KT-196-325

SHEEP PRODUCTION

Proceedings of Previous Easter Schools in Agricultural Science, published by Butterworths, London

*SOIL ZOOLOGY Edited by D. K. McL. Kevan (1955)
*THE GROWTH OF LEAVES edited by F. L. Milthorpe (1956)
*CONTROL OF THE PLANT ENVIRONMENT edited by J. P. Hudson (1957)
*NUTRITION OF THE LEGUMES edited by E. G. Hallsworth (1958)
*THE MEASUREMENT OF GRASSLAND PRODUCTIVITY Edited by J. D. Ivins (1959)
*DIGESTIVE PHYSIOLOGY AND NUTRITION OF THE RUMINANT Edited by D. Lewis (1960)
*NUTRITION OF PIGS AND POULTRY Edited by J. T. Morgan and D. Lewis (1961)
*ANTIBIOTICS IN AGRICULTURE Edited by A. M. Woodbine (1962)
*THE GROWTH OF THE POTATO Edited by J. D. Ivins and F. L. Milthorpe (1963)
*EXPERIMENTAL PEDOLOGY Edited by E. G. Hallsworth and D. V. Crawford (1964)
*THE GROWTH OF CEREALS AND GRASSES Edited by F. L. Milthorpe and J. D. Ivins (1965)
*REPRODUCTION IN THE FEMALE MAMMAL Edited by G. E. Lamming and E. C. Amoroso (1967)
*GROWTH AND DEVELOPMENT OF MAMMALS Edited by G. A Lodge and G. E. Lamming (1968)
*ROOT GROWTH Edited by W. J. Whittington (1968)
*PROTEINS AS HUMAN FOOD Edited by R. A. Lawrie (1970)
*LACTATION Edited by I. R. Falconer (1971)
*PIG PRODUCTION Edited by D. J. A. Cole (1972)
*SEED ECOLOGY Edited by W. Heydecker (1973)
HEAT LOSS FROM ANIMALS AND MAN: ASSESSMENT AND CONTROL Edited by J. L. Monteith and L. E. Mount (1974)
*MEAT Edited by D. J. A. Cole and R. A. Lawrie (1975)
*PRINCIPLES OF CATTLE PRODUCTION Edited by Henry Swan and W. H. Broster (1976)
*LIGHT AND PLANT DEVELOPMENT Edited by H. Smith (1976)
PLANT PROTEINS Edited by G. Norton (1977)
ANTIBIOTICS AND ANTIBIOSIS IN AGRICULTURE Edited by M. Woodbine (1977)
CONTROL OF OVULATION Edited by D. B. Crighton, N. B. Haynes, G. R. Foxcroft and G. E. Lamming (1978)
POLYSACCHARIDES IN FOOD Edited by J. M. V. Blanshard and J. R. Mitchell (1979)
SEED PRODUCTION Edited by P. D. Hebblethwaite (1980)
PROTEIN DEPOSITION IN ANIMALS Edited by P. J. Buttery and D. B. Lindsay (1981)
PHYSIOLOGICAL PROCESSES LIMITING PLANT PRODUCTIVITY Edited by C. Johnson (1981)
ENVIRONMENTAL ASPECTS OF HOUSING FOR ANIMAL PRODUCTION Edited by J. A. Clark (1981)
EFFECTS OF GASEOUS AIR POLLUTION IN AGRICULTURE AND HORTICULTURE Edited by M.H. Unsworth and D.P. Ormrod (1982)
CHEMICAL MANIPULATION OF CROP GROWTH AND DEVELOPMENT Edited by J. S. McLaren (1982)
CONTROL OF PIG REPRODUCTION Edited by D. J. A. Cole and G. R. Foxcroft (1982)

**These titles are now out of print but are available in microfiche editions*

Sheep Production

W. HARESIGN, PhD
University of Nottingham School of Agriculture

BUTTERWORTHS
London Boston Durban Singapore Sydney Toronto Wellington

All rights reserved. No part of this publication may be reproduced or transmitted in any form or by any means, including photocopying and recording, without the written permission of the copyright holder, application for which should be addressed to the Publishers. Such written permission must also be obtained before any part of this publication is stored in a retrieval system of any nature.

This book is sold subject to the Standard Conditions of Sale of Net Books and may not be re-sold in the UK below the net price given by the Publishers in their current price list.

First published 1983

© The several contributors named in the list of contents 1983

British Library Cataloguing in Publication Data

Sheep production.—(Nottingham Easter School proceedings; 35)
1. Sheep–Congresses
I. Haresign, W. II. Series
636.3 SF375

ISBN 0-408-10844-4

Typeset by Scribe Design, Gillingham, Kent
Printed and bound in Great Britain by Redwood Burn, Trowbridge, Wilts.

PREFACE

Many recent Easter Schools have studied in depth various specialized aspects relating to agricultural production. It was therefore thought appropriate that the 35th Easter School in Agricultural Science should provide an interdisciplinary platform for discussion of the many factors of importance in sheep production. In so doing it provided an opportunity to involve research workers in a number of disciplines (e.g. nutritionists, veterinarians, physiologists, geneticists, etc.), all of whom have a common aim of improving the efficiency of sheep production, and the quality of the end-product.

A major limitation to a meeting of this type is the time available, and an attempt was therefore made to select only the more important topics although it is realized that this inevitably means that some subject areas were not included.

The first session commenced with a paper outlining the factors affecting the economics of sheep production in the UK in an attempt to set the scene for the rest of the conference. The main sessions then considered carcass quality, sheep nutrition, forage production and utilization, sheep health, reproductive efficiency, wool production and genetic improvement. With the rapid progress taking place in these areas it is all too easy for research workers to lose sight of developments in subjects not immediately akin to their own, and for teachers and advisors to acknowledge the lack of a comprehensive up-to-date appraisal of the subject as a whole. The wide scope of the programme of this 35th Easter School in Agricultural Science was therefore an attempt to overcome these problems.

W. Haresign

ACKNOWLEDGEMENTS

The organizers of the 35th Easter School wish to thank the staff of the University of Nottingham, the Speakers and the Chairman who all contributed to the success of the meeting. The following organizations made donations to the School, without which the meeting could not have taken place:

J. Bibby Agriculture Ltd
BOCM Silcock Ltd
The British Council
Colborn Nutrition Ltd
Alfred Cox (Surgical) Ltd
Crown Chemical Co Ltd
Dalgety Spillers Agriculture Ltd
FBC Ltd
Peter Hand (GB) Ltd
Hoechst Pharmaceuticals
Intervet International B/V
Nutec Feed Supplements Ltd
Pauls Agriculture Ltd
Rumenco Ltd
Upjohn Ltd
Wellcome Foundation Ltd

CONTENTS

1

FACTORS AFFECTING THE ECONOMICS OF SHEEP PRODUCTION IN THE UK

J.B. KILKENNY
Meat and Livestock Commission, Bletchley, UK

This chapter is in two main sections; the first provides economic background information on UK sheep production and the second uses information from Meat and Livestock Commission (MLC) recorded flocks to highlight those factors having the biggest effect on profitability.

Sheep and farming systems

Sheep are the major enterprise on hill farms in the UK, very important enterprises on upland and permanent grassland farms and a secondary enterprise on a substantial number of dairy and mixed arable lowland farms. *Table 1.1* shows the distribution of ewes by type of farming.

Table 1.1 THE DISTRIBUTION OF BREEDING EWES BY TYPE OF FARMING

	Percentage of farms with breeding ewes	
Types of farming	*England and Wales*	*Scotland*
Hill	100	100
Livestock rearing	77	72
Lowland: dairy	27	32
mixed	51	51
cropping	21	27

From *The Changing Situation of Agriculture 1968–1975*, HMSO 1977

Sheep are often considered in lowland situations as an enterprise which makes a contribution to the farming system greater than that which would be assessed simply on the basis of the economics of the enterprise alone. They are utilisers of by-products, they 'put back fertility', and help in the control of grazing. Nevertheless with increasing working capital requirements (largely as a consequence of increases in the price of breeding ewes), the enterprise increasingly has needed to make a commensurate contribution to the total farm gross margin to offset its share of farm fixed costs. This means that the enterprise must be given the right order of management priority within the farming system to ensure satisfactory levels of performance and margins.

Table 1.2 shows the importance of sheep regionally in the United Kingdom and the competition or complementary relationship of the sheep and cattle enterprises. The contrast in this respect in England and Wales is between the relatively low sheep population in the eastern region with only 15 sheep livestock units (LU)/100 hectares of grassland and the 'sheep country' of northern England and Wales with 32 and 49 sheep LU/100 hectares of grass. In these areas, the ratio of sheep to cattle is narrowly in favour of catle compared with the ratio of 1:8 in the east and west of England. With the exception of the north-east of the country, Scotland shows up as a sheep-keeping area with high sheep stocking rates and a narrow ratio of sheep to cattle.

Table 1.2 THE DISTRIBUTION OF SHEEP REGIONALLY AND STOCKING RATES

	Percentage of UK total breeding ewes	*Sheep stocking rates* (LU/100 ha grassland)	*Ratio of sheep to cattle*
East England	10.4	15	1:8.3
West England	17.8	16	1:8.4
North England	20.7	32	1:3.0
	48.9		
Wales	23.3	49	1:2.0
Scotland	23.9	35	1:2.4
Northern Ireland	3.9	7	1:16.4

From Thomas and Kilkenny (1981)

In the hills and uplands sheep are the major enterprise, either alone or in association with cattle. The figures in *Table 1.3* show that 57% of the UK ewe flock is kept on hill and upland farms. In the lowlands, sheep must

Table 1.3 DISTRIBUTION OF BREEDING EWES BY ELEVATION OF FARM

	Lowland flocks (%)	*Upland flocks* (%)	*Hill flocks* (%)	*Totals* (%)
England	29.1	4.8	10.0	43.9
Wales	6.8	5.3	14.0	26.1
Scotland	6.0	4.9	15.6	26.5
Northern Ireland	1.1	0.6	1.8	3.5
United Kingdom	43.0	15.6	41.4	100.0

From MAFF, based on analysis of flocks eligible for Hill Livestock Compensatory Allowance, 1978

always compete on economic terms with other livestock and cropping enterprises and the success they have in this respect reflects back to the hill and upland sheep sector. This is because of the stratification system within the sheep industry by which culled breeding stock and store lambs from the hills come down to the lowlands for further breeding or for finishing.

The main contribution to a sheep producer's income is the returns received from the sale of lambs, and wool is much less important accounting for only about 6% of income (*Table 1.4*). The priority is

Table 1.4 COMPONENTS OF INCOME FROM DIFFERENT TYPES OF FLOCK

	1980			*1981 Estimate*		
	Lowland (%)	*Upland* (%)	*Hill* (%)	*Lowland* (%)	*Upland* (%)	*Hill* (%)
Lamb returns	82	75	56	80	72	55
Other stock sales	11	10	18	10	10	15
Wool	7	6	7	6	6	7
Ewe premium	0	0	0	4	4	6
Livestock compensatory allowance	0	9	19	0	8	16
Income (£/ewe)	46.7	43.6	28.9	53.5	51.7	36.1

From MLC

therefore to increase returns from lambs. The average return/kg of lamb in 1981 was 192p, the average return/kg of wool was 115p.

Factors affecting profitability

SEASONALITY OF PRODUCTION

Lambs from the annual lamb crop are sold throughout the year, starting with lambs born in December and January and sold in March, April and May, through to hoggetts sold in the spring the year after birth. There is a very marked seasonality of production which is illustrated in *Table 1.5*. The pattern of seasonal guide prices in the EEC Sheepmeat Regime has been designed in part to stimulate an increase in the proportion of lambs and hoggetts sold in the period January to June.

Table 1.5 SEASONALITY OF UK LAMB PRODUCTION (AVERAGE 1979 AND 1980)

	Lambs (%)	*Hoggetts*[a] (%)	*Total* (%)
January–March	1.2	20.9	22.1
April–June	2.4	13.6	16.0
July–September	28.6	0	28.6
October–December	33.3	0	33.3
	65.5	34.5	100

[a]Lambs sold in the year after the year in which they were born
From *MLC Commercial Sheep Production Yearbook 1980–81*

Table 1.6 shows the distribution of sales of slaughter lambs from MLC recorded lowland flocks according to lambing dates (these are the averages for the last five years). Clearly there is a relationship between lambing date and the proportion of lambs sold finished by the end of the grazing season, the earlier the lambing period the higher the proportion of lambs sold finished. For many flocks there is a very high proportion of sales in October, on average about a fifth of the lamb crop. Lambs not sold for slaughter by the end of October are either sold as stores for further feeding

Table 1.6 DISTRIBUTION OF LAMB SALES FOR SLAUGHTER IN MLC RECORDED LOWLAND FLOCKS BY LAMBING PERIOD

Lamb sales	*Average* (%)	*Lambing period*						
		December/ January (%)	*February*		*March*		*April*	
			1st Half (%)	*2nd Half* (%)	*1st Half* (%)	*2nd Half* (%)	*1st Half* (%)	*2nd Half* (%)
Up to end April	5	33	4	–	–	–	–	–
Breeding flock								
May	6	30	9	2	–	–	–	–
June	5	16	13	5	2	–	–	–
July	6	10	15	9	4	2	–	–
August	9	7	14	18	13	5	3	3
September	16	4	16	22	22	18	22	11
October	22	0	15	25	30	33	25	24
(stores end of October)	(31)	(0)	(14)	(19)	(29)	(42)	(50)	(62)
Feeding flock								
November	11		10	10	10	18	18	14
December	8		4	5	10	10	10	14
January	6			4	8	7	8	12
February	3				1	4	7	11
March	3					3	7	11

From *MLC Commercial Sheep Production Yearbook 1980–81*

on other farms or are transferred into a store feeding flock on the same farm. The choice of the right system for the farm largely depends upon lambing date, and the selling policy of the lambs is a major factor in determining the economics of production.

Table 1.7 shows that the average gross margin/hectare for different types of lowland flocks for the five-year period 1976–1980 were remarkably constant. The new seasonal scale of prices associated with the introduction of the EEC Sheepmeat Regime further altered the returns/lamb in favour of earlier season lambs and those fattened in the winter and carried through as hoggetts. However, the changes are not likely to dramatically change the average margins. For example in 1980 the average return/lamb was 26% higher for early lambs (sold March–May) than grass lambs (sold

Table 1.7 COMPARISON OF RESULTS FOR DIFFERENT TYPES OF LOWLAND FLOCKS (FIVE-YEAR AVERAGE 1976–80)

	Early lambs	*Grass lambs*	*Forage lambs*	*Hoggetts*
Lamb sales/ewe (£)	36.5	33.4	34.6	35.3
Feed cost/ewe (£)	14.1	9.8	11.0	11.6
Gross margin/ewe (£)	18.6	19.7	19.5	19.5
Gross margin/hectare (£)	228.3	213.4	196.8	191.0
Lambs reared/ewe	1.33	1.41	1.39	1.36
Stocking rate (ewes/hectare)	12.3	10.8	10.1	9.8
Average return/lamb (£)	27.4	23.7	24.9	26.0
Ratio of average lamb return (to grass lamb = 100)	117	100	105	110

From *MLC Commercial Sheep Production Yearbook 1980–81*

August–November), yet the difference in gross margins/hectare was only £6.00. The main reason why the extra returns/lamb have not significantly increased the gross margins of early lambing flocks is that they rear fewer lambs/ewe—on average 0.10 lambs less—equivalent to a loss in potential returns of about £4.00/ewe or £50.00/hectare. In addition early lambing flocks have higher costs because of a greater need for concentrate feeding. The economics of early lamb production would be substantially enhanced if the breeding performance of ewes mated in August/September could be significantly improved. Nevertheless, despite the small difference in margins/hectare over recent years, there is a case for more lowland flocks lambing a proportion of their ewes earlier both to increase their own total flock margin and to contribute to reducing the seasonality of production.

Analysis of results from recorded flocks shows that increased stocking rate is closely related to higher gross margins/hectare but this tends to be associated with a distinct fall in gross margin/ewe largely as a consequence of lower lamb returns because the lambs are lighter at the end of the grazing season and fewer are sold finished.

Table 1.6 showed that for flocks lambing in March and early April between a third and half of the lambs had not been sold finished by the end of October. A higher proportion of lambs could be sold off grass by pushing back the lambing date, by drawing lambs slightly younger and lighter (and leaner), by effective parasite control and a grazing management system that provides adequate grazing in the second half of the season. The greater flexibility in lambing date provided by the availability of winter housing is an important factor in making a sound economic case for erecting winter housing for sheep.

The new Sheepmeat Regime has substantially increased the returns to UK sheep producers, but taking full advantage of it (particularly in the longer term) means reducing the proportion of the lamb crop sold in October and November. For hill lambs there is only very limited ability to avoid selling at this time but most lowland and some upland flocks could make a small adjustment in their selling pattern which would have a significant effect on overall prices at this time of the year. This basically means selling grass lambs some time earlier than many are currently sold i.e. slightly lighter and younger. If this can be combined with pushing back the lambing date for a proportion of the flock the pattern of sale can be improved further. Ensuring an increased number of lambs sold in August and September also benefits grassland management by reducing stocking rate in line with the decline in grass production. The signs are that in the first year of operation of the Regime the producers have responded in this way. An alternative is to deliberately produce a higher proportion of store lambs for winter feeding by increasing stocking rates and possibly by lambing a bit later. The latter should allow for some saving in winter feed costs.

LAMB PRICES

Figure 1.1 compares the prices of slaughtered sheep, cattle and pigs during the 1970s. During this period sheep prices rose by 285%. The increase in

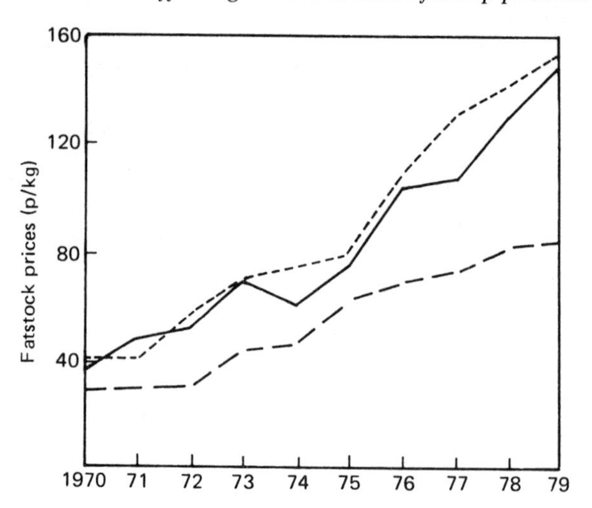

Figure 1.1 Fatstock prices in the 1970s: —— cattle (dw); ---- sheep (edcw); — — — pigs (dw). From *MLC Commercial Sheep Production Yearbook 1979–80.*

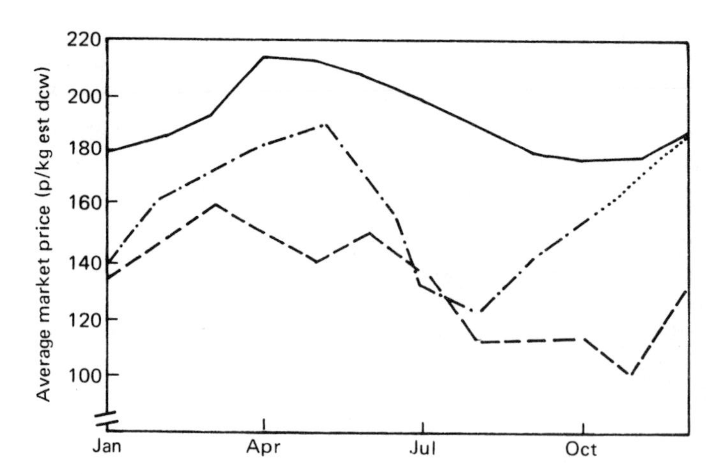

Figure 1.2 Average market price for finished sheep, 1980–81. — — — 1980; —·—·—· 1981; ...... forecast; —— Guide Price 1981. From *MLC Commercial Sheep Production Yearbook 1980–81.*

sheep prices was higher than the rate of inflation and when prices are deflated by the General Retail Price Index, an increase of 28% in real prices occurred. The corresponding increases in cattle prices was similar at 26%, but contrasts with a fall of 8% in the real price of pigs.

A subsidy scheme for sheepmeat prices operated during the 1970s and in October 1980 this was replaced by the EEC Sheepmeat Regime. *Figure 1.2* shows average lamb and hoggett prices for 1980 and 1981 and the Guide Price for 1981. Clearly there is a substantial gap between the average market price and the Guide Price which is made up by variable premium payments. In 1981 support accounted for nearly 20% of total returns from the sale of lambs and hoggetts.

PRODUCTION COSTS

Figure 1.3 shows changes in the costs of labour, feed, rent and fertilizer—major items in the cost structure of sheep production. *Figure 1.4* shows changes in the ratio of sheep returns to fertilizer prices. Although costs are important in the economics of production it is worth stressing that costs and capital investment tend to represent a lower proportion of total returns in sheep than in other livestock enterprises and the priority in most situations is to increase output rather than to reduce costs.

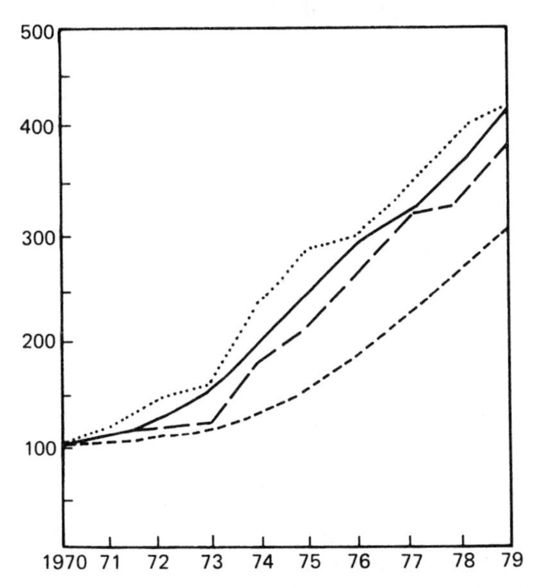

Figure 1.3 Agricultural input price changes in the 1970s (1970 = 100). —— labour; — — — fuel; - - - - rent; ...... fertilizer. From *MLC Commercial Sheep Production Yearbook 1979–80.*

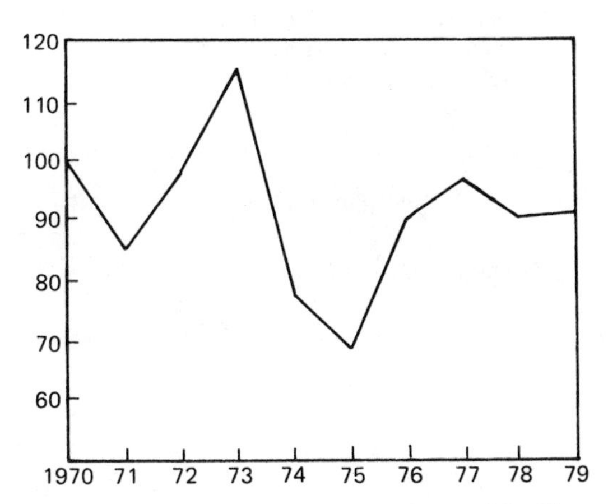

Figure 1.4 Comparison of sheep prices and fertilizer prices in the 1970s (index of sheep prices divided by index of fertilizer prices; 1970 = 100). From *MLC Commercial Sheep Production Yearbook 1979–80.*

One of the consequences of higher lamb returns referred to above is to generate an increase in the price of replacement breeding stock. This tends to 'share' the benefit of higher lamb returns between the lowland and hill sectors within the stratified system. Thus for example, the ratio of average lamb returns to flock replacement costs/ewe in lowland and upland flocks fell from 8.6:1 in 1976 to 5.0:1 in 1981.

Table 1.8 shows the make-up of fixed costs in sheep production from an MLC survey. The proportion of fixed costs accounted for by different items was very similar for both lowland and upland flocks. Labour was singly the most important item accounting for about 40% of fixed costs.

Table 1.8 STRUCTURE OF FIXED COSTS IN LOWLAND AND UPLAND FLOCKS

	Lowland (%)		*Upland* (%)	
Rent and buildings	22		24	
Machinery and equipment	18	24	17	23
Machinery running costs	6		6	
Labour: paid	18	39	15	43
family	21		28	
Other fixed costs	14		12	
Fixed costs as a proportion of total costs	51		56	

Based on three-year average 1977/79.
From *MLC Commercial Sheep Production Yearbook 1979–80*

The major items of variable costs are flock replacement costs, concentrate costs and forage costs, with each accounting for between a quarter and a third of variable costs both in lowland and upland flocks (*Table 1.9*).

MLC surveys of fixed costs associated with sheep production have revealed a close relationship between gross and net margins (*Figure 1.5*). About 78% of the variation in net margins was explained by variation in gross margins/hectare. Any factor influencing gross margin will affect net margin and a good level of gross margin is a prerequisite for a high net margin. The rest of the chapter therefore uses gross margin as an indicator of profitability.

Table 1.9 COMPONENTS OF PRODUCTION VARIABLE COSTS (FIVE-YEAR AVERAGE 1976–1980)

	Lowland (%)	*Upland* (%)	*Hill* (%)
Flock replacement cost	32	30	16[a]
Concentrates	30	29	29
Forage costs	27	29	23
Vet and medicine	7	8	11
Others	4	4	21[b]
	100	100	100

[a] Rams only
[b] Includes agistment
From MLC

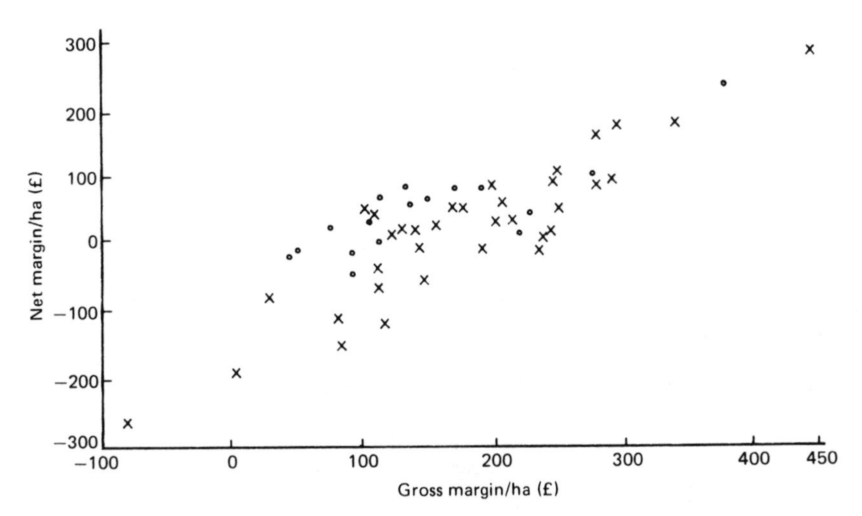

Figure 1.5 Relationship between gross margins and net margins. × = lowland; o = upland. From *MLC Commercial Sheep Production Yearbook 1979–80.*

TRENDS IN GROSS MARGINS

Figures 1.6 and *1.7* show average gross margins for different types of flocks from 1970–1981. Improvements in average margins are striking but high inflation through much of the period obscures the real trend. Adjusting previous years' figures for inflation reveals a different picture. Average gross margins in real terms for 1979–81 were, in fact, only between 18% and 25% higher (depending on the type of flock) than those for 1970–72. Nevertheless sheep improved their relative profitability to other livestock enterprises during the 1970s and improvements in margins in most years during the period were ahead of the general rate of inflation. The average gross margins in 1981 (the first full year of the CAP for sheepmeat) were

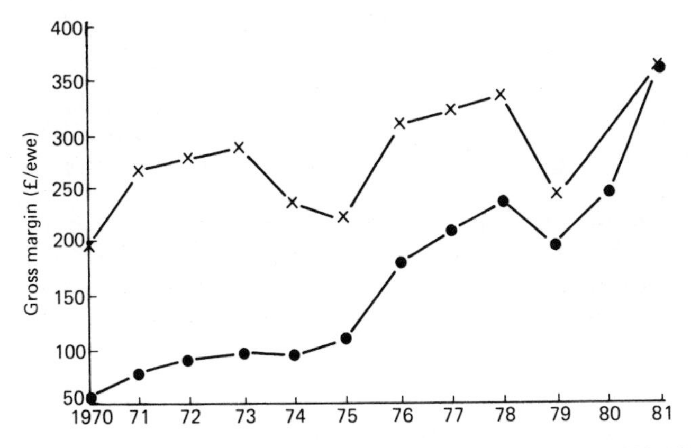

Figure 1.6 Average gross margins/hectare in lowland flocks 1970–1981. ×——× years prior to 1981 inflated to 1981 values using the General Retail Price Index; ●——● actual. From MLC.

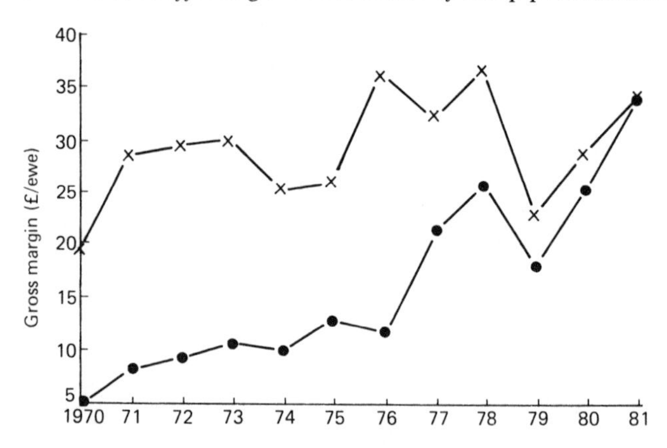

Figure 1.7 Average gross margins/ewe in upland flocks 1970–1981. ×——× years prior to 1981 inflated to 1981 values using the General Retail Price Index; ●——● actual. From MLC.

appreciably higher than for 1980 and above the general rate of inflation. All of the improvement in 1981 compared with 1980 was due to the higher lamb returns for slaughtered lambs because of the CAP and the effect of this on increasing the demand and prices for store lambs.

TRENDS IN PERFORMANCE

Despite the improved gross margins per ewe and per hectare there has been no consistent trend in lamb rearing percentage in recorded flocks. The average number of lambs reared/100 ewes overall averages around 140 in lowland flocks, 125 in upland flocks and 95 in hill flocks. But there has

Table 1.10 CHANGES IN STOCKING RATES, FERTILIZER NITROGEN USE AND FLOCK SIZE IN MLC RECORDED LOWLAND FLOCKS 1970–1980

	1970–75	*1976*	*1977*	*1978*	*1979*	*1980*	*1981*
Average no. ewes in flock	216	287	315	355	366	398	443
Stocking rate (ewes/ha)	9.6	10.4	10.4	10.6	11.1	11.5	13.0
Nitrogen use: (kg/ha)	82	94	98	115	138	151	154
(kg/ewe)	9	9	9	11	12	13	12

From *MLC Commercial Sheep Production Yearbook 1980–81*

been a consistent rise in stocking rates. This has been associated with both an increase in average flock size and higher application levels of fertilizer nitrogen/hectare (*Table 1.10*). Although nitrogen use/hectare increased by 72 kg (88%) between 1970–75 and 1981, the nitrogen use/ewe increased by only 3 kg (33%).

VARIATION IN RESULTS

MLC recording of commercial sheep enterprises emphatically demonstrates the wide variation in results achieved in practice and the close relationship between physical performance and gross margins and hence net profits. *Table 1.11* shows the difference in gross margins between average and top third enterprises. The differences are very significant in all cases and particularly marked in the case of sheep.

Table 1.11 DIFFERENCE BETWEEN AVERAGE AND TOP THIRD GROSS MARGINS/HECTARE (FIVE-YEAR AVERAGE)

		Superiority (%)	*Per 20 hectares*
Sheep			
Early lambing flocks		+59	3580
Grass flocks		+58	3060
Forage flocks		+63	
	Average	+60	
Beef			
18-month beef		+31	3140
20/24 month beef		+42	2650
Suckler herds		+47	1860
Suckler herds (selling finished cattle)		+48	2200
Store (finishing)		+45	1980
	Average	+42	
Dairying		+42	4580

From MLC

Variation in results between sheep enterprises is particularly marked. *Table 1.12* illustrates this; in lowland flocks 160 lambs reared/100 ewes is often quoted as a reasonable industry target, but in fact within the MLC recording scheme of 600 flocks only one in five flocks achieves this performance.

Table 1.12 EXAMPLE OF VARIATION IN PERFORMANCE BETWEEN LOWLAND FLOCKS (FIVE-YEAR AVERAGE)

	Average	*Top third*	*Percentage of flocks with values:*	
Lamb deaths			*>20*	
(% of live lambs born)	13	12	11	
Lambs reared			*>160*	*<120*
(per 100 ewes to ram)	139	154	18	20
Empty ewes			*>10*	
(per 100 ewes to ram)	6	4	22	
Dead ewes			*>10*	
(per 100 ewes to ram)	4	3	6	

From MLC

COMPONENTS OF PROFIT

Table 1.13 shows the key profit factors in sheep production. There is very little difference in costs between average and top third flocks. The major differences between the average and top third is due to stocking rate, number of lambs reared and the average sale value of the lamb.

Table 1.13 PERCENTAGE CONTRIBUTION TO TOP THIRD SUPERIORITY IN GROSS MARGINS/HECTARE

	Early lambing flocks (%)	*Lowland spring lambing flocks* (%)	*Upland flocks* (%)	*Overall* (%)
Lamb sale price/head	5	7	20	11
No. lambs reared	25	33	36	31
Flock replacement costs	5	11	4	7
Feed and forage costs	4	2	5	3
Stocking rate	52	37	25	38
Other factors	9	10	10	10
	100	100	100	100
Effect on gross margin/ewe of 0.1 increase in number of lambs reared/ewe (£)	+3.21	+2.56	+2.33	+2.70
Effect on gross margin/hectare of 0.1 increase in number of ewes/hectare (£)	+1.95	+1.89	+2.27	+2.04
Extra gross margin of top third/20 hectares (£)	3580	3060	2100	2910

From MLC

The differences in costs between average and top third flocks are relatively small and in addition the contribution made by the various items is also very similar (*Table 1.14*).

Table 1.14 COMPONENTS OF COSTS IN AVERAGE AND TOP THIRD LOWLAND FLOCKS

	Average (%)	*Top third* (%)
Fixed costs as % total costs	51	49
Components of variable costs		
Flock replacement cost	31	29
Concentrates	30	30
Forage costs	27	30
Vet and medicine	8	8
Others	4	4
	100	100

From MLC

The components of flock replacement costs are purchase price for replacements, returns for cull ewes, average flock life and ewe mortality. Top third flocks have lower flock replacement costs because of a lower ewe mortality (three fewer deaths/100 ewes), a slightly longer average flock life (+0.2 years), a higher proportion of ewe lambs in the flocks (three more/100 ewes), and a better return from cull ewes (+15%/cull ewe sold).

As shown in *Table 1.13*, differences in feed and forage costs accounted for little of the difference in gross margins between top third and average flocks. Top third flocks have a similar dependence on concentrates despite the substantially higher lamb output/ewe. The conclusion of this is that 'better feeding' is not synonymous with 'expensive feeding'.

LAMBS REARED

Top third flocks rear more lambs than average. This is because they have fewer empty ewes, a higher number of lambs born and lower lamb mortality. Differences in breed type and in the age structure of flocks account for some of the variation between flocks but of much greater significance are the differences in management and feeding. Ewes in top third flocks perform very differently to ewes of the same type in average flocks (*Table 1.15*).

Table 1.15 PERFORMANCE OF EWES IN AVERAGE, TOP THIRD[(a)] AND TOP 10% FLOCKS[(a)]

		No. live lambs born/100 ewes to ram		
Ewe type		*Average*	*Top third*	*Top 10%*
Scotch Halfbred	(Biggest)	169	180	195
Mule	↓	166	181	198
Masham		168	180	195
Welsh Halfbred	(Smallest)	150	160	175

[(a)]Based on gross margin/hectare
From MLC

There are important performance differences between ewe breed types and they vary markedly in size (with consequent implications for relative stocking rates) and the choice made by farmers will be influenced by many factors—particularly local availability and price, and the weight range of lambs required for sale. Whatever the choice, the priority is to provide adequate all-year round nutrition to the ewe.

Over recent years the routine use of body condition scoring of ewes has become established as a technique for assessing the adequacy of feeding and for drawing out ewes for preferential feeding. MLC records have demonstrated the close relationship between ewe body condition and lamb rearing percentage. *Table 1.16* shows lambing percentages of ewes in various body condition scores at mating in both lowland and hill situations.

As a result of research and development it is now possible to define the feeding of the flock in terms of achieving target body condition scores at

Table 1.16 LAMBING PERCENTAGES OF EWES IN VARIOUS BODY CONDITIONS AT MATING

	Body condition score at mating						
	1	*1½*	*2*	*2½*	*3*	*3½*	*4*
Hill ewes							
Blackface	–	79	–	–	162	–	–
Swaledale	–	78	133	140	156	–	–
Lowland							
Mule	–	–	149	166	178	194	192
Masham	–	–	–	167	181	215	–
Scotch Halfbred	–	–	148	170	183	217	202
Greyface	–	–	147	163	176	189	184

From *Feeding the Ewe*, MLC (1981)

Table 1.17 TARGET BODY SCORES

Mating	3½
Mid-pregnancy	3
Lambing	2½
2 months post-lambing	2
Weaning	2½

From MLC

the key times of the year (*Table 1.17*). The achievement of these targets by the drawing out of ewes and preferential management and feeding will ensure the optimum production for the type of ewe being kept.

Reduced lamb mortality (*Tables 1.18* and *1.19*) is also a major component of the flock's lamb rearing percentage responsive to improved nutrition and management e.g. the use of tubes for colostrum feeding of lambs. A high proportion of lambs that die at birth do so from starvation rather than a specific disease contact. The influence of proper feeding of the ewe in late pregnancy cannot be overstated to ensure good lamb birthweights and vitality, and to develop a strong maternal instinct in the ewe at lambing.

Other factors that influence the flock's lamb rearing percentage are the age structure of the flock and disease conditions. It is essential to have a detailed record of the components of a given rearing percentage to

Table 1.18 LAMB DEATHS IN LOWLAND FLOCKS

Proportion of lambs dying at different ages (%)			
0–7 days	*8–30 days*	*30+ days*	*Unknown*
64	12	3	21

Late pregnancy feeding and management at lambing time

	Causes of lamb deaths (%)	
Mismothering/starvation	41	Directly responsive to ewe nutrition
Physical damage	3	
E. coli scour/septicaemia	15	
Pneumonia	2	
Abscess/peritonitis/navel ill	5	
Miscellaneous	16	Proportion responsive to ewe nutrition
No cause found	20	

From MLC

Table 1.19 FACTORS AFFECTING LAMB DEATHS IN LOWLAND FLOCKS

Birthweight (kg)	*Percentage still births*	*Age of ewe*	*% lamb mortality*	*Litter size*	*% lamb mortality*
Under 3 (responsive to ewe nutrition)	9 Preferential treatment	Ewe lambs	11.8	1	4.5
3–5.5	2	Shearlings	8.4	2	6.0
Over 5.5	2	Ewes	7.1	3	13.5

From MLC

highlight where the weakness is so that the appropriate remedial action can be taken.

The components of the average and top third lambing percentages in 1980 for lowland and upland are shown in *Table 1.20.*

Table 1.20 COMPONENTS OF LAMB REARING PERCENTAGES (1980)

		Lowland		*Upland*	
		Average[a]	*Top third*[a]	*Average*[a]	*Top third*[a]
		(per 100 ewes to ram)			
No. ewes	Empty	7	5	6	5
	Died before lambing	1	2	2	1
	Lambed	92	93	92	94
No. lambs	Born dead	10	9	8	7
	Born alive	150	164	131	143
	Died after birth	9	7	9	7
	REARED	141	157	122	136
Total lambs died		19	16	17	14
Percentage of lambs born		12	9	12	9

[a]Based on gross margin/hectare
From MLC

LAMB RETURNS

Table 1.13 showed that the extra return/lamb achieved by the top third producers made a significant contribution to the extra gross margin. Differences in time of lambing and time of sale do not account for the difference, so the extra returns are a reflection of either superior marketing skill and/or the production of a more acceptable product.

As the requirements of both the UK home market and the export markets become more precise, producers need to adapt their production to meet more closely market requirements. These are defined primarily in terms of lamb carcass weight and fatness.

Weight requirements show quite wide variations. In Great Britain the main weight band is approximately 15–22 kg, and heavier carcasses normally incur a price penalty. The main European markets (Belgium, France and West Germany) require carcasses in the narrower 16–19 kg range and the requirements for major British multiples are similar. By exploiting the known variations in body size between the different terminal sire breeds and the main ewe crosses and breeds allied to careful selection of lambs for slaughter, it is possible to increase the proportion of lambs falling into the required or target weight range.

Meat trade requirements for fatness levels are more uniform and are essentially for MLC fat class 2 and 3L (scale 1 (lean) to 5 (fat)); the European preference for lambs is fat class 2 with good conformation. Producers should therefore aim to produce as many lambs as possible in this target area for fatness whilst providing their particular outlet with the weight range of carcass required. Top third flocks have a higher proportion of their lambs in the target area than average.

STOCKING RATE

Table 1.13 showed that stocking rate is a major component of profitability in lamb production. In 1980 on average, for each increase of 0.1 in the number of ewes/hectare the gross margin/hectare rose by £1.89.

Figure 1.8 gives the distribution of gross margins/hectare by fertilizer nitrogen usage/grass hectare, and *Figure 1.9* gives the distribution of gross margins per ewe and per hectare by stocking rate.

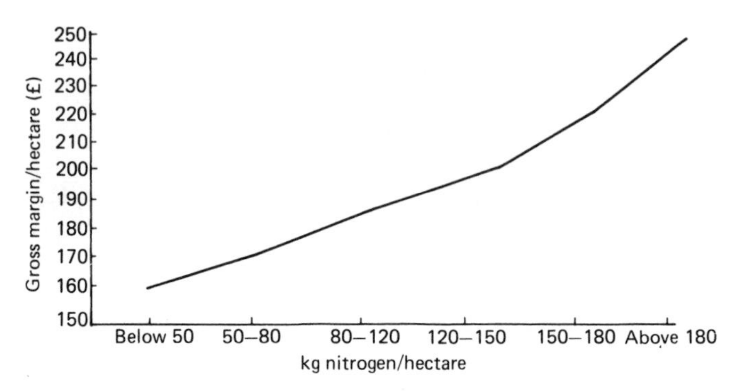

Figure 1.8 Distribution of gross margins/hectare by fertilizer nitrogen usage/grass hectare in 1979. From *MLC Commercial Sheep Production Yearbook 1979–80.*

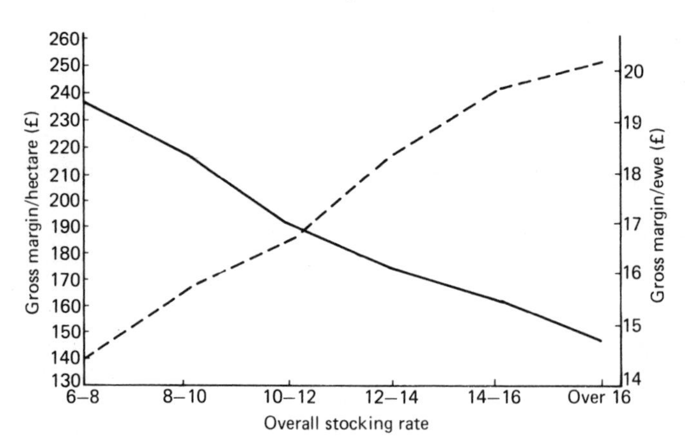

Figure 1.9 Distribution of gross margins/ewe (——) and /hectare (---) by overall stocking rate, 1979. From *MLC Commercial Sheep Production Yearbook 1979–80.*

A closer look at the relationship between gross margin/hectare and stocking rate reveals a fall in gross margin/ewe despite the higher gross margin/hectare. This is almost entirely because the average return/lamb falls with stocking rate; the lambs are lighter at the end of the grazing season and fewer are sold finished. These problems can be overcome by effective parasite control and the adoption of a grazing management system that provides adequate grazing in the second half of the season. Additionally *Table 1.6* showed that between a third and half of the lambs from flocks lambing in March and early April had not been sold finished by the end of October. A higher proportion could be sold off grass by pushing back the lambing date and by selling lambs slightly younger and lighter (and leaner).

Stocking rate potential depends on the inherent quality of the land and levels of fertilizer nitrogen use. The convention in costings is to express fertilizer use on a 'per hectare' basis. Such values invariably point to the fact that top third producers apply higher levels of fertilizer nitrogen/ hectare than average. This implies a 'more intensive' high-cost system. But if, instead, fertilizer use is expressed on a 'per head' basis, in common with other inputs such as concentrates, the conclusion is different. On this basis, fertilizer nitrogen use in top third flocks is hardly any different from average. This is because top third producers increase stocking rates to utilize the extra grass grown (*Table 1.21*). This concept has been extended to provide guidelines for fertilizer nitrogen use for different systems.

Table 1.21 FERTILIZER NITROGEN USAGE IN AVERAGE AND TOP THIRD FLOCKS (FIVE-YEAR AVERAGES)

	Average		*Top third*	
	kg N/ha	*kg N/head*	*kg N/ha*	*kg N/head*
Lowland early lambing	145	13	174	13
Spring lambing	138	12	159	12
Upland	90	9	110	9

From *MLC Commercial Sheep Production Yearbook 1980–81*

The clean grazing system developed at the East of Scotland College of Agriculture has made a major contribution to grassland management and increased sheep profitability. The adoption of this technique allows sheep stocking rates to be increased without reducing lamb performance (or cattle production) by effective parasite control. In essence the system means *never* grazing sheep on next year's sheep grazing. The production of clean grazing needs to be planned to suit the individual farm but there are few situations where it cannot be employed; it does not depend, for example, on an equal proportion of cattle and sheep. Full details of clean grazing systems will be given in later chapters.

References

MLC (1981). *Commercial Sheep Production Yearbook 1980–81*

MLC (1981). *Commercial Sheep Production Yearbook 1979–80*

MLC (1981). Sheep Improvement Services. *Feeding the Ewe*

THOMAS, W.J.K and KILKENNY, J.B. (1981). *The Role of Grassland in Sheep Production. Grassland in the British Economy.* Centre for Agricultural Strategy. Paper 10

I

Growth and Carcass Quality

2

GROWTH AND DEVELOPMENT OF LAMBS

J.L. BLACK
CSIRO Division of Animal Production, Prospect, Blacktown, New South Wales, Australia

Sheep raised for meat and wool production form a significant part of the economy of many countries of the world. There is a large variation between these countries in climatic conditions, in management procedures and in the genotypes available. Some animals are grazed continuously outdoors with only limited supplementation of available pasture whereas others are housed from birth and fed carefully formulated diets. In addition, the desires of consumers vary widely, with some preferring heavy, fat carcasses whereas others are attracted to leaner meat. Despite this diversity, it should be possible from a better understanding of the factors affecting the growth and development of sheep to devise management procedures which will improve the efficiency of production in each situation and develop products more suited to the demands of particular consumers.

This chapter reviews factors that affect the growth and development of sheep from fertilization to maturity. Many of the concepts discussed have been incorporated into a computer programme that simulates the growth and body composition of sheep from conception. It is believed that with programmes of this type, the advantages or disadvantages of alternative management procedures can be readily assessed and the most efficient methods of production devised.

Prenatal growth and development

Although growth of the prenatal lamb is a continuous process, it is often divided into three periods. The period of the ovum lasts from ovulation until attachment of the blastocyst to the endometrium about 10 days later. The initial cell division takes place midway through the second day after mating and by the fourth day the morula stage with 16 to 64 cells is reached (Rowson and Moor, 1966). The first sign of cell differentiation occurs at the 16-cell stage (Calarco and McLaren, 1976). By the sixth day, the blastocyst is formed, and the zona pellucida is lost from 90% of the ova by the end of the ninth day. At this stage the embryo is only a few millimetres in length (Rowson and Moor, 1966), but the chorion is well developed and by day 10 has made a loose attachment to the endometrium (Bryden, Evans and Binns, 1972a).

The embryonic period commences when the blastocyst becomes associated with the endometrium and continues until about day 34 of gestation. It covers the period of genesis of the main organ systems. The embryo undergoes rapid elongation early in this period increasing in length from about 10 mm at day 12 to 100 mm by day 14 of gestation. Growth rates of 10 mm/hour are achieved during parts of this period (Rowson and Moor, 1966). Morphological changes which occur during the embryonic period, and development of the alimentary tract and associated organs have been described by Bryden, Evans and Binns (1972a,b). The neural plate and head folds are apparent by day 15 of gestation and the mesoderm shows segmentation. Heart and liver bulges are clearly seen by day 18 and the first limb buds appear on day 20. Major organs of the digestive tract can also be distinguished at this time. By days 25 and 26 of gestation, limb buds have differentiated into proximal and distal segments, olfactory pits are deep, eyelid grooves clearly visible, salivary glands have started to develop, the reticulo-rumen has separated from the remainder of the stomach with the omasal bulge apparent, and the intestine has elongated. Development continues rapidly so that most of the major organ systems have formed by day 34 and many of the cell types have differentiated. The chorion develops villi about day 31 and the foetal placenta becomes firmly attached to the uterus at this stage.

The foetal period extends from about day 34 of gestation until birth and covers mainly the growth and development of organ systems but also includes further differentiation of some tissues. For example, the pituitary gland develops between days 45 and 55 of gestation and pituitary hormones can be isolated from the foetal system at this time (Hopkins, 1975). Ossification centres in foetal limbs first become radio-opaque around day 50 with bones in the limb being clearly visible by day 60 (Hopkins, 1975). Also by day 60 of gestation primary wool follicles have initiated in all areas of the body, but secondary wool follicles do not start to form until the 80th to 90th day (Carter and Hardy, 1947). Similarly the growth of dissectible adipose depots in the abdomen does not commence until about day 70 of gestation, whereas subcutaneous adipose tissue is not visible until closer to day 90 of gestation (Alexander, 1978a).

FACTORS AFFECTING FOETAL GROWTH

The foetus doubles in weight every four days during early pregnancy, but its absolute rate of gain is extremely low being only 0.06 g/day between days 20 and 30 of gestation (Joubert, 1956). Forty days after fertilization it still weighs only about 4.5 g. As pregnancy proceeds, the relative growth rate of the foetus declines but its absolute growth rate increases markedly. From the 455 observations collated by Joubert (1956), it can be seen (*Figure 2.1*) that the average foetus took 25 days to double in weight towards the end of pregnancy and that its growth rate continued to increase throughout pregnancy, exceeding 100 g/day near birth. However, there was a large variation between foetuses in growth such that birthweight ranged from about 2 kg to over 7 kg.

Many factors are known to affect foetal growth and lamb birthweight (Alexander, 1974). Most of these act by altering the rate of nutrient or

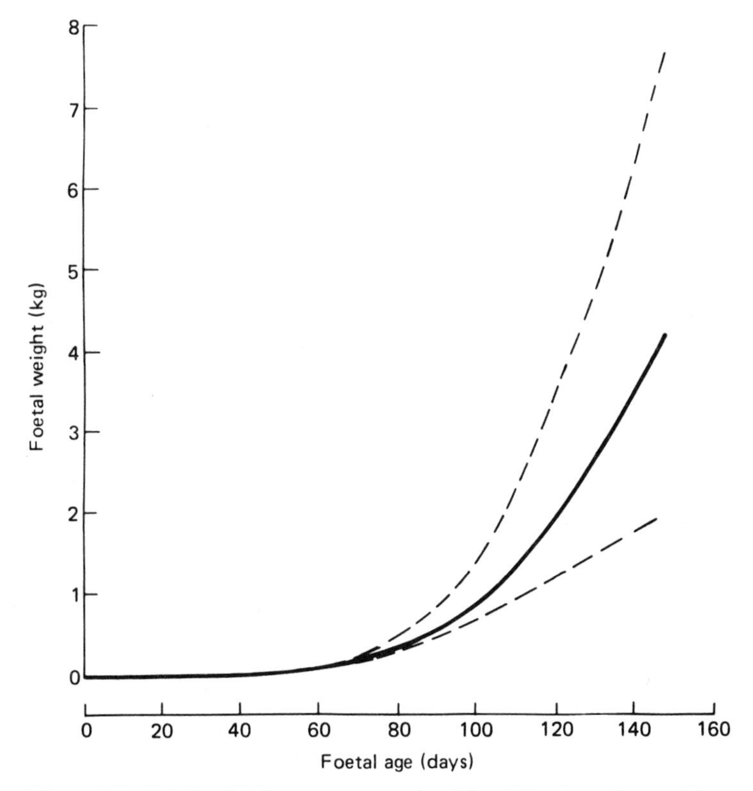

Figure 2.1 Relationship between age and weight of the sheep foetus. The continuous line represents the mean of 455 observations (Joubert, 1956) and the dotted lines indicate the approximate upper and lower range.

oxygen uptake by the foetus. Delivery of nutrients to the foetus depends upon the concentration of required substrates in maternal arterial circulation, the rate of blood flow to that part of the uterus in contact with the foetal placenta and the utilization of nutrients within the utero-placenta (Battaglia and Meschia, 1981).

In sheep, attachment of the foetal placenta to the uterus occurs at specific sites or caruncles where a button-shaped cotyledon of both maternal and foetal tissue develops. A strong correlation (r>0.8) has been observed between the weight of cotyledons and the birthweight of lambs from both normal ewes and others induced to produce small lambs either through under-feeding or heat stress (Alexander, 1974). Although cotyledon weight provides only a crude estimate of nutrient exchange, it is useful for assessing the relative nutrition of foetuses at the same stage of gestation. In well-fed ewes, the weight of cotyledons remains static or even falls during the last third of pregnancy, but surface villi, vascularity and blood flow continue to increase (Alexander, 1978b). Surgical removal of 60–84 caruncles prior to fertilization significantly reduces both the number of cotyledons formed and lamb birthweight (Alexander, 1974). Direct evidence that a reduction in blood flow to the uterus affects foetal growth has been obtained by Creasy *et al.* (1972) who embolized uterine arterial vessels of sheep with 15 μ microspheres at day 109 of gestation and observed a fall of 30% in the weight of foetuses 30 days later.

Litter size and sex

As litter size increases there is a decrease in the birthweight of individual lambs. Donald and Russell (1970) calculated from published information that the mean birthweight of twins was 80% of singles, and triplets 77% of twins. Robinson *et al.* (1977) observed successive reductions in lamb birthweight of 19%, 20% and 14% as litter size increased in the prolific Finn-Dorset ewe from 1 to 2, to 3 and finally to 4. However, there was considerable variation from time to time in these values.

Differences in foetal weight due to litter size become progressively greater as pregnancy proceeds. Twin and single foetuses are of similar weight at days 25 or 40 of gestation (El-Sheikh *et al.*, 1955; Foote *et al.*, 1959) but Robinson *et al.* (1977) calculated for the Finn-Dorset ewe that the weight of a foetus at 60 days was reduced by 5% for each increase in litter size.

The number of cotyledons attached to each foetus is affected by litter size and also the number of ovulations (Rhind, Robinson and McDonald, 1980). In ewes having 4–5 ovulations, the mean number of cotyledons/foetus was 49, 27 and 18 for 1, 3 and 5 viable foetuses in each uterus, respectively. Associated with the fall in cotyledons attached to each foetus, was a compensatory rise in the weight of individual cotyledons. The overall effect was a decline of only 12% in total cotyledon weight for each increase in litter size. The corresponding decrease in lamb birthweight was 11%. The number of cotyledons/foetus also declined substantially as the ovulation rate increased. For example, in ewes maintaining two foetuses, the number of cotyledons/foetus fell from 40 to 21 as ovulation rate increased from 2–3 to 6–9. This suggests that the effect of competition for caruncles during implantation remains even when some foetuses fail to survive. Foetal weight was reduced by 7% in ewes having six or more ovulations.

Male lambs are generally from 5 to 12% heavier than female lambs (Hunter, 1957; Everitt, 1964; Alexander, 1974; Robinson *et al.*, 1977). Contrary to other reports, Bradford *et al.* (1974) found in ova transfer experiments that females were heavier than males when litter size exceeded two. Rhind, Robinson and McDonald (1980) observed no difference between male and female foetuses in the number of cotyledons/foetus, but the cotyledons were 10.5% heavier for males than for their female litter mates. However, even when the effects of cotyledon weight were considered, male foetuses were still 4.9% heavier than female foetuses. Similar observations have been made by Everitt (1964).

Parental characteristics

There is a positive relationship between parental size and birthweight of lambs both between and within breeds (Donald and Russell, 1970). The embryo transplant experiments of Hunter (1957) and Dickinson *et al.*

(1962) clearly demonstrate that the genotype of lamb and ewe both affect birthweight. In the experiment of Hunter (1957), Border Leicester embryos carried by the small Welsh Mountain ewe instead of their own breed were 17% lighter at birth, indicating that the maternal environment can be insufficient to allow the foetus to grow to its genetic potential. When embryos of the Welsh Mountain breed were carried by Border Leicester ewes, birthweight increased by 14%, but was still only 70% of that achieved by Border Leicester embryos in the same environment. This result shows the importance of the growth potential of the foetus and Dickinson *et al.* (1962) concluded that the genotype of the lamb has a greater effect on birthweight than the genotype of the ewe.

Birthweight of lambs is strongly correlated with ewe liveweight at mating. From the equations developed by Donald and Russell (1970), it can be calculated that a change in ewe mating weight from 45 to 75 kg would result in an increase in lamb birthweight from 3.6 to 5.3 kg. A similar, but less pronounced, effect of mating weight on lamb birthweight has been observed in Finn-Dorset ewes (Robinson *et al.*, 1977). However, nutrition of the ewe during pregnancy can alter the relationship between ewe mating weight and lamb birthweight (Russel *et al.*, 1981).

The birthweight of lambs from maiden ewes is also less than from older ewes (Bradford *et al.*, 1974). The few studies on development of the placenta suggest that the total weight of cotyledon tissue is influenced by genotype and increases with age and parity of the ewe (Alexander, 1974).

Maternal nutrition

Effects of maternal nutrition on foetal growth and lamb birthweight have been reviewed (Everitt, 1968; Robinson, 1977 and Chapter 6; Robinson and McDonald, 1979). The major influence is during the last eight weeks of gestation when foetal growth is most rapid. However, there are conflicting reports regarding the effect of nutrient intake prior to about day 100 of gestation. In various experiments, nutrient restriction during mid pregnancy has been found either to have no effect (Wallace, 1948; Stern *et al.*, 1978), to reduce (Bennett, Axelsen and Chapman, 1964; Everitt, 1964; Curll, Davidson and Freer, 1975; Russel *et al.*, 1981) or to stimulate (Faichney, 1981; Russel *et al.*, 1981) foetal growth and lamb birthweight.

In the experiment of Russel *et al.* (1981), ewes selected from two flocks had vastly different body weights when mated at 18 months of age (*Table 2.1*). During days 30–98 of pregnancy, animals from each flock were fed either a high plane of nutrition sufficient to maintain maternal body weight or a low plane which resulted in an estimated maternal weight loss of 5–6 kg. All ewes were then fed adequately during the last phase of pregnancy. Birthweight of the lambs from the lighter ewes was significantly reduced by poor nutrition during mid pregnancy, whereas the reverse was seen in the heavier ewes.

Everitt (1968) noted that severe under-nutrition during early and mid pregnancy reduced the number of cotyledons attached to single foetuses. However, by day 91 of gestation there was considerable compensation in the size of the attached cotyledons, although the total weight of cotyledon

Table 2.1. EFFECT OF EWE LIVEWEIGHT AND FEEDING LEVEL FROM DAY 30 TO DAY 98 OF GESTATION ON THE BIRTHWEIGHT OF LAMBS

Flock	*Mating weight* (kg)	*Nutrition in mid pregnancy*[(a)]	*Lamb birthweight* (kg)
A	42.5	High	3.83
		Low	3.32
B	54.5	High	4.23
		Low	4.95

[(a)]High level of nutrition sufficient to maintain conceptus-free, ewe body weight; low level resulted in an estimated loss of 5–6 kg in ewe body weight.
From Russel *et al.* (1981)

tissue was still less than in well-nourished animals. Growth of the placenta in sheep often ceases around days 80 to 100 of gestation (Alexander, 1974; Robinson *et al.*, 1977). Nevertheless, in the experiment of Rattray *et al.* (1974), weight of the placenta, including the cotyledons, of twin foetuses, but not single foetuses, continued to increase between days 124 and 140 of gestation.

Faichney (1981) has also shown that moderate nutritional restriction between days 50 and 100 of pregnancy can stimulate foetal weight at day 135 when compared with animals well-fed throughout (*Table 2.2*). Moderate nutrient restriction after day 100 did not affect foetal weight at day 135 when compared with well-fed animals, but reduced it significantly from that found in animals restricted during mid pregnancy. Weights of the placentae of these animals (*Table 2.2*) show clearly that moderate nutrient restriction during pregnancy stimulates placental growth and the effect continues after day 100 of gestation.

Table 2.2. EFFECT OF VARYING FEED INTAKE OF EWES DURING GESTATION ON FOETAL AND PLACENTAL WEIGHT AT 135 DAYS AFTER FERTILIZATION

Treatment	*Feed intake*[(a)] (g/day)		*Foetus weight* (kg)	*Placenta weight* (g)
	Day of gestation			
	50–99	*100–135*		
MM	900	900	3.3^{ab}	321^{a}
MR	900	500	3.3^{ab}	437^{ab}
RM	500	900	3.7^{a}	463^{b}
RR	500	500	3.0^{b}	413^{ab}

[(a)]A pelleted lucerne hay, oat grain diet (3:2). M = 900 g/day sufficient to maintain conceptus-free ewe body weight; R = 500 g/day restricted.
^{a,b}Means with different letters within columns are different at $P<0.05$.
From Faichney (1981)

Foetal growth and lamb birthweight appear to be differentially affected depending upon the severity, duration and timing of a nutritional restriction during pregnancy. The critical times are probably around the period of implantation and also late in pregnancy when the placenta loses some of its ability to compensate for a limited nutrient supply. At other times, moderate nutrient restriction stimulates cotyledon growth but severe restriction has been shown to cause involution of the cotyledons (Mellor and Murray, 1981). The duration of a period of under-nutrition is likely to affect the extent of change, either compensation or involution, in the cotyledons.

Thus, moderate nutrient restriction after implantation would appear to stimulate cotyledon development in an attempt to compensate for the lower nutrient supply from maternal blood. In some cases, the compensation may be sufficient to maintain foetal growth rate (Faichney, 1981). Subsequent realimentation to a higher level is then likely to stimulate nutrient transfer across the placenta to a greater extent than would have occurred without the period of nutrient restriction. Hence the growth rate of the foetus is raised above normal. On the other hand, with progressive under-nutrition, the ability of the cotyledons to compensate declines and eventually their partial involution results. In situations where no compensation occurs, refeeding would not stimulate foetal growth above that of continuously well-fed animals, and smaller foetuses would result.

Examination of the results of Russel *et al.* (1981) show that the light animals were more severely treated, losing 3–3.5 kg during the first six weeks of under-feeding compared with a loss of only 1.5 kg in the heavier animals. This may have caused a differential response in placental growth and thus explain their apparently contradictory results.

The influence of maternal nutrition on foetal growth during late pregnancy has been summarized by Robinson (1977 and Chapter 6) and Robinson and McDonald (1979). Foetal growth appears not to be affected greatly until energy intake falls below that required for maintenance of the ewe's tissue. Indeed, some evidence suggests that submaintenance feeding may be tolerated in animals of fat condition before foetal growth rate declines (Robinson, 1977). Nevertheless, Rattray *et al.* (1974) observed a 10% difference in the birthweight of lambs from ewes fed at either 1.5 or 2.0 times maintenance.

Environmental conditions

Exposure of pregnant sheep to ambient temperatures sufficient to raise deep body temperature for several hours daily, has been shown to reduce the birthweight of lambs (Alexander, 1974). Although heat stress affects the appetite of ewes, the effects on foetal growth of heat exposure are greater than those associated with the reduced intake. Experimentally imposed heat stress in ewes during both mid and late pregnancy reduced lamb birthweight to between 1.5 to 2.0 kg. Associated with heat stress is a reduction in both the weight (Alexander, 1974) and blood flow (G. Alexander, J.R.S. Hales, D. Stevens and J.B. Donnelly, unpublished observations) of the cotyledon tissue.

Various diseases, such as ovine brucellosis, affect foetal growth probably by reducing the effectiveness of nutrient exchange (Alexander, 1974). Alterations to day length also affect foetal growth (Hulet, Foote and Price, 1969).

PREDICTION OF FOETAL GROWTH RATE

Robinson and McDonald (1979) examined a number of relationships which have been used to describe growth of foetal lambs. They conclude, as do

Geisler and Jones (1979), that the Gompertz equation best describes foetal growth when nutrient supply is adequate because it extrapolates to realistic values in early pregnancy and incorporates an exponentially decreasing specific growth rate.

Geisler and Jones (1979) use the equations established by Dickinson *et al.* (1962) or by Donald and Russell (1970) for a variety of breeds to predict lamb birthweight from the mating weight of the ewe and litter size. They then use a Gompertz equation established from a range of data to predict foetal weight at any time from conception. Robinson *et al.* (1977), on the other hand, have modified the Gompertz equation to include terms which account for the effect of litter size, weight of ewe at mating and sex of lamb. Both approaches give similar estimates of foetal weight when the birthweights of lambs are identical. However the equation of Robinson *et al.* (1977) predicts a smaller effect of both litter size and ewe mating weight on lamb birthweight and may apply more specifically to Finn-Dorset ewes (Geisler and Jones, 1979).

Normal growth rate of the foetus at any stage of gestation can be obtained by selecting the most appropriate equation for the specific situation and differentiating it with respect to time. There is insufficient information to predict precisely the effect of nutrient restriction on foetal growth rate. However, foetal growth rate appears to fall curvilinearly as metabolizable energy intake is reduced below that required for maintenance of the ewe's body tissue (Robinson, Chapter 6). This is currently represented in our computer model of sheep growth (Graham *et al.*, 1976) by the following equation

$$G = P(2E_mE - E^2)/E_m^2$$

where G and P are the actual and normal growth rate of the foetus, respectively, E_m is the metabolizable energy intake required for maintenance of the ewe and E metabolizable energy intake. When E is above E_m, actual foetal growth is equal to normal growth. This representation is considered to be more appropriate than the methods used earlier (Graham *et al.*, 1976; Geisler and Jones, 1979), where foetal growth was only reduced when estimated body energy, nitrogen or weight loss exceeded predetermined values. There is at present insufficient information to predict with confidence the compensatory stimulus to foetal growth following refeeding after moderate nutrient restriction.

ORGAN DEVELOPMENT AND BODY COMPOSITION

Development of individual organs, skeletal tissue and muscles of the foetal lamb has been examined (Wallace, 1948; Joubert, 1956; Thurley, Revfeim and Wilson, 1973; Gabbedy, 1974; McDonald, Wenham and Robinson, 1977; Alexander, 1978a; Richardson and Hebert, 1978). The pattern of growth varies between tissues (*Figure 2.2*). Relative to the rate of change in foetal weight, some organs such as the liver, thymus, brain and lungs develop rapidly in early gestation and may even fall in weight towards birth. Others, like the adrenals, skeleton, muscles and fat depots tend to

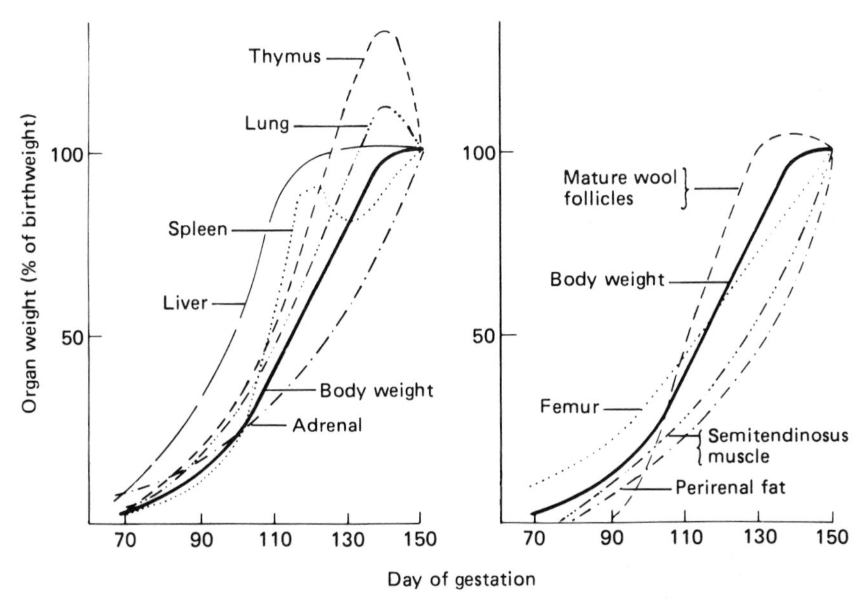

Figure 2.2 Growth of foetal organs and tissues relative to their weight at birth. Internal organs after Thurley, Revfeim and Wilson (1973), wool follicles (Carter and Hardy, 1947), femur (McDonald, Wenham and Robinson, 1977), semitendinosus muscle (Gabbedy, 1974) and perirenal fat (Alexander, 1978a).

develop most rapidly late in gestation. Wool follicles mature mainly between days 90 and 130 of gestation.

The differential pattern of organ growth influences the way data should be examined when attempting to determine the consequences of a retardation in foetal growth on the relative development of individual organs. Creasy *et al.* (1972) and Alexander (1974) concluded from a comparison of organ weight with body weight at a specific gestational age that the growth of some organs but not others was affected by foetal growth retardation. However, if lambs that are small at birth have identical proportions to normal foetuses of the same weight, the differential pattern of organ development throughout gestation would lead to this conclusion, even though relative to body weight there may be no differential development. Everitt (1968) used the allometric equation to compare organ weight with birthweight in lambs receiving different nutrition during foetal development and concluded that malnourished lambs appear disproportioned at birth only because they are small. However, if this were true, the liver for example, should be larger relative to body weight not smaller as observed by Alexander (1974).

McDonald, Wenham and Robinson (1977) outlined the limitations of the allometric equation for describing organ development during foetal growth. They concluded that both foetal weight and gestational age should be considered and used a modified Gompertz equation to compare organ weight with that expected for the particular gestational age. These analyses show that foetal growth retardation can cause differential development of some tissues (McDonald, Wenham and Robinson, 1977; Robinson, 1981).

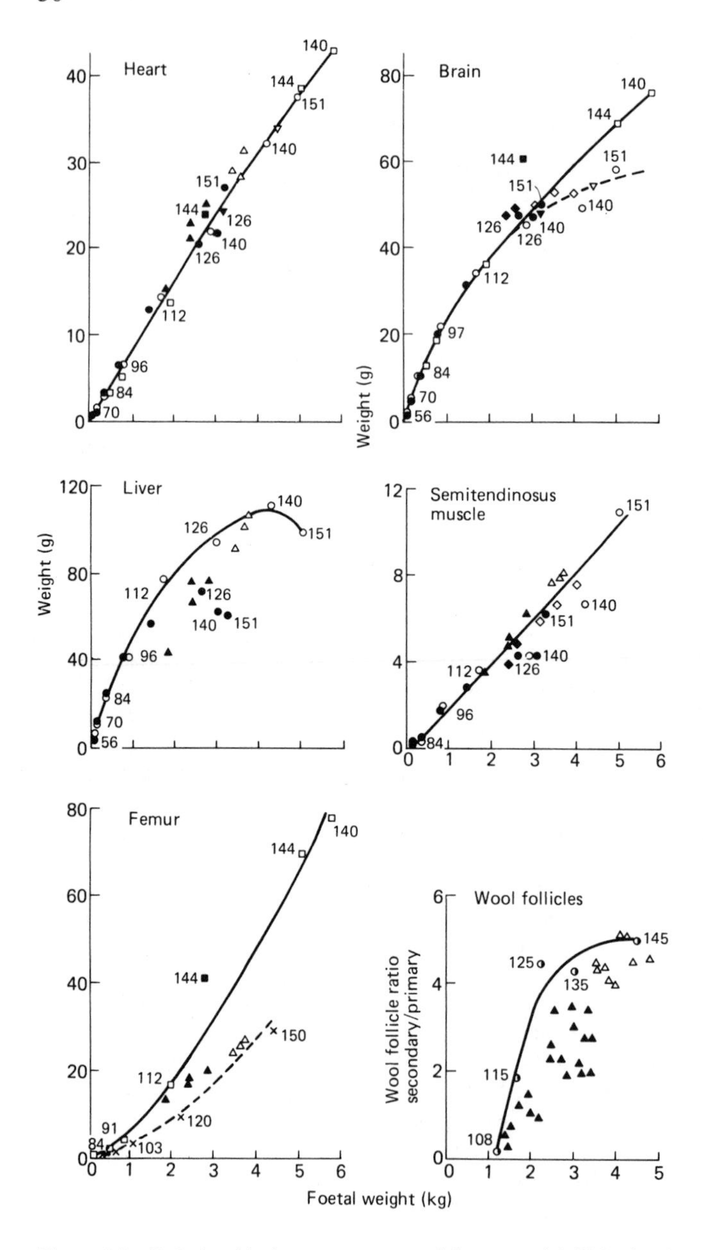

Figure 2.3 Relationship between organ weight or wool follicle development and weight of foetuses of different gestational age and treatment. Open symbols, crosses and half closed symbols for wool follicles represent normal animals. Closed symbols represent treatments in which foetal growth rate has been restricted. Numbers indicate gestational age (days) for data sets in which it varies. △, all at birth, Alexander (1974); ◑, variable age, Carter and Hardy (1947); ▽, day 139 of gestation, Creasy *et al.* (1972); ○, variable age, Gabbedy (1974); ×, variable age, Joubert (1956); ◇, all at birth, Sykes and Field (1972); □, variable age, Wallace (1948). Lines indicate the pattern of normal growth (drawn by eye). Separate lines are drawn for brain and brain plus spinal cord. Data for femur are separated between those of Joubert and Wallace.

Weights of several tissues from foetuses of different gestational age and growth rate are graphed against body weight in *Figure 2.3*. Although variation exists among data sources, a comparison between foetuses of the same weight but vastly different ages indicates that there is little effect of growth retardation on development of the heart or skeletal muscle, but that light animals at birth have smaller livers and heavier bones than normal foetuses of the same weight. The liver of lambs weighing about 2 kg at birth appears to be little over half the weight of normal foetuses, whereas the femur is nearly twice as heavy. Although less conclusive, the data indicate that the brain and spinal cord may be more developed at birth in growth-retarded animals.

Development of secondary, but not primary, wool follicles is adversely affected by growth retardation during the last third of pregnancy (Corbett, 1979). When compared with normal foetuses, those of light birthweight have a lower ratio of secondary to primary follicles. This could have serious consequences on further follicle development because follicle initiation is completed by birth, even in severely undernourished animals (Schinckel and Short, 1961).

Changes in chemical composition of the developing foetus (*Table 2.3*) observed by McDonald *et al.* (1979) agree closely with other studies

Table 2.3. CHANGE IN THE CHEMICAL COMPOSITION (g/kg) OF FOETUSES WITH GESTATIONAL AGE

Component	*Age of foetus* (days)			
	60	*88*	*116*	*144*
Dry matter	91	125	164	206
Crude protein	64	86	113	145
Fat	8	12	18	24
Ash	14	23	29	34
Calcium	2.2	5.0	7.7	9.7
Phosphorus	1.5	3.4	4.8	5.4
Magnesium	0.11	0.18	0.24	0.28
Sodium	2.2	2.0	1.9	1.9
Potassium	1.5	1.7	1.7	1.7

By courtesy of McDonald *et al.* (1979)

(Langlands and Sutherland, 1968; Sykes and Field, 1972; Gabbedy, 1974; Rattray *et al.*, 1974). Dry matter content of the foetus increases from under 10% at day 60 of gestation to over 20% at term and is associated with increases in all major constituents except sodium. Unfortunately the chemical composition of severely retarded foetuses has not been recorded. Rattray *et al.* (1974) found no difference between twin and single foetuses in composition at the same gestational age, but Gabbedy (1974) observed a larger ash and smaller protein and fat content in twin than in single foetuses near term. The analyses of McDonald *et al.* (1979) indicate that light for age foetuses have an immature composition more appropriate to their weight.

There is some evidence that the fat content of growth-retarded foetuses is reduced. Alexander (1974, 1978a) noted that the proportion of fat in new born lambs ranges widely from 1.5 to 4.5% and tends to be lower in light birthweight animals. Gabbedy (1974) observed that the fat content of twin, but not single, foetuses declined from day 126 of gestation and at term was almost 25% less than in single foetuses of the same weight. In addition, Robinson (1981) calculated that foetuses 40% above their expected weight for gestational age contained substantially more fat relative to skeletal size than did foetuses 40% below expected weight.

CONSEQUENCES OF PRENATAL GROWTH RETARDATION

Lamb survival

The major consequence of prenatal growth retardation is on lamb survival. Neonatal mortality of lambs increases markedly in many environments as birthweight falls below about 3 kg. Compared with normal lambs, low birthweight animals have reduced insulation due to the smaller number of wool fibres, greater relative heat loss because of their larger surface area/unit of body weight and a reduced capacity to maintain heat production because of their lower fat and energy reserves (Alexander, 1974). All these factors increase their susceptibility to environmental stress and reduce their ability to compete with normal sized siblings.

Growth rate, body composition and ultimate body size

There are few studies of the consequences of prenatal growth retardation on the subsequent performance of sheep. A reduction in growth rate of lambs has been observed following restricted maternal nutrition during pregnancy (Wallace, 1948; Short, 1955; Curll, Davidson and Freer, 1975; Tissier and Theriez, 1979). However, it is not possible to distinguish the effects of a lower milk production of the ewe from a reduced capacity of the lamb to grow. When Taplin and Everitt (1964) artificially fed lambs whose birthweights had been manipulated by maternal nutrition during pregnancy, they found that the absolute growth rate of light lambs was reduced, but their relative growth rate was higher than for heavier lambs. Nevertheless, there were still substantial differences in liveweight when the experiment finished 20 weeks after birth. There is evidence from some experiments with lambs reared normally (Schinckel and Short, 1961; Everitt, 1967) but not others (Short, 1955) that severe prenatal growth retardation in late pregnancy results in a reduction of up to 10% in the liveweight of animals at 2–2.5 years of age. Severe early postnatal growth retardation also appears to have a small effect on mature body size (Allden, 1979). Gunn (1977) observed that moderate under-nutrition during the last six weeks of pregnancy and the first 12 months of life reduced liveweight of Scottish Blackface ewes by 5–6 kg at maturity.

There is little evidence that prenatal growth retardation has a permanent effect on organ size or chemical composition of lambs. Differences noted by Taplin and Everitt (1964) in length of the cannon bone at 6 kg liveweight between lambs from ewes either well-fed or poorly-fed during pregnancy had disappeared by the time the animals weighed 16 kg. Similarly, differences between twin and single lambs in liver weight and fat content relative to body weight observed by Gabbedy (1974) near birth were no longer apparent when the animals weighed 6–10 kg. Nevertheless, McClelland, Bonaiti and Taylor (1976) found that 3.3% of the variance in the total fat content of lambs from several breeds and slaughtered at different stages of maturity was attributable to litter size which was manipulated from one to five by ova transfer.

Wool growth

Growth retardation during late pregnancy can affect wool production of adult sheep from some breeds. Schinckel and Short (1961) and Everitt (1967) found that wool production of Merino sheep at 2–2.5 years of age was about 8% less in animals born of ewes under-fed during late pregnancy. Alternatively, Short (1955) showed that, although the number of secondary follicles was reduced by prenatal growth restriction in Merinos, a compensatory increase in the growth of each fibre meant that there was no difference in wool production at 200 days of age between lambs from ewes fed differentially during pregnancy. In a recent review, Corbett (1979) suggests that the lifetime wool production of twin lambs is about 4% less and of lambs born to maiden ewes about 6% less than from single lambs of multiparous Merino ewes. Although early postnatal nutrition can affect subsequent wool production (Schinckel and Short, 1961; Allden, 1979), the wool growth of twin lambs reared as singles is still often less than for single lambs (Corbett, 1979). The effects of prenatal nutrient restriction on subsequent wool growth appear to be restricted to sheep breeds, like the Merino, which have a high ratio of secondary to primary follicles (Allden, 1970).

Postnatal growth and development

During postnatal life, growth rate of the lamb is primarily determined by energy intake relative to liveweight. Under ideal conditions, growth rate tends to remain relatively constant from soon after birth until the animal reaches about half mature weight when it then progressively declines to zero at maturity. Rate of growth and the relative development of tissues can be modified by composition of the diet, breed and sex of animal and environmental conditions such as disease and extremes in ambient temperatures.

POSTNATAL GROWTH OF ORGANS AND TISSUES

Differences in the relative growth rate of organs and tissues continues throughout postnatal life (*Figure 2.4*). By convention, tissues which are a greater proportion of their mature weight than is body weight or which increase in weight at a slower relative rate than body weight over the postnatal period are classified as early maturing. Tissues which have the

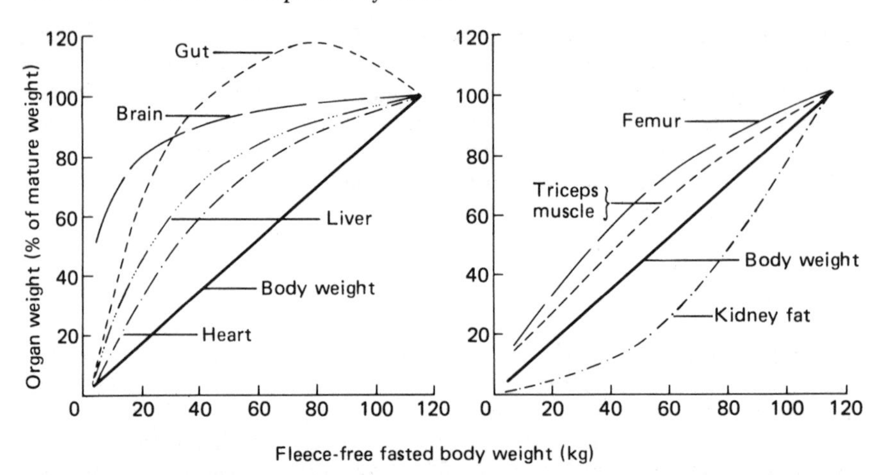

Figure 2.4 Postnatal growth of organs and tissues relative to their weight at maturity. Data above 20 kg from R.M. Butterfield (unpublished) for Large Merino rams fed a good quality diet *ad libitum* until they ceased to increase in weight and reached a mean fleece-free fasted liveweight of 116 kg. Data at birth: brain, femur, heart (Alexander, 1974); kidney fat (Alexander, 1978a); liver (Gabbedy, 1974); triceps muscle (Lohse, Moss and Butterfield, 1971); gut (Wardrop, 1960).

converse characteristics are classified as late maturing. In many studies, allometric relationships between tissue weight and body weight have been used to make these classifications. However, the allometric constant changes throughout the postnatal period for many tissues (Benevent, 1971; Lohse, Moss and Butterfield, 1971) and overall relationships need to be interpreted carefully.

Most internal organs in sheep are early maturing (Benevent, 1971; Kirton, Fourie and Jury, 1972). Brain represents the extreme, having reached about 90% of its maximum size by the time the animal is 35% of mature weight. Major exceptions are some parts of the digestive tract and organs associated with sexual development (Kirton, Fourie and Jury, 1972). The small intestine is early maturing, whereas the large intestine and abomasum mature at a rate similar to the whole body (Kirton, Fourie and Jury, 1972). The rumen, reticulum and omasum increase in weight extremely rapidly early in postnatal life and there is some indication that the total weight of the digestive tract falls as maturity is approached (*Figure 2.4*). Small differences between breeds and sexes have been observed in the allometric growth coefficients for some organs, but the general classification of maturation rate remains similar (Kirton, Fourie and Jury, 1972). Sex differences are not apparent in lambs weighing less than 10 kg (Benevent, 1971).

Of the major body components, the order of maturation is skeleton, muscle and fat, with only fat being classified as late maturing (Fourie, Kirton and Jury, 1970; Butterfield *et al.*, 1983). Studies of the growth patterns of individual bones (Prud'hon *et al.*, 1978; Thompson, Atkins and Gilmour, 1979b), muscles (Lohse, Moss and Butterfield, 1971; Thompson, Atkins and Gilmour, 1979b) and fat depots (Thompson, Atkins and Gilmour, 1979a; Wood *et al.*, 1980) in the whole body or carcass of sheep indicate a general disto-proximal pattern of growth. Tissues of the limbs tend to be a greater proportion of mature weight in early life than do tissues of the trunk. Although all fat depots are classified as late maturing, they vary in their pattern of development, maturing in the order: intermuscular, channel, kidney, subcutaneous and omental (Kirton, Fourie and Jury, 1972; Wood *et al.*, 1980).

Differences occur between genotypes in the pattern of development of major body components. Males have a greater postnatal development of the bones (Prud'hon *et al.*, 1978) and muscles (Lohse, 1973) of the head and neck region than do females. Likewise, the muscles of the proximal pelvic limb mature later in males (Lohse, 1973). As a result of these differences in muscle distribution, females have a greater proportion of their muscle tissue in the prime commercial meat cuts than do males (Taylor, Mason and McClelland, 1980). Although there are differences between breeds compared at the same weight in carcass composition and the relative distribution of muscles and bones, these differences largely, but not entirely, disappear when comparisons are made at the same percentage of mature weight (Taylor, Mason and McClelland, 1980) and reflect the relative rate of maturation of different tissues.

The patterns of organ development described above do not necessarily hold when growth rate diverges markedly from normal or when the introduction of solid feed to lambs is delayed. Robinson (1948) grew adult sheep rapidly either from 32 to 68 kg or in the reverse direction and observed that most organs and tissues followed similar growth patterns relative to body weight during both the increasing and decreasing weight phases. However, the weight of some organs such as brain and eyes remained unchanged throughout the experiment, and weight of the skeleton changed only slightly. Other organs like the liver, kidneys and digestive tract showed marked divergence of growth patterns, increasing rapidly in weight when animals were first introduced to supermaintenance diets and showing a marked atrophy following the initial feeding of submaintenance diets. Organs associated with digestion and metabolism fluctuate widely in weight in relation to nutritional and physiological status. For example, the reticulo-rumen of lambs does not undergo rapid development until solid feed is eaten (Wardrop, 1960) and shows substantial atrophy and loss of papillae if adult sheep are given a liquid diet per abomasum (Black, Robards and Thomas, 1973). From a review of published data, Baldwin and Black (1979) showed that there was not a close relationship between the weight of these organs and body weight in sheep. Reported weights of liver range from less than 100 g to over 1600 g for sheep weighing between 40 and 50 kg. The growth pattern of muscles also appears to be affected by nutritional treatment, but to a smaller extent. Murray and Slezacek (1975) found that lambs grown slowly had greater development of leg muscles and lesser development of abdominal muscles than did animals grown to the same weight more rapidly.

Because the growth pattern of many organs is affected by the nutritional and physiological status of an animal and by its genotype, Baldwin and Black (1979) developed a computer simulation model of mammalian tissue growth which was based on describing the changes which occur in cell number and cell size of organs in relation to genetic and nutritional factors.

Details of this model are not described, but limited data on the cellularity of organs in sheep are reviewed.

CHANGES IN TISSUE CELLULARITY DURING GROWTH

Tissue growth occurs by an increase in both cell number (hyperplasia) and cell size (hypertrophy). Although the number of muscle fibre cells in mammals is fixed near birth, the number of nuclei in each cell continues to increase into postnatal life through transfer from satellite cells outside the fibre (Allen, Merkel and Young, 1979). Cheek *et al.* (1971) suggested that each nucleus has the capacity to support a certain amount of cell tissue. Thus to avoid problems with polyploid tissues, it is convenient to regard hyperplasia as an increase in the amount of deoxyribonucleic acid (DNA) in each organ and hypertrophy as a change in the mass of tissue/unit DNA.

N.M. Tulloh and B.J. Gabbedy (personal communication) obtained comprehensive data on changes in cellularity of several organs from day 56 of gestation to 10 weeks after birth for both single and twin lambs. Data for brain, kidney and quadriceps muscle are presented in *Figure 2.5*. Although birthweight of the twin lambs was only 67% of the single lambs, litter size had no effect on cellularity when animals were compared at the same weight. However, there were substantial differences between organs in the

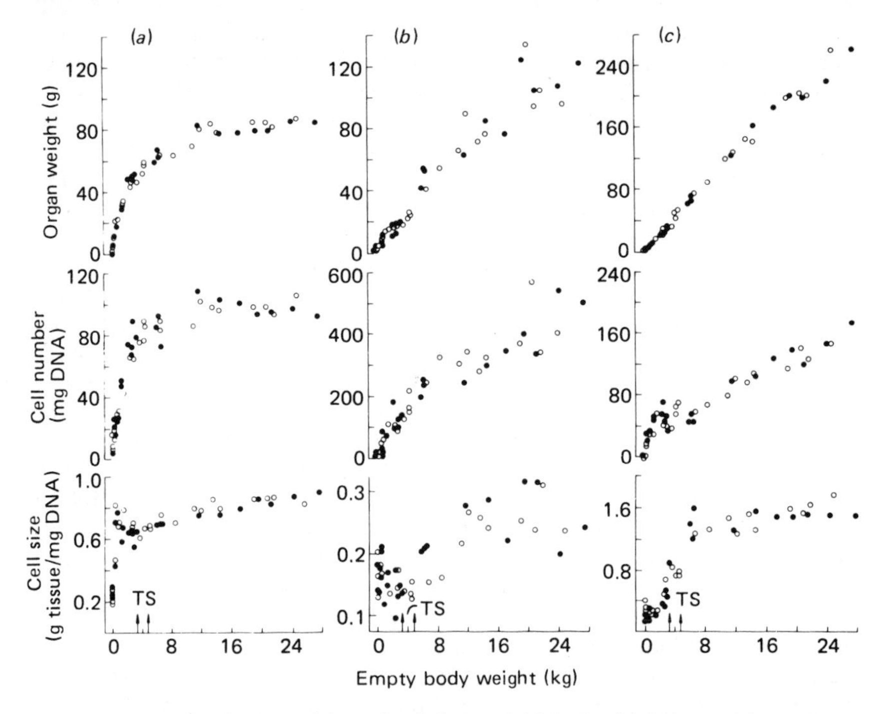

Figure 2.5 Changes in the weight and cellularity of (a) brain, (b) kidneys, (c) quadriceps muscle of twin (●) and single (○) lambs from day 56 of gestation to 10 weeks after birth. Arrows indicate birth weight of twin (T) and single (S) lambs. Lambs were all artificially fed after birth. From N.M. Tulloh and B.J. Gabbedy (unpublished).

pattern of cell growth. Brain showed rapid hypertrophy in early life with cell size increasing four-fold by the time the foetus weighed 500 g; a slight increase in cell size then continued until the end of the experiment. Thus, the marked increase in brain weight in late pregnancy was mainly associated with hyperplasia. However, there was no increase in cell number after the lambs weighed about 10 kg and the subsequent increase in brain weight was entirely the result of hypertrophy. With kidneys, cells were much smaller and the overall increase in size was only two-fold, but continued more or less uniformly throughout the experiment. Although the rate of hyperplasia appeared to decline after birth, growth of the kidneys resulted from substantial hyperplasia and hypertrophy. Cell size in the quadriceps muscle increased rapidly during foetal and early postnatal life and showed only a slight increase after the animals weighed 6 kg. However, hyperplasia in this muscle continued rapidly throughout the experiment and was responsible for most of the increase in tissue weight during the postnatal period.

These results, like those of Sands, Dobbing and Gratrix (1979) do not support the widely held contention that tissues grow first by increasing cell number and subsequently by increasing the size of these cells. Clearly, hypertrophy contributed significantly to the early growth of all organs. However, evidence from brain suggests that hyperplasia ceases before maximum organ size is achieved. The results of Johns and Bergen (1976) who studied the cellularity of the liver and gastrocnemius muscle in lambs from birth to 45 kg also indicate that there was no increase in cell number after the animals weighed 35 kg and that continuing organ growth resulted entirely from hypertrophy. Similar conclusions have been drawn by Hood and Thornton (1979) for adipose tissue. In continuously growing lambs, these workers observed a two-fold increase in adipose cell number as animals grew from 7–11 months of age (28–45 kg), but no change subsequently. In contrast, adipose cell size increased linearly with increasing weight of body fat. Thus, in animals weighing more than 45 kg, the entire increase in body fat was attributable to increases in cell size.

The cessation in cellular hyperplasia is one feature of growth which distinguishes land-dwelling mammals and birds from many other forms of animal life. In some of these mammals, nutrient restriction during the period of hyperplasia can result in reduced cell division and a permanent reduction in organ size. However, this has not been adequately examined in sheep. Hood and Thornton (1979) and Haugebak, Hedrick and Asplund (1974) found only small differences in adipose cell numbers between lambs rehabilitated after periods of weight loss or weight maintenance and normally grown animals, although there were substantial differences in fat weight and cell size. Similarly, Johns and Bergen (1976) did not observe any difference in muscle cell number of lambs given low protein or normal diets. There is insufficient evidence to assess whether the reduction in adult size of sheep following severe malnutrition in prenatal and early postnatal life is due to a reduction in hyperplasia, but it is a plausible explanation.

Changes in organ weight in adult sheep result almost entirely from changes in cell size. Masters (1963) fed 3–4 year old ewes for eight weeks an amount of poor quality diet sufficient only to prevent death and found that the DNA content of heart, liver, kidneys, spleen, pancreas, lungs and brain was unaltered despite marked reductions in the weight of some organs. Similarly, Symons and Jones (1975) found that infestation of sheep with intestinal nematodes resulted in substantial reductions in the size of cells in both liver and muscle tissue, but did not alter their DNA content. The same observation has been made for adipose tissue in lambs in which adipose hyperplasia has ceased (Hood and Thornton, 1979). Probable exceptions to the above pattern occur in gut and skin tissues. Significant cell turnover continues throughout life in these tissues and a decrease in both cell number and cell size would be expected during weight loss, but this has not been examined in sheep.

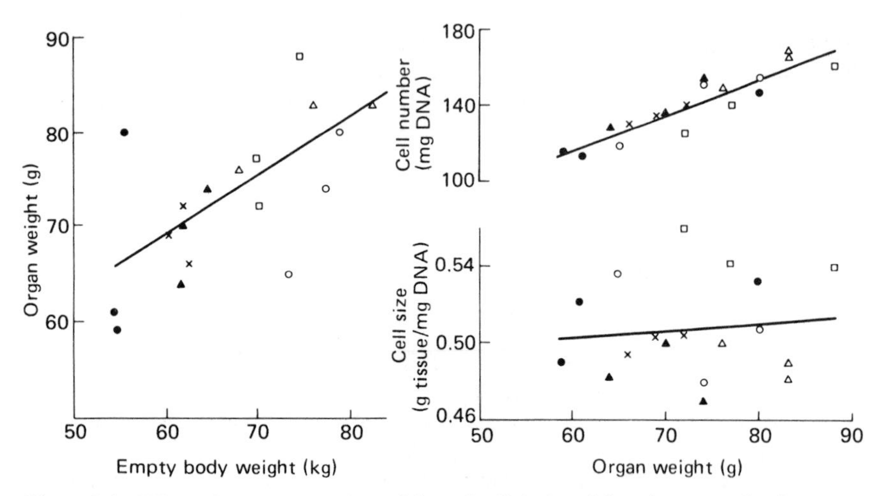

Figure 2.6 Effect of genotype on the weight and cellularity of the triceps muscle of mature, 3–6 year old rams. ○, Large Merino strain; ×, Small Merino strain; ●, Camden Park Merino; □, Corriedale; △, Dorset Horn; ▲, Cheviot. The Large and Small Merino strains are from the same flock as those described by Butterfield *et al.* (1983). From W.F. Colebrook, J.L. Black and T.W. Searle (unpublished).

Baldwin and Black (1979) postulated that the major determinant of ultimate organ size in mammals was the genetic control of cell number, with smaller differences resulting from the genetic control of potential cell size. To examine this postulate in sheep the cellularity of 3–6 year old rams from several genotypes (Cheviot, Dorset Horn, Corriedale and three strains of Merino) ranging in body weight from 54–90 kg was recently studied (Colebrook, Black and Searle, unpublished). All animals were offered a good quality diet *ad libitum* for at least 12 weeks before slaughter to reduce the effect of previous nutrition on cell size. Results for the triceps muscle (*Figure 2.6*) confirm that organ weight and cell number are closely correlated ($r = 0.91$) and that there was little overall effect of organ weight on cell size. Nevertheless, the cells of Corriedales appear consistently larger than the cells of Dorset Horns. Results from the mouse selection experiment of Falconer, Gauld and Roberts (1978) also indicate that both cell number and size vary with genotype. In contrast to differences between selection lines, these authors noted that cell size contributed more to differences between the sexes than did cell number. This has not been examined in sheep. However, testosterone has been shown to increase both cell proliferation and protein synthesis in muscle cells in tissue culture (Allen, Merkel and Young, 1979).

FACTORS AFFECTING CHEMICAL COMPOSITION OF THE WHOLE BODY

Age and body weight

There is no intrinsic relationship between age and chemical composition of sheep (*Figure 2.7(a)*). The figure contains results for sheep of several breeds and sexes subjected to various nutritional treatments including prolonged milk feeding, periods of starvation or submaintenance feeding

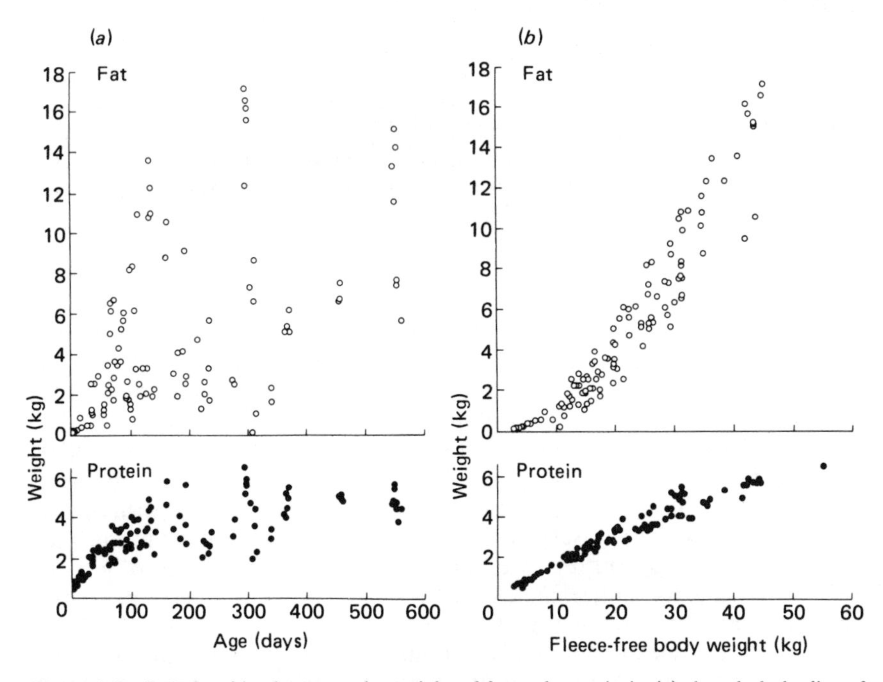

Figure 2.7 Relationships between the weight of fat and protein in (a) the whole bodies of sheep and age or (b) fleece-free empty body weight for animals of different genotypes and given a range of nutritional treatments.

and intakes of balanced diets which produced a range of positive growth rates. At 300 days of age, the fat content of the animals ranged from 100 g to 17 kg. However, as was originally shown by Tulloh (1964) and Reid *et al.* (1968), there is a relatively close relationship between body composition and body weight (*Figure 2.7(b)*). Light animals contain mostly lean and little fat, but as they become heavier the proportion of fat in the body increases.

The rate of fat accretion increases markedly at some empty body weight between 10 and 35 kg. Beyond this weight has been termed the fattening phase of growth and the relationship between fat content and empty body weight remains substantially linear (Searle, Graham and O'Callaghan,

1972). When nutrient intake is not limiting growth, the relationship between body fat and empty body weight can be described with a two-phase linear regression, joined by a smooth transition (Searle and Griffiths, 1976a). Nevertheless, the body composition of lambs weighing less than 40 kg is known to be influenced by a variety of factors. The fat content of sheep weighing 30 kg ranged from 5–9 kg or from 17–30% of body weight in *Figure 2.7(b)*.

Most of the variation in body composition of sheep appears to be associated with the amount of fat. Searle and Griffiths (1983) showed that the amounts of protein, water and ash were linearly related to weight of the fat-free body for animals of different genotypes subjected to the range of nutritional treatments described for *Figure 2.7*. The slopes of these relationships were unaffected by treatment, but there were small differences in intercept such that prolonged milk feeding produced animals with a slightly higher proportion of water and there was more ash in older animals of low body weight. The overall equations suggest that as fat-free empty body weight changes from 5 to 40 kg the proportions of water, protein and ash change from 0.795, 0.171, 0.041 to 0.730, 0.210, 0.053, respectively. Corresponding asymptote values are 0.721, 0.215, 0.055.

Sex and breed

When comparisons are made at the same weight, genotypes which are heavier at maturity generally grow faster, contain less fat and more protein and bone in their whole bodies and carcasses than do animals of smaller mature size (McClelland, Bonaiti and Taylor, 1976; Searle and Griffiths, 1976a,b; Thompson, Atkins and Gilmour, 1979a; Wood *et al.*, 1980; Theriez, Tissier and Robelin, 1981). Differences due to sex do not become apparent until the commencement of the fattening phase of growth. Searle and Griffiths (1976a) observed that the rate of change in body fat with body weight during both the pre-fattening and fattening phases of growth were the same for entire males and females, but the transition to the fattening phase was later in males (*Figure 2.8(a)*). Castrated males have also been reported to contain less fat and more bone than females of the same weight (Thompson, Atkins and Gilmour, 1979a), but the differences are smaller than between entire males and females.

When comparing two strains and a cross of the Merino breed, Searle and Griffiths (1976b) also found that the slopes of regressions relating body fat to body weight were the same for each genotype during both the pre-fattening and fattening phases of growth. Again, the transition to the fattening phase occurred at a heavier weight in the genotypes of larger mature size (*Figure 2.8(b)*). In this experiment, the intercepts of the relationships in the pre-fattening phase also varied with heavier breeds having less fat. However, this may not be a true genotype effect because the animals were treated differently during early life and were 3–4 months old before the experiment commenced.

Differences between breeds are not always as clearly defined as those in the experiment of Searle and Griffiths (1976b). McClelland, Bonaiti and Taylor (1976) found that although Finnish Landrace and Southdown

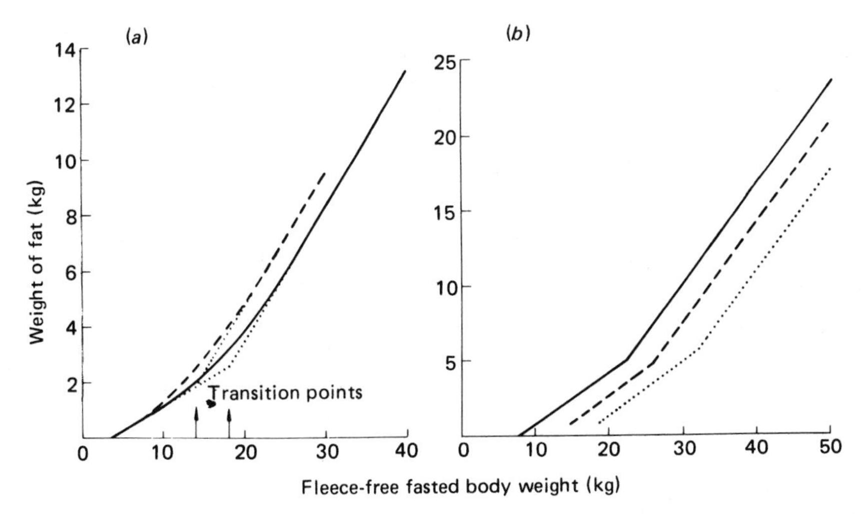

Figure 2.8 Relationship between amount of body fat and the fasted fleece-free body weight of sheep of different genotypes. (a) Effect of sex in cross-bred lambs fed milk diet *ad libitum*: —— male; --- female. Dotted lines indicate the extrapolation of the linear regression for each sex and the arrows the body weight at which the change in slope occurs. From Searle and Griffiths (1976a). (b) Effect of different strains of Merinos and their cross: —— Camden Park Merino; ---- Medium Peppin Merino; .... Border Leicester × Merino. From Searle and Griffiths (1976b).

breeds had similar mature weights, the Finn sheep were considerably fatter, particularly at empty body weights between 20 and 30 kg. However, it is possible that this difference resulted from the lower food intakes and digestive upsets observed in the Southdowns. Wood *et al.* (1980) also noted that the carcasses of Colbred sheep contained less fat in comparison with other breeds than would be expected from their mature body size.

When McClelland, Bonaiti and Taylor (1976) compared the body composition of breeds of vastly different mature sizes at the same percentage of mature weight, all the differences due to sex and most of those due to breed were removed. Thus most of the differences in body composition observed between genotypes compared at the same weight simply reflect differences in relative maturity of these breeds. Nevertheless, the Soay breed, which is a wild breed from uninhabited islands off the west coast of Scotland, had substantially less fat and more bone when compared at the same percentage of mature liveweight as the domesticated breeds. Thompson (1982) has recently suggested that the age of selection for weight may affect the relative body composition of sheep breeds. He has evidence indicating that animals selected for heavy weights at weaning have similar body composition when compared at the same weight as animals selected for light weaning weight, but at maturity they are heavier and have a greater percentage of fat.

Concept of a potential rate of energy deposition and body composition for each genotype. In several experiments, different sheep genotypes have been fed high quality diets *ad libitum* until they cease to increase in weight (Blaxter, 1976; Butterfield *et al.*, 1983 and unpublished). Provided the diets supplied all essential nutrients and feed intake was not limited by the capacity of the digestive tract, these experiments give an estimate of the potential rate of energy deposition for different genotypes. Estimates of the rate of energy deposition for the genotypes studied by Butterfield and his colleagues in combination with the results of Hodge (1974) for lambs fed milk *ad libitum* suggest that the potential daily rate of energy deposition increases as empty body weight increases to about 30 kg and then declines to maturity (*Figure 2.9*). Clearly there are differences between genotypes, with those of smaller mature weight having lower rates of energy deposition.

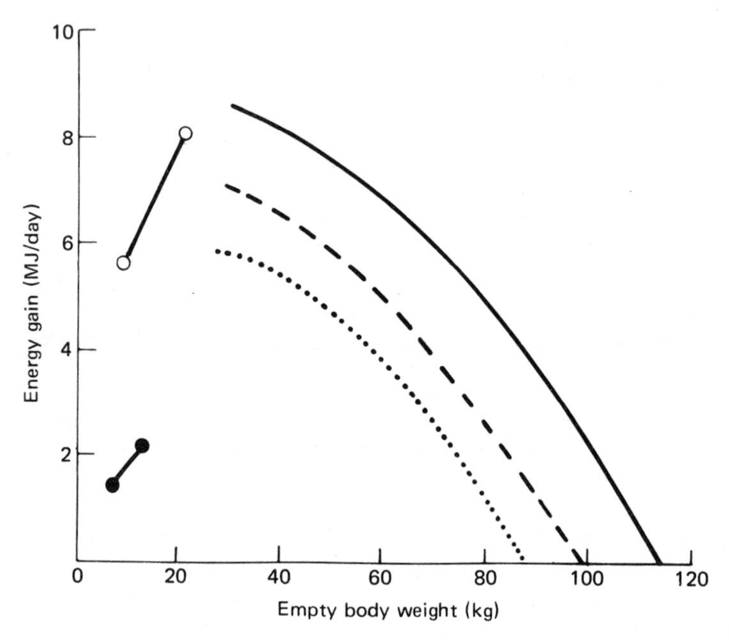

Figure 2.9 Relationship between daily energy gain and empty body weight for different genotypes fed high quality diets *ad libitum*. ●, Superfine wool Saxon Merino wethers fed reconstituted cow's milk from birth (McLaughlin, 1971); ○, Dorset Horn × (Border Leicester × Merino) ram lambs fed reconstituted cow's milk (Hodge, 1974); —— Large Merino rams; ---- Dorset Horn rams; ...... Dorset Horn wethers and Small Merino rams fed a high quality diet from about 25 kg weight (Butterfield *et al.*, 1983, unpublished). Energy content of empty liveweight gain calculated from Searle, Graham and O'Callaghan (1972).

Experiments which describe the interaction between genotype and body composition provide limited data to speculate on partition of the potential energy gain between fat and protein for different genotypes at various stages of maturity. Although the chemical composition of sheep grown from birth to maturity under ideal conditions has not been examined, it can be postulated from the experiment of Searle and Griffiths (1976a) that there would be only two phases to the relationship between fat and body weight which are joined by a smooth transition. Further, evidence from Searle and Griffiths (1976a,b) indicates that there is little difference between genotypes in the slope of the body fat/body weight relationship during either the pre-fattening or fattening phases of growth. It is

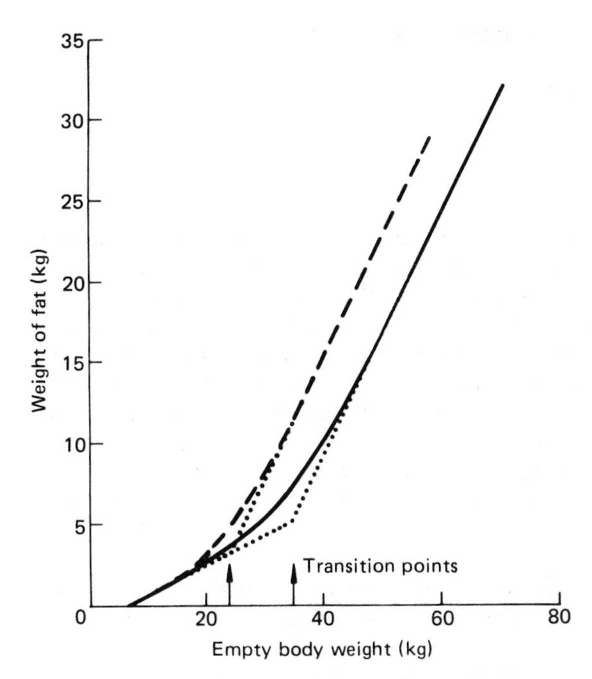

Figure 2.10 Proposed description of the relationship between body fat and empty body weight for sheep of different genotypes: —— large genotype; ---- small genotype. The slope and intercept of the relationship is the same for all genotypes during the pre-fattening phase of growth, whereas the slopes are also identical during the fattening phase but the transition to this phase occurs at heavier body weights in animals of larger mature size. Dotted lines indicate the extrapolation of the linear regression for each genotype.

proposed, therefore, that under unlimited nutritional and environmental conditions, both the slope and the intercept of the body fat/body weight relationship during the pre-fattening phase of growth are identical for different genotypes but transition to the fattening phase occurs at different body weights (*Figure 2.10*). Such a proposition needs to be examined by appropriate experimentation.

The following equation was developed by Searle and Griffiths (1976a) to describe the two-phase relationship between body fat and body weight.

$$F_w = \alpha_0 - \alpha_2 W_p + \alpha_1 W + \alpha_2[(W - W_p) + \gamma]^{0.5}$$

where F_w is weight of body fat, W is empty body weight, γ is a curvature parameter, W_p, α_0, α_1 and α_2 are parameters which describe the linear asymptotes such that α_0 is the intercept of the asymptote for the first phase, $\alpha_1 - \alpha_2$ is the slope of this line, $\alpha_1 + \alpha_2$ is the slope of the second phase asymptote and W_p represents the point of transition between the two linear phases. A similar equation can be developed to describe the relationship between body protein and body weight.

By differentiating these equations with respect to body weight and converting the weight of fat and protein to an equivalent amount of energy, it is possible to predict the partition of potential energy gain between fat and protein for different genotypes at different body weights. However, achievement of this potential composition can be influenced by nutritional and environmental conditions.

Nutrition

The effects of energy intake, protein content of the diet and the interactions between protein and energy intake on the growth and body composition of lambs have been reviewed (Black 1974, 1981b). Although there are many contradictory reports regarding the effects of nutrition on the body composition of sheep, most of these can be reconciled through an understanding of the partition of metabolizable energy between the needs of maintenance, protein synthesis, lipogenesis and the heat loss associated with nutrient utilization. The following description of energy partition is based on studies of protein and energy utilization in milk-fed lambs (Black, 1974), but the principles apply to weaned animals (Black, Faichney and Graham, 1976).

Level of feeding. The effects of increasing the intake of a protein adequate diet on the partition of energy is illustrated for a 5 kg lamb in *Figure 2.11(a)*. When the animal is in energy balance, that is at maintenance, there is a gain in body protein and a loss of body fat. Further, once

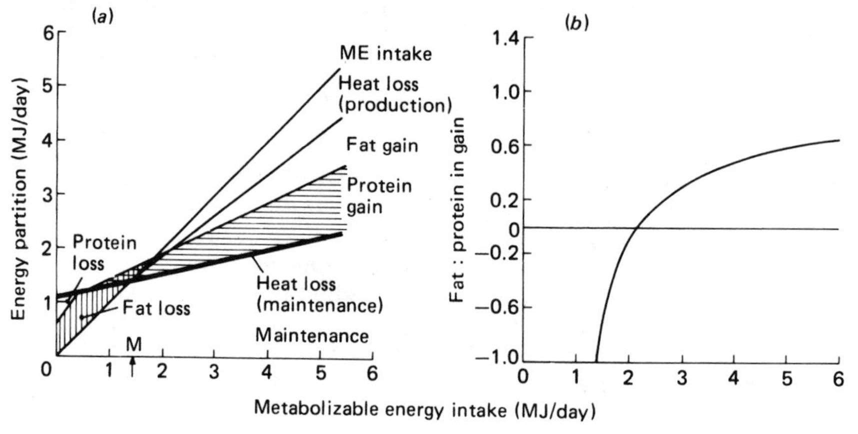

Figure 2.11 Effect of increasing the intake of metabolizable energy (ME) on (a) its partition between various body functions, and (b) the ratio of fat to protein in the energy gain for a lamb weighing 5 kg. Arrow (M) indicates point of energy balance. From Black (1974)

energy intake is raised above the level where fat deposition commences, the ratio of the gain in fat to the gain in protein is constant for each increment in energy intake. Consequently, as energy intake is raised above maintenance the proportion of fat to protein deposited in the body increases exponentially from —1 to an asymptote equivalent to the constant ratio of fat to protein for increments of energy intake above that at which fat is deposited (*Figure 2.11(b)*). An increase in the intake of a well balanced diet above maintenance, therefore, results in a faster rate of growth and an increase in the fat content of the gain. However, the effects on body composition are greatest when intake is raised from near maintenance rather than when it is increased above an already high level. An increase in fat content of the energy gain means that animals have more fat when compared at the same weight as animals receiving less energy. The curvilinear relationship between intake of a well balanced diet and fat deposition can explain why an increase in fat content is sometimes observed whereas in other experiments there is no effect of increasing the level of feeding; the outcome depends upon the levels of feeding relative to maintenance.

When intake falls below about half maintenance, there is a marked increase in the rate of protein catabolism (Black and Griffiths, 1975). Thus, the composition of body weight loss is influenced by the severity of intake restriction. With intakes near maintenance, there is considerable loss of body fat but little change in body protein (*Figure 2.11(a)*). However, when intake is reduced to below half maintenance, the amount of energy lost as protein increases substantially. Therefore, when compared with normal growth, animals that are fed near maintenance will tend to be leaner, but those placed on a severely restricted diet will initially be fatter. A number of experiments support this contention (Black, 1974). However, with prolonged under-feeding it is probable that the relationship between protein deposition and energy intake shown in *Figure 2.11(a)* will gradually alter such that, at maintenance, there would ultimately be no gain in body protein nor loss in fat. Body composition thus gradually returns towards that characteristic of normal growth as the period of undernutrition is prolonged, except that the body will tend to contain more bone (Searle, Graham and Smith, 1979). Thus, the effect of submaintenance feeding on body composition depends upon the severity and duration of the nutrient restriction.

Protein absorption. The effect of increasing protein absorption on the partition of energy is illustrated in *Figure 2.12(a)* for a 5 kg lamb given a

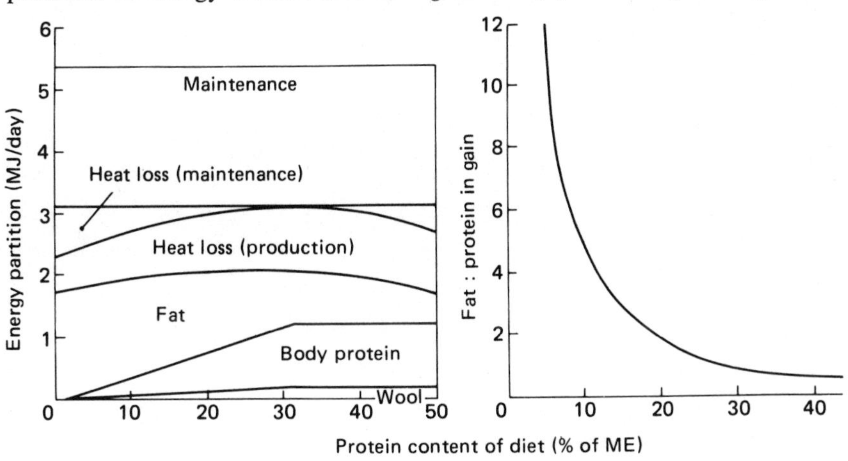

Figure 2.12 Effect of increasing protein content of diet on (a) the partition of metabolizable energy (ME) and (b) the ratio of fat to protein in the energy gain for a lamb weighing 5 kg and fed 5.3 MJ metabolizable energy/day. From Black (1974).

constant intake of metabolizable energy. As protein absorption increases, the proportion of energy deposited in protein increases until protein availability is no longer limiting. Over the same range in protein absorption, the amount of energy deposited in fat decreases. Thus, the ratio of fat to protein in the body gain decreases markedly as protein absorption increases from deficient levels (*Figure 2.12(b)*). Moderate increases in protein absorption above the animal's requirements appear to have little further effect on the fat to protein content of the gain. Thus lambs fed diets in which protein absorption is below requirement, grow more slowly and contain more fat than animals of the same weight receiving adequate protein (Ørskov *et al.*, 1976).

There is evidence from rats and lambs that, when protein absorption is considerably in excess of the amount needed for maximum protein synthesis, heat production is increased and the efficiency of utilization of metabolizable energy is reduced (Hartsook and Hershberger, 1971; Walker and Norton, 1971). Under these conditions, it is probable that less energy is available for lipogenesis and the ratio of fat to protein in the gain falls below that which occurs in animals given better balanced diets. Diets which provide protein well in excess of requirement could be expected to produce leaner animals. There is some evidence to support this contention in weaned lambs (Searle, Graham and Donnelly, 1982).

Protein–energy interaction. The decline in protein synthesis which occurs when protein absorption falls below requirement means that the net gain in

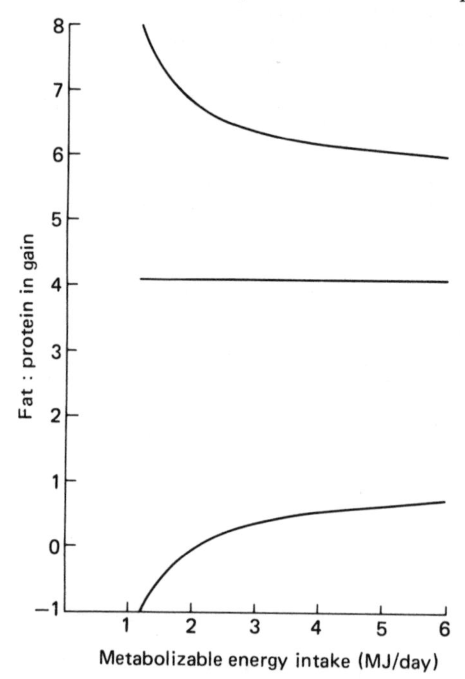

Figure 2.13 Calculated effect of increasing the intake of metabolizable energy on the ratio of fat to protein in the energy gain of a 5 kg lamb given diets supplying different amounts of protein: (a) grossly deficient in protein, (b) deficient, (c) adequate.

body protein seen at energy balance in animals receiving adequate protein (*Figure 2.11(a)*) gradually turns into a net loss as protein absorption becomes more inadequate. Animals receiving diets grossly inadequate in protein, gain fat and lose protein when in energy balance. Thus, the effect of increasing feed intake on the ratio of fat to protein in the gain depends upon the adequacy of dietary protein (*Figure 2.13*). In contrast to animals fed adequate protein, a severe protein deficiency results in a decline in the ratio of fat to protein in the gain as feed intake increases. At some intermediate protein intake, there will be no effect of feeding level on body composition.

The interaction between protein and energy is more complex in weaned animals. The amount of protein absorbed often bears little relationship to the amount eaten because of the activity of rumen microbes. Although the principles outlined above still apply, a clear understanding of the absorption of protein and energy must be obtained to adequately predict the effects of nutrition on the body composition of ruminant lambs. An increase in the intake of many diets results in an increase in the absorption of protein relative to energy because of an increase in the rate of passage of feed particles from the rumen which decreases microbial degradation of dietary protein. A change in feed intake in ruminants may therefore alter the ratio of protein to energy available to the animals as well as total energy intake.

Refeeding after periods of nutrient restriction. The growth rate of lambs is normally greater following refeeding after periods of prolonged under-nutrition than is observed in animals of the same weight given unlimited access to feed. There are conflicting reasons given for this observation. In some experiments (Elliott and O'Donovan, 1969; Allden, 1970; Graham and Searle, 1979) but not others (Meyer and Clawson, 1964; Drew and Reid, 1975; Ørskov *et al.*, 1976; Murray and Slezacek, 1980), appetite was increased. Allden (1970) suggested that the time and duration of the growth stress may affect the appetite response. He observed an increase in appetite only in lambs suffering severe nutritional restriction in the second six months of life. However, this explanation does not apply to all experiments.

Compensatory growth can occur without an increase in feed intake. Although the maintenance requirement of lambs falls during periods of submaintenance feeding (Graham and Searle, 1979) there is little evidence of a substantial improvement in the efficiency of utilization of metabolizable energy except perhaps for the first one or two weeks of realimentation (Drew and Reid, 1975; Graham and Searle, 1975; Murray and Slezacek, 1980). The most consistent explanation for the increase in growth rate following refeeding after a period of under-nutrition is a change in the chemical composition of the liveweight gain. An increase in the protein and water content and a decrease in the fat content of the liveweight gain has been commonly observed (Drew and Reid, 1975; Ørskov *et al.*, 1976; Searle, Graham and Smith, 1979; Murray and Slezacek, 1980). This results in a marked increase in liveweight gain compared with an equivalent retention of energy during normal growth.

Climatic conditions

The effect of cold stress on energy utilization and its partition between body fat and protein in animals has been reviewed (Graham *et al.*, 1976; Mount, 1980; Black, 1981a). Moderate cold stress does not affect protein deposition but reduces growth rate and fat synthesis. Although the rate of protein synthesis can be depressed in animals subjected to cold stress, this occurs only when conditions are so severe that survival is threatened.

Relationship between growth rate and body composition

The relationship between growth rate and body composition of lambs for the conditions discussed above is summarized in *Table 2.4*. Clearly, there is

Table 2.4. A SUMMARY OF THE EFFECT OF GENOTYPE, NUTRITION AND ENVIRONMENT ON THE RELATIONSHIP BETWEEN GROWTH RATE AND BODY COMPOSITION OF SHEEP

Treatment[(a)] *comparisons*	*Conditions*	*Body composition of faster growing animal when treatments are compared at same weight*
Large–small genotype	Balanced diet *ad libitum*	Leaner
High–low intake	(a) Balanced diet, positive growth	
	(i) near maintenance	Fatter
	(ii) near *ad libitum*	Marginally fatter
	(b) Moderate protein deficiency, positive growth	Not different
	(c) Gross protein deficiency positive growth	Leaner
Adequate–grossly excess protein	Energy intake constant	Fatter
Adequate–inadequate protein	Energy intake constant	Leaner
Thermoneutral–cold exposed	Energy intake constant	Fatter
Compensatory gain–normal growth	Energy intake constant	Leaner

[(a)]Left hand treatment produces faster growth rate

no simple association between these characteristics. Many factors interact to affect them both and assessment of the relationship is extremely difficult unless the partition of absorbed nutrients is fully understood.

Application of principles

Many of the concepts discussed in this chapter have been incorporated into a computer programme which simulates the growth of lambs from conception to maturity (Graham *et al.*, 1976). The programme deals with the intake and utilization of protein and energy, and their partition between various body functions. The ability of the programme to predict the effects of altering protein and energy intake, changing animal activity or imposing cold stress on the growth and body composition of lambs has been considered previously (Black, 1974, 1981a). Recently, the concepts outlined in this chapter regarding the effect of genotype have been included in the programme.

The potential energy gain for a genotype is first calculated and then partitioned between protein and fat depending upon empty body weight. Although the energy gain of sheep given unlimited nutrients from birth to maturity has not been examined, the following equation is used to describe the relationship between potential energy gain and empty body weight for different genotypes (*Figure 2.9*):

$$E_g = kW^a((E_c - E_w)/E_c)$$

where E_g is the potential daily gain in body energy, W is empty body weight, E_c the maximum energy content of the genotype at maturity, E_w is the energy content at W, k is a rate constant representing the maximum daily rate of energy gain per unit W^a and a is a constant. An approximate fit to the data of Hodge (1974) and Butterfield *et al.* (1983) for large Merino rams provided estimates of 1.94 for k and 0.45 for a when energy is in MJ and weight in kg. The results of McLaughlin (1971) suggest that k is influenced by genotype (*Figure 2.9*), but currently a is held constant between genotypes.

Potential energy gain is partitioned between fat and protein on the basis outlined (*Figure 2.10*). From a survey of the literature (Searle, Graham and O'Callaghan, 1972; Blaxter, 1976; Searle and Griffiths, 1976a,b; Searle, Graham and Smith, 1979), the slope for the relationship between body fat and empty body weight was set at 0.18 and 0.65 for the pre-fattening and fattening phases of growth respectively. Corresponding values for protein were 0.132 and 0.085. The gain in fat (F_g) and protein (P_g) for each unit increase in empty body weight is calculated from the differentiated equations of Searle and Griffiths (1976a).

$$F_g = 0.415 + 0.235\,((W - W_p)/((W - W_p)^2 + 33)^{0.5})$$
$$P_g = 0.1085 - 0.0235\,((W - W_p)/((W - W_p)^2 + 33)^{0.5})$$

where W_p represents the body weight at the transition point between the lines describing the pre-fattening and fattening phases of growth. The protein to energy ratio of the potential gain is then determined and the potential deposition of protein calculated for the particular body weight and genotype. When energy intake is below the potential required to satisfy all body functions, protein deposition is decreased using the relationship shown in *Figure 2.11(a)* and described by Black and Griffiths (1975). The amount of body fat deposited is determined from the energy remaining after all other body functions have been satisfied. The amount of water and ash deposited are calculated from relationships with the fat-free tissue. Thus, the effects of genotype on the potential growth and body composition of lambs can be obtained once values for k, E_c and W_p

Table 2.5. ESTIMATES[a] OF THE MAXIMUM DAILY RATE OF ENERGY GAIN (k), MAXIMUM ENERGY CONTENT AT MATURITY (E_c) AND THE BODY WEIGHT REPRESENTING THE TRANSITION POINT BETWEEN THE PRE-FATTENING AND FATTENING PHASES OF GROWTH (W_p) FOR THREE DIFFERENT GENOTYPES

Genotype	k (MJ/$W^{0.45}$/d)	E_c (MJ)	W_p (kg)
Large Merino ram	1.94	2525	30
Small Merino ram	1.60	1970	25
Small Merino ewe	1.30	1609	20

[a]Based on results from McLaughlin (1971); Searle, Graham and O'Callaghan (1972); Hodge (1974); Blaxter (1976); Searle and Griffiths (1976a,b); Butterfield *et al.* (1983).

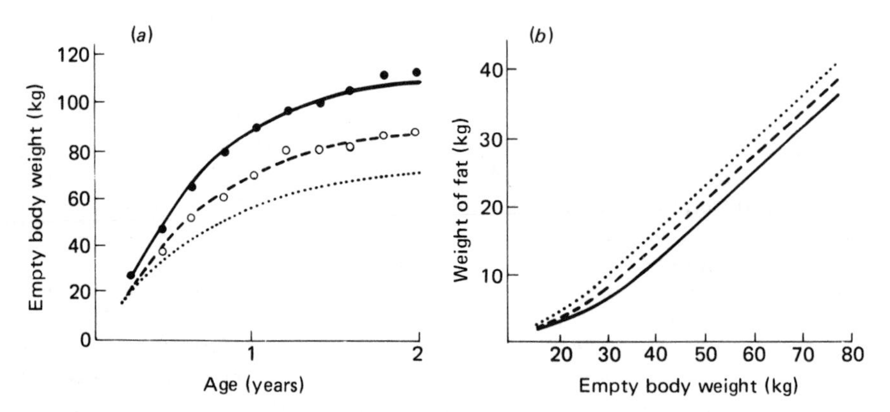

Figure 2.14 Predicted effect of growth (a) and body composition (b) of three genotypes (—— Large Merino ram, --- Small Merino ram, ···· Small Merino ewe) given a subterranean clover containing 26% protein and with a digestibility of 74%. Intake was limited by energy demand. Observed body weight of Large (●) and Small Merinos (○) fed high quality diet *ad libitum* (Butterfield *et al.*, 1983).

Table 2.6. PREDICTED EFFECT OF GENOTYPE AND LEVEL OF FEEDING ON THE GROWTH OF LAMBS FROM 15–35 KG GIVEN A SUBTERRANEAN CLOVER DIET CONTAINING 26% PROTEIN AND WITH AN ORGANIC MATTER DIGESTIBILITY OF 74%

Genotype	*Feeding level*	*Mean growth rate* (g/day)	*Total feed intake* (kg)	*Body fat at 35 kg*	
				(kg)	(% of body)
Large Merino ram	*ad libitum*[a]	398	109	8.9	25.5
Small Merino ram	*ad libitum*[a]	272	135	11.0	31.3
Small Merino ewe	*ad libitum*[a]	178	168	13.0	37.0
Small Merino ram	0.8 *ad libitum*	201	142	10.4	29.6
	0.5 *ad libitum*	96	181	8.9	25.4

[a]Feed intake limited by energy demand of the animal, not rumen fill.

are available. Although these values cannot be set with confidence, estimates for three genotypes are given in *Table 2.5*. Experiments are required to provide accurate estimates of these parameters for different genotypes. Predicted growth rates and body composition for animals fed *ad libitum* (*Figure 2.14*) show the behaviour of the model when these concepts are included.

With models of the type described, current concepts and information can be rapidly used to assess the practical benefits of alternative management strategies. For example, *Table 2.6* shows the predicted effects of sheep genotype on the growth rate, total feed intake and body fat content of lambs grown from 15–35 kg empty body weight when nutrient intake is limited by energy demand. Similar results are shown for one genotype given various levels of intake. Predictions can be made for any variations in diet composition, nutrient intake or cold exposure. This type of information provides a sound basis for producers to consider the consequences and profitability of alternative farm practices.

Conclusion

Foetal growth is determined primarily by the rate of nutrient exchange across the utero-placenta. It is influenced by the sex and genotype of the foetus, litter size, body weight and condition of the ewe, maternal nutrition during mid and late pregnancy and environmental factors such as high ambient temperature and disease. Small deviations in tissue development result from severe foetal growth retardation, such that small lambs at birth tend to have lighter livers, less fat and fewer wool follicles but larger skeletons relative to their weight than do normal foetuses.

Although foetal growth retardation greatly influences lamb survival, its effects on productivity in later life are generally small. Slight reductions at maturity resulting from restricted prenatal growth have been reported in body weight and wool production.

Both hyperplasia and hypertrophy contribute to organ growth in sheep. The major change in cell size occurs during foetal life in several organs but not in adipose tissue. In continuously growing animals, hyperplasia ceases only when most organs approach mature size. Except for gut and skin, variations in organ weight within an adult sheep appear to be almost entirely due to changes in cell size, whereas differences between genotypes result mainly from variations in cell number.

During the postnatal period, growth rate and the relative distribution of fat and lean within the body are affected by the sex, breed and stage of maturity of the lamb as well as by energy intake, composition of the diet and climatic conditions. Within each genotype, there appears to be a maximum rate of energy deposition with the partition between protein and fat varying with body weight. There is no simple relationship between growth rate and body composition of sheep. However, from an understanding of the partition of absorbed nutrients between various body functions, it is possible to explain the association between these characteristics in specific situations.

Many of the concepts discussed within this chapter have been incorporated into a computer programme which simulates growth and body composition of sheep from conception to maturity. With such a programme much of the current information on growth and development of sheep can be used to investigate the practical significance of varying the genotype or feeding and environmental regimes on performance.

Acknowledgements

I am particularly grateful to Professor N.M. Tulloh, Faculty of Agriculture and Forestry, University of Melbourne, and Professor R.M. Butterfield, Department of Veterinary Anatomy, University of Sydney, and their colleagues for allowing me to use their unpublished data. I also thank Dr. I.H. Williams, Department of Animal Science, University of Western Australia for his collaboration during development of the concepts regarding the influence of genotype on potential energy gain and its partition.

References

ALEXANDER, G. (1974). Birth weight of lambs: influences and consequences. In *Size at Birth* (K. Elliott and J. Knight, Eds.), pp. 215–245. Amsterdam, Elsevier

ALEXANDER, G. (1978a). Quantitative development of adipose tissue in foetal sheep. *Australian Journal of Biological Sciences* **31**, 489–503

ALEXANDER, G. (1978b). Factors regulating the growth of the placenta: with comments on the relationship between placental weight and foetal weight. In *Abnormal Fetal Growth: Biological Bases and Consequences* (F. Naftolin, Ed.), pp. 149–164. Berlin, Dahlen Konferenzen

ALLDEN, W.G. (1970). The effects of nutritional deprivation on the subsequent productivity of sheep and cattle. *Nutrition Abstracts and Reviews* **40**, 1167–1184

ALLDEN, W.G. (1979). Undernutrition of the Merino sheep and its sequelae. V. The influence of severe growth restriction during early post-natal life on reproduction and growth in later life. *Australian Journal of Agricultural Research* **30**, 939–948

ALLEN, R.E., MERKEL, R.A. and YOUNG, R.B. (1979). Cellular aspects of muscle growth: myogenic cell proliferation. *Journal of Animal Science* **49**, 115–127

BALDWIN, R.L. and BLACK, J.L. (1979). Simulation of the effects of nutritional and physiological status on the growth of mammalian tissues: description and evaluation of a computer program. *Animal Research Laboratories Technical Paper No. 6*, pp. 1–35. Australia, CSIRO

BATTAGLIA, F.C. and MESCHIA, G. (1981). Foetal and placental metabolisms: their interrelationship and impact upon maternal metabolism. *Proceedings of the Nutrition Society* **40**, 99–113

BENEVENT, M. (1971). Croissance relative pondérale postnatale, dans les deux sexes, des principaux tissus et organes de l'agneau Mérinos d'Arles. *Annales de Biologie Animale Biochimie Biophysique* **11**, 5–39

BENNETT, D., AXELSEN, A. and CHAPMAN, H.W. (1964). The effect of nutritional restriction during early pregnancy on numbers of lambs born. *Proceedings of the Australian Society of Animal Production* **5**, 70–71

BLACK, J.L. (1974). Manipulation of body composition through nutrition. *Proceedings of the Australian Society of Animal Production* **10**, 211–218

BLACK, J.L. (1981a). Nutritional needs of lambs for growth and prediction of growth rates. In *Sheep Nutrition* (G.J. Tomes and I.J. Fairnie, Eds.), pp. 117–132. Perth, Western Australian Institute of Technology

BLACK, J.L. (1981b). Manipulation of body composition in growing and mature sheep. In *Sheep Nutrition* (G.J. Tomes and I.J. Fairnie, Eds.), pp. 133–147. Perth, Western Australian Institute of Technology

BLACK, J.L. and GRIFFITHS, D.A. (1975). Effect of live weight and energy intake on nitrogen balance and total N requirements of lambs. *British Journal of Nutrition* **33**, 399–413

BLACK, J.L., FAICHNEY, G.J. and GRAHAM, N.McC. (1976). Future role of computer simulation in research and its application to ruminant protein nutrition. In *Protein Metabolism and Nutrition* (D.J.A. Cole *et al.*, Eds.), pp. 477–491. London, Butterworths

BLACK, J.L., ROBARDS, G.E. and THOMAS, R. (1973). Effects of protein and energy intakes on the wool growth of Merino wethers. *Australian Journal of Agricultural Research* **24**, 399–412

BLAXTER, K.L. (1976). Experimental obesity in farm animals. *European Association of Animal Production, Publication No. 19*, 129–132

BRADFORD, G.E., TAYLOR, St.C.S., QUIRKE, J.F. and HART, R. (1974). An egg-transfer study of litter size, birth weight and lamb survival. *Animal Production* **18**, 249–263

BRYDEN, M.M., EVANS, H.E. and BINNS, W. (1972a). Embryology of the sheep. I. Extraembryonic membranes and the development of body form. *Journal of Morphology* **138**, 169–186

BRYDEN, M.M., EVANS, H.E. and BINNS, W. (1972b). Embryology of the sheep. 2. The alimentary tract and associated glands. *Journal of Morphology* **138**, 187–206

BUTTERFIELD, R.M., GRIFFITHS, D.A., THOMPSON, J.M., ZAMORA, J. and JAMES, A.M. (1983). Changes in body composition relative to weight and maturity of large and small strains of Australian Merino rams. I. Muscle, bone and fat. *Animal Production* **36**, (In press)

CALARCO, P.G. and McLAREN, A. (1976). Ultrastructural observations of preimplantation stages of the sheep. *Journal of Embryology and Experimental Morphology* **36**, 609–622

CARTER, H.B. and HARDY, M.H. (1947). Studies in the biology of the skin and fleece of sheep. 4. The hair follicle group and its topographical variations in the skin of the Merino foetus. *CSIRO, Bulletin No. 215*, 1–41. Melbourne, CSIRO

CHEEK, D.B., HOLT, A.B., HILL, D.E. and TALBERT, J.L. (1971). Skeletal muscle mass and growth: the concept of the deoxyribonucleic acid unit. *Pediatric Research* **5**, 312–328

CORBETT, J.L. (1979). Variations in wool growth with physiological state. In *Physiological and Environmental Limitations to Wool Growth* (J.L. Black and P.J. Reis, Eds.), pp. 79–98. Armidale, Australia, University of New England Press

CREASY, R.K., BARRETT, C.T., SWIET, de.M., KAHANPAA, K.V. and RUDOLPH,

A.M. (1972). Experimental intrauterine growth retardation in sheep. *American Journal of Obstetrics and Gynecology* **112**, 566–573

CURLL, M.L., DAVIDSON, J.L. and FREER, M. (1975). Efficiency of lamb production in relation to the weight of the ewe at mating and during pregnancy. *Australian Journal of Agricultural Research* **26**, 553–565

DICKINSON, A.G., HANCOCK, J.L., HOVELL, G.J.R., TAYLOR, St.C.S. and WIENER, G. (1962). The size of lambs at birth—a study involving egg transfer. *Animal Production* **4**, 64–79

DONALD, H.P. and RUSSELL, W.S. (1970). The relationship between live weight of ewe at mating and weight of newborn lamb. *Animal Production* **12**, 273–280

DREW, K.R. and REID, J.T. (1975). Compensatory growth in immature sheep. III. Feed utilization by sheep subjected to feed deprivation followed by realimentation. *Journal of Agricultural Science, Cambridge* **85**, 215–220

ELLIOTT, R.C. and O'DONOVAN, W.M. (1969). Compensatory growth in Dorper sheep. *Proceedings of Second Symposium on Animal Production* (J. Oliver, Ed.), pp. 41–51. Salisbury, University College of Rhodesia

EL-SHEIKH, A.S., HULET, C.V., POPE, A.L. and CASIDA, L.E. (1955). The effect of level of feeding on the reproductive capacity of the ewe. *Journal of Animal Science* **14**, 919–929

EVERITT, G.C. (1964). Maternal undernutrition and retarded foetal development in Merino sheep. *Nature, London* **201**, 1341–1342

EVERITT, G.C. (1967). Residual effects of prenatal nutrition on the postnatal performance of Merino sheep. *Proceedings of the New Zealand Society of Animal Production* **27**, 52–68

EVERITT, G.C. (1968). Prenatal development of uniparous animals, with particular reference to the influence of maternal nutrition in sheep. In *Growth and Development of Mammals* (G.A. Lodge and G.E. Lamming, Eds.), pp. 131–157. London, Butterworths

FAICHNEY, G.J. (1981). Amino acid utilization by the foetal lamb. *Proceedings of the Nutrition Society of Australia* **6**, 48–53

FALCONER, D.S., GAULD, I.K. and ROBERTS, R.C. (1978). Cell numbers and cell sizes in organs of mice selected for large and small body size. *Genetical Research* **31**, 287–301

FOOTE, W.C., POPE, A.L., CHAPMAN, A.B. and CASIDA, L.E. (1959). Reproduction in the yearling ewe as affected by breed and sequence of feeding levels. II. Effects on foetal development. *Journal of Animal Science* **18**, 463–474

FOURIE, P.D., KIRTON, A.H. and JURY, K.E. (1970). Growth and development of sheep. II. Effect of breed and sex on the growth and carcass composition of the Southdown and Romney and their cross. *New Zealand Journal of Agricultural Research* **13**, 753–770

GABBEDY, B.J. (1974). A growth comparison of single and twin lambs during prenatal and early postnatal life. Master of Agricultural Science Thesis. University of Melbourne

GEISLER, P.A. and JONES, C.M. (1979). A model for calculation of the energy requirements of the pregnant ewe. *Animal Production* **29**, 339–355

GRAHAM, N.McC. and SEARLE, T.W. (1975). Studies of weaner sheep during and after a period of weight stasis. I. Energy and nitrogen utilization. *Australian Journal of Agricultural Research* **26**, 343–353

GRAHAM, N.McC. and SEARLE, T.W. (1979). Studies of weaned lambs before, during and after a period of weight loss. I. Energy and nitrogen utilization. *Australian Journal of Agricultural Research* **30**, 513–523

GRAHAM, N.McC., BLACK, J.L., FAICHNEY, G.J. and ARNOLD, G.W. (1976). Simulation of growth and production in sheep—model 1: a computer program to estimate energy and nitrogen utilization, body composition and empty liveweight change day by day for sheep of any age. *Agricultural Systems* **1**, 113–138

GUNN, R.G. (1977). The effects of two nutritional environments from 6 weeks prepartum to 12 months of age on lifetime performance and reproductive potential of Scottish Blackface ewes in two adult environments. *Animal Production* **25**, 155–164

HARTSOOK, E.W. and HERSHBERGER, T.V. (1971). Interactions of major nutrients in whole-animal energy metabolism. *Federation Proceedings* **30**, 1466–1473

HAUGEBAK, C.D., HEDRICK, H.B. and ASPLUND, J.M. (1974). Adipose tissue accumulation and cellularity in growing and fattening lambs. *Journal of Animal Science* **39**, 1016–1025

HODGE, R.W. (1974). Efficiency of food conversion and body composition of the preruminant lamb and the young pig. *British Journal of Nutrition* **32**, 113–126

HOOD, R.L. and THORNTON, R.F. (1979). The cellularity of ovine adipose tissue. *Australian Journal of Agricultural Research* **30**, 153–161

HOPKINS, P.S. (1975). The development of the foetal ruminant. In *Digestion and Metabolism of Ruminants* (I.W. McDonald and A.C.I. Warner, Eds.), pp. 1–14. Armidale, Australia, University of New England Press

HULET, C.V., FOOTE, W.C. and PRICE, D.A. (1969). Factors affecting growth of the ovine foetuses during early gestation. *Animal Production* **11**, 219–224

HUNTER, G.L. (1957). The maternal influence on size in sheep. *Journal of Agricultural Science, Cambridge* **48**, 36–60

JOHNS, J.T. and BERGEN, W.G. (1976). Growth in sheep. Pre- and post-weaning hormone changes and muscle and liver development. *Journal of Animal Science* **43**, 192–200

JOUBERT, D.M. (1956). A study of pre-natal growth and development in the sheep. *Journal of Agricultural Science, Cambridge* **47**, 382–427

KIRTON, A.H., FOURIE, P.D. and JURY, K.E. (1972). Growth and development of sheep. III. Growth of the carcass and non-carcass components of the Southdown and Romney and their cross and some relationships with composition. *New Zealand Journal of Agricultural Research* **15**, 214–227

LANGLANDS, J.P. and SUTHERLAND, H.A.M. (1968). An estimate of the nutrients utilized for pregnancy by Merino sheep. *British Journal of Nutrition* **22**, 217–227

LOHSE, C.L. (1973). The influence of sex on muscle growth in Merino sheep. *Growth* **37**, 177–187

LOHSE, C.L., MOSS, F.P. and BUTTERFIELD, R.M. (1971). Growth patterns of muscles of Merino sheep from birth to 517 days. *Animal Production* **13**, 117–126

McCLELLAND, T.H., BONAITI, B. and TAYLOR, St.C.S. (1976). Breed differences in body composition of equally mature sheep. *Animal Production* **23**, 281–293

McDONALD, I., WENHAM, G. and ROBINSON, J.J. (1977). Studies on reproduction in prolific ewes. 3. The development in size and shape of the foetal skeleton. *Journal of Agricultural Science, Cambridge* **89**, 373–391

McDONALD, I., ROBINSON, J.J., FRASER, C. and SMART, R.I. (1979). Studies on reproduction in prolific ewes. 5. The accretion of nutrients in the foetuses and adnexa. *Journal of Agricultural Science, Cambridge* **92**, 591–603

McLAUGHLIN, J.W. (1971). Study on the nutritional efficiency of the Merino lamb. PhD. Thesis. University of Melbourne

MASTERS, C.J. (1963). Nucleic acids and protein stores in the Merino sheep. *Australian Journal of Biological Sciences* **16**, 192–200

MELLOR, D.J. and MURRAY, L. (1981). Effects of placental weight and maternal nutrition on the growth rates of individual fetuses in single and twin bearing ewes during late pregnancy. *Research in Veterinary Science* **30**, 198–204

MEYER, J.H. and CLAWSON, W.J. (1964). Undernutrition and subsequent realimentation in rats and sheep. *Journal of Animal Science* **23**, 214–224

MOUNT, L.E. (1980). Growth and the thermal environment. In *Growth in Animals* (T.L.J. Lawrence, Ed.), pp. 47–63. London, Butterworths

MURRAY, D.M. and SLEZACEK, O. (1975). The effect of growth rate on muscle distribution in sheep. *Journal of Agricultural Science, Cambridge* **85**, 189–191

MURRAY, D.M. and SLEZACEK, O. (1980). Growth pattern and its effect on feed utilization of sheep. *Journal of Agricultural Science, Cambridge* **95**, 349–355

ØRSKOV, E.R., McDONALD, I., GRUBB, D.A. and PENNIE, K: (1976). The nutrition of the early weaned lamb. IV. Effects on growth rate, food utilization and body composition of changing from a low to a high protein diet. *Journal of Agricultural Science, Cambridge* **86**, 411–423

PRUD'HON, M., BENEVENT, M., VEZINHET, A. and DULOR, J.P. (1978). Croissance relative du squelette chez l'agneau. Influence du sexe et de la race. *Annales de Biologie Animale Biochimie Biophysique* **18**, 5–9

RATTRAY, P.V., GARRETT, W.N., EAST, N.E. and HINMAN, N. (1974). Growth, development and composition of the ovine conceptus and mammary gland during pregnancy. *Journal of Animal Science* **38**, 613–626

REID, J.T., BENSADOUN, A., BULL, L.S., BURTON, J.H., GLEESON, P.A., HAN, I.K., JOO, Y.D., JOHNSON, D.E., McMANUS, W.R., PALADINES, O.L., STROUD, J.W., TYRRELL, H.F., NIEKERK, Van B.D.H. and WELLINGTON, G.W. (1968). Some peculiarities in the body composition of animals. In *Body Composition in Animals and Man,* pp. 19–44. National Academy of Science, Publication No. 1598

RHIND, S.M., ROBINSON, J.J.and McDONALD, I. (1980). Relationships among uterine and placental factors in prolific ewes and their relevance to variations in foetal weight. *Animal Production* **30**, 115–124

RICHARDSON, C. and HEBERT, C.N. (1978). Growth rates and patterns of organs and tissues in the ovine foetus. *British Veterinary Journal* **134**, 181–189

ROBINSON, J.J. (1977). The influence of maternal nutrition on ovine foetal growth. *Proceedings of the Nutrition Society* **36**, 9–16

ROBINSON, J.J. (1981). Prenatal growth and development in the sheep and its implications for the viability of the newborn lamb. *Livestock Production Science* **8**, 273–281

ROBINSON, J.J. and McDONALD, I. (1979). Ovine prenatal growth, its mathematical description and the effects of maternal nutrition. *Annales de Biologie Animale Biochimie Biophysique* **19**, 225–234

ROBINSON, J.J., McDONALD, I., FRASER, C. and CROFTS, R.M.J. (1977). Studies on reproduction in prolific ewes. I. Growth of the products of conception. *Journal of Agricultural Science, Cambridge* **88**, 539–552

ROBINSON, P. (1948). The effect of supermaintenance and submaintenance diets on mature Border Leicester–Cheviot ewes. *Journal of Agricultural Science, Cambridge* **38**, 345–353

ROWSON, L.E.A. and MOOR, R.M. (1966). Development of the sheep conceptus during the first fourteen days. *Journal of Anatomy* **100**, 777–785

RUSSEL, A.J.F., FOOT, J.Z., WHITE, I.R. and DAVIES, G.J. (1981). The effect of weight at mating and of nutrition during mid pregnancy on the birth weight of lambs from primiparous ewes. *Journal of Agricultural Science, Cambridge* **97**, 723–729

SANDS, J., DOBBING, J. and GRATRIX, C.A. (1979). Cell number and cell size: organ growth and development and the control of catch-up growth in rats. *Lancet* **8141**, 503–505

SCHINCKEL, P.G. and SHORT, B.F. (1961). The influence of nutritional level during pre-natal and early post-natal life on adult fleece and body characters. *Australian Journal of Agricultural Research* **12**, 176–202

SEARLE, T.W. and GRIFFITHS, D.A. (1976a). The body composition of growing sheep during milk feeding, and the effect on composition of weaning at various body weights. *Journal of Agricultural Science, Cambridge* **86**, 483–493

SEARLE, T.W. and GRIFFITHS, D.A. (1976b). Differences in body composition between three breeds of sheep. *Proceedings of the Australian Society of Animal Production* **11**, 57–60

SEARLE, T.W. and GRIFFITHS, D.A. (1983). The composition of the fat-free empty body of growing sheep and the effect of weight stasis, weight loss and compensatory growth. *Journal of Agricultural Science, Cambridge* (in press)

SEARLE, T.W., GRAHAM, N.McC. and DONNELLY, J.B. (1982). The effect of plane of nutrition on the body composition of two breeds of weaner sheep fed a high protein diet. *Journal of Agricultural Science, Cambridge* **98**, 241–245

SEARLE, T.W., GRAHAM, N.McC. and O'CALLAGHAN, M. (1972). Growth in sheep. I. The chemical composition of the body. *Journal of Agricultural Science, Cambridge* **79**, 371–382

SEARLE, T.W., GRAHAM, N.McC. and SMITH, E. (1979). Studies of weaned lambs before, during and after a period of weight loss. II. Body composition. *Australian Journal of Agricultural Research* **30**, 525–531

SHORT, B.F. (1955). Developmental modification of fleece structure by adverse maternal nutrition. *Australian Journal of Agricultural Research* **6**, 863–872

STERN, D., ADLER, J.H., TAGARI, H. and EYAL, E. (1978). Responses of dairy ewes before and after parturition, to different nutritional regimes during pregnancy. I. Ewe body weight, uterine contents, and lamb birth weight. *Annales de Zootechnie* **27**, 317–333

SYKES, A.R. and FIELD, A.C. (1972). Effect of dietary deficiencies of energy, protein and calcium on the pregnant ewe. II. Body composition and mineral content of the lamb. *Journal of Agricultural Science, Cambridge* **78**, 119–125

SYMONS, L.E.A. and JONES, W.O. (1975). Skeletal muscle, liver and wool protein synthesis by sheep infected by the nematode *Trichostrongylus colubriformis*. *Australian Journal of Agricultural Research* **26**, 1063–1072

TAPLIN, D.E. and EVERITT, G.C. (1964). The influence of prenatal nutrition on postnatal performance of Merino lambs. *Proceedings of the Australian Society of Animal Production* **5**, 72–81

TAYLOR, St.C.S., MASON, M.A. and McCLELLAND, T.H. (1980). Breed and sex differences in muscle distribution in equally mature sheep. *Animal Production* **30**, 125–133

THERIEZ, M., TISSIER, M. and ROBELIN, J. (1981). The chemical composition of the intensively fed lamb. *Animal Production* **32**, 29–37

THOMPSON, J.M. (1982). Genetic manipulation of fatness in lamb carcases. *Proceedings of the Australian Society of Animal Production* **14**, 54–57

THOMPSON, J.M., ATKINS, K.D. and GILMOUR, A.R. (1979a). Carcass characteristics of heavy weight crossbred lambs. II. Carcass composition and partitioning of fat. *Australian Journal of Agricultural Research* **30**, 1207–1214

THOMPSON, J.M., ATKINS, K.D. and GILMOUR, A.R. (1979b). Carcass characteristics of heavy weight crossbred lambs. III. Distribution of subcutaneous fat, intermuscular fat, muscle and bone in the carcass. *Australian Journal of Agricultural Research* **30**, 1215–1221

THURLEY, D.C., REVFEIM, K.J.A. and WILSON, D.A. (1973). Growth of the Romney sheep foetus. *New Zealand Journal of Agricultural Research* **16**, 111–114

TISSIER, M. and THERIEZ, M. (1979). Influence du niveau des apports énergétiques distribués à la brebis pendant la gestation sur le poids à la naissance et la croissance des agneaux. *Annales de Biologie Animale Biochemie Biophysique* **19**, 235–240

TULLOH, N.M. (1964). The carcase compositions of sheep, cattle and pigs as functions of body weight. In *Carcase Composition and Appraisal of Meat Animals* (D.E. Tribe, Ed.), pp. 5–1, 5–30. Melbourne, CSIRO

WALKER, D.M. and NORTON, B.W. (1971). The utilization of the metabolizable energy of diets of different protein content by the milk-fed lamb. *Journal of Agricultural Science, Cambridge* **77**, 363–369

WALLACE, L.R. (1948). The growth of lambs before and after birth in relation to level of nutrition. I, II, III. *Journal of Agricultural Science, Cambridge* **38**, 93–153, 243–302, 367–401

WARDROP, I.D. (1960). The post-natal growth of the visceral organs of the lamb. II. The effect of diet on growth rate, with particular reference to the parts of the alimentary tract. *Journal of Agricultural Science, Cambridge* **55**, 127–131

WOOD, J.D., MacFIE, H.J.H., POMEROY, R.W. and TWINN, D.J. (1980). Carcass composition in four sheep breeds: the importance of type of breed and stage of maturity. *Animal Production* **30**, 135–152

3

CARCASS QUALITY AND ITS MEASUREMENT IN SHEEP

A.J. KEMPSTER
Meat and Livestock Commission, Bletchley, UK

Sheep meat production is dominated by two factors. First there are the biological and agricultural considerations of resource use, in particular the efficiency with which feed is converted into lean meat. Secondly, there are marketing considerations, in particular the ability of farmers to adapt output to conform more closely with changing market demand.

Progress in both factors is needed urgently in Britain if sheep meat is to continue as an important component of the diet. Since the war, consumption of mutton and lamb has fallen from near 12 kg/head/year to about 7 kg; it now constitutes 10% of total meat consumption compared with 24% in the 1940s. Attitude research among housewives purchasing meat, conducted by the Meat Promotion Executive of MLC, shows that lamb is seen to have significant disadvantages in comparison with other meats; it is thought of, especially by young housewives, as the least versatile, fattest and the most wasteful meat.

The carcass is central to the various activities upon which improvement is based. Carcass quality provides the primary measure of output and is a key selection criterion for genetic improvement. The carcass is the focus of trade and of new marketing efforts by farmers. It provides a convenient defined point of conversion of an animal to meat. It is handled in bulk at a relatively small number of slaughtering points; small, that is, in relation to the number of farms from which the animals come and to the number of retail shops and other trading points through which the meat is sold.

Factors influencing carcass quality

The value of sheep carcasses depends on several factors, namely weight, conformation, proportion of the main tissues (muscle, fat and bone), distribution of these tissues through the carcass, muscle thickness and meat quality.

The weight and size of the carcass has a major influence, not only on the quantity of the various tissues, but also on the size of the muscles exposed on cutting and of the individual joints prepared from it. This is of importance particularly in relation to a retailer's ability to provide cuts of suitable size for customer requirements. Over generations, wholesalers and retailers in different regions of the country have become accustomed to handle certain weight ranges of carcasses; abattoir practices and cutting methods have been developed accordingly. Most wholesalers state desired weight ranges in buying schedules and apply discounts to carcasses outside these ranges. In some cases, these discounts are severe and are a constraint on the use of improved breeds and production systems. There has, for example, been a demand historically for small legs suitable for the Sunday roast, which has slowed the movement to heavier carcasses which are likely to be more efficient to produce and process. This demand has also had a major effect on the type of carcass produced in New Zealand for the British market.

Although, as will be indicated later, conformation is a poor index of carcass composition, it is still regarded by many in the meat industry as valuable in this respect. Carcasses with good conformation normally command higher prices and most national classification and grading schemes include conformation as a factor. It is particularly important in some European markets and good conformation carcasses are in demand by British meat exporters.

Among carcasses of similar weight, the percentage formed by each tissue varies considerably depending on breed type and level of feeding. The proportion of lean meat in the carcass is of major importance since this is a prime determinant of yield and commercial value. The modern housewife increasingly judges quality and value for money by the lean content of the meat she buys. Taken as a generalized ideal, the best carcasses should have an optimum level of fatness, sufficient to ensure that carcasses do not dry out and to ensure good eating quality, and minimum bone. Fat class 2 in MLC's Sheep Carcass Classification Scheme is considered optimum for fatness.

Production has been slow to respond to the demand for leaner carcasses. *Table 3.1* shows the proportion of carcasses falling into different fat classes of the MLC Scheme and their composition. The failure to reduce fatness levels has almost certainly been the major factor contributing to the fall in the per capita consumption of sheep meat compared with other meats. Other factors are the limited processing market for sheep meat due largely to the unattractive characteristics of the fat, the swing away from the weekend joint and the limited demand for sheep meat in the catering trade.

Lack of response to consumer demand stems to some extent from traditional meat trade attitudes to fatness: some wholesalers and retailers still lay great emphasis on the importance of fat as an indicator of meat quality. However, there is growing evidence which fails to support the belief that fatness significantly improves the eating quality of lean meat once a minimum level is present.

The distribution of tissues through the carcass is potentially important because there are large differences between cuts in their retail value; for example loin chops are worth three times as much as breast. However, muscle weight distribution is a fairly constant characteristic and there is little variation to exploit commercially. *Table 3.2* indicates the differences found between breeds in major trials carried out in Britain. On the other hand the distribution of fat between different depots in the carcass and in the body cavity does show important variation (*Table 3.2*). The position of

Table 3.1 PERCENTAGE OF CARCASSES IN BRITAIN FALLING INTO EACH FAT CLASS OF THE MLC CLASSIFICATION SCHEME AND THEIR AVERAGE COMPOSITION

	Fat class					
	1	*2*	*3L*	*3H*	*4*	*5*
Percentage in each class	0.8	25.4	38.9	16.4	12.6	1.5
Separable fat in carcass (%)	14	21	24	29	32	37
Separable lean in carcass (%)	64	60	57	55	52	48
Typical saleable meat yield in carcass (%)[a]	96	93	91	90	89	86

[a]Corresponds approximately to the meat sold over the retail counter. Includes bone which is sold at retail.
From Kempster (1979)

Table 3.2 BREED DIFFERENCES IN LEAN DISTRIBUTION AND FAT PARTITION (RESULTS ARE EXPRESSED AS A PERCENTAGE OF THE SUFFOLK-CROSS RESULTS)

Reference	*Proportion of the total lean distributed in the higher-priced joints*[a]		*Fat depots as a proportion of carcass weight*[b]				
	(1)	*(2)*	*(1)*		*(2)*		*(3)*
			KKCF	*IMF*	*KKCF*	*IMF*	*TF*
Suffolk-cross	100	100	100	100	100	100	100
Purebred							
Welsh Mountain	102		160	98			
Blackfaced Mountain	100		121	90			
British Longwool	97		150	93			
Crossbred with sires							
Border Leicester		98			90	103	
British Longwool	99		118	87			
Dorset Down	} 100	100	} 114	94	98	103	101
Hampshire Down		100			93	104	
Ile de France		100			98	101	101
North Country Cheviot		97			103	98	
Oldenburg							99
Oxford Down		99			98	103	99
Southdown	102	100	132	98	95	105	
Texel		98					98
Wensleydale		99			103	95	

[a]Proportion of total lean which occurred in the leg, chump, loin and best-end neck.
[b]Results are adjusted to equal carcass subcutaneous fat percentage.
KKCF = perinephric and retroperitoneal fat.
IMF = intermuscular fat.
TF = total dissectable fat.
(1) Kempster and Cuthbertson (1977).
(2) Kempster, Croston and Jones (1981).
(3) Wolf, Smith and Sales (1980).

fat is important for several reasons. Subcutaneous fat can be trimmed more easily than intermuscular fat and is, therefore, preferable in carcasses containing fat in excess of consumer requirements; indeed, excess intermuscular fat cannot be trimmed from some lamb joints without mutilating them. Fat in the body cavity is waste even in the leanest carcasses. The evenness of fat distribution is also important because wedges or bulges of fat in some joints can lead to excess trimming or devaluation.

The importance of differences in muscle thickness is less obvious. Meat traders tend to prefer blockier carcasses with thicker muscles because the appearance of the joints prepared from them is better. There may also be advantages in terms of increased tenderness and reduced weight loss in the preparation and cooking of cuts, but the size of these advantages and their commercial significance is not clear.

Meat texture (degree of toughness or tenderness) is the most important factor contributing to eating quality. Texture is influenced not only by the state of the contractile components of muscle but by the quantity and chemical nature of the connective tissue. As the animal grows older the connective tissue within each muscle becomes tougher, primarily because the collagen is more cross-linked and does not dissolve so easily on cooking. Toughness can also occur (known as cold shortening) if carcasses are rapidly chilled immediately after slaughter. However, recent developments in electrical stimulation which speed up physiological changes in the muscle can overcome this problem to a large extent. Lamb exported from New Zealand is nearly all electrically stimulated.

The chemical composition of carcasses does not usually have a direct bearing on their commercial value (except insofar as it reflects physical composition). However, chemical composition may be important in relation to a number of factors including eating quality of the meat itself, the processing characteristics, the propensity to lose water between slaughter and consumption, the keeping qualities and the nutritive value. The chemical composition may also have an influence on eating quality. For example, the highly saturated nature of lamb fat leads to it solidifying quickly during cooling after cooking, making it less palatable than more unsaturated fats.

Techniques for carcass evaluation

Most carcass evaluation work is carried out with an economic objective ultimately in mind and is concentrated on those characteristics which have most influence on carcass retail value. Setting aside carcass weight, description of carcass leanness is of most importance and the remainder of the chapter will concentrate on this characteristic.

The first consideration is how carcass leanness is to be defined: on what definition of overall carcass leanness (baseline definition) should we decide, for example, that one breed of sheep or one individual sire should be used in preference to others. The distinction between the baseline and carcass evaluation technique used to predict it should be clearly understood. The baseline is the ideal evaluation which would be applied to all animals if cost were no obstacle. When estimating carcass composition from simpler measurements one is trying to get as close as possible to the results which would have been obtained had the baseline itself been used on all animals.

A range of possible baseline techniques is available from fat trimming as in commercial practice, through more standardized cutting and tissue separation to whole body chemical analysis. For most purposes, a baseline in terms of cutting seems preferable to chemical analysis since value judgements by consumers are made on the basis of the appearance or physical composition of the carcass and joint. Considering the various cutting techniques available, the division of the carcasses into standardized commercial joints and their separation into component tissue and different fat depots using a butcher's knife is recommended (Cuthbertson, Harrington and Smith, 1972). This has provided a satisfactory baseline in a whole range of MLC studies and is also used effectively in a number of other countries. More detailed techniques involving scalpel and scissors are rarely cost effective; the money is often spent more effectively by increasing sample sizes.

If resources allow, the most comprehensive procedure is to apply tissue separation after dividing the carcass into cuts followed by chemical analysis of the resulting tissues. Such an approach is particularly valuable in growth and nutritional studies when it is important to determine the extent to which specific nutrients are retained in the carcass. Apart from providing information on the chemical composition of tissues, such a procedure also enables the standard of physical separation to be monitored.

The dissection of one side of the carcass only is adequate for most baseline work. Significant differences have been recorded between sides in tissue percentages but it is generally accepted that the improvement in precision obtained from dissecting both sides does not justify the considerable increase in costs involved. Most side to side differences occur in bone content because of inaccuracies in centre splitting. If there is a problem, the vertebral column can be removed from both sides to obtain vertebral bone weight for the dissected side. Real biological bilateral asymmetry in carcasses is of negligible importance.

The problem of selecting the most suitable carcass evaluation technique (prediction of the baseline technique) for a given application can only be solved satisfactorily with reference to the particular circumstances, labour force, resources available and the required sensitivity of the work in question. There are three main areas where carcass evaluation is required; in the commercial classification and grading of carcasses, in breeding schemes, and in experiments and population studies.

COMMERCIAL CLASSIFICATION AND GRADING

The main reason for the development of carcass classification and grading schemes is to provide better communication of consumer requirements back to the producer. Given clear price differentials related to carcass characteristics of commercial importance, the producer can exploit existing knowledge of breeding and management to provide what is required. Without such price differentials, there can be no incentive to improve carcass quality unless the improvement is related positively to production efficiency.

The distinction between classification and grading is an important one. 'Classification' covers the description of carcasses by characteristics which are of prime importance to meat traders; the characteristics are described separately without attributing relative importance or cash value differences to them. The buyer creates grades by combining the classification data to form a summarized assessment to which prices are attached. It is, therefore, for the individual trader to decide what is good and what is poor in relation to his own business.

A classification scheme is operated in Britain by MLC and this will be discussed in detail. Some schemes do grade carcasses directly, for example those operated by the United States Department of Agriculture and by the New Zealand Meat Producers Board (the features of the main classification and grading schemes around the world have been discussed by Kempster, Cuthbertson and Harrington, 1982). The MLC classification scheme describes carcasses by four key characteristics—weight, category, fatness and conformation.

Weight

'Weight' is the weight determined after the carcass has been dressed in a specified manner; an independently recorded weight linked to a firm dressing specification is seen as an important feature of successful deadweight marketing. A fundamental difference between this classification scheme and grading schemes like that operated in New Zealand, is that the latter include weight ranges; thus grade YM relates only to lambs in the carcass weight range 13.0–16.0 kg.

Category

The term 'category' is used to distinguish lambs from older groups such as hoggets (lambs born before the 1st October in any one year are classed as hoggets at the beginning of the following year).

Fatness

Fatness is described by a visual judgement of the level of external fat cover on a five class scale ranging from 1 (very lean) to 5 (very fat): the middle fat class is divided into 3L (low fat) and 3H (high fat). Carcasses with excessive perinephric and retroperitoneal fat (KKCF) are also identified. Although the classification for subcutaneous fat development is carried out subjectively, there is an underlying objective baseline for each class defined in terms of a range of subcutaneous fat in carcass (Kempster, 1979). This baseline can be used for standardization purposes.

Selection of the visual fat score was based on a national survey and comparison of different predictors suitable for use on the hot intact carcass under normal slaughter line conditions (Kempster *et al.*, 1976; Kempster and Cuthbertson, 1977). The visual fat score was found to be the best single predictor among those tested. Addition of a visual assessment of KKCF development or a measure of the circumference of buttocks to fat score provided a useful increase in precision. Individual linear measurements taken on the intact carcass had no worthwhile predictive value, confirming the results of earlier studies (see *Table 3.4*).

The development of probing instruments for measuring fat thickness. Visual assessments are not easy to standardize and there are advantages in using measurements of fat thickness in classification and grading schemes. The problem is how to take such measurements effectively on the intact carcass. MLC has been testing simple steel rules with caliper for probing fat thickness but are not yet convinced that they should replace the visual fat score. A simple measuring tool of this type is now used in the New Zealand Lamb Grading Scheme to identify the overfat (F) grade, which must be cut and trimmed before export. This measurement is referred to as GR and is the total tissue thickness between the surface of the lamb carcass and the rib at a point 11 cm from the dorsal mid line in the region of the twelfth rib. GR has been shown to be as accurate as fat thickness measurements over the *M. longissimus* taken on the cut carcass in predicting carcass fat content in New Zealand lambs (Kirton and Johnston, 1979), but there are difficulties in taking the measurement at fast line speeds.

In recent years there have been major developments in automatic recording probes for measuring fat and muscle thicknesses. The main thrust of the work has been for pig carcass classification and grading where the use of such probes is simpler, but interest in their use is now being extended to beef and sheep. Probes are essentially of two types: those which identify muscle and fat boundaries using differences in electrical conductivity and those which identify the differences in colour (reflectance) between muscle and fat (Kempster, 1981a). Research with sheep is not very advanced and it is too early to say such probes will make a major contribution to classification and grading schemes in the next five years.

Conformation

The other factor in the MLC Sheep Carcass Classification Scheme is conformation. The main reason for its inclusion is to identify carcasses with thicker muscles and possible advantages in meat to bone ratio. The description of carcass conformation is based on four classes—extra (E), average, poor (C) and very poor (Z) but consideration is now being given to extend the number of classes. Conformation is assessed as the thickness of lean and fat in relation to skeletal size, fatness being allowed to play its full part in influencing the score given. This approach to conformation is considered preferable to attempts to adjust the conformation assessment for fatness.

There is still much debate about the value of conformation as an indicator of carcass composition and uncertainty about the emphasis that should be given to it in commercial classification and grading schemes. Most scheme conformation studies have been carried out within breed and without adjustment to equal fatness. Predictably the general conclusion has been that lambs with good conformation tend to be fatter than poorer conformation carcasses, with little difference in muscle thickness or proportion of higher-priced cuts (e.g. Kirton and Pickering, 1967; Jackson and Mansour, 1974). An association between fatness and conformation is expected when carcasses with thicker fat cover are judged to have better conformation.

Detailed information on the value of conformation in the British sheep population has been obtained from two large-scale ram breed trials (Kempster, Croston and Jones, 1981). In these trials, conformation was examined as a predictor of carcass composition when variation in fatness was eliminated. 'Fat-corrected' conformation was not found to be a valuable predictor of composition in sheep carcasses (*Table 3.3*). At best, it

Table 3.3 RESIDUAL STANDARD DEVIATIONS FOR THE PREDICTION OF CARCASS LEAN PERCENTAGE FROM CONFORMATION SCORE IN DIFFERENT CIRCUMSTANCES

	Overall	*Within sire breed*	*Within year, dam breed, sire breed and sex*
Standard deviation of carcass lean percentage	3.83	3.58	3.30
Residual S.D. for:			
Carcass weight (W)	3.63	3.32	3.01
W + SF_5[(a)]	2.97	2.88	2.67
W + SF_5 + C_4[(b)]	2.96	2.88	2.67
W + SF_e[(a)]	2.61	2.56	2.38
W + SF_e + C_{15}[(b)]	2.61	2.56	2.37

[(a)] SF_5 = visual assessment of carcass subcutaneous fat percentage on a 5-point scale.
SF_e = visual assessment of carcass subcutaneous fat percentage to the nearest percentage point.
[(b)] C_4 = visual assessment of conformation on a 4-point scale.
C_{15} = visual assessment of conformation on a 15-point scale.
Sire breeds are as shown in *Table 3.2*; dam breeds were Scottish Blackface, Border Leicester × North Country Cheviot (Scotch Halfbred) and Bluefaced Leicester × Swaledale (Mule).
From Kempster, Croston and Jones (1981)

only contributed marginally to prediction and was sensitive to the accuracy with which fatness was controlled. Indeed, poor control of fatness, as would occur in commercial classification when a visual score into one of a limited number of fat classes is used, could mean that conformation would identify fatter rather than leaner carcasses.

The marginal advantage of conformation as a predictor of leanness appears to lie in the identification of differences in lean to bone ratio, although examination of breed differences indicates that breeds with better conformation do not necessarily have higher lean to bone ratios. Suffolk crosses, for example, have relatively low lean to bone ratios in relation to their conformation whereas Texel crosses have a high lean to bone ratio but do not have sufficiently high conformation scores to identify their advantage.

The results in *Table 3.3* indicate that sire breed is more effective than conformation in improving precision of carcass lean prediction and it can be argued that breed should be used as an alternative to conformation (or used in association with it) for commercial classification and grading schemes where this is possible. Identification difficulties preclude its use in most circumstances but some wholesalers purchasing sheep from well-known suppliers may be able to include breed in their grading specifications. The same conclusion has been reached for other species (Kempster, 1981b). The importance of breed in these relationships indicates that the most suitable measuring techniques may change with time. As populations of sheep change in their breed structure, so there is a need for continued reappraisal of the measuring techniques used.

BREEDING SCHEMES, POPULATION STUDIES AND EXPERIMENTS

The selection of predictors for use in breeding schemes, population studies and experiments presents quite a different problem to that in commercial classification and grading. Much more is known about the animals, in particular the breed, system of production, age, liveweight and so on. Results obtained in a mixed sample of carcasses, as one would have in commercial classification and grading, might not apply within breed and system of feeding.

Although there are many published papers on techniques for predicting sheep carcass composition, the selection of a technique for a particular application is rarely straightforward and often made with uncertainty. The literature provides general guidance on the limit of precision of various techniques but many of the trials appear to lack precise objectives and often fail to show any real practical value in the techniques studied. In the following section, the relative value of predictors is considered in relation to the prediction of carcass lean content.

Criterion for selecting techniques

Three criteria are involved in the selection of the most suitable technique for a particular application:

1. the precision with which the technique is expected to predict carcass lean content;
2. the cost of taking the predicting measurements; this will reflect the ease, speed and accuracy with which the measurements can be recorded, and the carcass depreciation involved; and
3. the stability of the prediction equations to treatment differences or differences between the types of lamb being compared.

Relative precision of different predicting measurements. There have been many trials to compare the precision of different predictors of sheep carcass composition. The early trials assessed techniques over as great a range of material as possible so the results are of limited value except possibly in the context of commercial classification. Harrington (1966) drew attention to the deficiencies of this approach. Since then a number of trials have been carried out within breed and sex and within other factors normally controlled in breeding schemes, population studies and experiments.

Table 3.4 PRECISION WITH WHICH DIFFERENT MEASUREMENTS AND COMBINATIONS OF MEASUREMENTS HAVE BEEN FOUND TO PREDICT CARCASS LEAN PERCENTAGE IN DIFFERENT TRIALS (RESIDUAL STANDARD DEVIATIONS)

Trial	*(1)*	*(2)*	*(3)*	*(4)*	*(5)*
Standard deviation of carcass lean percentage	3.29	3.1	3.6	3.70	3.50
Residual S.D. for:					
M. longissimus dimensions	2.83	2.7	3.3	3.65	3.48
Carcass dimensions		2.9		3.33	3.01
Visual fat scores		2.3		2.82	2.60
KKCF in carcass (%)	2.81		3.2	2.98	2.60
Fat thickness over *M. longissimus*	2.71	1.7	2.9	2.80	2.60
Specific gravity	2.83	1.7			
Combination of the above		1.6	2.1	2.40	
Lean in joints (%)					
leg (27)[a]	1.63	1.6	1.8	2.15	2.34
breast (8)	1.88	1.7		1.82	2.15
loin (14)	1.83	1.2	2.2	1.81	1.80
ribs (13)	1.83	1.2	1.9	1.53	1.92
shoulder (16)	1.58	1.4	1.3	1.62	1.74
legs + loin (41)		0.9		1.23	
loin + ribs (27)		1.0		1.32	
ribs + shoulder (29)				1.04	

[a]Cost as a percentage of side dissection cost under MLC conditions.
(1) Field, Kemp and Varney (1963): 165 Southdown × Western lambs.
(2) Timon and Bichard (1965): 83 Clun Forest lambs.
(3) Latham, Moody and Kemp (1966): 121 Southdown crossbred lambs.
(4) Kempster *et al.* (1976): 424 lambs, pooled within breed groups.
(5) Kempster and Jones (unpublished data): 894 lambs pooled within year, dam breed, sire breed and sex groups.

Results for the more important published trials are shown in *Table 3.4*, together with some unpublished results from recent MLC studies. The results as a whole present a fairly constant picture indicating that:

1. carcass dimensions and *M. longissimus* dimensions are poor individual predictors;
2. visual fat assessments, perinephric and retroperitoneal fat (KKCF) percentages, fat thicknesses and specific gravity determinations provide increased precision, possibly in that order;
3. sample joints are the most precise predictors.

When the simpler measurements not involving dissection are combined in multiple regression, their precision approaches that achieved by dissection of the leg (and is better than that achieved by dissection of some smaller joints, such as the neck, which is not shown in the table). However, for greater precision, dissection of other joints and combination of joints is necessary.

Table 3.4 also shows the relative costs of the sample joints under the circumstances of MLC's Central Carcass Evaluation unit. The leg does not appear to offer value for money, whereas the breast offers quite good precision in relation to its cost. Besides the more common predictors considered in *Table 3.4*, several others have been evaluated, for example fat to lean areas of cut surfaces, Electronic Meat Measuring Equipment (EMME), but not in comparison with a range of other methods. These techniques have been discussed by Kempster, Cuthbertson and Harrington (1982).

Use of prediction equations and their relative stability for different predictors. The most satisfactory approach to prediction is to construct equations within the study where they are to be used. Equations can be constructed separately for each treatment, or individual treatment intercepts can be used together with pooled within treatment slopes as in the double sampling with regression procedure (Conniffe and Moran, 1972; Evans and Kempster, 1979). However, this approach requires some baseline dissection, which is costly to perform, and it is often necessary to use equations developed elsewhere. In such cases, it may be best to use equations developed in circumstances as close as possible to those in the proposed trial. Alternatively, it may be decided to select prediction equations which have been shown to be stable over a range of circumstances. In neither case is there a guarantee that the estimates of carcass composition will not be biased to some extent.

The selection of predictors having stable equations is particularly important in trials without baseline evaluation, since no amount of replication will compensate for biases. It is also worthwhile emphasizing that it is not necessary actually to apply a regression equation to occasion a bias; by assuming that the ranking and the relative differences in carcass composition between treatments are the same as for the predicting measurements, an equation is effectively being applied.

Although there are many reports of analyses to compare predictors within a single group of lambs, few workers have examined the same predictors over a series of groups differing in breed type, sex, etc. to determine the stability of prediction equations; sample sizes have generally been too small or the data insufficiently variable in origin to do this. A number of trials have, however, provided indirect evidence of instability in prediction relationships (e.g. Kirton and Barton, 1958; Oliver *et al.*, 1968).

More specific indications emerged from a large-scale breed comparison trial carried out in Britain by the Animal Breeding Research Organization in collaboration with MLC (*Table 3.5*). This involved the dissection of 894 crossbred lambs out of two dam types by Dorset Down, Ile de France, Oldenburg, Oxford Down, Suffolk and Texel sires. The results indicate that bias between breeds can still be a problem even when breeds are compared within flock and year. The Texel cross appeared quite different from other breeds in several of the prediction relationships; they had a significantly higher carcass lean percentage than would be predicted, due largely to their high lean to bone ratio and low percentage of perinephric and retroperitoneal fat.

The more precise predictors tended to be among the more stable

Table 3.5 ESTIMATES OF BIAS FROM THE ASSUMPTION THAT PREDICTION EQUATIONS ARE THE SAME FOR ALL BREEDS[a]

Sire breed (common dam breed)	*Number of carcasses*	*Predictor*						
		Visual external fat score	*Visual[b] internal fat score*	*Fat[c] thickness*	*Percentage lean in*			
					best-end neck	*shoulder*	*loin*	*leg*
Dorset Down	154	+1.7	+2.0	+1.6	+0.7	+0.6	+1.0	+0.9
Ile de France	165	+0.7	+0.8	+0.6	−0.1	0.0	−0.1	+0.9
Oldenburg	138	−0.4	−0.2	−0.4	0.0	+0.1	−0.4	−0.8
Oxford Down	156	+0.7	+0.7	+0.3	+0.4	+0.1	+0.2	−0.3
Suffolk	158	+0.6	+0.4	+0.6	+0.4	+0.1	+0.4	−0.4
Texel	190	−3.3	−3.7	−2.8	−2.0	−0.7	−1.3	−0.5

[a]Predicted minus actual carcass lean percentage.
[b]Visual assessment of the amount of perinephric and retroperitoneal fat in the carcass.
[c]Measured over the *M. longissimus* at the 12th rib.
From Kempster and Jones (unpublished)

although some of the sample joints were exceptions to this rule. On the basis of these results, dissection of the shoulder joint would be recommended for use in breed comparison trials when the full dissection of some carcasses was impossible. Trials involving the Texel or similar breeds require special attention when selecting predictors.

The occurrence of sample joints with unstable prediction equations is of particular concern because sample joints are primarily research tools and used in trials where small treatment of genetic differences may be of critical interest. In such trials, some baseline dissection may, therefore, be essential and careful consideration needs to be given to whether the use of sample joints represents an optimal use of resources: it may, for example, be more cost effective to dissect a subset of carcasses and take simple measurements on the remainder. The point was discussed in detail by Evans and Kempster (1979) in the context of using a double sampling with regression procedure for pigs.

General considerations

On the basis of available information on lamb carcass evaluation, consideration should be given to the following points when choosing prediction methods.

1. Wherever possible develop prediction equations or intercepts within the trial where they are to be used.
2. Find the balance between baseline dissection and predicting measurements which is most cost-effective, even if this involves taking only the simplest measurement on some animals.
3. If full dissection is impossible:
   (a) use predictors which have been shown to have the most stable prediction equations;
   (b) use two or more predictors which are based on the measurements of different characteristics in the animal body; and
   (c) change the prediction measurements from time to time, for example in different replicates of the same experiment or population study. Such a policy may be particularly important in different generations of breeding schemes since reliance on a single predictor might lead to accumulated biases.

Since no carcass measurements or sample joint dissections have yet been found which are of outstanding value in relation to their cost, it is unlikely that such a biological panacea exists. It is, therefore, probably futile to waste resources in haphazard attempts to find it. It is much more important to examine the underlying features of growth and development in relation to the prediction problem and attempt to determine why some predictors are more precise than others and why some predictors have more stable regression relationships. This point has been discussed in detail by Kempster (1981c).

References

CONNIFFE, D. and MORAN, M.A. (1972). Double sampling with regression in comparative studies of carcass composition. *Biometrics* **28**, 1011–1023

CUTHBERTSON, A., HARRINGTON, G. and SMITH, R.J. (1972). Tissue separation to assess beef and lamb variation. *Proceedings of the British Society of Animal Production (New Series)* **1**, 113–122

EVANS, D.G. and KEMPSTER, A.J. (1979). A comparison of different predictors of the lean content of pig carcasses. 2. Predictors for use in population studies and experiments. *Animal Production* **28**, 97–108

FIELD, R.A., KEMP, J.D. and VARNEY, W.Y. (1963). Indices of lamb carcass composition. *Journal of Animal Science* **22**, 218–221

HARRINGTON, G. (1966). The relative accuracy and cost of alternative methods of pig carcass evaluation. *Zeitschrift fur Tierzuchtung und Zuchtungsbiologie* **82**, 187–198

JACKSON, T.H. and MANSOUR, Y.A. (1974). Differences between groups of lamb carcasses chosen for good and poor conformation. *Animal Production* **19**, 93–105

KEMPSTER, A.J. (1979). Variation in the carcass characteristics of commercial British sheep with particular reference to over-fatness. *Meat Science* **3**, 199–208

KEMPSTER, A.J. (1981a). Recent developments in measuring techniques for use in pig carcass classification and grading. *Pig News and Information* **2**, 145–148

KEMPSTER, A.J. (1981b). The problem of breed bias in commercial carcass classification and grading. *Animal Production* **32**, 360–361 (Abstract)

KEMPSTER, A.J. (1981c). The indirect evaluation of sheep carcass composition in breeding schemes, population studies and experiments. *Livestock Production Science* **8**, 263–271

KEMPSTER, A.J. and CUTHBERTSON, A. (1977). A survey of the carcass characteristics of the main types of British lamb. *Animal Production* **25**, 165–179

KEMPSTER, A.J., CROSTON, D. and JONES, D.W. (1981). Value of conformation as an indicator of sheep carcass composition within and between breeds. *Animal Production* **33**, 39–49

KEMPSTER, A.J., CUTHBERTSON, A. and HARRINGTON, G. (1982). *Carcase Evaluation in Livestock Breeding, Production and Marketing*. Granada Publishing, St. Albans

KEMPSTER, A.J., AVIS, P.R.D., CUTHBERTSON, A. and HARRINGTON, G. (1976). Prediction of the lean content of lamb carcasses of different breed types. *Journal of Agricultural Science*, Cambridge **86**, 23–34

KIRTON, A.H. and BARTON, R.A. (1958). Specific gravity as an index of the fat content of mutton carcasses and various joints. *New Zealand Journal of Agricultural Research* **1**, 633–641

KIRTON, A.H. and JOHNSON, D.L. (1979). Interrelationships between GR and other lamb carcass fatness measurements. *Proceedings of the New Zealand Society of Animal Production* **39**, 194–201

KIRTON, A.H. and PICKERING, F.S. (1967). Factors associated with differences in carcass conformation in lamb. *New Zealand Journal of Agricultural Research* **10**, 183–200

LATHAM, S.D., MOODY, W.G. and KEMP, J.D. (1966). Techniques for estimating lamb carcass composition. *Journal of Animal Science* **25**, 492–496

OLIVER, W.M., CARPENTER, Z.L., KING, G.T. and SHEPTON, M. (1968). Predicting cutability of lamb carcasses from carcass weights and measures. *Journal of Animal Science* **27**, 1245–1260

TIMON, V.M. and BICHARD, M. (1965). Quantitative estimates of lamb carcass composition. *Animal Production* **7**, 173–182, 183–187, 189–201

WILSON, A. (1975). Carcass studies in crossbred lambs. PhD. Thesis. University of Newcastle upon Tyne

WOLF, B.T., SMITH, C. and SALES, D.I. (1980). Growth and carcass composition in the crossbred progeny of six terminal sire breeds of sheep. *Animal Production* **31**, 307–313

4

EVALUATION OF CARCASS QUALITY IN THE LIVE ANIMAL

J.C. ALLISTON
Animal Breeding Research Organisation, Edinburgh, UK

The need for accurate assessment of the body composition of live sheep is very important if selection for improved carcass quality is to be effective. Recent developments in techniques available has meant that there are now several methods that are both accurate and repeatable. Some of the machines are quick and easy to use under field conditions whilst being strong and readily transportable. Examples of these techniques are visual assessments, body measurements and ultrasonic scannings with machines based on a simple principle. Other more advanced techniques are the use of deuterium oxide dilution and computerized tomography from X-ray transmission data. These tend to be slower to use and require the animals to be brought to the assessment area.

The techniques available have been reviewed by Houseman (1972), Cuthbertson (1975), Wright (1982) and Miles (1981). As there are many varied methods of determining body composition only a few selected techniques will be outlined in this chapter.

Techniques suitable for commercial application

LIVE ANIMAL ASSESSMENT

Live visual assessments of sheep have for a long time been the only means used for the selection of breeding stock. The advantages of the methods are that the performance of the animals is not affected and the costs are low.

Kallweit (1976), writing about the use of visual assessments for beef production, concluded that in spite of their inherent subjectivity, these criteria have been successfully employed in the establishment of large numbers of rather well defined breeds, differing widely in specialization and performance. The variability in special breeds and lines has been minimized to such an extent, that still more refined breeding accomplishments can hardly be expected without resorting to objective evaluation. However, because of past success and low cost, visual assessments will continue to be practised. Cuthbertson (1975) suggested that the simplest technique for assessing the degree of fat development and conformation was to handle the animal.

At a recent CEC workshop on '*in vivo* estimation of body composition in beef', results presented showed that visual assessment of animals are used in progeny and performance tests for cattle in West Germany, France, Belgium and Holland (Kallweit, 1981; Rehben, 1981; Verbeke, 1981; Jansen, 1981).

BODY CONDITION SCORING IN SHEEP

Jefferies (1961) first described a system of scoring body condition in which the fat cover on the back of a sheep, as measured by the cover on the lumbar vertebrae, is given a quantitative value. The technique was used by Everitt (1962) but it was concluded that condition score in both sheep and cattle was not a good indicator of either total chemical composition or dissectable fatty tissues. Russel, Doney and Gunn (1969), however, developed a grading system at the Hill Farming Research Organisation which consisted of a six-point scale. Condition score gave a better estimate (r = 0.94 versus 0.81) of chemical fat in the body of mature Scottish Blackface ewes than liveweight (*Figures 4.1* and *4.2*). It was therefore concluded that body condition scoring could give an acceptable and useful estimate of the proportion of fat in the body.

The Meat and Livestock Commission recommend the use of a five-grade body condition scoring system in the booklet *Lamb Carcass Production* published in 1981. The grades are defined in *Table 4.1*. As with all subjective techniques the skill of the assessor is vital if any sort of consistency is to be achieved. It is now possible to calibrate individual estimators against a more objective technique such as ultrasonic measurements and in this way minimize trends over time and differences between estimators. Other methods involving the incision into the fat and muscle area under the skin are outlined in *Table 4.2*. Whilst having some value as a predictor, these techniques obviously cause some discomfort to the animal and are not easy to carry out in the field.

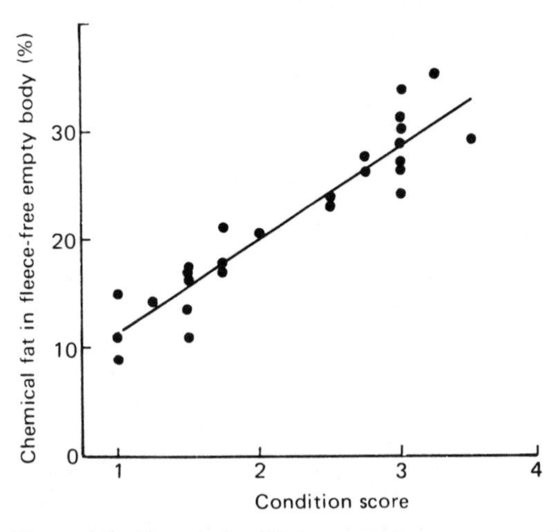

Figure 4.1 The relationship between the percentage of chemical fat in the fleece-free empty body and subjectively assessed condition score. By courtesy of Russel, Doney and Gunn (1969).

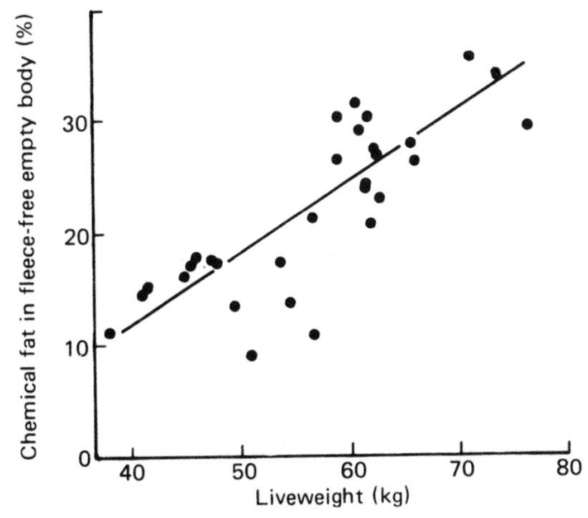

Figure 4.2 The relationship between the percentage of chemical fat in the fleece-free empty body and liveweight. By courtesy of Russel, Doney and Gunn (1969).

Table 4.1 MEAT AND LIVESTOCK COMMISSION CLASSIFICATION OF LAMBS INTO FAT CLASSES

Fat class	*Dock*	*Loin*
1	Fat cover very thin. Individual bones very easy to detect	Spinous process very prominent. Individual processes felt very easily. Transverse processes prominent. Very easy to feel between each process
2	Fat cover thin. Individual bones detected easily with light pressure	Spinous processes prominent. Each process is felt easily. Transverse processes: each process felt easily
3	Individual bones detected with light pressure	Spinous and transverse processes: tips rounded. With light pressure individual bones felt as corrugations
4	Fat cover quite thick. Individual bones detected only with firm pressure	Spinous processes: tips of individual bones felt as corrugations with moderate pressure. Transverse processes: tips detected only with firm pressure
5	Fat cover thick. Individual bones cannot be detected even with firm pressure	Spinous and transverse processes: individual bones cannot be detected even with firm pressure

Table 4.2 PREDICTION OF CARCASS COMPOSITION OF SHEEP BY PROBING THE LIVE ANIMAL

Authors	*No. of animals*	*Measurements*	*Correlations*
Stouffer *et al.* (1958)	34	Thickness of loin muscle and fat. Last rib	+0.58 carcass loin depth +0.62 carcass rib eye width +0.42 rib eye area
Spurlock *et al.* (1966)	31	Liveweight (fasted) skin fold behind front leg needle behind the shoulder probe 7th rib fat depth 12th rib fat depth. Combination of liveweight live conformation grade probe seventh rib fat depth	+0.98 cold carcass weight +0.73 fat depth in carcass +0.75 fat depth in carcass +0.77% trimmed cuts

CONFORMATION

Conformation can be defined as the visual shape of the body of an animal, particularly the relationship between the skeleton and the covering of muscle and fat. Most of the work relating conformation to body composition has been carried out on carcasses and was briefly reviewed by Kempster, Croston and Jones (1981). They found that most of the published studies have been carried out within breed, with the general conclusion that lamb carcasses with good conformation tended to be fatter than poorer conformation carcasses and had little advantage in muscle thickness or proportion of higher price cuts.

The results of Kempster, Croston and Jones (1981) with many breeds emphasized that conformation gave a poor indication of both carcass lean content and the proportion of higher priced joints in carcasses of the same weight and fatness, and confirmed that in animals of variable fatness conformation is likely to identify fatter rather than leaner carcasses.

The importance of conformation in beef cattle has probably been emphasized by the presence or absence of double muscling. In sheep no such comparable phenomenon occurs and as a result conformation has not dominated visual assessment of animals in the same way. However, the recent importation of the Texel with its 'blocky' appearance has given new interest in the assessment of conformation in British sheep breeding.

So long as conformation is a factor in classification and grading schemes then it will be used in the selection of breeding stock. The problem then arises as to how to maintain and improve conformation whilst at the same time reducing the fat content of the animals. The conclusion from the evidence must be that in the long term, selection for leaner carcasses based on the visual assessment of conformation is unlikely to be effective.

LIVEWEIGHT

Tulloh (1963) re-examined the work of Hammond (1932), Wallace (1948a,b), Palsson and Verges (1952a,b) and Wardop (1957) and computed the regression equations required for estimating the body composition of sheep from empty liveweight (that is, the weight of the live animal at slaughter minus the contents of its alimentary tract).

The equations indicate that as a sheep increases in empty liveweight the bone, fat and muscle tissues also increase in weight. However, the slopes of these lines show that as empty liveweight increases, the percentage of carcass fat also increases, but the percentage of bone decreases, whereas the percentage carcass muscle remains almost constant (*Figure 4.3*). Tulloh (1963) found that the deviation from the regression lines are large enough to suggest that no equation is likely to be sufficiently accurate in its predictive value to adequately represent all sheep for any one carcass component. More accurate regression equations will have to be obtained within breeds in various environments.

Unfortunately liveweight is a less useful indicator of body composition in ruminants than in monogastrics. Indeed, gut fill in sheep can vary considerably and ranges of 7–21% are reported in the literature (Tayler,

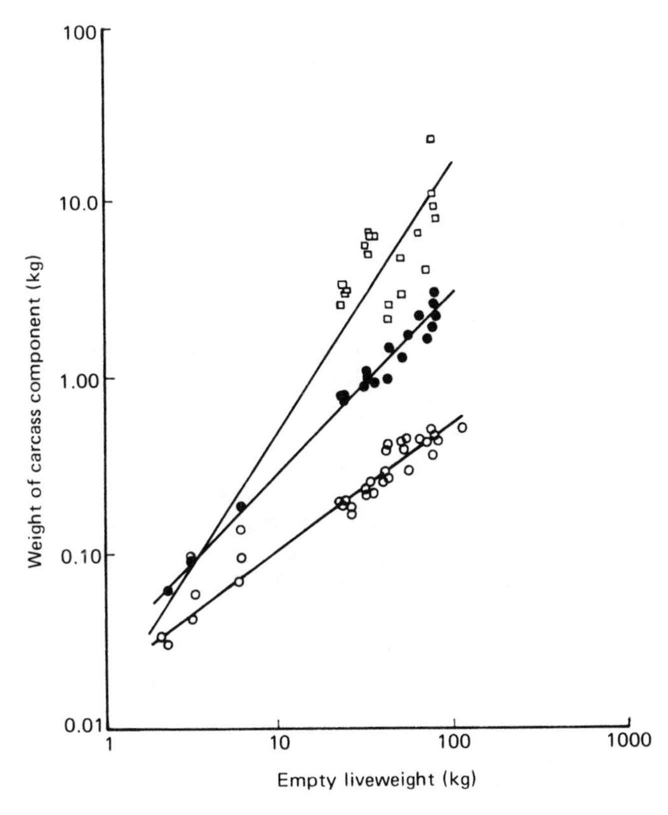

Figure 4.3 Weight of dissected carcass bone (○), muscle (●) and fat (□) compared with empty liveweight of sheep. By courtesy of Tulloh (1963)

1954; Kirton and Barton, 1958). At present there is no practical means of estimating the weight of gastrointestinal contents and liveweight measurements will therefore at best give only an indication of body composition. Spurlock, Bradford and Wheat (1966) working with 31 ram lambs found that liveweight was an important single measure of weight of trimmed cuts (r = 0.96).

ULTRASONICS

There are now several ultrasonic machines in operation that are able to give two-dimensional scans of body tissues in animals. The most common of these have been reviewed by Miles (1978) and Andersen *et al.* (1981).

The machines work on the principle that high frequency sound signals pass through the tissues of the body but when an interface between two tissues is encountered some sound is reflected back. A pulse generator transmits electric pulses which are converted into sound signals in the transmitter. These signals are then passed through the tissues until reflected at an interface. The reflected signals are then picked up by the receiver and can be amplified and shown in a visual form by an oscilloscope. Variations in the time taken for the reflected signals to return to the transmitter–receiver are used to measure variations in the distance of boundaries between tissues. The principles are outlined by Andersen *et al.* (1981), Miles (1978), Wells (1969) and Simm (unpublished review, 1982). 'A' mode ultrasonic machines have a display of echo amplitude against time. The appearance on the screen is of peaks superimposed on a time baseline. The distance between the peaks represents the thickness of the tissues being measured (e.g. Sonatest). For 'B' mode machines the signals are shown on a cathode ray tube as a series of bright spots. The thickness of tissues is represented by the distance between successive bright spots. Generally speaking these machines have a single transducer which moves across the body of the animal on a track. As the transducer moves a picture is built up either on polaroid film or on a cathode ray screen (e.g. Scanogram).

Real-time machines produce a practically instantaneous picture by rapid electronic switching from element to element. The principle involved is similar to that already described except that movements of the tissues of the animal can be seen because of the continuous nature of the picture (e.g. Danscanner). The principles are illustrated in *Figures 4.4* and *4.5*.

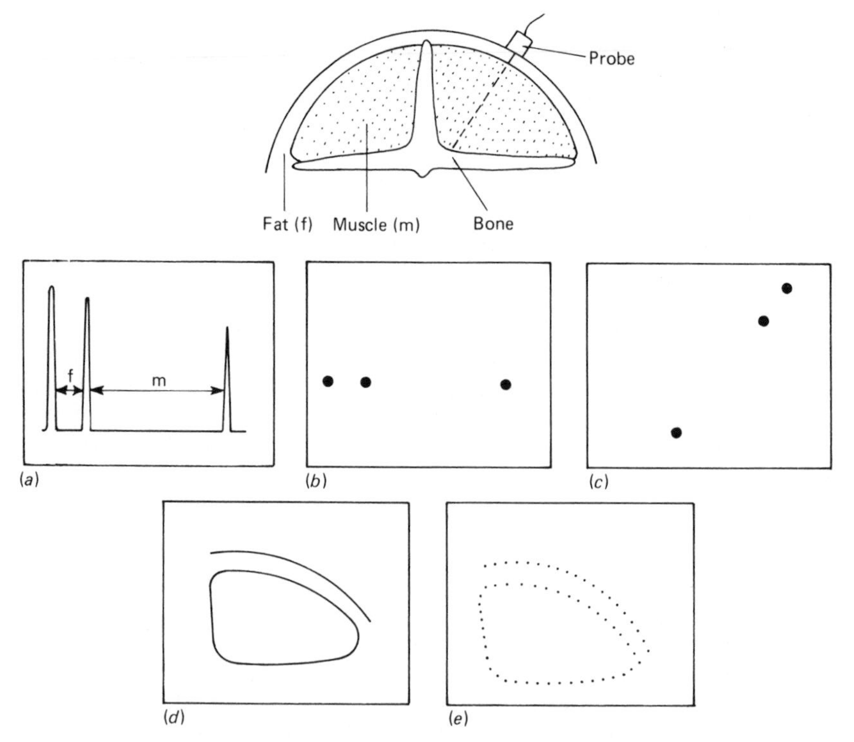

Figure 4.4 Schematic diagram of a cross section through an animal and different types of ultrasonic scan. (a) 'A' mode presentation; (b) 'B' mode presentation; (c) 'B' mode with direction of the timebase linked to the direction of the ultrasonic beam; (d) 'B' mode presentation built up as the probe moves across the back of the animal; (e) 'B' mode presentation from a multi-element transducer. From Simm (unpublished, 1982).

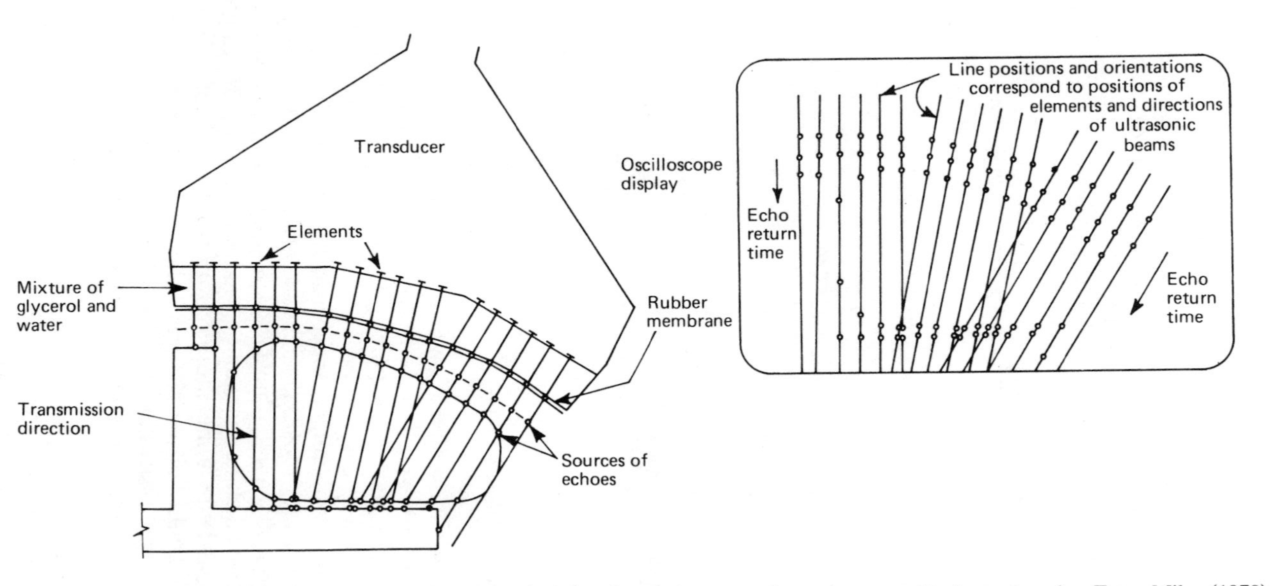

Figure 4.5 A schematic drawing illustrating the principle of real-time scanning using a multi-element probe. From Miles (1978).

The interpretation of results for 'B' mode machines and real-time scanning usually requires the tracing of depths and areas from pictures. This can now be done by using planimeters linked to microprocessors or computers.

Miles and Fursey (1974) reported on a technique which measures the velocity of ultrasound in the soft tissue of the limbs of living meat animals. Sheep were measured across the leg region away from the bone with a caliper incorporating two transducers (*Figure 4.6*). Acoustic velocity alters as the proportions of fat and lean tissues through which the sound travels changes. Fat/lean ratios can therefore be estimated.

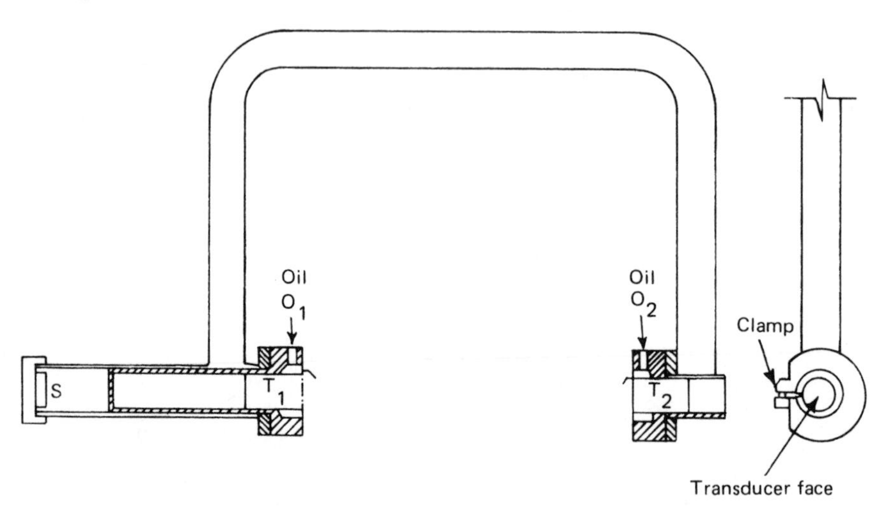

Figure 4.6 Apparatus used to measure the velocity of sound through the limbs of living animals. T_1, T_2, transducers; O_1, O_2, holes for introduction of oil; S, spacing piece. By courtesy of Miles and Fursey (1974).

Most trials scanning sheep have centred on the loin and back regions. Skeletal shapes are well defined in these areas and the positioning of the equipment on a particular point of the body can then be precise. Ultrasonics are very easy to use in the field as the machines are usually robust, portable and quick. Sheep can be held for scanning or can be restrained in a crate. Most machines require that a strip of wool is clipped at the position of scanning but Gooden, Beach and Purchas (1980) have reported a new machine that can be used without clipping. Liquid paraffin or acoustic gel helps eliminate air from under the transducers and this improves the quality of the pictures. All the machines require calibrating before use. This is easily done by scanning a suitable block of material with a change of medium at a known depth. The machine is then adjusted to read the correct depth.

Scanning sheep is facilitated by the animal standing still with its weight evenly distributed on its four legs. Sheep that are handled regularly will be easy to scan. Miles, Pomeroy and Harries (1972) used ultrasonics on cattle and reported a number of ways that errors could occur in interpretation of pictures. These were misidentification of boundaries due to incomplete signals, the lack of recognition of multiple signals and the omission of data

Table 4.3 CORRELATION COEFFICIENTS BETWEEN ULTRASONIC AND CARCASS MEASUREMENTS

Reference	*Type of scan*	*Number of animals*	*Position*	*Fat depth correlation*	*Eye muscle area correlation*	*Fat area correlation*
Moody *et al.* (1965)	A scan	235	13th rib		0.52**–0.66** (estimated)	
Campbell *et al.* (1959)	A scan	65	13th rib		0.44**–0.68** (depths)	
Pattie *et al.* (1975)	B scan	23	12th rib	0.74*	0.29	
Andersen *et al.* (1977)	Real-time Scanner	18	Unspecified		0.83–0.85	0.62–0.87
Jensen (1977)	Real-time Scanner	38	1st lumbar	0.74–0.82	0.72–0.78	0.81–0.83
Alliston *et al.* (1981)	Real-time	254	13th rib	0.50***	0.62***	0.68***
	B scan			0.71***	0.71***	0.75***
Gooden *et al.* (1980)	A scan	106	13th rib	0.72		
		97		0.81		
		30		0.91		
		32		0.88		
Shelton *et al.* (1977)	B scan	102	12–13th rib	0.77*	0.69*	

*$P<0.05$
**$P<0.01$
***$P<0.001$

on the scans when the ultrasonic beam is not perpendicular to the anatomical boundary.

Several of the ultrasonic machines have variable scaling factors. There appears to be only a small improvement, however, in the accuracy obtained by the use of the 1:1 as opposed to the 2:1 scales (Alliston *et al.*, 1981; Kempster *et al.*, 1982).

In cattle it is often possible to distinguish between fat and hide when interpreting pictures (Andersen *et al.*, 1981), but with sheep skin and fat are interpreted together because the fat layer can be very thin and also because the measurable areas are then larger. The number of pictures taken of each animal and the number of interpretations taken of each picture will affect the accuracy of the results. In practice two pictures of each animal, each interpreted twice, gives a satisfactory result.

The correlation coefficients between ultrasonic measurements and carcass composition range between –0.9 to 0.63 for percentage lean, 0.27 to 0.76 for percentage total fat and 0.46 to 0.77 for percentage subcutaneous fat. These also vary from the non-significant to the highly significant. The results are summarized in *Tables 4.3* and *4.4*. Measurements showing the repeatability of ultrasonic techniques in the field are not common but some results are given in *Table 4.5* and these do show ultrasonic measurements to be highly repeatable.

Kempster *et al.* (1977, 1982), working with the Scanogram found residual standard deviations of 2.47, 3.20 and 2.40 percentage units for

Table 4.4 CORRELATION COEFFICIENTS BETWEEN ULTRASONIC AND CARCASS MEASUREMENTS

References	*Fat measurements with percentage lean*	*Fat measurements with percentage subcutaneous fat*	*Fat measurements with percentage total fat*
Moody *et al.* (1965)			0.27–0.34 (fat depth)
Pattie *et al.* (1975)	–0.09		0.55*
Alliston *et al.* (1981)	0.54*** to –0.63**	0.61*** to 0.77***	
Gooden *et al.* (1980)	0.61*** to 0.77***		0.76
Shelton *et al.* (1977)	–0.59* (fat depth) –0.42 (eye muscle area)	0.55* (fat depth) 0.46* (eye muscle area)	0.47* (fat depth) 0.45* (eye muscle area)

*$P<0.05$
**$P<0.01$
***$P<0.001$

Table 4.5 REPEATABILITIES OF MEASUREMENTS USING SCANNING OF THE LIVE ANIMAL

Reference	*Type of scan*	*Position*	*Fat depth*	*Fat area*	*Eye muscle area*
Pattie *et al.* (1975)	B scan	12th rib	0.78		0.73
Jensen (1977)	Real-time	1st lumbar	0.79–0.93	0.91–0.95	0.83–0.88

prediction of percentage lean in the carcass (standard deviations of percentage lean in the carcass at constant liveweight were 2.90, 3.85 and 2.89 respectively; *Table 4.7*). Results from another trial with the Scanogram gave a residual standard deviation of 2.08 for prediction of percentage boneless cuts using ultrasonic measurements (S.D. of 2.31 at constant liveweight) (Fortin, 1980; *Table 4.6*). Reported residual standard deviations of ultrasonic measurements are given in *Table 4.7*.

Table 4.6 STANDARD DEVIATION OF CARCASS COMPONENTS AND RESIDUAL STANDARD DEVIATIONS OF ULTRASONIC MEASUREMENTS AS PREDICTORS OF CARCASS COMPOSITION

Number of animals	*Percentage boneless cuts*		*Percentage trimmed cuts*	
33	(2.31)[a]	2.08[b]	(2.00)[a]	1.93[b]
		2.15[c]		1.99[c]
		2.22[d]		1.96[d]

[a] Corrected to constant liveweight
[b] Scanogram
[c] Krautkrumer
[d] Scanoprobe
Standard deviation given in parentheses
From Fortin (1980)

The application of ultrasonics in animal science is reviewed by Stouffer and Westervelt (1977). Although most of the ultrasonic data for sheep suggests that the precision achieved with scanning machines is low, the equipment and techniques are improving rapidly and many more developments are likely in the near future.

Techniques suitable for research application

SPECIFIC GRAVITY

Body fat is considerably less dense than the other body tissues. Many researchers have therefore studied the relationship between body specific gravity and body fat content. Timon and Bichard (1965) reviewed the literature on investigations on lamb and mutton carcasses and concluded that an inverse relationship between fat and carcass specific gravity existed with correlation coefficients ranging from –0.56 to –0.88. Working with 83 lambs Timon and Bichard (1965) found that carcass specific gravity accounted for 86.1% and 78.1% of the respective variances in carcass fat percentage and muscle percentage.

With live animals, body volume is estimated but there are problems associated with lung volume and the contents of the digestive system that makes these techniques inaccurate.

DILUTION TECHNIQUES

The dilution technique involves the introduction of a known amount of tracer which will become uniformly distributed throughout a compartment

Table 4.7 STANDARD DEVIATION OF CARCASS COMPONENTS AND RESIDUAL STANDARD DEVIATIONS OF ULTRASONIC MEASUREMENTS AS PREDICTORS OF CARCASS COMPOSITION

Reference	*Number of animals*	*Percentage total fat*	*Percentage subcutaneous fat*	*Percentage lean*	*Fat thickness*
Kempster *et al.* (1977)	51		(2.42)[d] 1.93[a] fat area	(2.90)[d] 2.47[a] fat area	
			1.86[a] *M. long dorsi*	2.39[a] *M. long dorsi*	
Pattie *et al.* (1975)	23	(4.55)[d] 3.97[a]			
Kempster *et al.* (1982)	254		(3.15)[d] 2.18[a]	(3.85)[d] 3.20[a]	(2.42)[d] 1.80[a]
			2.79[b]	3.58[b]	(2.43)[d] 2.20[b]
	147		(2.25)[d] 1.86[a]	(2.89)[d] 2.40[a]	(1.64)[d] 1.42[a]
Gooden *et al.* (1980)	106				(1.67) 1.16[c]
	97				(1.04) 0.61[c]
	30				(1.52) 0.63[c]
	32				(0.78) 0.38[c]

[a]Scanogram
[b]Danscanner
[c]Prototype pulse echo instrument
[d]Corrected to constant liveweight
Standard deviation given in parentheses

in the animal body (Houseman, 1972; Cuthbertson, 1975; Wright, 1982). After the concentration of the tracer is at equilibrium a sample of the compartment is taken and the concentration of the tracer is measured. A tracer should not be toxic, must not be metabolized, should be easily measurable, and must diffuse homogeneously into all the volume to be measured (Robelin, 1981).

The different tracers used in sheep include antipyrene and N-acetyl-antipyrene (reviewed by Panaretto and Till, 1963), tritiated water (TOH) or deuterium oxide (D_2O) (Foot and Greenhalgh, 1970; Searle, 1970). Of these deuterium oxide (D_2O) would seem to be most suitable as it is accurate and not radioactive. Results from work with antipyrene generally demonstrate that antipyrene dilution was too variable to give good estimations of total body water (Panaretto and Till, 1963).

Foot and Greenhalgh (1970), working with seven Blackface ewes, estimated deuterium oxide concentration by infrared spectroscopy. The amount of body fat in each ewe was estimated from the deuterium oxide dilution using a prediction equation relating body fat to body water. Body fat predicted from deuterium oxide dilution differed from values obtained by analysis of the slaughtered animals by –0.8 to +1.7 kg in seven ewes containing 5.2 to 21.4 kg fat. The standard deviation from regression of percentage of body fat estimated from deuterium oxide dilution on percentage body fat measured after slaughter was ±1.2 percentage units. Both Cuthbertson (1975) and Robelin (1981) highlighted the problem associated with the estimation of body water in ruminants, caused by the variation in the water content of the alimentary tract.

The results of Searle (1970) and Panaretto (1968) were compared by Searle (1970) and presented as a relationship between fat predicted by the tritiated water space and the fat found by dissection. *Figure 4.7* gives the results for 76 sheep aged from three days to adults. Cuthbertson (1975)

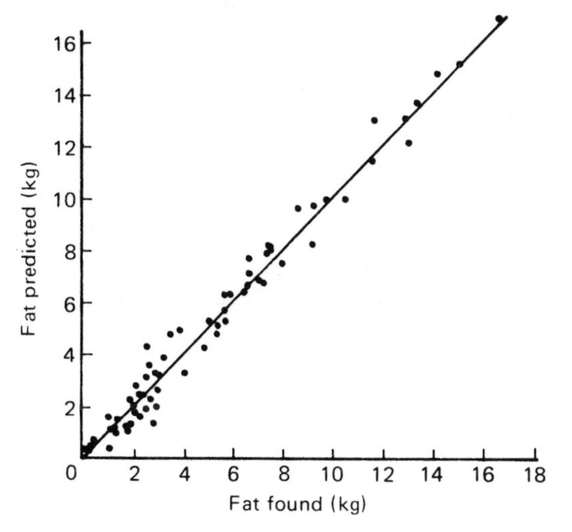

Figure 4.7 Relationship between fat determined by analysis and fat predicted for sheep aged from three days to adults. The line of equality is indicated (two high values have been omitted with actual fat 36.5 and 39.6 kg and predicted fat 37.3 and 38.1 kg respectively). By courtesy of Searle (1970).

dismissed dilution techniques because body water estimations cannot provide information on the distribution of fat between carcass and non-carcass parts, nor of the distribution and thickness of the tissues in different parts of the carcass.

CREATININE EXCRETION

The metabolism of creatine in the body of animals ends with the production of creatinine. Borsook and Dubnoff (1947) found that as 98% of the creatine reserves of the body exist in the muscles it is possible to predict body composition as it is assumed that a fixed amount of the creatine of the body is converted each day into creatinine which is excreted in the urine. Van Niekerk *et al.* (1963b), working with 65 sheep of differing fatness, found that the protein and water content and the fat-free mass of the ingesta free body were highly correlated with the amount of creatinine excreted in the urine (r = 0.98). However it was also reported that a significant fraction of creatinine may be voided from the body via such routes as the sweat, saliva and the digestive tract (Van Niekerk *et al.*, 1963a). The evidence available led Wright (1982) to conclude that it is doubtful if accurate prediction of body composition could be achieved from measurements of creatinine excretion in animals over a small range of liveweight or fatness.

COMPUTERIZED TOMOGRAPHY

Computed tomography was first developed for medical use (Cormack, 1980; Houndsfield, 1980). An X-ray tube rotates around an object and by the use of a computer it is possible to reconstruct from a series of pictures a slice through the object. By this technique the density (CT number) of different body tissues at different distances from the X-ray tube can be calculated (Vangen *et al.*, 1981).

Skjervold *et al.* (1981) used CT to estimate body composition in 23 pigs of variable fatness and the results are given in *Table 4.8*. The results obtained were promising and it may be possible to improve the technique by combining information from two or more tomographic slices and by

Table 4.8 PREDICTION OF BODY COMPOSITION OF PIGS ON BASIS OF THE RELATIVE DISTRIBUTION OF CT NUMBERS FROM ONE TOMOGRAPHIC PLAN

	Prediction of body composition based on CT numbers (R^2 values)
Percentage fat in slice	0.89
Carcass	0.89
Percentage protein in slice	0.80
Carcass	0.83
Percentage water in slice	0.85
Carcass	0.82
Energy content slice	0.85
kJ/g carcass	0.85

From Skjervold *et al.* (1981)

removing from the tomographic slice CT values which describe intestines, stomach and organs not in the carcass.

The main drawbacks to these machines are the expense and the complication that the animals have to be brought to the scanning area and anaesthetized for scanning. Even considering this, however, it would seem that improved technique coupled with computer tomographs more suited for scanning animals will give great possibilities for a future development of a new field in animal breeding (Skjervold, 1981).

OTHER METHODS

Miles (1981) reviewed several other methods including dielectric methods, methods using X-ray or gamma radiation, hormone measurements, nuclear magnetic resonance, potassium-40 and neutron activation analysis for determining body composition.

Of the techniques mentioned the electronic meat measuring equipment (EMME) developed in America appears to be interesting. The principle works on eddy currents being induced in the animal by an alternating magnetic field produced by a current passing through a coil surrounding the animal. The eddy currents generate a magnetic field which can be picked up by a change of impedance in the coil. The conductivity of muscle is much higher than that of fat.

Domermuth *et al.* (1976) used 44 pigs of varying breeds and weight to compare the EMME machine with lean body composition. The results showed that EMME in combination with fasted body weight was useful in predicting the weight of carcass protein (R^2 = 0.78) and lean cuts (R^2 = 0.80).

Fredeen, Martin and Sather (1979) found correlations of –0.70 between EMME values and percentage total fat for 130 pigs and 0.79 and 0.40 for percentage lean for 130 and 228 pigs respectively. The results obtained with EMME in general, however, were too variable to be acceptable at present.

Each EMME machine has to be calibrated and formulae worked out for the specific machine. Several exterior factors also affect the working of the machine. These are the weight of the animal, the movement of the animal as it passes through the machine, temperature and humidity of the air (Skelley *et al.*, 1977, 1978). At present, therefore, the EMME machines show some potential when used to determine body composition in pigs but more substantial results are required before it can be used extensively.

The use of *in vivo* techniques in breeding

In California a selection of breeders' flocks have been selected using a combination of a fasted liveweight, a live conformation score and a seventh rib probe (R = 0.77 for predicting percentage trimmed cuts) (Spurlock, Bradford and Wheat, 1966). This technique has proved economical and easy to use in the field. In 1966 there were few techniques available and Spurlock, Bradford and Wheat concluded that it seems advisable to use the best combination of live animal measures available, recognizing that this may be considerably less than ideal.

At ABRO a project designed to select rams with a high rate of lean tissue growth and a reduction in fat is underway (C. Smith, personal communication). Selection is on estimated lean growth rate to 100 days of age with adjustment for dam age, type of birth and day of birth. Lean growth rate (LGR) is measured as:

$$LGR = \frac{(100 \text{ day weight} - \text{birthweight})}{100} \text{ (estimated lean percentage)}$$

The parameters for the estimation of lean percentage in live rams are derived from MLC dissection results and are measured by ultrasonics.

Performance testing of ram lambs is now a part of the breeding plans for sheep in Denmark. During the test period of 60 days growth rate, feed consumption and carcass quality are assessed. An index is constructed that relates the performance of the ram to the breed average. The carcass quality is estimated from ultrasonic measurements of muscle and fat. The degree of fatness is regarded as being too high if the fat thickness is more than 9 mm (skin and fat) (Jensen and Hansen, 1981).

The use of *in vivo* methods will depend on the relative economic importance of the carcass criteria and the returns from their improvement. Selection aims will usually include several characteristics such as growth rate and food conversion efficiency. A selection index, therefore, aimed at improving a number of traits will mean that *in vivo* measurements are less critical as more emphasis will be put on the characteristics that can be measured directly (King, 1981).

A development for the future might be the manipulation of developing embryos to produce identical twins to order (Willadsen, 1979). It would then be possible to slaughter one of the twins and assess its carcass quality whilst still preserving an identical genotype.

The use of *in vivo* techniques for estimating body composition in sheep breeding will become significantly more important if and when carcass payments reflect lean meat content more and carcass weight less.

Conclusions

Serial slaughtering techniques might soon be substituted by *in vivo* techniques. None of the tested techniques available at present appear to give accurate indications of body composition. There is, however, a need for quick estimators for use in the commercial field situation and a combination of liveweight, body condition score and a simple ultrasonic measurement appear to be both easy to obtain and better than visual assessment alone. A more accurate assessment is required for experimental determination of body composition changes in nutrition, growth, and environmental trials and in selection experiments within breeds over a small liveweight range. Modern techniques such as two-dimensional ultrasonics and computerized tomography are likely to be used in the future and further improvements will occur in the equipment.

Further developments in techniques may make it possible to follow the development of various vital organs and the deposition of energy.

Practicability, portability, cost and public acceptability must be balanced with the accuracy required and obtained with the various techniques. The pig industry has been revolutionized by body composition information and it is only a matter of time before sheep breeding objectives lead to improved carcasses.

References

ALLISTON, J.C., BARKER, J.D., KEMPSTER, A.J. and ARNALL. D. (1981). The use of two ultrasonic machines (Danscanner and Scanogram) for the prediction of body composition in crossbred lambs. *Animal Production* **32**, 375 (Abstract)

ANDERSEN, B.B., BUSK, H. and JENSEN, N.E. (1977). Provisional results of Danscanner ultrasonic measurement on sheep, pigs and cattle. National Institute of Animal Science, Copenhagen

ANDERSEN, B.B., BUSK, H., CHADWICK, J.P., CUTHBERTSON, A., FURSEY, G.A.J., JONES, D.W., LEWIN, P., MILES, C.A. and OWEN, M.G. (1981). Ultrasonic techniques for describing carcass characteristics in live cattle. CEC Publication of the Beef Production Research Committee

BORSOOK, H. and DUBNOFF, J.W. (1947). The hydrolysis of phosphocreatine and the origin of urinary creatinine. *Journal of Biological Chemistry* **168**, 493–510

CAMPBELL, D., STONAKER, H.H. and ESPLIN, A.L. (1959). The use of ultrasonics to estimate the size of the longissimus dorsi muscle in sheep. *Journal of Animal Science* **18**, 1483 (Abstract)

CORMACK, A.M. (1980). A presentation of anatomical information by computed synthesis of an image from X-ray transmission data obtained in many different directions through the plane under consideration. Nobel Lecture. *Journal of Computer Assisted Tomography* **4**, 658–664

CUTHBERTSON, A. (1975). Carcass quality. In *Meat* (D.J.A. Cole and R.A. Lawrie, Eds.), Chapter 8, pp. 147–181. London, Butterworths

DOMERMUTH, W., VEUM, T.L., ALEXANDER, M.A., HEDRICK, H.B., CLARK, J. and EKLUND, D. (1976). Prediction of mean body composition of live market weight swine by indirect methods. *Journal of Animal Science* **43**, 966–976

EVERITT, G.C. (1962). On the assessment of body composition in live sheep and cattle. *Proceedings of the Australian Society of Animal Production* **4**, 79–89

FOOT, J.Z. and GREENHALGH, J.F.D. (1970). The use of deuterium oxide space to determine the amount of body fat in pregnant Blackface ewes. *British Journal of Nutrition* **24**, 815–825

FORTIN, A. (1980). Fat thickness measured with three ultrasonic instruments on live ram lambs as predictors of cutability. *Canadian Journal of Animal Science* **60**, 857–867

FREDEEN, H.T., MARTIN, A.H. and SATHER, A.P. (1979). Evaluation of an electronic technique for measuring lean content of the live pig. *Journal of Animal Science* **48**, 536–540

GOODEN, J.M., BEACH, A.D. and PURCHAS, R.W. (1980). Measurement of subcutaneous backfat depth in live lambs with an ultrasonic probe. *New Zealand Journal of Agricultural Research* **23**, 161–165

HAMMOND, J. (1932). *Growth and Development of Mutton Qualities in the Sheep*. Edinburgh, Oliver and Boyd

HOUNDSFIELD, G.M. (1980). Computed medical imaging. Nobel Lecture. *Journal of Computer Assisted Tomography* **4**, 665–674

HOUSEMAN, R.A. (1972). Studies of methods of estimating body composition in the living pig. PhD Thesis. University of Edinburgh

JANSEN, J. (1981). *In vivo* estimation of body composition in beef in the Netherlands. CEC Workshop on '*In vivo* estimation of body composition in beef'. National Institute of Animal Science, Copenhagen

JEFFERIES, B.C. (1961). Body composition scoring and its use in management. *Tasmanian Journal of Agriculture* **32**, 19–21

JENSEN, N.E. (1977). Preliminary results from Danscan ultrasonic measurements of fat thickness and loin area of lamb. *Proceedings of European Association for Animal Production, 28th Annual Meeting*

JENSEN, N.E. and HANSEN, K. (1981). Performance test of ram lambs 1981. Report of Performance Testing Stations. National Institute of Animal Science, Copenhagen

KALLWEIT, E. (1976). Visual assessments. EEC Seminar on criteria and methods for assessment of carcass and meat characteristics in beef. *Production Experiments* 81–89

KALLWEIT, E. (1981). Review of practical use and experimental results of *in vivo* techniques for the estimation of body composition in beef cattle in the Federal Republic of Germany. CEC Workshop on '*In vivo* estimation of body composition in beef'. National Institute of Animal Science, Copenhagen

KEMPSTER, A.J., CROSTON, D. and JONES, D.W. (1981). Value of conformation as an indicator of sheep carcass composition within and between breeds. *Animal Production* **33**, 39–50

KEMPSTER, A.J., ARNALL, D., ALLISTON, J.C. and BARKER, J.D. (1982). An evaluation of two ultrasonic machines (Scanogram and Danscanner) for predicting the body composition of live sheep. *Animal Production* (In press)

KEMPSTER, A.J., CUTHBERTSON, A., JONES, D.W. and OWEN, M.G. (1977). A preliminary evaluation of the 'Scanogram' for predicting the carcass composition of live lambs. *Animal Production* **24**, 145–146

KING, J.W.B. (1981). Potential use of *in vivo* techniques for breeding purposes and their limitations. CEC Workshop on '*In vivo* estimation of body composition in beef'. National Institute of Animal Science, Copenhagen

KIRTON, A.H. and BARTON, R.A. (1958). Liveweight loss and its components in Romney ewes subjected to L-thyroxine therapy and a low plane of nutrition. *Journal of Agricultural Science, Cambridge* **51**, 265–281

MEAT AND LIVESTOCK COMMISSION PUBLICATION (1981). *Lamb Carcass Production*

MILES, C.A. (1978). A note on recent advances in ultrasonic scanning of animals. Proceedings of the 24th European Meat. Research Workers Congress, Kulbach, pp. W13.3–W13.6

MILES, C.A. (1981). *In vivo* techniques for the estimation of body composition in beef: other techniques and future possibilities. CEC Workshop on '*In vivo* estimation of body composition in beef'. National Institute of Animal Science, Copenhagen

MILES, C.A. and FURSEY, G.A.J. (1974). A note on the velocity of ultrasound in living tissue. *Animal Production* **18**, 93–96

MILES, C.A., POMEROY, R.W. and HARRIES, J.M. (1972). Some factors affecting reproducibility in ultrasonic scanning of animals. *Animal Production* **15**, 239–249

MOODY, W.G., ZOBRISKY, S.E., ROSS, C.V. and NAUMANN, H.D. (1965). Ultrasonic estimates of fat thickness and longissimus dorsi area in lambs. *Journal of Animal Science* **24**, 364–367

PALSSON, H. and VERGES, J.B. (1952a). Effect of the plane of nutrition on growth and the development of carcass quality in lambs. *Journal of Agricultural Science, Cambridge* **42**, 1–92

PALSSON, H. and VERGES, J.B. (1952b). Effects of the plane of nutrition on growth and the development of carcass quality in lambs. *Journal of Agricultural Science, Cambridge* **42**, 93–149

PANARETTO, B.A. (1968). Body composition *in vivo*. IX. The relation of body composition to the tritiated water space of ewes and wethers fasted for short periods. *Australian Journal of Agricultural Research* **19**, 267–272

PANARETTO, B.A. and TILL, A.R. (1963). Body composition *in vivo*. II. The composition of mature goats and its relationship to the antipyrine, tritiated water and N-acetyl-4-amine antipyrine spaces. *Australian Journal of Agricultural Research* **14**, 926–943

PATTIE, W.A., THOMPSON, J.M. and BUTTERFIELD, R.M. (1975). An evaluation of the Scanogram as an ultrasonic aid for assessing carcass composition in live sheep. A report submitted to the Australian Meat Board (1975)

REHBEN, E. (1981). *In vivo* estimation of body composition in beef. CEC Workshop on '*In vivo* estimation of body composition in beef'. National Institute of Animal Science, Copenhagen

ROBELIN, J. (1981). Measurement of body water in living cattle. By dilution technique. CEC Workshop on '*In vivo* estimation of body composition in cattle'. National Institute of Animal Science, Copenhagen

RUSSEL, A.J.F., DONEY, J.M. and GUNN, R.G. (1969). Subjective assessment of body fat in live sheep. *Journal of Agricultural Science, Cambridge* **72**, 451–454

SEARLE, T.W. (1970). Body composition in lambs and young sheep and its prediction *in vivo* from tritiated water space and body weight. *Journal of Agricultural Science, Cambridge* **74**, 357–362

SHELTON, M., SMITH, G.C. and ORTS, F. (1977). Predicting carcass cutability of rambouillet rams using live animals traits. *Journal of Animal Science* **44**, 333–337

SKELLEY, G.C., SLIGH, C.R., HANDLIN, D.L. and McCONNELL, J.C. (1977). The use of the EMME as a method for measuring carcass composition of live swine. Animal Science Research Series 33, South Carolina University

SKELLEY, G.C., SLIGH, C.R., HANDLIN, D.L. and TERLIZZIE, F.M. (1978).

Additional observations with the EMME in live swine evaluation. Animal Science Research Series 35, South Carolina University

SKJERVOLD, H. (1981). Estimation of body composition in live animals by use of computerised tomography. CEC Workshop on '*In vivo* estimation of body composition in beef'. National Institute of Animal Science, Copenhagen

SKJERVOLD, H., GRONSETH, K., VANGEN, O. and EVENSEN, A. (1981). *In vivo* estimation of body composition by computerised tomography. *Zeitschrift fur Tierzuchtung und Zuchtungsbiologie* **98**, 77–79

SPURLOCK, G.M., BRADFORD, G.E. and WHEAT, J.D. (1966). Live animals and carcass measures for the prediction of carcass traits in lambs. *Journal of Animal Science* **25**, 454–459

STOUFFER, J.R. and WESTERVELT, R.G. (1977). A review of ultrasonic applications in animal science. Reverberations in echo cardiograms. *Journal of Clinical Ultrasound* **5**, 124–128

STOUFFER, J.R., HOGUE, D.E., MARDEN, D.M. and WELLINGTON, G.H. (1958). Some relationships between live animal measurements and carcass characteristics of lamb. *Journal of Animal Science* **17**, 1151 (Abstract)

TAYLER, J.C. (1954). Technique of weighing the grazing animal. *Proceedings of British Society of Animal Production* 3–16

TIMON, V.M. and BICHARD, M. (1965). Quantitative estimates of lamb carcass composition. *Animal Production* **7**, 183–187

TULLOH, N.M. (1963). Relation between carcass composition and liveweight of sheep. *Nature, London* **197**, 809–810

VANGEN, O., GRONSETH, K., EVENSEN, A. and SKJERVOLD, H. (1981). Estimation of body composition in live animals by use of computerized tomography. *Proceedings of European Association for Animal Production, 32nd Annual Meeting*

VAN NIEKERK, B.D.H., BENSADOUN, A., PALADINES, O.L. and REID, J.T. (1963a). A study of some of the conditions affecting the rate of excretion and stability of creatine in sheep urine. *Journal of Nurition* **79**, 373–380

VAN NIEKERK, B.D.H., REID, J.T., BENSADOUN, A. and PALADINES, O.L. (1963b). Urinary creatine as an index of body composition. *Journal of Nutrition* **79**, 463–473

VERBEKE, R. (1981). Beef production research outline in Belgium. CEC Workshop on '*In vivo* estimation of body composition in beef'. National Institute of Animal Science, Copenhagen

WALLACE, L.R. (1948a). The growth of lambs before and after birth in relation to the level of nutrition. *Journal of Agricultural Science, Cambridge* **38**, 243–302

WALLACE, L.R. (1948b). The growth of lambs before and after birth in relation to the level of nutrition. *Journal of Agricultural Science, Cambridge* **38**, 367–401

WARDROP, I.D. (1957). Body composition studied in sheep. PhD Thesis, University of Melbourne

WARDROP, I.D. and COOMBE, J.B. (1961). The development of rumen function in the lamb. *Australian Journal of Agricultural Research* **12**, 661–680

WELLS, P.N.T. (1969). *Physical Principles of Ultrasonic Diagnosis.* London, Academic Press

WILLADSEN, S.M. (1979). A method of culture for micro-manipulated sheep embryos and its use to produce monozygotic twins. *Nature, London* **277**, 298–300

WRIGHT, I.A. (1982). Studies on the body composition of beef cows. PhD Thesis, University of Edinburgh

II

Nutrition

5

THE INFLUENCE OF NUTRITION ON THE REPRODUCTIVE PERFORMANCE OF EWES

R.G. GUNN
Hill Farming Research Organisation, Bush Estate, Penicuik, Midlothian, UK

The influence of nutrition on the reproductive performance of ewes has long been a matter for concern to both sheep farmers and research workers throughout the world. With the wide range of environmental conditions within which sheep are to be found and the seasonality of breeding which most sheep breeds exhibit, it is clear that the relationship between nutritional provision and requirement for optimum reproductive performance is seldom ideal. Where nutrition is 'adequate', whatever that might mean, reproductive performance may be expected also to be adequate but this tells little about the specific nutritional requirements of the different phases of the reproductive cycle and it is information about these that is vital to the optimization of reproductive performance in nutritionally fluctuating environments. Only when this information is available can grazing and feeding management be adjusted in the necessary tactical manner.

Much relevant work has been carried out over many years on the relationships between nutrition and reproductive performance in the sheep but a great deal of this has been observational and differences in detail of both method and results have led to variable interpretation. Some differences in response are undoubtedly genetic in origin but even these would probably fit a general pattern if the fundamental mechanisms and interactions of nutritional state with reproductive physiology were better understood.

Knowledge has certainly improved over the years but there are still aspects about which there is little understanding and without which it is not possible to impose the nutritional fine tuning to give control at specific levels of response to suit different objectives. The state of knowledge and understanding has been well reviewed by others (e.g. Edey, 1976a; Lindsay, 1976; Rattray, 1977; Cockrem, 1979) and it is not intended in this chapter to carry out an exhaustive review of the considerable amount of detailed work that has been put into clarifying those areas in which there is unequivocal understanding but, instead, to look at those areas which are still obscure, in an attempt to stimulate thought and discussion as to how on the one hand the research needs may be indicated and on the other the problems may be overcome in practice.

Firstly, it is necessary to set the scene and to define the parameters by which responses will be recognized. Nutrition can, for convenience, be defined in terms of energy, protein and specific dietary components such as vitamins and minerals. Measurement of nutritional adequacy in the animal is, in practice, usually based on the relatively insensitive parameters of liveweight and body condition change, which together express nutritional accumulation or loss, mainly of energy. It can, in an experimental context, be measured in terms of intake or absorbed nutrients but such techniques have little applicability to the practical scene. The development of easy and rapid techniques to improve the measurement of nutritional status and short-term change in the ewe at critical stages of the reproductive cycle would be of immense value as an aid to the definition of nutritional requirements for specific reproductive responses.

Measurement of reproductive response is in terms of both fertility, defined as the ability to breed, and fecundity, defined as breeding rate. The latter can be expressed in terms of litter size, which is the number of lambs produced/ewe lambing. This is a consequence of ovulation rate or the number of ova shed by the ovaries in the oestrous cycle at which mating occurs, minus ova, embryo and foetal wastage expressed as the number of ova not represented by viable lambs at parturition. The effect of wastage is to reduce litter size or create barrenness. Where total wastage occurs early enough, remating and a viable pregnancy are still possible although the size of the litter may be less than from an earlier pregnancy. Where partial wastage of multiples occurs, pregnancy continues but at a lower litter size and, in practice, there is no way of knowing in specific cases whether this is derived from a high ovulation rate and high wastage or a low ovulation rate and a low wastage.

Three major areas of interest can be identified:

1. When are the nutritionally critical phases of the life and reproductive cycles?
2. What are the nutritional requirements for these?
3. How can nutritional adequacy in the animal be measured or predicted?

Nutritionally critical phases and their nutritional requirements

Areas 1 and 2 are so closely related that they will be discussed together under the following three headings: energy nutrition; protein nutrition; and specific nutritional components.

ENERGY NUTRITION

Energy nutrition is clearly of major importance to reproductive performance and there are three aspects to be considered:

1. Long-term energy intake from the foetal stage throughout the growth and development of the animal until sexual and body maturity;
2. Medium-term energy intake in a given annual cycle in the mature animal; and
3. Short-term energy intake in the immediate pre-mating and mating periods.

Long-term energy effects

Nutrition of the female from the foetal stage until it reaches maturity may influence its subsequent reproductive performance by affecting the time or age of onset of the first viable oestrus, by affecting the fertility and fecundity at this first oestrus, or by residual effects on reproductive performance during the remainder of the reproductive life. Since the first two areas will be discussed in later chapters, observations here will be limited to the final alternative.

It is known that severe undernourishment of ewes during late pregnancy can depress birth weight and vigour of lambs, irrespective of the number carried, and that lambs born as twins are likely to be more undernourished and have smaller birth weights than singles, even when ewes have been similarly nourished. These differences are also known to remain in early life, associated with lower milk intake of the smaller lambs and the inadequate ewe nutrition which is likely in such circumstances.

The few reports on this subject to date (Allden, 1979) mainly suggest that there are no long-term effects of restricted *in utero* or early postnatal nutrition on subsequent reproductive performance when removal of the restriction has permitted compensatory growth. When, however, the restriction is severe and of long duration, as was the case in the study of Allden (1979) and also in a study of Gunn (1977), compensatory growth did not overcome the initial restriction in growth and mature size, either because the adult high plane of nutrition was inadequate or because the low plane of nutrition during rearing had produced animals incapable of compensating.

Table 5.1 MEAN LIVEWEIGHTS AT 12 AND 54 MONTHS OF AGE AND LAMBING PERCENTAGES IN FIVE SUCCESSIVE YEARS OF SCOTTISH BLACKFACE EWES IN DIFFERENT NUTRITIONAL ENVIRONMENTS DURING REARING AND ADULT LIFE

Nutrition level		*Liveweight* (kg)		*Lambing % in five successive years*					
Rearing (0–12 months)	*Adult life (12–78 months)*	*12 months*	*54 months*	*2 years*	*3 years*	*4 years*	*5 years*	*6 years*	*Mean*
H	H	38.5	65.7	169	176	167	161	177	170
L	H	27.5	60.1	114	143	129	167	147	140
H	L	38.7	53.1	96	129	139	138	122	124
L	L	27.2	46.3	84	124	163	127	92	118

H = high; L = low plane of nutrition

In the study of Gunn (1977), the early limitations had long-term persistent effects on adult reproductive rate in both 'high' and 'low' adult nutritional environments (*Table 5.1*; R.G. Gunn, unpublished data). When the shorter-term nutritional limitations were removed in later life, however, (*Table 5.2*; Gunn and Doney, 1979), although liveweight and ovulation rate at comparable levels of body condition were still less in the LH ewes (low plane of nutrition during rearing, high in adult life) than in the HH ewes (high level of nutrition during both rearing and adult life), the

Table 5.2 PRE-MATING LIVEWEIGHT (kg) AND OVULATION RATE OF SCOTTISH BLACKFACE EWES IN FAT CONDITION AT MATING WHICH WERE DRAWN FROM ONE GENETIC SOURCE AND WHICH HAD BEEN IN DIFFERENT NUTRITIONAL ENVIRONMENTS DURING REARING AND ADULT LIFE

Nutritional environment during		*Condition score*	*Pre-mating liveweight* (kg)	*Ovulation rate*
Rearing (0–12 months)	*Adult life (12–66 or 78 months)*			
High (H)	High (H)	3	66.4	2.24
	Low (L)	3	61.2	2.27
Low (L)	High (H)	3	60.1	2.00
	Low (L)	3	56.5	1.68

From Gunn (1977)

ovulation rate difference was not statistically significant. Re-analysis of Allden's (1979) data (analysis of variance of 0, 1 and 2 lambs born/ewe over a five-year productive life, derived from the total numbers of lambing opportunities, lambings and lambs) did indicate a significantly lower reproductive performance from the ewes on a restricted plane of nutrition from birth to 14 months, although the chi-square analysis used by Allden did not show significance (*Table 5.3*). These results suggest that body size and maturity are important in some way and the provision of a high plane of nutrition in adult life, although probably producing some positive compensatory effect, does not seem to permit the expression of genetic reproductive potential since long-term nutritional limitations are still evident.

However, body size and maturity may not be the whole story. In the study of Allden (1979), an unrestricted plane of nutrition for as short a period as the first eight weeks of life prior to a lengthy period of restriction was clearly effective, when compared with restricted nutrition from birth, in establishing a high reproductive potential. The potential shown by group HL in *Table 5.2* (Gunn and Doney, 1979) would also appear to indicate the importance of nutrition in early life, a conclusion which may be similarly

Table 5.3 REPRODUCTIVE PERFORMANCE OF MERINO EWES OVER FIVE LAMBING YEARS FOLLOWING DIFFERENT LEVELS OF REARING NUTRITION DURING THE FIRST 14 MONTHS OF LIFE (L = RESTRICTED NUTRITION, H = UNRESTRICTED NUTRITION)

Group	*Treatment*	*Lambing opportunities*	*Ewes lambed*	*Lambs born*	*No. of times 0, 1 and 2 lambs born*			*Lambing (%)*
					0	*1*	*2*	
1	L (0–14 months)	37	32	33	5	31	1	89^{bc}
2	H (0–2 months) L (2–14 months)	34	33	35	1	31	2	103^{ab}
3	H (0–14 months)	70	62	76	8	48	14	109^{ac}

Significance of difference between values with same superscript: ^{a} = NS, ^{b} = $P<0.1$, ^{c} = $P<0.05$
From Allden (1979)

derived from Reardon and Lambourne (1966). These results suggest that there may therefore be a critical development period which is highly sensitive to nutrition and which markedly affects the extent to which the genetic potential is achieved in later life.

Present evidence is largely based on the effects of treatments imposed later than the period of greatest sensitivity to permanent stunting which has been predicted to lie in the immediate pre- and postnatal periods (Schinckel, 1963). There appears to be little or no information available on the reproductive performance of offspring derived from ewes ill-fed during pregnancy. Since cells, tissues, organs and systems are all initiated during gestation in a sequential developmental pattern, it has been suggested (Everitt, 1967) that the factors operating during these intrauterine formative stages may well modify subsequent productive possibilities. Concern about the long-term effects of maternal undernutrition in early pregnancy on the adult productive life of the progeny has also recently been expressed by Parr, Cumming and Clarke (1982). Existing knowledge of the consequences of early growth patterns is clearly inadequate for informed decisions to be made concerning practical management at this very important stage.

Medium-term energy effects

Fluctuations in energy intake during the annual cycle will lead to accumulation or loss of energy stored in the body, mostly in the form of body fat. Since requirements for pregnancy and lactation are additional to those for maintenance of the maternal body, and since the provision of energy in a grazing situation may often fail to match requirements, there are considerable differences in storage within the body according to the stage of the cycle. Liveweight and body condition are the criteria used to define medium-term energy effects. While there may well be interactions with long-term energy effects and with genotype, the most important period in relation to reproductive performance is that between the end of one pregnancy and lactation and the start of the next. This is often referred to as the recovery period, since it is the time that body tissue reserves, utilized during the previous pregnancy and lactation, are replenished.

The relationship between liveweight and/or body condition and reproductive performance is clearly established as a positive one. Only at the extremes of liveweight or condition is there any doubt and in these circumstances there are probably additional hormonally-based metabolic factors present which affect the relationship. The component of reproductive performance which has the clearest positive relationship with liveweight and body condition is ovulation rate and here the evidence is unequivocal. The evidence is less certain in respect of the effect of liveweight and condition on wastage of ova but this is an aspect which is subject to interactive effects with non-nutritional factors such as stress, not only during the recovery period but also during pregnancy.

Short-term energy effects

Current energy intake, in the period immediately prior to and at mating, is a short-term energy effect. Much confusion has existed in the past on the relative importance of medium- and short-term energy intake, due to the almost inevitable confounding of high levels of nutrition, or flushing as it is commonly called, with the degree of body condition achieved. Interpretation has also been difficult on account of either cumulative, or possibly different, effects on ovulation rate and wastage.

Ovulation rate appears to respond to short-term energy intake only within a specific intermediate range of body condition. Above and below this range, which appears to vary with genotype, it is the condition achieved that matters and there is no additional positive or negative effect of the energy intake currently applying (*Figure 5.1*). Much difficulty has

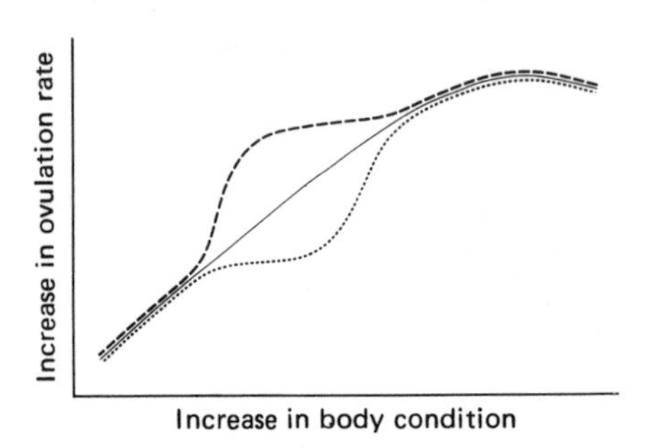

Figure 5.1 Effect of body condition and feed intake on ovulation rate. Current feed levels at mating: --- high; —— maintenance; ········ low.

been experienced in determining optimum durations of the flushing period in grazing ewes, since responses do not seem very predictable. These results have suggested a need to attempt a different interpretation of the nature of the response to flushing. Body condition manipulation and experimental treatments involving several weeks of flushing, have demonstrated ovulation rate responses in groups of ewes on many occasions. But variation in the response of groups, as well as individual variation within groups, is quite high and recent studies lead to the suggestion that positive group responses to flushing are only achieved where the ewes have a greater appetite drive over mating. Ewes which are not in such a physiological state, and which have a lower appetite drive, will not respond to flushing, no matter for how long this is provided.

Estimates of dry matter intake of ewes on good ryegrass and clover pasture (*Table 5.4*; R.G. Gunn, unpublished data) have indicated an inverse relationship between intake at the time of mating and body condition at four weeks before mating. Those ewes which had been in lean condition were eating up to 35% more at mating than those which had been in fat condition. On the basis of their liveweight at the time of mating, intakes per kg liveweight of those ewes which were originally in leaner condition were up to 55% greater. Ovulation rates of ewes initially in intermediate condition were as good as those of ewes initially in fat condition which were still heavier and fatter at mating (*Table 5.5*). Such a response to flushing may be interpreted as being associated with the increase in voluntary intake shown by the ewes in poorer, but not poor, body condition. This interpretation would perhaps account for the old idea of reducing condition for a period prior to flushing. Like so many of the old

Table 5.4 MEAN DAILY DRY MATTER (DM) INTAKE AT MATING OF NORTH COUNTRY CHEVIOT EWES IN DIFFERENT LEVELS OF BODY CONDITION AT FOUR WEEKS BEFORE MATING AND ALL GRAZING GOOD QUALITY RYEGRASS/CLOVER PASTURE FOR THE FOUR WEEKS TO MATING[(a)]

Condition score at 4 weeks before mating	*No. of ewes*	*Mean daily DM intake during week prior to and including mating* (g)			
		Total		*Per kg liveweight at mating*	
		Actual	*As % of ewes in 3+/3½*	*Actual*	*As % of ewes in 3+/3½*
3+/3½	8	1090	100	14.2	100
3–/3	11	1301	119	16.7	118
2+/2½	9	1282	118	17.6	124
2–/2	7	1473	135	22.0	155

[(a)]Based on faecal nitrogen estimations derived from grab samples of chromic oxide dosed ewes and the ryegrass/clover regression of R.H. Armstrong (personal communication)

Table 5.5 MEAN LIVEWEIGHT (kg) AND CONDITION SCORE AT FOUR WEEKS BEFORE MATING AND AT MATING, OVULATION RATE AND LAMBING RATE OF NORTH COUNTRY CHEVIOT EWES IN DIFFERENT LEVELS OF BODY CONDITION AT FOUR WEEKS BEFORE MATING AND ALL GRAZING GOOD QUALITY RYEGRASS/CLOVER PASTURE FOR THE FOUR WEEKS TO MATING

Condition score at 4 weeks before mating	*No. of ewes*	*Mean condition score at*		*Mean liveweight at*		*Ovulation rate*	*Lambing rate*
		4 wks before mating	*Mating*	*4 wks before mating*	*Mating*		
⩾3	73	3.10	3.15	73.1	74.8	2.23(17)	1.29(56)
2½/3–	71	2.64	2.92	65.3	70.7	2.21(24)	1.53(47)
⩽2+	59	2.08	2.60	56.5	66.1	1.87(23)	1.33(36)

Figures in parentheses indicate the number of ewes

shepherds' ideas, it was probably soundly based, provided it wasn't overdone. Appetite in grazing ewes in the pre-mating period is clearly an aspect which requires further study in relation to ovulation rate.

The effect of short-term energy nutrition on wastage is also an area which is insufficiently understood. Part of the total wastage is inherent or basal, being of genetic origin, and is unlikely to be influenced by nutritional factors. The remainder may be referred to as induced wastage, which could be influenced by nutritional factors. There is, however, little clear-cut evidence of nutrition consistently affecting wastage. The issue is complicated by a confounding with ovulation rate and the interaction of pre- and post-mating levels of nutrition. During the most sensitive first 30–40 days, a whole range of wastage responses can be identified and the level can be very high indeed. Anything up to 50% is not out of the ordinary but 25–30% is closer to the norm. Although nutrition appears to influence wastage, it is not a predictable or simple effect and does not lend itself to the sort of repeatable control which is required in practice. Wastage is also dependent on genotype and on complex interactions with a variety of stress factors (Edey, 1976b; Doney, Gunn and Horák, 1982). It is not even clear whether the relationship between the level of nutrition before and/or after mating and wastage is positive or negative since increased wastage has been associated with high levels of nutrition as well as with low (Cumming *et al.*, 1975; Gunn, Doney and Smith, 1979), which gives some idea of the difficulty that exists in clarifying this area. Nutritional factors may operate by disturbing the critical endocrine pattern that is important to the establishment and maintenance of pregnancy, particularly in the early stages, and the studies which have been started in this area clearly require extending.

Although the evidence on the effects of short-term post-mating energy nutrition on wastage is largely equivocal, it is clear that ewes should not be losing weight at the time of mating. If they are well enough nourished to at least maintain weight, they will produce more lambs than if they had been losing weight. This has been shown quite clearly over several years study of the consequences of the individual direction of liveweight change about the

Table 5.6 THE EFFECTS OF DIRECTION OF LIVEWEIGHT CHANGE ABOUT MATING ON LAMB PRODUCTION IN GREYFACE (BORDER LEICESTER × BLACKFACE) EWES

	Direction of liveweight change about mating		
	Gaining	*Maintaining*	*Losing*
No. of ewes	114	339	221
Lambs born/100 ewes alive at lambing	196	178	158
Liveweight at mating (kg)	70.4	67.9	66.6

From Gunn and Maxwell (1978)

time of mating in Greyface (Border Leicester × Blackface) ewes (*Table 5.6*; Gunn and Maxwell, 1978) and could be due to nutritional effects on both ovulation rate and wastage. It is also further evidence of the individual variation in response at this time which was discussed earlier.

PROTEIN NUTRITION

Much less is known about protein nutrition in relation to reproductive performance than is known about energy nutrition. This is to a large extent due to the problem of degradation of ingested protein by rumen organisms and the consequent uncertainty about how much of and in what form the protein is actually absorbed by the animal. The matter is further complicated by interaction with energy intake.

In recent years, much interest has been aroused by the increased ovulation rate response that has followed the relatively short duration of feeding of lupin grain in Australia, particularly in Western Australia (Lindsay, 1976). This has, until quite recently, been considered to be due to the high protein content of this grain (>30%) and many field scale studies have been carried out to determine the optimum usage of this protein supplement. Unfortunately, results have been very variable and many of the studies have not been able to supply the sort of basic physiological information that is required to describe the underlying mechanisms. Where attempts have been made to look at protein supplied in other forms of supplement, response has either been nil or very much less than with lupins (Fletcher, 1981). Whether the response to lupins is due to the low protein status of the feed that was being supplemented in the experiments or to the very high protein content or some other component of the lupin grain is as yet unclear. Studies by Corbett and Edey (1977), using rations containing 60 g casein either treated or untreated with formaldehyde, have shown that the treated casein diet provided the ewe with greater quantities of amino acids than did the untreated but there was no difference in ovulation rate. The evidence on protein/fecundity relationships seems to suggest that protein could be limiting at very low levels of protein intake but that provision in excess of maintenance produces no greater response (Morley *et al.*, 1978). Much work needs to be done on protein nutrition in the ewe and on its effects on reproductive performance.

SPECIFIC NUTRITIONAL COMPONENTS

Although there are a number of specific nutritional components which may affect reproductive performance in the ewe, these appear to operate indirectly through their effect on the general health of the animal. Direct effects of specific nutritional components are less obvious and response to them may be little more than a bonus from improving the overall standard of nutrition.

Mineral elements whose functions are interrelated and which may have some influence on reproductive performance in the ewe are copper, cobalt, molybdenum and sulphur but there is still work to be done to clarify their relationships and effects. Selenium appears to be important with regard to embryo survival and interacts in some way with vitamin E. There is little evidence of other vitamins affecting reproductive performance in the ewe. Substances of plant origin, in particular the plant oestrogens present in legumes, can upset the normal hormone patterns associated with reproduction.

Measurement or prediction of nutritional adequacy

Earlier in this chapter, three major areas of interest were listed, two of which were concerned with identification of the nutritionally critical phases of the life and reproductive cycles and with determination of the nutritional requirements for these. Before the information derived in these areas can be applied in practice, the third major area of interest becomes important; namely, how can nutritional adequacy be measured or predicted in the animal on the farm? Without some means of doing this, the information of when, what and how much to feed is difficult to apply.

Current techniques are relatively insensitive, being based on liveweight and body condition measurements. Liveweight is an objective measurement but is subject to so many influences which are unrelated to nutritional adequacy, such as gut-fill and skeletal size, that it is, on its own, of only limited value. Body condition, on the other hand, is a subjective measurement, being usually based on the feel of tissue cover in the loin area, but may also be measured using ultrasonic techniques (e.g. 'Danscan'; Jensen, 1977). The rate of change in body condition is, however, too slow to indicate short-term change in nutritional state. The relative merit of liveweight and body condition as predictors of ovulation rate was examined by Doney, Gunn and Horák (1982) and it was concluded that, in a diverse population of a given breed, liveweight may be the most useful parameter for prediction while in a uniform flock with known history, body condition may be the most useful.

It is important to recognize that body condition, although recorded individually, cannot be used as a measure of the individual ewe's ovulatory response and its effect can only be considered on a flock basis. It cannot be assumed that an individual ewe will produce a specific number of ova because she has been graded at a specific level of condition. With the ratio of single- to twin-shedding ewes varying according to the flock mean condition level, there will still be ewes at any one level which shed a single ovum and others which shed two or more. Increasing the condition of a ewe increases her likelihood of shedding another one or two ova but does not guarantee it. Even with a mean ovulation rate of 2.00 there will be some individuals which shed only one matched by others shedding three.

Body size measurements can, within a population, be related to ovulation rate, irrespective of associated body condition, but are unlikely to be of much use as predictors of differences between populations, nor of changes in nutritional adequacy in the short-term.

Nutritional adequacy in terms of energy and protein status can be measured experimentally from analyses of various blood parameters e.g. cholesterol, free fatty acids, 3-hydroxybutyrate, blood urea (Cockrem, 1979; Russel, 1978; Sykes and Russel, 1979) but the techniques require complex analytical facilities and the results of some are sometimes of doubtful value when derived from untrained animals in the field. There is clearly a need for the development of simple measures of nutritional adequacy, and of its short-term change, which have application in commercial practice as well as in an experimental context. Ideally, but perhaps unrealistically, these should be based on a single measurement of an individual animal on any one particular occasion. In the meantime, however, practical measurement of nutritional adequacy is likely to remain based on liveweight and body condition.

In summary the reproductive performance of ewes is strongly influenced by nutrition, particularly energy nutrition, and although ovulation rate is under predictable nutritional control, ova wastage does not lend itself to an acceptable level of control. For optimization of reproductive performance, nutrition during the developmental stages of life must be adequate to ensure that the later potential is not restricted by long-term limitations. Nutrition must also be managed during the recovery period in each year to achieve and maintain good body condition at the time of mating and extremes of, or changes in, nutritional level must be avoided at this critical stage in the reproductive cycle.

References

ALLDEN, W.G. (1979). Undernutrition of the Merino sheep and its sequelae. V. The influence of severe growth restriction during early post-natal life on reproduction and growth in later life. *Australian Journal of Agricultural Research* **30**, 939–948

COCKREM, F.R.M. (1979). A review of the influence of liveweight and flushing on fertility made in the context of efficient sheep production. *Proceedings of the New Zealand Society of Animal Production* **39**, 23–42

CORBETT, J.L. and EDEY, T.N. (1977). Ovulation in ewes given formaldehyde-treated or untreated casein in maintenance-energy rations. *Australian Journal of Agricultural Research* **28**, 491–500

CUMMING, I.A., BLOCKEY, M.A. de B., WINFIELD, C.G., PARR, R.A. and WILLIAMS, A.H. (1975). A study of relationships of breed, time of mating, level of nutrition, live weight, body condition, and face cover to embryo survival in ewes. *Journal of Agricultural Science, Cambridge* **84**, 559–565

DONEY, J.M., GUNN, R.G. and HORÁK, F. (1982). Reproduction. In *Sheep and Goat Production* (I.E. Coop, Ed.), pp. 57–80. Amsterdam, Elsevier

EDEY, T.N. (1976a). Nutrition and embryo survival in the ewe. *Proceedings of the New Zealand Society of Animal Production* **36**, 231–239

EDEY, T.N. (1976b). Embryo mortality. In *Sheep Breeding* (G.J. Tomes, D.E. Robertson and R.J. Lightfoot, Eds.), pp. 400–410. Proceedings of the 1976 International Congress, Muresk, Western Australian Institute of Technology

EVERITT, G.C. (1967). Prenatal development of uniparous animals, with particular reference to the influence of maternal nutrition in sheep. In *Growth and Development of Mammals* (G.A. Lodge and G.E. Lamming, Eds.) pp. 131–157. London, Butterworths

FLETCHER, I.C. (1981). Effects of energy and protein intake on ovulation rate associated with the feeding of lupin grain to Merino ewes. *Australian Journal of Agricultural Research* **32**, 79–87

GUNN, R.G. (1977). The effects of two nutritional environments from 6 weeks prepartum to 12 months of age on lifetime performance and reproductive potential of Scottish Blackface ewes in two adult environments. *Animal Production* **25**, 155–164

GUNN, R.G. and DONEY, J.M. (1979). Ewe management for control of reproductive performance. *ADAS Quarterly Review* No. 35, 231–245

GUNN, R.G. and MAXWELL, T.J. (1978). The effects of direction of liveweight change about mating on lamb production in Greyface ewes. *Animal Production* **26**, 392 (Abstract)

GUNN, R.G., DONEY, J.M. and SMITH, W.F. (1979). Fertility in Cheviot ewes. 3. The effect of level of nutrition before and after mating on ovulation rate and early embryo mortality in South Country Cheviot ewes in moderate condition at mating. *Animal Production* **29**, 25–31

JENSEN, N.E. (1977). Preliminary results from Danscan ultrasonic measurements of fat thickness and loin area of lamb. *Proceedings of European Association for Animal Production, 28th Annual Meeting*

LINDSAY, D.R. (1976). The usefulness to the animal producer of research findings in nutrition on reproduction. *Proceedings of the Australian Society of Animal Production* **11**, 217–224

MORLEY, F.H.W., WHITE, D.H., KENNEY, P.A. and DAVIS, I.F. (1978). Predicting ovulation rate from liveweight in ewes. *Agricultural Systems* **3**, 27–45

PARR, R.A., CUMMING, I.A. and CLARKE, I.J. (1982). Effects of maternal nutrition and plasma progesterone concentrations on survival and growth of the sheep embryo in early gestation. *Journal of Agricultural Science, Cambridge* **98**, 39–46

RATTRAY, R.V. (1977). Nutrition and reproductive efficiency. In *Reproduction in Domestic Animals* (H.H. Cole and P.T. Cupps, Eds.) pp. 553–575. New York, San Francisco and London, Academic Press

REARDON, T.F. and LAMBOURNE, L.J. (1966). Early nutrition and lifetime reproductive performance of ewes. *Proceedings of the Australian Society of Animal Production* **6**, 106–108

RUSSEL, A.J.F. (1978). The use of measurement of energy status in pregnant ewes. In *The Use of Blood Metabolites in Animal Production* (D. Lister, Ed.) pp. 31–40. British Society of Animal Production, Occasional Publication No. 1

SCHINCKEL, P.G. (1963). The potential for increasing efficiency of feed utilization through newer knowledge of animal nutrition: (c) Sheep and goat. *Proceedings of the 1st World Conference on Animal Production, Rome*, pp. 199–218

SYKES, A.R. and RUSSEL, A.J.F. (1979). Seasonal variation in plasma protein and urea nitrogen concentrations in hill sheep. *Research in Veterinary Science* **28**, 223–229

6

NUTRITION OF THE PREGNANT EWE

J.J. ROBINSON
Rowett Research Institute, Bucksburn, Aberdeen, UK

Both the mode of action and the final outcome to reproduction of alterations in maternal nutrition vary with stage of gestation. For this reason it is convenient to divide the 21-week gestation into a number of periods and consider the effects of nutrition on production within each period. Most experimental observations on nutrition are restricted to one of three periods. To begin with there is the first month in which embryo loss is the main sequel to incorrect nutrition (Edey, 1976). During this period it is generally recommended that the body condition of the ewe is maintained, thus minimizing embryo and early foetal losses (Meat and Livestock Commission, 1981). This is followed by a period of two months in which there is rapid growth of the placenta but in which growth of the foetus in absolute terms is very small (Robinson *et al.*, 1977). Over this period it is normal to advocate that losses in body weight do not exceed 5% (Meat and Livestock Commission, 1981). Finally there is the phase from 90 days to parturition in which the gain in mass of the foetus amounts to 85% of its birthweight (Robinson *et al.*, 1977). For this period of pregnancy there is general agreement that nutrient intakes should be increased.

With one exception it is proposed to adhere to these three periods for a discussion of the effects of plane of nutrition and alterations in the intake of specific dietary nutrients on reproduction and where possible to consider the practical implications of carry-over effects from one period to another. The one exception is the first month of pregnancy which will be considered as two separate periods. The reason for splitting the first month into two is not to suggest alterations in feeding levels in this period. Indeed this would be both impractical and unnecessary. Rather it is to draw attention to the fact that embryo loss can have quite different effects on reproduction, depending on when it occurs.

The first period to consider is from mating until day 15 when, as a result of embryo migration, a balance in the distribution of the embryos between the two uterine horns is achieved in ewes with multiple ovulations and the first tenuous links that initiate the implantation process are established between the rapidly growing trophoblast and the epithelial cells of the maternal caruncles (Boshier, 1969). This is the *preimplantation phase*. The second is the ensuing two weeks which are characterized by a progressive strengthening of the bond between specialized areas on the chorionic membrane of the trophoblast i.e. the cotyledons, and the maternal caruncles situated on the uterine epithelium. This is the *implantation phase.*

Preimplantation

PLANE OF NUTRITION EFFECTS

Losses of fertilized ova in the first 12 days of pregnancy manifest themselves as a high incidence of repeat oestrous cycles at the normal interval and/or a reduced lambing percentage. Although losses in this period are frequently well above the 6–8% (Long and Williams, 1980) that can be attributed to chromosomal abnormalities and cracked zona pellucida, it is often difficult to attribute a significant proportion of them to current nutrition except in those ewes that are in poor condition at mating (Gunn, Doney and Russel, 1972) or those subjected to severe undernutrition (Edey, 1976). At the other extreme, i.e. very high levels of nutrition, there is also evidence of detrimental effects on embryo survival (reviews by Edey, 1976; Robinson, 1977; Doney, 1979).

In view of the extremely small requirements of embryos for nutrients it is perhaps not surprising that only extreme nutritional regimes affect their survival. Such nutritional effects probably operate by altering the endocrine balance of the ewe, for a link between the concentration of progesterone in maternal plasma in early pregnancy and plane of nutrition has been demonstrated by Cumming *et al.* (1971) and Parr, Cumming and Clarke (1982). Furthermore it has been shown that the growth of sheep embryos in the first two weeks of pregnancy is influenced by the progesterone concentration in maternal plasma (Wintenberger-Torrès, 1976; Lawson, 1977). The mode of action of the shift in progesterone appears to be through an alteration in the balance of amino acids in the uterine fluid (Ménézo and Wintenberger-Torrès, 1976). Recent confirmation of the importance to blastocyst development of the correct protein content of the uterine fluid and the influence of progesterone/oestrogen ratios in maternal plasma on fluid composition can be found in a review by Beier (1982).

Implantation

PLANE OF NUTRITION EFFECTS

In comparison to the preimplantation phase, nutritionally-induced deaths of embryos during the implantation phase have a wider range of effects on pregnancy. These effects include a higher than normal proportion of the flock returning to the ram at more than 19 days after a fertile mating, a reduced lambing percentage and the birth of smaller than average lambs for their particular litter size. The latter effect is based on the results of Rhind, Robinson and McDonald (1980) and McDonald, Robinson and Fraser (1981), which show that the death of embryos in the third and fourth weeks of pregnancy tends to disturb the balance in the distribution of the foetuses between the two uterine horns. In doing so it not only increases

Table 6.1 SOME EXAMPLES OF THE EFFECTS OF PLANE OF NUTRITION IN EARLY PREGNANCY ON THE SURVIVAL AND GROWTH OF EMBRYOS

Reference	*Breed of ewe*	*Age* (years)	*Body condition at mating*[a]	*Plane of nutrition* (× maintenance, M)	*Stage of gestation*	*Summary of main findings*
El-Sheikh *et al.* (1955)	Shropshire, Hampshire, Oxford	1	Medium	High 2.0 M	0–40 days	High-plane feeding reduced embryo survival.
Edey (1970)	Merino	4–7	Fat	0.5 M	0–37 days	Extended interval to repeat oestrus suggesting embryo mortality.
Gunn, Doney and Russel (1972)	Scottish Blackface	5–6	Condition scores of 1.5 and 3.0	0.4 M v. 2.5 M	0–26 days	Embryo mortality higher for ewes in lower body condition (score 1.5); suggestion that high-plane feeding reduced embryo survival.
Cumming (1972)	Perendale	4–5	Fat	0.2 M	7, 14 or 21 days on 0.2 M in first 21 days	Embryo survival decreased as period of restricted intake was extended.
Blockey, Cumming and Baxter (1974)	Merino	Mature	Not given	Fasting	Fasted for 3 days starting on day 1, 5, 8, 10 or 12	Inconclusive; reduction in pregnancy rate in single- but not twin-ovulators
MacKenzie and Edey (1975)	Merino	1.5 (Primiparous)	Not given	0.3 M	0–14 days	Extended interval to repeat oestrus. This suggests embryo mortality after day 12.
	Merino	3.5–5.5		0.15 M	0–14 days	
Cumming *et al.* (1975)	Merino and Border Leicester × Merino	5	Condition score 2.5	0.25, 1.0 and 2.0 M	2–16 days	Embryo survival highest on 1.0 M
Gunn, Doney and Smith (1979)	South County Cheviot	6	Condition score 2.0	0.7 M v. 1.75 M	0–30 days	Embryo survival improved by high-plane feeding particularly in ewes underfed before mating.
Parr, Cumming and Clarke (1982)	Merino	Mature	Not given	0.25 M v. 1.0 M	0–21 days	Retarded embryo development.

[a] Description of the procedures for the subjective assessment of body condition score given by Meat and Livestock Commission (1981)

the within-litter variability in foetal growth but, as a result of the inability of the surviving embryos to utilize the maternal cotyledons vacated by those that died, it also reduces the birthweight of those that survive.

Many of the nutritionally-induced deaths that occur between days 15 and 30 after mating undoubtedly arise from on-going nutritional defects that start soon after mating or even earlier. From the many experiments carried out on the effects of nutrition in the immediate post-mating period a small number have been selected to illustrate the range of nutritional regimes employed and the effects they had on embryo survival and growth. These are given in *Table 6.1*. It would appear that the embryos of young ewes and older ewes that are in poor condition at mating are most at risk. Extremes of nutrition appear to be detrimental to embryo survival. From this observation comes the recommendation that ewes should be kept at maintenance levels of feeding in the first month of pregnancy. There is a paucity of data on highly prolific breeds but higher embryo loss might be expected on the basis that nutritionally-induced losses appear to be more prevalent in twin- than single-ovulating ewes (Edey, 1976).

Another interesting aspect of the information summarized in *Table 6.1* is the evidence that restrictions in food intake in the first three weeks of pregnancy can, in the absence of embryo mortality, retard embryo development, as assessed by the size of the pharangeal lobe and the fore and hind limb buds on day 21 (Parr, Cumming and Clarke, 1982). Whether or not these observations could explain an apparent carry-over effect on the birthweight of single lambs from Merino ewes on 0.5 × maintenance regime between days 7 and 37 of pregnancy (Edey, 1970) is uncertain, for a similar effect could arise through death of one of a pair of embryos during the implantation phase.

EFFECTS OF SPECIFIC NUTRIENTS

Corrections of dietary deficiencies of protein (Fletcher, 1981), and of certain trace elements, notably copper (Annenkov, 1981), manganese and zinc (Egan, 1972), iron and cobalt (Hidiroglou, 1979), and selenium (Hartley and Grant, 1961; Annenkov, 1981), have been shown to improve lambing percentages. However, it is only for selenium that there is strong evidence that at least part of the improvement comes from a reduction in embryo mortality in the third and fourth weeks of pregnancy (Hartley, 1963; Piper *et al.*, 1980). Provided there is not a concurrent vitamin E deficiency in the diet then the supplementation of selenium-deficient diets with 0.1 mg selenium/kg dry matter will reduce embryo mortality arising from selenium deficiency.

SPECIFIC FEEDS THAT REDUCE EMBRYO SURVIVAL

The most notable example of a specific feed causing embryo mortality during the implantation phase is kale (Williams, Hill and Alderman, 1965), but it is not known if the increased mortality was the result of a goitrogenic effect, anaemia or a reduced copper status, as all three factors were significantly altered by kale feeding. If it resulted from a general anaemia then the use of varieties low in the anaemia factor, S-methyl-cysteine sulphoxide (Smith, 1974) might overcome the problem.

Other feeds that are a common cause of reduced lambing percentages are the oestrogenic forages, in particular red clover, with its high content of the isoflavone, formononetin, which is converted in the rumen into the phyto-oestrogen, equol, or 7,4-dihydroxyisoflavan (Shutt, Weston and Hogan, 1970). Although it is generally agreed that the adverse effect of the phyto-oestrogen on lambing percentages is the result of a reduction in fertilization rate there is evidence that prolonged grazing of oestrogenic forages can lead to cystic endometrial hyperplasia and reduced embryo survival during the latter half of the implantation period (Turnbull, Braden and George, 1966).

The second and third months

This is the period over which it is generally assumed that plane of nutrition is relatively unimportant to the success of the pregnancy. No doubt the early studies of Wallace (1948) played a significant role in arriving at this view for, in comparison with the dramatic effects of undernutrition in late pregnancy on birthweight, his results showed that a 7% decrease in liveweight in the first three months had no detrimental effects on foetal weights at 90 days. Since then there have been a number of studies which indicate that losses of this magnitude may be approaching the upper limit. The results of these studies are summarized in *Table 6.2* and show clearly that the important cotyledonary component of the placenta is much more sensitive than the foetus to undernutrition during this period.

The extent to which detrimental effects on placental and foetal weights at 90 days can be counteracted by a high plane of nutrition in late pregnancy is of considerable practical importance. In reviewing the studies

Table 6.2 THE EFFECTS OF PLANE OF NUTRITION IN THE SECOND AND THIRD MONTHS OF PREGNANCY ON PLACENTAL AND FOETAL WEIGHTS AT AROUND 90 DAYS

Reference	*Breed of ewe*	*Weight* (kg)	*Litter size*	*Plane of nutrition defined as % loss(–) or % gain(+) in liveweight*		*Cotyledon weights* (Low as % of high)	*Foetal weights* (Low as % of high)
				Low	*High*		
Everitt (1964)	Merino	Not given	1	–12	+12	69.6	89.5
Rattray, Trigg and Urlich (1980)	Coopworth	58	2	–11	+16	Not given	87.3
Clark and Speedy (1980)	Border Leicester × Cheviot	85[a]	2	–9	+6	90.3	97.5
		69[a]	2	–7	+16	89.9	98.8

[a]Difference in liveweight arising from difference in plane of nutrition before mating.

that immediately followed the observations of Everitt (1964; *Table 6.2*), Robinson (1977) concluded that there was evidence for at least a partial compensation in foetal growth when ewes that were under-nourished in mid pregnancy were subsequently well fed. Also young ewes that had not reached their mature size at mating appeared less capable of compensating than mature ewes. The more recent observations of Rattray, Trigg and Urlich (1980) suggest that even in mature ewes complete compensation is unlikely. When ewes that were on low and high planes up to 95 days (see *Table 6.2*) were given the same high level of feeding for the next 40 days, the low plane ewes partitioned more of their available nutrients to replenishing their maternal reserves and less to foetal growth than those on the high plane up to day 95.

In addition to the degree of maturity of the ewe, many other factors undoubtedly influence the extent to which late pregnancy nutrition can compensate for undernutrition in mid pregnancy. These will include the condition of the ewe at the end of the first month of gestation and litter size, but there is insufficient information on these factors to provide quantitative relationships.

The last two months

PLANE OF NUTRITION EFFECTS

Since the gain in mass of the foetus in the last 8, 4 and 2 weeks of gestation is equivalent to 85, 50 and 25% of its birthweight, it is reasonable to expect a relationship between plane of nutrition in late pregnancy and lamb birthweight. Although a relationship exists, a review of the literature reveals that for a given reduction in birthweight there is wide variation in metabolizable energy (ME) intakes between data sources (see *Figure 6.1*). Some of this variation may be attributed to differences between studies in

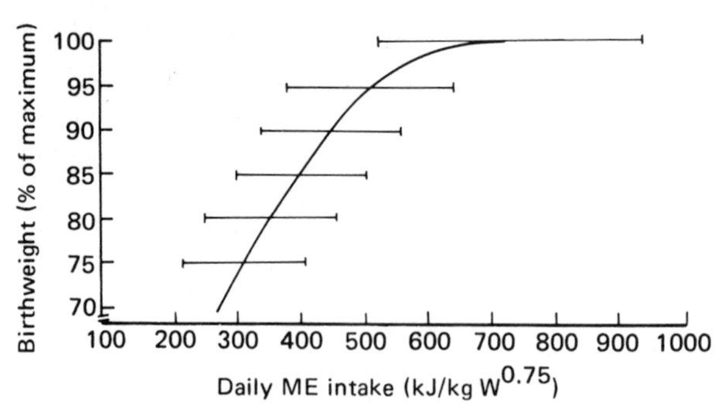

Figure 6.1 Ranges in metabolizable energy (ME) intake in late pregnancy that are associated with given reductions in lamb birthweight. Data sources used in defining the ranges are: Wallace (1948); Russel, Doney and Reid (1967); Robinson and Forbes (1968); Robinson, Brown and Lucas (1973); Shevah, Black and Land (1975); Prior and Christenson (1976); Russel *et al.* (1977); Valdez Espinosa, Robinson and Scott (1977) and Sheehan, Lawlor and Bath (1977). Plots of relationship for each data source between the daily ME intake and the reduction in birthweight are given by Robinson (1982).

genotype and maternal reserves (Robinson and McDonald, 1979) and some to the specific effects of a deficiency of dietary protein. This aspect will be discussed in detail later.

These effects of chronic undernutrition influence birthweight by a gradual slowing down of prenatal growth (Rattray *et al.*, 1974b; Mellor and Murray, 1981, 1982a). In contrast, a severe and sudden restriction in food intake at the beginning of the final month of pregnancy can reduce foetal growth by 30–40% within three days or in some instances cause a complete cessation in growth, as determined by *in vivo* estimates of the daily changes in the curved crown-rump lengths of individual foetuses (Mellor and Matheson, 1979). Severe restrictions in food intake can also give rise to lines of arrested growth in the bones of the foetal skeleton, notably the scapula, os coxae and talus (Wenham, 1981). For relatively short periods of severe undernutrition of one week duration there is evidence of a partial compensation in growth rate when food intake is increased (Mellor and Matheson, 1979). For longer periods this does not appear to be so and in a recent study Mellor and Murray (1982b) concluded that after 16 days of severe undernutrition, characterized by pre-feeding levels of glucose in maternal plasma which were only 50% of those of well-fed ewes, foetuses lacked the ability to return to normal growth rates when food intake was increased.

EFFECTS OF DIETARY PROTEIN

Although 70% of the total energy in the newborn lamb is in the form of protein, it is only in the last 15 years that researchers have become interested in studying the specific effects of dietary protein in late pregnancy on lamb birthweight. A quick glance at the literature on the

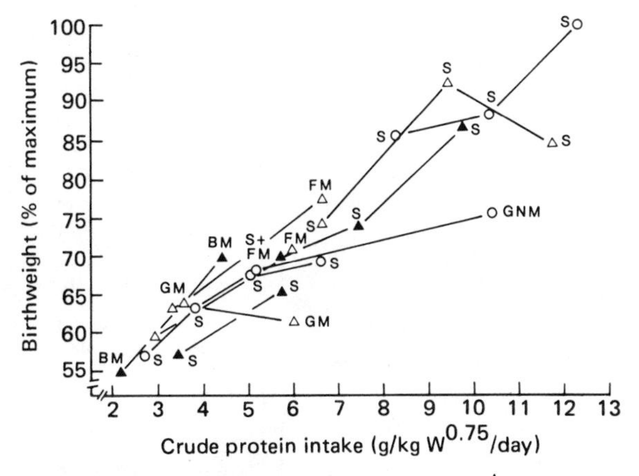

Figure 6.2 The effect of crude protein intake in late pregnancy on lamb birthweight. Symbols with no letters refer to basal diets. Letters refer to those with supplements viz. S, soya bean; FM, fish meal; GM, grass meal; GNM, ground nut meal; BM, blood meal. The data sources are: Forbes and Robinson (1967); Robinson and Forbes (1968); McClelland and Forbes (1968); Sykes and Field (1972a); Parkins *et al.* (1974); Clark and Speedy (1981). Daily ME intakes/kg $W^{0.75}$ of 300–400 kJ, ▲; 400–500 kJ, △; and >500 kJ, ○.

subject reveals that in many instances the effects of dietary protein have been tested at suboptimal intakes of energy. In order to summarize the information, the birthweights for each level of protein intake were expressed as percentages of maximum values with further classifications based on protein source and level of ME intake. The results are presented in *Figure 6.2.* Clearly protein intake can have a marked effect on birthweight but there are insufficient observations to disentangle the effects of protein source and energy status. By analogy with other physiological states it seems reasonable to expect the highest response in birthweight to dietary protein to occur at low ME intakes and with protein sources that are fairly resistant to degradation in the rumen (Ørskov and Robinson, 1981).

The net accretion of nutrients

Estimates of the net accretion rates of nutrients in the products of conception together with efficiency coefficients for the utilization of dietary nutrients for net accretion form the basis of the factorial method for estimating requirements. This approach is used in the subsequent calculations. Rather than expressing accretion rates in terms of ewe body weight

Table 6.3 DAILY RATES OF ACCRETION OF ENERGY AND PROTEIN IN THE GRAVID UTERUS AND OF MACRO MINERALS IN THE FOETUSES IN RELATION TO LITTER SIZE AND STAGES OF GESTATION (VALUES EXPRESSED/kg LAMB BIRTHWEIGHT)

	Number of foetuses[a]	*Stage of gestation* (days)						
		60	*74*	*88*	*102*	*116*	*130*	*144*
Energy (kJ)	1	14	22	34	55	82	111	141
	2				54	80	106	131
	3				53	78	101	123
	4				52	76	98	117
Protein (g)	1	0.49	0.75	1.14	1.82	2.69	3.67	4.64
	2				1.80	2.64	3.50	4.34
	3				1.78	2.59	3.37	4.12
	4				1.76	2.54	3.27	3.97
Calcium (mg)	1	5	21	58	110	171	224	257
	2				111	170	217	243
	3				111	168	211	231
	4				112	166	204	218
Phosphorus (mg)	1	3.5	15	38	67	96	116	123
	2				68	96	113	118
	3				69	96	111	112
	4				70	95	108	107
Magnesium (mg)	1	0.21	0.77	1.86	3.42	5.16	6.59	7.52
	2				3.39	5.01	6.29	7.00
	3				3.36	4.87	5.99	6.50
	4				3.34	4.74	5.69	6.02

[a] Values for singles were obtained by extrapolation from the values for twins, triplets and quadruplets.
From McDonald *et al.* (1979)

and predicted averages for birthweight, which is the usual method of presentation (Agricultural Research Council, 1980), they are expressed/kg of birthweight in *Table 6.3*. It is hoped that presentation of the information in this form will make it relatively easy for users to relate it to their own circumstances, in which effects on birthweight, of breed (Dickinson *et al.*, 1962) and certain management factors such as shearing in mid pregnancy (Rutter, Laird and Broadbent, 1971) may cause significant deviations from the average.

Up to the end of the third month of pregnancy the differences between litter sizes in accretion rates/kg of birthweight are so small that only single values are given in *Table 6.3*. Thereafter the values diverge and for this reason they are presented for each litter size.

Efficiency of energy utilization for conceptus gain (k_c)

In a review of the literature Robinson *et al.* (1980) took the six main publications which provided estimates of k_c (viz. Graham, 1964; Sykes and Field, 1972b; Lodge and Heaney, 1973; Rattray *et al.*, 1973; Heaney and Lodge, 1975 and Rattray *et al.*, 1974a) and together with their own estimates from a comparative slaughter experiment put forward the hypothesis that k_c is positively related to the ME concentration of the diet. For diets with an ME concentration of 10.5 MJ/kg dry matter the value for k_c was 0.145 and the slope of the regression of k_c on ME concentration (MJ/kg dry matter) was 0.029.

Energy requirements

Division of the daily rates of energy deposition in the gravid uterus (*Table 6.3*) by the coefficient, 0.145 for k_c, provides estimates of the ME requirements for conceptus growth in ewes receiving a diet containing 10.5 MJ of ME/kg dry matter. By way of example these estimates are given in *Table 6.4*.

If diets with ME concentrations above or below 10.5 are being used the estimates can be obtained by appropriate adjustment of the coefficient using the value 0.029/MJ of ME. For example a diet with an ME concentration of 9.5 MJ/kg dry matter would have an efficiency coefficient

Table 6.4 ESTIMATES OF THE DAILY METABOLIZABLE ENERGY REQUIREMENTS ABOVE MAINTENANCE FOR CONCEPTUS GROWTH (ME_p) IN RELATION TO LITTER SIZE AND STAGE OF GESTATION. (VALUES EXPRESSED AS MJ/kg LAMB BIRTHWEIGHT)

ME concentration of diet (MJ/kg dry matter)	*Number of foetuses*	*Stage of gestation* (days)						
		60	*74*	*88*	*102*	*116*	*130*	*144*
10.5	1	0.10	0.15	0.23	0.38	0.57	0.77	0.97
	2				0.37	0.55	0.73	0.90
	3				0.37	0.54	0.70	0.85
	4				0.36	0.52	0.68	0.81

of 0.116 which translates to a daily ME requirement at 144 days of gestation of 1.13 MJ/kg lamb birthweight for a twin-bearing ewe. Since these estimates refer only to the energy costs of pregnancy they must be added to a daily ME requirement for maternal maintenance of 0.42 MJ/kg $W^{0.75}$ (Robinson *et al.*, 1980) in order to arrive at overall needs.

Those familiar with recommended feeding standards will note that the estimates in *Table 6.4* are higher than those advocated in Bulletin 33 (MAFF, DAFS and DANI, 1975) but they are in close agreement with more recent recommendations (Agricultural Research Council, 1980) and experimental results (Corbett *et al.*, 1980).

Protein requirements

The transformation of the net accretion rates for protein in the gravid uterus into dietary amounts requires an understanding of protein digestion and utilization. What follows therefore is an attempt to use recent information on the synthesis of microbial protein and on the subsequent digestion and utilization of both microbial and undegraded dietary protein to translate the net requirements for protein into dietary amounts. In doing this it is useful first of all to consider how well the microbial and undegraded dietary protein meet the protein needs for maintenance.

It is now generally accepted that basal feeds, i.e. roughages and cereal grains, must contain a minimum of about 10 g of crude protein (CP)/MJ of ME to provide the maximal synthesis of microbial protein, i.e. 8 g/MJ of ME, from the 80% of their protein that is degraded to ammonia in the rumen (Agricultural Research Council, 1980). About 80% of this microbial protein is in the form of amino acids and these together with the 20% undegraded protein in the basal feed are available for absorption in the small intestine. Taking coefficients for true digestibility and subsequent utilization of 0.85 and 0.80 respectively (Storm and Ørskov, 1982) gives an estimate for net protein synthesis of 5.7 g/MJ of ME. Assuming a daily maintenance requirement for energy of 0.42 MJ of ME/kg $W^{0.75}$, as in the preceding section, then the net synthesis of protein at energy maintenance is 2.4 g/kg $W^{0.75}$. With a net protein requirement for tissue maintenance of 2.2 g/kg $W^{0.75}$ (Ørskov, MacLeod and Grubb, 1980), or perhaps slightly more, as seems probable for a pregnant animal, and a net daily accretion of protein in wool of approximately 0.25 g/kg $W^{0.75}$ (Agricultural Research Council, 1965), then the combined net requirements for maintenance and wool production are very similar to the net amount produced from microbial and undegraded dietary protein.

The preceding conclusion has a number of important practical implications. Firstly it means that during the first three months of pregnancy, when it is usual to recommend maintenance levels of energy intake, basal feeds which contain less than 10 g CP/MJ of ME should be supplemented with urea in order to meet the minimum requirements for rumen-degradable protein. Secondly, if energy intake falls below maintenance in the mid pregnancy period then it is necessary to provide a supplement of dietary protein of low degradability in the rumen in order to prevent a loss of protein from the maternal body. Finally it means that production requirements for protein in late pregnancy can be considered solely in relation to production requirements for energy.

NET PROTEIN REQUIREMENTS PER MJ OF ME

In addition to the protein accretion in the gravid uterus there is the udder and its secretions. Estimates of the latter have been derived from the data of Rattray *et al.* (1974b) and Robinson *et al.* (1977), and added to the values for the gravid uterus (*Table 6.3*) before obtaining the values given in *Table 6.5* which are the net protein requirements for pregnancy (NPR_p) expressed in relation to the corresponding ME requirements (ME_p).

This exercise produces some interesting results. On the basis of the conclusion arrived at earlier, that diets containing adequate protein for maximal synthesis of microbial protein supply a net protein requirement of 5.7 g/MJ of ME, then pregnant ewes that are meeting all their energy requirements from the diet only require an additional supplement of undegraded dietary protein in the last three weeks of pregnancy.

Table 6.5 THE NET ACCRETIONS OF PROTEIN IN THE GRAVID UTERUS AND UDDER, i.e. NET PROTEIN REQUIREMENTS FOR PREGNANCY (NPR_p) IN RELATION TO THE CORRESPONDING ME REQUIREMENTS ME_p (g/MJ)

Number of foetuses	*Stage of gestation* (days)						
	60	*74*	*88*	*102*	*116*	*130*	*144*
1				5.2	5.2	6.0	11.3
2	5.2	5.2	5.2	5.3	5.3	6.1	11.7
3				5.3	5.3	6.2	12.1
4				5.3	5.4	6.2	12.6

The other interesting feature of the values in *Table 6.5* is the extremely rapid increase in the net protein requirements at the end of pregnancy to twice the value of 5.7 g that is provided by microbial protein. This arises from the large influx of colostrum to the udder. Clearly it would be extremely difficult to supply these additional amounts of protein which, for a twin-bearing ewe, are equivalent to 6 g of net protein/MJ of ME_p or 8.8 g of undegraded dietary protein/MJ of ME_p (i.e. assuming a true digestibility of 0.85 and a utilization of 0.80). In terms of dietary amounts this is equivalent to 12.6 g of protein/MJ of ME_p from a source that has a degradation coefficient in the rumen of only 0.3. For a 70 kg ewe producing twin lambs of total birthweight 10 kg, ME_p = 9 MJ/day (see *Table 6.4*). Thus the additional protein requirement = 113 g/day. This is in addition to the 90 g required to meet the microbial needs of the 9 MJ of ME for pregnancy and another 100 g for the microbial needs of the 10 MJ of ME (i.e. 0.42 MJ/kg $W^{0.75}$) required for maternal maintenance.

Since redistribution of body protein in order to ensure adequate udder development and colostrum production has been shown to occur in the last few weeks of pregnancy (Robinson *et al.*, 1978), it is therefore more practical to replace the rapid increase in the final week of pregnancy (*Table 6.5*) with a more even pattern. A suitable alternative to emulating the

Table 6.6 CALCULATION OF THE AVERAGE ADDITIONAL REQUIREMENTS FOR NET PROTEIN IN THE LAST THREE WEEKS OF PREGNANCY AND THEIR TRANSFORMATION TO DIETARY AMOUNTS

Number of foetuses	*Mean daily*[a] *requirement for net protein* (g/MJ of ME_p)	*Net protein*[b] *supplied by basal diet* (g/MJ of ME_p)	*Net protein*[b] *requirement above that supplied by the basal diet* (g/MJ of ME_p)	*Requirements for additional protein* (g/MJ of ME_p)		
				Truly absorbed(C)	*Available for absorption*	*Dietary amount* (assuming a degradability coefficient of 0.3)
	(A)	(B)	(A – B)	(A – B)/0.80	C/0.85	
1	7.7	5.7	2.0	2.5	2.9	4.1
2	7.8	5.7	2.1	2.6	3.1	4.4
3	8.0	5.7	2.3	2.9	3.4	4.9
4	8.2	5.7	2.5	3.1	3.6	5.2

[a]Taken from *Table 6.5*.
[b]Basal diet containing 10 g CP/MJ of ME (see text, p.120)

actual pattern is to give constant daily amounts of supplementary dietary protein (expressed/MJ of ME_p) in the last three weeks of pregnancy so that the same overall accretion of net protein is achieved. The various steps involved in calculating these amounts are set out in *Table 6.6*. The dietary amounts, which are expressed in g/MJ of ME_p, are easily transformed into absolute daily amounts. For example a twin-bearing ewe has a daily ME_p requirement of 0.9 MJ/kg lamb birthweight at the end of pregnancy (see *Table 6.4*), which is equivalent to 9 MJ of ME_p for a total lamb birthweight of 10 kg. Since the requirement for an additional protein supplement with a degradability coefficient of 0.3 is 4.4 g/MJ of ME_p (*Table 6.6*), the overall amount of supplementary dietary protein, above that required for maximal microbial protein synthesis, is $4.4 \times 9 = 40$ g. This is equivalent to about 64 g of fish meal/day. The corresponding amount at two weeks before lambing would be $0.73 \times 10 \times 4.4 = 32$ g of dietary protein or 50 g of fish meal.

It will be noted that this approach to the determination of protein requirements gives higher values than those recommended by the Agricultural Research Council (1980). This is understandable for ARC made no allowance for the additional needs of protein for udder development and colostrum production.

THE IMPLICATIONS OF AN ENERGY DEFICIT TO PROTEIN NEEDS

In most commercial systems of sheep production it is extremely difficult to meet the entire requirements for energy in late pregnancy from dietary sources alone. The immediate implications of a reduced intake of energy is a concomitant reduction in the synthesis of microbial protein and thus a lower contribution of microbial protein to the net protein needs of the ewe. To correct this deficit in net protein, diets will require supplementation with proteins of low degradability in the rumen at an earlier stage of pregnancy and at higher amounts than those given in the preceding section. By way of example, the daily amounts of dietary protein supplied by basal diets containing the minimum amount of protein for maximal microbial protein production (10 g/MJ of ME) and given at energy intakes equivalent to maintenance (M) plus 1.0, 0.8 or 0.6 ME_p have been calculated for different stages of pregnancy for a twin-bearing ewe of 70 kg at mating, and producing twin lambs of 10 kg total birthweight. These are presented in *Figure 6.3*. In addition the daily requirements for supplementary protein have been calculated using the values given in the preceding section. The data are expressed/kg lamb birthweight for ease in application to different birthweights. The data are also readily transformed to other sizes of ewe by reducing each of the values (i.e. the *y*-axis of *Figure 6.3*) by 10 g, this being the amounts of dietary protein (g/kg birthweight) supplied by the maintenance component of the diet of a 70 kg ewe. In each case the remaining value is the dietary protein supplied by the ME_p portion of the basal diet. This is multiplied by the appropriate value for total lamb birthweight and added to the contribution of protein from the maintenance component of the diet (i.e. $10 \times 0.42\ W^{0.75}$). The sum of these two values gives the contribution of protein from the basal diet. The necessary amounts of

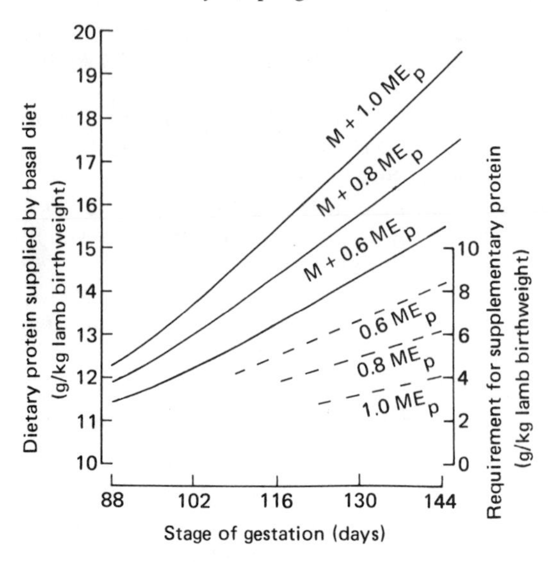

Figure 6.3 The daily amounts of dietary protein supplied at different stages of pregnancy by a basal diet containing 10 g CP/MJ of ME and given at energy intakes of maintenance (M) plus 1.0, 0.8 and 0.6 ME_p to a twin-bearing ewe of 70 kg at mating and producing a total lamb birthweight of 10 kg. Also included are the daily requirements for supplementary protein assuming the use of a protein source with a degradability coefficient of 0.3.

supplementary protein are easily obtained by multiplying the values in *Figure 6.3* by the appropriate total birthweight. For supplements differing in degradability from the value of 0.3 used in *Figure 6.3*, the amounts are calculated by multiplying the estimates illustrated by the broken lines in *Figure 6.3* by the ratio of the degradability coefficients.

For those concerned with the practical application of recent research on protein digestion and utilization to the feeding of the pregnant ewe, the preceding exercise is of more value in the principles that it embraces than in the precision of the final estimates. Some of the factors used in the calculations may require numerical modification when more data become available. Indeed it will be noted that in the present exercise many of the factors have been modified from those used by the Agricultural Research Council (1980). Even so no attempt has been made to accommodate the recent finding that pregnancy reduces protein degradation in the rumen through its effects on the retention times of food in the rumen (Thompson, Robinson and McHattie, 1978; Weston, 1979; Faichney and White, 1980). Nevertheless a knowledge of the principles involved in protein digestion and utilization provides a useful framework for making recommendations for a wide range of flock types and management systems.

Mineral requirements

It is not the purpose of this chapter to deal in detail with the mineral needs of the pregnant ewe, for meeting the mineral requirements of sheep is the subject of Chapter 9. Rather it is briefly to draw attention to the high rates of net accretion of calcium and phosphorus in the foetuses in the late stages of pregnancy (see *Table 6.3*). Despite these high rates, equivalent in a twin-bearing ewe of 70 kg, to about 2.5 g daily for calcium, Sykes and Field (1972a) found that protein undernutrition in late pregnancy had a much greater effect on foetal size than the reduction of the intake of dietary calcium to around 1.0 g/day, i.e. 18% of ARC (1965) recommendations, in hill ewes carrying one and two foetuses. The reason for this appears to be the direct effects of a protein deficiency on the development of the bone matrix (Sykes, Nisbet and Field, 1973), although the possibility of a specific effect of protein on the absorption of calcium was not ruled out (Sykes and Field, 1972c).

Linking the findings on maternal nutrition to foetal metabolism

Recent advances in technique for studying the metabolism of the gravid uterus and individual foetuses show some interesting relationships between maternal nutrition and foetal metabolism. For example in the foetus of the well-nourished ewe in late pregnancy the proportion of oxygen consumption accounted for by the metabolism of the various substrates was as follows: glucose, 0.50; lactate, 0.25; amino acids, 0.20; acetate, 0.05 and glycerol, 0.01 (Girard, Pintado and Ferre, 1979). In contrast when ewes were severely undernourished the values were: glucose, 0.30; lactate, 0.15; amino acids, 0.60; acetate, 0; and glycerol, 0.01. The larger contribution, in relative terms, of amino acids to foetal oxidative metabolism in the undernourished ewe is of particular interest in view of the beneficial effects of dietary protein on lamb birthweight (see *Figure 6.2*). Similarly in prolonged undernutrition, Faichney (1981) has shown that there is a substantial increase in urea synthesis by the foetus thus indicating that amino acids are diverted from protein synthesis to the provision of metabolic fuel. Turning to the energy metabolism of the conceptus the low value for k_c, while largely the result of expressing it as an overall efficiency akin to a food conversion efficiency, may well have a component attributable to the high turnover rate of tissue proteins in the foetal lamb (Noakes and Young, 1981). Despite these very tangible links between maternal nutrition, foetal metabolism and production responses, many practical observations still defy explanation. The most obvious of these is the observation that overnutrition in mid pregnancy can reduce birthweight particularly in young ewes (Robinson, 1977; Russel *et al.*, 1981). Similarly the mechanisms whereby shearing ewes in mid pregnancy increases lamb birthweight have not been elucidated. Whether or not they could arise from a specific effect of the management procedures on uterine blood flow or an alteration in a 'specific uterine mechanism for the maintenance of glucose uptake' (Oddy *et al.*, 1981) remains to be proven.

References

AGRICULTURAL RESEARCH COUNCIL (1965). *The Nutrient Requirements of Farm Livestock, No. 2 Ruminants.* London, Agricultural Research Council

AGRICULTURAL RESEARCH COUNCIL (1980). *The Nutrient Requirements of Farm Livestock, No. 2 Ruminants, Second edition.* Slough, Commonwealth Agricultural Bureaux

ANNENKOV, B.N. (1981). Mineral feeding of sheep. In *Mineral Nutrition of Animals* (V.I. Georgievski, B.N. Annenkov and V.I. Samokhin, Eds.), pp. 321–354. London, Butterworths

BEIER, H.M. (1982). Uteroglobin and other endometrial proteins; biochemistry and biological significance in beginning pregnancy. In *Proteins and Steroids in Early Pregnancy* (H.M. Beier and P. Karlson, Eds.), pp. 39–71. Berlin, Heidelberg and New York, Springer Verlag

BLOCKEY, M.A. DE B., CUMMING, I.A. and BAXTER, R.W. (1974). The effect of short term fasting in ewes on early embryonic survival. *Proceedings of the Australian Society of Animal Production* **10**, 265–269

BOSHIER, D.P. (1969). A histological and histochemical examination of implantation and early placentome formation in sheep. *Journal of Reproduction and Fertility* **19**, 51–61

CLARK, C.F.S. and SPEEDY, A.W. (1980). The effects of pre-mating and early pregnancy nutrition on foetal growth and body reserves in Scottish half-bred ewes. *Animal Production* **30**, 485

CLARK, C.F.S. and SPEEDY, A.W. (1981). Effect of mating condition and protein/energy nutrition on foetal growth and maternal reserves in crossed ewes. *Animal Production* **32**, 364

CORBETT, J.L., FURNIVAL, E.P., INSKIP, M.W., PICKERING, F.S. and PLAZA, J. (1980). Pasture intake and heat production of breeding ewes. In *Energy Metabolism* (L.E. Mount, Ed.), pp. 319–323. London, Butterworths

CUMMING, I.A. (1972). The effect of nutritional restriction on embryonic survival during the first three weeks of pregnancy in Perendale ewes. *Proceedings of the Australian Society of Animal Production* **9**, 199–203

CUMMING, I.A., BLOCKEY, M.A. DE B., WINFIELD, C.G., PARR, R.A. and WILLIAMS, A.H. (1975). A study of relationships of breed, time of mating, level of nutrition, live weight, body condition and face cover to embryo survival in ewes. *Journal of Agricultural Science, Cambridge* **84**, 559–565

CUMMING, I.A., MOLE, B.J., OBST, J., BLOCKEY, M.A. DE B., WINFIELD, C.G. and GODING, J.R. (1971). Increase in plasma progesterone caused by undernutrition during early pregnancy in the ewe. *Journal of Reproduction and Fertility* **24**, 146–147

DICKINSON, A.G., HANCOCK, J.L., HOVELL, G.J.R., TAYLOR, St. C.S. and WIENER, G. (1962). The size of lambs at birth—a study involving egg transfer. *Animal Production* **4**, 64–79

DONEY, J.M. (1979). Nutrition and the reproductive function in female sheep. In *The Management and Diseases of Sheep*, pp. 152–160. Slough, Commonwealth Agricultural Bureaux

EDEY, T.N. (1970). Nutritional stress and preimplantation mortality in Merino sheep, 1967. *Journal of Agricultural Science, Cambridge* **74**, 193–198

EDEY, T.N. (1976). Embryo mortality. In *Sheep Breeding* (G.J. Tomes, D.E. Robertson and R.J. Lightfoot, Eds.), pp. 400–410. Armidale, New England University

EGAN, A.R. (1972). Reproductive responses to supplemental zinc and manganese in grazing Dorset Horn ewes. *Australian Journal of Experimental Agriculture and Animal Husbandry* **12**, 131–135

EL-SHEIKH, A.S., HULET, C.V., POPE, A.L. and CASIDA, L.E. (1955). The effect of level of feeding on the reproductive capacities of the ewe. *Journal of Animal Science* **14**, 919–929

EVERITT, G.C. (1964). Maternal undernutrition and retarded foetal development in Merino sheep. *Nature, London* **201**, 1341–1342

FAICHNEY, G.J. (1981). Amino acid utilization by the foetal lamb. *Proceedings of the Nutrition Society of Australia* **6**, 48–53

FAICHNEY, G.J. and WHITE, G.A. (1980). Mean retention time of markers in the rumen of pregnant sheep. *Proceedings of the Australian Society of Animal Production* **13**, 455

FLETCHER, I.C. (1981). Effects of energy and protein intake on ovulation rate associated with the feeding of lupin grain to Merino ewes. *Australian Journal of Agricultural Research* **32**, 79–87

FORBES, T.J. and ROBINSON, J.J. (1967). The effect of source and level of dietary protein on the performance of in-lamb ewes. *Animal Production* **9**, 521–530

GIRARD, J., PINTADO, E. and FERRE, P. (1979). Fuel metabolism in the mammalian fetus. *Annales de Biologie Animale, Biochimie, Biophysique* **19**(1B), 181–197

GRAHAM, N.McC. (1964). Energy exchanges of pregnant and lactating ewes. *Australian Journal of Agricultural Research* **15**, 127–141

GUNN, R.G., DONEY, J.M. and RUSSEL, A.J.F. (1972). Embryo mortality in Scottish Blackface ewes as influenced by body condition at mating and by post mating nutrition. *Journal of Agricultural Science, Cambridge* **79**, 19–25

GUNN, R.G., DONEY, J.M. and SMITH, W.F. (1979). Fertility in Cheviot ewes. 3. The effect of level of nutrition before and after mating on ovulation rate and early embryo mortality in South Country Cheviot ewes in moderate condition at mating. *Animal Production* **29**, 25–31

HARTLEY, W.J. (1963). Selenium and ewe fertility. *Proceedings of the New Zealand Society of Animal Production* **23**, 20–27

HARTLEY, W.J. and GRANT, A.B. (1961). A review of selenium responsive diseases of New Zealand livestock. *Federation Proceedings* **20**, 679–688

HEANEY, D.P. and LODGE, G.A. (1975). Body composition and energy metabolism during late pregnancy in the *ad libitum* fed ewe. *Canadian Journal of Animal Science* **55**, 545–555

HIDIROGLOU, M. (1979). Trace element deficiencies and fertility in ruminants: A review. *Journal of Dairy Science* **62**, 1195–1206

LAWSON, R.A.S. (1977). Research application of embryo transfer in sheep and goats. In *Embryo Transfer in Farm Animals* (K.J. Betteridge, Ed.), pp. 72–78. Monograph 16. Canada Department of Agriculture

LODGE, G.A. and HEANEY, D.P. (1973). Energy cost of pregnancy in single and twin-bearing ewes. *Canadian Journal of Animal Science* **53**, 479–489

LONG, S.E. and WILLIAMS, C.V. (1980). Frequency of chromosomal abnormalities in early embryos of the domestic sheep (*Ovis aries*). *Journal of Reproduction and Fertility* **58**, 197–201

MacKENZIE, A.J. and EDEY, T.N. (1975). Short-term undernutrition and prenatal mortality in young and mature Merino ewes. *Journal of Agricultural Science, Cambridge* **84**, 113–117

McCLELLAND, T.H. and FORBES, T.J. (1968). A study of the effect of energy and protein intake during late pregnancy on the performance of housed Scottish Blackface ewes. *Record of Agricultural Research, N. Ireland* **17**, 131–138

McDONALD, I., ROBINSON, J.J. and FRASER, C. (1981). Studies on reproduction in prolific ewes. 7. Variability in the growth of individual foetuses in relation to intra-uterine factors. *Journal of Agricultural Science, Cambridge* **96**, 187–194

McDONALD, I., ROBINSON, J.J., FRASER, C. and SMART, R.I. (1979). Studies on reproduction in prolific ewes. 5. The accretion of nutrients in the foetuses and adnexa. *Journal of Agricultural Science, Cambridge* **92**, 591–603

MEAT AND LIVESTOCK COMMISSION (1981). *Feeding the Ewe. Revised Edition.* Bletchley, Meat and Livestock Commission

MELLOR, D.J. and MATHESON, I.C. (1979). Daily changes in the curved crown-rump length of individual sheep fetuses during the last 60 days of pregnancy and effects of different levels of maternal nutrition. *Quarterly Journal of Experimental Physiology* **64**, 119–131

MELLOR, D.J. and MURRAY, L. (1981). Effects of placental weight and maternal nutrition on the growth rates of individual fetuses in single and twin bearing ewes during late pregnancy. *Research in Veterinary Science* **30**, 198–204

MELLOR, D.J. and MURRAY, L. (1982a). Effects of long term undernutrition of the ewe on the growth rates of individual fetuses during late pregnancy. *Research in Veterinary Science* **32**, 177–180

MELLOR, D.J. and MURRAY, L. (1982b). Effects on the rate of increase in fetal girth of refeeding ewes after short periods of severe undernutrition during late pregnancy. *Research in Veterinary Science* **32**, 377–382

MÉNÉZO, Y. and WINTENBERGER-TORRÈS, S. (1976). Free amino acid content of the ewe uterine fluid under various hormonal treatments during early pregnancy. *Annales de Biologie Animale, Biochimie, Biophysique* **16**, 537–543

MINISTRY OF AGRICULTURE, FISHERIES AND FOOD (1975). *Energy Allowances and Feeding System for Ruminants.* Technical Bulletin 33. London, Her Majesty's Stationery Office

NOAKES, D.E. and YOUNG, M. (1981). Measurement of fetal tissue protein synthetic rate in the lamb *in utero. Research in Veterinary Science* **31**, 336–341

ODDY, V.H., GOODEN, J.M., TELENI, E. and JONES, A.W. (1981). The effect of diet and stage of pregnancy on glucose uptake and extraction by the uterus and leg muscle of sheep. *Proceedings of the Nutrition Society of Australia* **6**, 107

ØRSKOV, E.R. and ROBINSON, J.J. (1981). The application of modern concepts of ruminant protein nutrition to sheep production systems. *Livestock Production Science* **8**, 339–350

ØRSKOV, E.R., MacLEOD, N.A. and GRUBB, D.A. (1980). New concepts of N metabolism in ruminants. In *Proceedings of the Third European Association for Animal Production* (H.J. Oslage and K. Rohr, Eds.), pp. 451–457. EAAP Publication No. 27

PARKINS, J.J., FRASER, J., RITCHIE, N.S. and HEMINGWAY, R.G. (1974). .Urea as a protein source for ewes with twin lambs in late pregnancy and early lactation. *Animal Production* **19**, 321–329

PARR, R.A., CUMMING, I.A. and CLARKE, I.J. (1982). Effects of maternal nutrition and plasma progesterone concentrations on survival and growth of the sheep embryo in early gestation. *Journal of Agricultural Science, Cambridge* **98**, 39–46

PIPER, L.R., BINDON, B.M., WILKINS, J.F., COX, R.J., CURTIS, Y.M. and CHEERS, M.A. (1980). The effect of selenium treatment on the fertility of Merino sheep. *Proceedings of the Australian Society of Animal Production* **13**, 241–244

PRIOR, R.L. and CHRISTENSON, R.K. (1976). Influence of dietary energy during gestation on lambing performance and glucose metabolism in Finn-cross ewes. *Journal of Animal Science* **43**, 1114–1124

RATTRAY, P.V., TRIGG, T.E. and URLICH, C.F. (1980). Energy exchanges in twin-pregnant ewes. In *Energy Metabolism* (L.E. Mount, Ed.), pp. 325–328. London, Butterworths

RATTRAY, P.V., GARRETT, W.N., EAST, N.E. and HINMAN, N. (1973). Net energy requirements of ewe lambs for maintenance, gain and pregnancy and net energy values of feedstuffs for lambs. *Journal of Animal Science* **37**, 853–857

RATTRAY, P.V., GARRETT, W.N., EAST, N.E. and HINMAN, N. (1974a). Efficiency of utilization of metabolizable energy during pregnancy and the energy requirements for pregnancy in sheep. *Journal of Animal Science* **38**, 383–393

RATTRAY, P.V., GARRETT, W.N., EAST, N.E. and HINMAN, N. (1974b). Growth, development and composition of the ovine conceptus and mammary gland during pregnancy. *Journal of Animal Science* **38**, 613–626

RHIND, S.M., ROBINSON, J.J. and McDONALD, I. (1980). Relationships among uterine and placental factors in prolific ewes and their relevance to variations in foetal weight. *Animal Production* **30**, 115–124

ROBINSON, J.J. (1977). The influence of maternal nutrition on ovine foetal growth. *Proceedings of the Nutrition Society* **36**, 9–16

ROBINSON, J.J. (1982). Pregnancy. In *Sheep and Goat Production*, Volume C1 (I.E. Coop, Ed.), pp. 103–118. Amsterdam, Elsevier

ROBINSON, J.J. and FORBES, T.J. (1968). The effect of protein intake during gestation on ewe and lamb performance. *Animal Production* **10**, 297–309

ROBINSON, J.J. and McDONALD, I. (1979). Ovine prenatal growth, its mathematical description and the effects of maternal nutrition. *Annales de Biologie Animale, Biochimie, Biophysique* **19**(1B), 225–234

ROBINSON, J.J., McDONALD, I., FRASER, C. and CROFTS, R.M.J. (1977). Studies on reproduction in prolific ewes. 1. Growth of the products of conception. *Journal of Agricultural Science, Cambridge* **88**, 539–552

ROBINSON, J.J., McDONALD, I., FRASER, C. and GORDON, J.G. (1980). Studies on reproduction in prolific ewes. 6. The efficiency of energy utilization for conceptus growth. *Journal of Agricultural Science, Cambridge* **94**, 331–338

ROBINSON, J.J., McDONALD, I., McHATTIE, I. and PENNIE, K. (1978). Studies on reproduction in prolific ewes. 4. Sequential changes in the maternal body during pregnancy. *Journal of Agricultural Science, Cambridge* **91**, 291–304

ROBINSON, W.I., BROWN, W. and LUCAS, I.A.M. (1973). Effects of shelter, exposure and level of feeding over winter on the productivity of Welsh Mountain ewes and lambs. *Animal Production* **17**, 21–32

RUSSEL, A.J.F., DONEY, J.M. and REID, R.L. (1967). The use of biochemical parameters in controlling nutritional state in pregnant ewes, and the effect of undernourishment during pregnancy on lamb birth weight. *Journal of Agricultural Science, Cambridge* **68**, 351–358

RUSSEL, A.J.F., FOOT, J.Z., WHITE, I.R. and DAVIES, G.J. (1981). The effect of weight at mating and of nutrition during mid-pregnany on the birth weight of lambs from primiparous ewes. *Journal of Agricultural Science, Cambridge* **97**, 723–729

RUSSEL, A.J.F., MAXWELL, T.J., SIBBALD, A.R. and McDONALD, D. (1977). Relationships between energy intake, nutritional states and lamb birth weight in Greyface ewes. *Journal of Agricultural Science, Cambridge* **89**, 667–673

RUTTER, W., LAIRD, T.R. and BROADBENT; P.J. (1971). The effects of clipping pregnant ewes at housing and of feeding different basal roughages. *Animal Production* **13**, 329–336

SHEEHAN, W., LAWLOR, M.J. and BATH, I.H. (1977). Energy requirements of the pregnant ewe. *Irish Journal of Agricultural Research* **16**, 233–242

SHEVAH, Y., BLACK, W.J.M. and LAND, R.B. (1975). Differences in feed intake and the performance of Finn × Dorset ewes during late pregnancy. *Animal Production* **20**, 391–400

SHUTT, D.A., WESTON, R.H. and HOGAN, J.P. (1970). Quantitative aspects of phyto-oestrogen metabolism in sheep fed on subterranean clover (*T. subterraneum cult. clare*) or red clover (*Trifolium pratense*). *Australian Journal of Agricultural Research* **21**, 713–722

SMITH, R.H. (1974). Kale poisoning. In *Annual Report of Studies in Animal Nutrition and Allied Sciences* **30**, 112–131. Rowett Research Institute

STORM, E. and ØRSKOV, E.R. (1982). Biological value and digestibility of rumen microbial protein in lamb small intestine. *Proceedings of the Nutrition Society* **41**, 78A

SYKES, A.R. and FIELD, A.C. (1972a). Effects of dietary deficiencies of energy, protein and calcium on the pregnant ewe. II. Body composition and mineral content of the lamb. *Journal of Agricultural Science, Cambridge* **78**, 119–125

SYKES, A.R. and FIELD, A.C. (1972b). Effects of dietary deficiencies of energy, protein and calcium on the pregnant ewe. III. Some observations on the use of biochemical parameters in controlling energy under nutrition during pregnancy and on the efficiency of utilization of energy and protein for foetal growth. *Journal of Agricultural Science, Cambridge* **78**, 127–133

SYKES, A.R. and FIELD, A.C. (1972c). Effects of dietary deficiencies of energy, protein and calcium on the pregnant ewe. I. Body composition and mineral content of the ewes. *Journal of Agricultural Science, Cambridge* **78**, 109–117

SYKES, A.R., NISBET, D.I. and FIELD, A.C. (1973). Effects of dietary deficiencies of energy, protein and calcium on the pregnant ewe. V. Chemical analyses and histological examination of some individual bones. *Journal of Agricultural Science, Cambridge* **81**, 433–440

THOMPSON, J.L., ROBINSON, J.J. and McHATTIE, I. (1978). An effect of physiological state on digestion in the ewe. *Proceedings of the Nutrition Society* **37**, 71A

TURNBULL, K.E., BRADEN, A.W.H. and GEORGE, J.M. (1966). Fertilization and early embryonic losses in ewes that had grazed oestrogenic pastures for six years. *Australian Journal of Agricultural Research* **17**, 907–917

VALDEZ ESPINOSA, R., ROBINSON, J.J. and SCOTT, D. (1977). The effect of different degrees of food restriction in late pregnancy on nitrogen metabolism in ewes. *Journal of Agricultural Science, Cambridge* **88**, 399–403

WALLACE, L.R. (1948). The growth of lambs before and after birth in relation to the level of nutrition. *Journal of Agricultural Science, Cambridge* **38**, 93–153; 367–401

WENHAM, G. (1981). A radiographic study of the changes in skeletal growth and development of the foetus caused by poor nutrition in the pregnant ewe. *British Veterinary Journal* **137**, 176–187

WESTON, R.H. (1979). Digestion during pregnancy and lactation in sheep. *Annales de Recherches Veterinaires* **10**, 442–444

WILLIAMS, H.Ll., HILL,R. and ALDERMAN, G. (1965). The effects of feeding kale to breeding ewes. *British Veterinary Journal* **121**, 2–17

WINTENBERGER-TORRÈS, S. (1976). Action de la progésterone et des stéroides ovariens sur la segmentation des oeufs chez la brebis. *Annales de Biologie Animale, Biochimie, Biophysique* **7**, 391–406

7

NUTRIENT REQUIREMENTS FOR LACTATION IN THE EWE

T.T. TREACHER
Grassland Research Institute, Hurley, Maidenhead, Berkshire, UK

In the majority of sheep production systems lactating ewes are used solely to rear their lambs, but in some countries, particularly in southern and eastern Europe and the Middle East, ewe's milk is an important direct source of animal protein in the human diet (Gall, 1975). In dairy systems, lactation is generally two or three months longer than in suckling systems, but, as milking is usually preceded by a period of suckling, there is no clear demarcation between suckling and dairy systems (Flamant and Casu, 1978). The lengths of the suckling and milking periods vary considerably from one month of suckling and five or six months milking in the traditional Mediterranean systems, to three months of suckling and about a month of milking in central and eastern Europe. In the eastern Mediterranean, in Cyprus and, particularly, Israel, a system of simultaneous suckling and milking is common with lambs being allowed to suck after each milking until they are weaned about two months after lambing. The main exception appears to be in the small population of East Friesland ewes in northern Germany where the lambs are removed very soon after birth and the ewes milked for six or seven months.

Very few studies of nutrition have been made on milking ewes, where research has concentrated on selection, breed improvements and milking characteristics, but there appears to be no reason why the results of nutritional studies on suckling ewes should not be applied to milking ewes, if suitable adjustments are made for the level of milk yield.

The importance of milk for lamb growth

In the majority of sheep production systems throughout the world lambs obtain nutrients from only two sources, their mother's milk and grazed herbage. Milk is essential in the first 3–4 weeks of the lamb's life and, in this period, correlations between milk intake and liveweight gain are approximately 0.9 (e.g. Wallace, 1948). Although the necessity for milk then ceases, it is still a source of highly digestible energy and of high quality protein which is utilized very efficiently. The level of milk intake strongly influences the pattern of herbage intake. The overall importance of milk intake is indicated by correlations between milk intake and liveweight of lambs of approximately 0.7 over four months of lactation (e.g. Wallace, 1948).

Many estimates have been made of the efficiency of conversion of milk to gain in liveweight of lambs (Boyazoglu, 1963). In the first six weeks of lactation, when intake of solid food is negligible, the mean value of 18 estimates of conversion efficiency was 0.167 kg liveweight gain/kg of fresh milk intake with individual values ranging from 0.238 to 0.138 kg of liveweight gain/kg of milk.

Part of this variation arises from differences in the composition of milk as a result of breed and nutrition of the ewe, stage of lactation and number of lambs suckled. The energy concentration of ewe's milk varies quite widely from 4.10–5.7 MJ/kg fresh milk with a mean value of 4.53 MJ/kg. As there is a tendency for there to be an inverse relationship between milk yield and energy concentration, expression of the gain in relation to intake of energy, or even dry matter, would be preferable. The difficulty of sampling accurately to determine the composition of milk actually consumed by the lambs generally makes it impossible to express efficiency in these ways, unless the ewes are milked and samples of the milk are analyzed prior to it being fed to the lambs (Joyce and Rattray, 1970; Penning and Gibb, 1979).

Experiments with milk substitutes have shown that, as the proportion of energy derived from protein (protein:energy ratio) falls below 0.30, the efficiency of conversion of the lamb decreases (Penning, Penning and Treacher, 1978). Lamb growth and the conversion of milk to gain in suckling ewes may therefore be affected by variation in protein:energy ratio of ewe's milk, which ranges from 0.25–0.32 with a mean of 0.28.

High milk intake increases growth rate of lambs and their efficiency of conversion as at low intakes a greater proportion of the intake is used for maintenance. Similarly, at a particular level of intake, large lambs and those with low growth potential have lower efficiencies of conversion (Bonsma, 1939; Robinson, Foster and Forbes, 1969). Conversion of milk to gain in body energy will always increase as intake of milk increases above maintenance, but there are indications that conversion to liveweight gain reaches a maximum and declines as composition of the gain changes. Boccard (1963) showed that a maximum conversion efficiency of 0.2 kg liveweight gain/kg milk occurred in lambs growing at 200 g/day and that greater gains in lambs older than 20 days resulted in slight reductions in efficiency of conversion.

The pattern and level of intake of solid food is strongly influenced by the milk consumption of the lambs. Lambs with low milk intakes start to consume appreciable amounts of solid food at an earlier age and then increase intake faster (Joyce and Rattray, 1970; Penning and Gibb, 1979). Lambs on *ad libitum* milk eat negligible amounts of solid food (Penning, Bradfield and Treacher, 1971). It seems possible that the decline in intake of milk that occurs after the peak of lactation is the stimulus for the lambs to start eating solid food. This decline in milk intake when expressed relative to liveweight of the lamb is large.

When herbage is available *ad libitum* substantial increases in intake by lambs of dried grass (Hodge, 1966a), frozen grass (Spedding, Brown and Large, 1963) and grazed herbage (Hodge, 1966b; Penning and Gibb, 1979) occur when milk intake is reduced. Although intakes relative to liveweight are higher in lambs receiving less milk, more rapid growth by lambs receiving more milk can result in them having higher absolute intakes by seven or eight weeks of age (Penning and Gibb, 1979). Even large differences in milk intake between single and twin lambs can result in only small differences in absolute intake in the second month of lactation (Gibb and Treacher, 1982).

In the situation where low milk intake results from poor milk yields of ewes caused by low herbage availability, lambs will have great difficulty in making compensatory increases in herbage intake. Williams, Geytenbeck and Allden (1976) showed that low herbage availability on short pastures caused decreases in the intake of lambs before milk production by the ewes was affected. In a comparison of ewes and lambs grazing at two herbage allowances Gibb, Treacher and Shanmugalingam (1981) found that a reduction in milk intake of 30% on the lower allowance caused a slight increase in intake of herbage relative to lamb liveweight initially, but the absolute intakes were higher by 10% at five weeks and 47% at 13 weeks of age, in lambs at the high allowance that were receiving more milk.

Penning and Gibb (1979) calculated, using efficiencies of utilization of metabolizable energy (ME) for growth and fattening (k_f) of 0.33 for herbage and 0.71 for milk, that intake of herbage has to increase by 4.7 g for each 1 g reduction in milk to maintain the same total net energy intake for gain. This did not occur in their experiment until 11 weeks of age and indicates the importance of high milk yield for maintaining high lamb growth rates in the second and third months of the lamb's life.

The low growth of lambs weaned on to grass swards confirms the importance of milk production in mid or late lactation in maintaining lamb growth rate (e.g. Langlands and Donald, 1975). Gibb, Treacher and Shanmugalingam (1981) found that lambs weaned before 10 weeks of age on to a high herbage allowance could not increase intake sufficiently to increase growth rate even when lamb growth prior to weaning had been reduced to 170 g/day, about 60% of potential growth rate, by low herbage availability, which also reduced milk yield by 30%.

The problem of measuring milk yield in suckling ewes

There are considerable problems in measuring yield responses in suckling ewes. Changes in lamb growth rate, while ultimately of practical importance, may not reflect the full change in milk yield of the ewe mainly because of differing proportions of the intake being used for maintenance, changes in intake of solid food by the lamb and differing levels of parasitism. The accuracy and limitations of the direct methods of measuring milk yield of suckling ewes by weighing lambs before and after suckling (test weighing method), by measuring secretion rate over short intervals with the aid of oxytocin (oxytocin method) or by body water dilution technique have been reviewed by Boyazoglu (1963) and Doney *et al.* (1979). Comparison of the more common test weighing and oxytocin methods suggests that the oxytocin method gives higher estimates

(Coombe, Wardrop and Tribe, 1960; Moore, 1962). However, a thorough comparison by Doney *et al.* (1979) found that the oxytocin method gave higher estimates of yield in the first week of lactation, especially in ewes suckling single lambs, but, by the third week, there were no significant differences between estimates made by the two methods, and they were not affected by level of milk production, number of lambs suckled or the genotype of the ewe.

NON-NUTRITIONAL FACTORS INFLUENCING YIELD

The response of the lactating ewe to nutrition can be greatly modified by physiological factors. The most important of these is the withdrawal stimulus (suckling or milking stimulus) but the ewe's parity and age at first lambing can also modify the potential milk yield.

Most studies of the yields of ewes suckling different numbers of lambs have shown that ewes rearing twin lambs have considerably higher yields than ewes under the same management but suckling singles. This effect can be greater than the response to nutrition (Wallace, 1948; Gardner and Hogue, 1964). The differences range from negligible ones in the studies by Louda and Doney (1976) and Burris and Baugus (1955) to increases of about 70% (Wallace, 1948; Kovnerev, 1974). However, the majority of the increases lie between 30–50% with an unweighted mean of 24 estimates in the literature of 41% (Treacher, 1978).

These differences in yield are accompanied by changes in the lactation curve and milk composition. Ewes with twin lambs normally reach their peak yield in the second or third week of lactation compared to the third to fifth week in ewes with singles (Wallace, 1948; Hunter, 1957; Ricordeau and Denamur, 1962; Gibb and Treacher, 1982). However, persistency is slightly lower in ewes with twins and, by 12 weeks of lactation, the differences in yield between ewes with one and with two lambs are generally negligible. An increase in the fat percentage of milk from ewes with twins has been found in experiments by Gardner and Hogue (1964), Moore (1966), Peart, Edwards and Donaldson (1972) where yield was measured using the oxytocin technique. Peart, Edwards and Donaldson (1972) also found a slight decrease in protein percentage in ewes with twins. Similar changes in composition have not been found when yield has been estimated by the lamb weighing technique, probably because of the difficulty of obtaining samples of milk for analysis that accurately represent the total yield.

There is very little information on yields of ewes suckling more than two lambs. Wallace (1948) and Guyer and Dyer (1954) found only slight increases in yield of ewes with triplets. However, more recent work on Romanov and Finnish Landrace crosses has found large increases in yield of ewes with triplets compared to ewes with singles. Flamant and Labussiere (1972) found an increase of 62%, Peart, Edwards and Donaldson (1972, 1975) 90% and Kovnerev (1974) 103%. Kovnerev (1974) obtained a 155% increase in yield of ewes with four lambs over those with singles but Nikolaev and Magomedev (1976) found no difference in yield between ewes suckling two, three or four lambs.

Although the level of yield appears to be mainly determined by the number of lambs actually suckled, and not the number of foetuses carried in pregnancy (Barnicoat, Logan and Grant, 1949), Alexander and Davies (1959) found a 12% but non-significant increase in yield of ewes that suckled one lamb after they had carried two foetuses in pregnancy. An increase in yield of 11% occurred in ewes that carried twin foetuses and were then milked from four days after lambing (Stern *et al.*, 1978). The importance of hormones produced by the placenta, notably oestrogens, in the development of the udder during pregnancy in the ewe suggests a mechanism by which the number or genotype of foetuses can affect udder development and hence yield potential (Delouis, 1981).

Flamant and Casu (1978) have discussed the problem of assessing genetic potential of breeds for milk production under suckling conditions and the need to separate the genetic potential of the ewe to produce milk, from the potential of the lambs to obtain milk. It is clear that similar considerations affect the design of studies to assess the response of ewes to nutrition. Slen, Clark and Hironaka (1963) considered that the quantity of milk obtained by twins estimated the mother's milk production potential whereas the amount obtained by the single lamb measures the lamb's voluntary intake of milk. This view is supported by the differences in yields found by Moore (1966), Langlands (1972) and Peart, Doney and MacDonald (1975) between ewes suckling single lambs of different genotypes, as a result of the use of different breeds of ram or cross-fostering breeds. Part of this effect may be mediated through differences in birthweight. For example, the differences between genotypes in the study of Moore (1966) were generally non-significant after using birthweight as a covariate. However, in the experiments of Langlands (1972) and Peart, Doney and MacDonald (1975) the differences in birthweights between genotypes were not significant and the yield differences appear to result from differences in appetite, but as the genotype with the greater appetite grew more rapidly, weight of the lambs may have had an effect.

These results all suggest that, in most breeds, ewes suckling single lambs will be lactating at considerably below their potential. Therefore, the use of ewes with singles in nutritional studies will considerably limit the range of response. The effects of lamb appetite on milk withdrawal potential can be minimized by using ewes suckling twin lambs. The further increases in yield in ewes rearing triplets or quads suggest that, at least in theory, the range of response could be extended by using ewes with multiples. The practical difficulties of managing ewes with multiples are great with a high risk of udder damage and mastitis as the lambs reach 4–6 weeks of age, particularly if the animals are in individual pens.

Attempts have been made to eliminate the direct effects of the lambs on milk yield by using machine milking, but in most breeds the withdrawal stimulus of the milking machine appears to limit milk yield considerably. Treacher (1970a) found that non-dairy breeds machine milked twice daily from the second day of lactation gave yields similar to those of ewes suckling single lambs. In dairy flocks, the change from suckling to milking 4–6 weeks after lambing is accompanied by a rapid reduction in yield of 30–40% (Labussiere, 1981). Clearly the removal of the lamb and the starting of milking after the maternal behaviour pattern has been fully established is very stressful, but Labussiere and Petrequin (1969) found that reducing the suckling period to a few days only improves adaptation to milking to a limited extent. Morag, Raz and Eyal (1970) suggest that there is a critical period of attachment between ewe and lamb of about six hours after which the removal of the lamb causes stress. Thus selection for high yield under milking conditions cannot be made efficiently when a period of suckling precedes milking. The East Friesland breed, which has an exceptional yield potential when milked, has been selected under a management in which the lambs are removed within a few hours of birth.

EFFECTS OF PARITY AND AGE AT FIRST LAMBING

Lactation yields in successive lactations are influenced by both parity and age at first lambing. In ewes lambing for the first time at two years of age, yields increase from the first to third lactations by 5–40%, then remain relatively constant until the sixth lactation, after which a decline occurs except under very good management and feeding (Mason and Dassat, 1954). This pattern is found in studies of both groups of ewes in different lactations recorded in one year, and therefore possibly affected by management in previous years (Barnicoat, Logan and Grant, 1949; Owen, 1957), and in studies following the same ewes in successive lactations, with year to year variations in environment and some influence of culling of low yielders (Barnicoat *et al.*, 1957; Bonsma, 1939). It is possible that unimproved breeds reach their peak lactation yield later, in the fourth rather than the third lactation, as slow growth increases the time taken to reach the ewe's mature weight (Starke, 1953). Information on yields of ewe lambs only exists for dairy breeds (e.g. Mason and Dassat, 1954). Yield in the first lactation is lower, partly as a result of a shorter lactation, than in ewes that lamb for the first time at two years of age, but the subsequent lactation yields are similar.

THE IMPORTANCE OF THE LATE PREGNANCY PERIOD

Measurements of DNA content indicate that almost all the development of secretory tissue in the ewe's udder occurs in the last third of pregnancy. At parturition, DNA levels are 95% of the level found at day 30 of lactation. A slight synthesis of lactose and casein occurs from about day 90 of pregnancy but the onset of lactation at parturition is accompanied by cellular hypertrophy and a rapid increase in RNA levels (Delouis *et al.*, 1980).

Oestrogens secreted by the placenta play an important role in the complex of hormones that control development of secretory tissue in the udder during late pregnancy. Although the plasma concentration of oestrogen is correlated with size of the placenta (Delouis, 1981), no studies have been made of the quantities of secretory tissue in ewes' udders following undernutrition, that has resulted in reduced foetal weight and placenta size. In most investigations of the effects of nutrition in pregnancy on the subsequent lactation, no attempt has been made to separate the direct effects of pregnancy nutrition on potential yield from the effects on withdrawal stimulus (suckling stimulus) resulting from differences in weight and vigour of lambs. Thomson and Thomson (1953) attempted to do this by cross-fostering lambs from different pregnancy treatments and Treacher (1970b) used machine milking.

Severe undernutrition in late pregnancy results in small udders with little or no secretion being present prior to lambing and a delay of several hours in the onset of copious lactation after lambing (Thomson and Thomson, 1953; McCance and Alexander, 1959). Reductions in milk production over the whole lactation have occurred in a few instances where ewes carrying twin foetuses have lost (Wallace, 1948) or only just maintained body weight (Guyer and Dyer, 1954; Butterworth and Blore, 1969; Treacher, 1970b) in the last 6–8 weeks of pregnancy. These patterns of weight change resulted in major mobilization of body reserves of the ewes and were associated with reductions in lamb birthweights of 17–32% and subsequent reductions in milk yield under suckling of 7–35%. The reduction in yield was greater (55%) in the study by Treacher (1970b) using machine milking. Wallace (1948), Thomson and Thomson (1953) and Butterworth and Blore (1969) found similar reductions in milk yield or lamb growth from underfeeding ewes with single foetuses in late pregnancy.

In experiments by Peart (1967) and Maxwell *et al.* (1979) no reduction in milk production followed severe underfeeding of ewes in late pregnancy, although similar reductions in lamb birthweight occurred. Lamb growth rate was, however, significantly lower in the study of Peart (1967). The biochemical parameters of blood non-esterified fatty acids or ketone levels were used to adjust undernourishment of individual ewes precisely in relation to the total weight of foetus carried. In each case the severely undernourished ewes were given slight increases in intake and made slight gains (6–8%) in body weight in the last weeks of pregnancy. This contrasts with more severe undernourishment in the experiments of Wallace (1948), Thomson and Thomson (1953) and Treacher (1970b) in which the low feeding levels were then reduced still further in successive weeks in late pregnancy in order to just maintain or reduce the weight of the pregnant ewe.

If direct reductions in milk production only occur in occasional extreme situations of underfeeding in late pregnancy, the body condition of the ewe at the start of lactation may influence its subsequent performance. Increases in body fatness can be accompanied by declines in feed intake in cows (Bines, Suzuki and Balch, 1969) and in wether sheep (Allden and Scott Young, 1964; Langlands, 1968; Arnold and Birrell, 1977). A similar response occurred in lactating ewes (Cowan *et al.*, 1980); intake was 10% lower in ewes with 33% fat compared to those with 17% fat in the empty body.

The reanalysis by Cowan, Robinson and McDonald (1982) of the relationship between body fat content after lambing, energy intakes and rate of loss of body fat during early lactation leads to the conclusion that the rate of fat loss in fat ewes is very responsive to changes in ME intake but restriction in energy intake in thin ewes has much less effect on the rate of fat loss. Therefore a high level of body fatness at parturition may lead to a considerable increase in milk yield if intake in lactation is low but will have less effect if ME intake is high, partly because of reductions in voluntary intake in fat ewes.

Studies of lactating ewes at pasture have not found significant differences in intake between ewes differing in body condition at lambing over the range 1.9–3.2 (Gibb and Treacher, 1980; 1982) or the range 1.6–2.2 (Maxwell *et al.*, 1979). Although these studies showed no effect of body condition on intake, they indicated some effect of body condition on performance in lactation. Growth rate was signiicantly higher in lambs sucking ewes in condition score 3.2 and these ewes lost more weight (18.4 v. 15.2 kg) during lactation (Gibb and Treacher, 1980). The milk production of ewes with twins in better body condition was consistently slightly but not significantly higher in the study of Maxwell *et al.* (1979).

Nutrition during lactation

A large number of experiments have shown the importance of nutrition during lactation. The variations in response of milk yield to change in intake of energy and protein have been very wide, partly because of the accompanying changes in the liveweight, which are often very much larger relative to body weight than in dairy cows (Cowan *et al.*, 1980). A sound theoretical framework for examining the response of the lactating ewe to current nutrient inputs has been provided only in the last five years by Robinson and his colleagues, drawing on the approach proposed by Ørskov (1977), Roy *et al.* (1977) and eventually published in *Nutrient Requirements of Ruminant Livestock* (ARC, 1980).

Robinson (1978) described a model (*Figure 7.1*) derived from data of Robinson and Forbes (1970), which demonstrates three important principles of the lactating ewe's response to variation in intake of both metabolizable energy and protein:

1. At a particular level of energy intake there is a minimum protein intake,

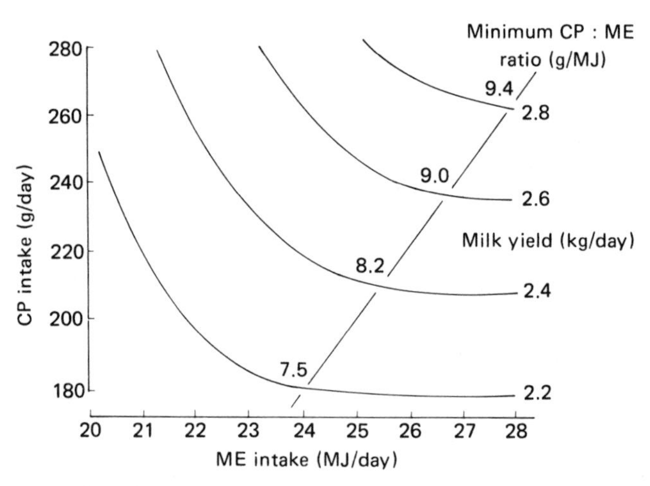

Figure 7.1 Response of milk yield to change in intake of metabolizable energy (ME) and crude protein (CP). From Robinson (1978)

and reduction in protein intake below this level will cause a reduction in milk yield.

2. This minimum ratio of crude protein (CP) to ME increases with increasing level of milk yield.
3. An increase in dietary CP concentration without a change in ME intake will increase milk production if the ewe has not reached her potential yield.

The validity of this theoretical model is confirmed by the results of a number of experiments in which crude protein intake was altered to ewes receiving energy intakes below their requirements. Robinson *et al.* (1974) demonstrated a curvilinear increase in milk yield from 2.4 to 3.1 kg/day as CP:ME ratio increased from 10.5 to 16.6 g/MJ ME when ME intake was 25 MJ/day. At the same time weight loss by the ewe increased from 120–260 g/day. The response of milk yield to change in protein intake is very rapid. Robinson *et al.* (1979) found responses were established in three days from changes in protein:energy ratio in the first three weeks of lactation but did not occur after 30 days of lactation when potential yield had declined.

Although there is general acceptance of the principles for calculating energy and protein requirements proposed in ARC (1980), the particular values to be used in the calculations are still being discussed (Robinson, 1978; 1980; Maxwell *et al.*, 1979) and this will continue as new estimates are made.

Individual estimates of the efficiency of use of ME above maintenance for milk production (k_1) in the ewe vary from 0.47 on ground and pelleted dried grass diets (Peart, 1968; Peart, Edwards and Donaldson, 1972) to 0.9 for diets with a high proportion of concentrates (Gardner and Hogue, 1966). Robinson (1978) derived a mean value of 0.63±0.03 by adjusting data from nine experiments for changes in body weight assuming a value of 25.5 MJ ME/kg loss in weight and an efficiency of utilization of body reserves for milk production of 0.84. This lies within the range of values of k_1 for both the dairy cow and the ewe in ARC (1980) which vary from 0.56 to 0.68 as the concentration of ME in the gross energy (GE) of the diet (q) varies from 0.4 to 0.75. A value of 0.63 is predicted at a q of 0.6.

The maintenance requirements for ME in lactating ewes derived from ARC (1980) assuming an efficiency of conversion of 0.63 are 0.34 MJ/kg $W^{0.75}$ indoors and 0.39 MJ/kg $W^{0.75}$ out of doors. Much higher estimates of 218 kJ/kg liveweight, equivalent to 0.56 MJ/kg $W^{0.75}$ at the mean weight of the ewes (Langlands, 1977) and 242 kJ/kg, equivalent to 0.70 MJ/kg $W^{0.75}$ (Maxwell *et al.*, 1979), have been found in grazing lactating ewes. These values lie within the range of 212–244 kJ/kg liveweight for maintenance requirements of grazing dry sheep (Young and Corbett, 1972; Langlands and Bennett, 1973). In each case the maintenance requirement obtained is very dependent on the accuracy of estimation of changes in body energy; very considerable errors can arise from inaccuracies in measurements of liveweight changes and the use of standard values for the energy content of liveweight change.

As large liveweight losses are very common in the lactating ewe in early lactation, the problem of estimating the contribution of body reserves to milk production is of particular importance. ARC (1980) suggest that weight loss has an energy value of 26 MJ ME/kg change in empty body weight and that this energy is used for the production of milk energy with an efficiency of 0.84. The results of Cowan *et al.* (1979; 1980; 1981) indicate that large errors will arise in many cases from the use of these values. In three slaughter experiments the energy content of liveweight loss ranged from 24 to 90 MJ ME/kg, as losses in body fat were often accompanied by increases in gut and body water. Estimates of the efficiency of use of energy from body tissue for synthesis of milk energy also varied widely from 0.5 to 0.8.

As very little protein is available from the ewe's body to supply amino acids for milk secretion, the efficiency of utilization of body tissue for milk production appears to be closely linked to the level of intake of dietary protein (Cowan *et al.*, 1980). Cowan *et al.* (1979; 1980) found protein losses were negligible in ewes losing 4–8 kg liveweight in the first six weeks of lactation. In a later report Cowan *et al.* (1981) suggest, from comparisons of the milk protein output and efficiency of utilization of energy for milk production (k_1) in successive weeks of lactation on high and low protein diets, that labile body protein can contribute to maintaining milk production in the first weeks of lactation. The ewes in this experiment lost an average of 4.3 kg of liveweight and 800 g of protein between days 6 and 42 of lactation. This was approximately 10% of total body protein and sufficient for the production of about 10 kg of milk over the period.

The contribution of labile body protein to the protein requirements for milk production in high yielding ewes with twin lambs is small and calculations using the values proposed by Ørskov (1977) and Roy *et al.* (1977) for quantities of microbial and undegraded protein arriving at the abomasum and their subsequent utilization for synthesis of milk protein, give an adequate indication of the milk production of ewes that are not receiving their full requirements for energy (Robinson, 1980).

These calculations suggest that diets with approximately 11 g protein/MJ of ME—the minimum needed to meet the nitrogen requirements of the rumen bacteria—will sustain milk yields of up to approximately 2 kg/day if the intake is sufficient to supply the ewe's energy requirements. If the yield is higher, or the ewe is not eating sufficient food to meet its energy requirements, a higher concentration of protein in the diet will be needed. The exact amounts of extra protein will depend on the extent to which the particular protein escapes degradation in the rumen.

The investigation by Gonzalez *et al.* (1982) of the response of milk production to level and source of dietary protein, in general, showed greater yields of milk and increases in protein content of the milk on protein sources, such as fish meal and blood meal, which are not readily degraded in the rumen. The exact extent of degradation in the rumen is affected by the rate of passage of the diet through the rumen (Ørskov, Hughes-Jones and McDonald, 1980). This is greatly modified by the level of intake. Thus inherently highly degradable protein sources may avoid degradation at a high level of intake and supplements generally not degraded may be largely degraded if intake is low and they remain in the rumen for a long time. The allowances in MLC (1981) have been adjusted for these levels of intake effects.

MINERAL REQUIREMENTS

The requirements by lactating ewes for most major and trace minerals are given in ARC (1980). The most common mineral disorder in lactating ewes is hypomagnesaemia. Although the overall incidence of hypomagnesaemia is low it can be a major problem in individual flocks. It occurs mainly in the first six weeks of lactation, particularly in flocks on heavily fertilized grass swards. Prevention of hypomagnesaemia poses considerable problems (Kelly, 1979) as low body magnesium levels may arise from insufficient magnesium in the diet or because its absorption is reduced by other factors, notably potassium in spring grass. So potash fertilizers should not be applied in the spring to sheep pastures. An adequate intake of magnesium in early lactation can be achieved by including 6 g/ewe of magnesium oxide (calcined magnesite) in the daily concentrate ration. Where concentrates are not fed, free access to high magnesium mineral mixtures offers some protection but some ewes will not take these minerals.

THE LACTATING EWE AT PASTURE

The amounts of energy and protein needed by the ewe in the first months of lactation are considerably greater than the amounts needed in late pregnancy to achieve maximal birthweight in twin lambs (*Table 7.1*). Allowances from MLC (1981) indicate that a 70 kg ewe producing 3 kg/day of milk and maintaining weight, requires 75% more metabolizable energy and 55% more protein than in the last weeks of pregnancy. It is usual for the lactating ewe to utilize a considerable amount of body reserves, but even so, the requirement for energy is still much greater than in pregnancy. The MLC allowances suggest that, if a low degradable protein source is not used, a sharp rise in the protein:energy ratio of the diet from 12–19.5 g CP/MJ of ME is necessary, if production of milk is to be maintained in ewes that are losing weight. The allowances in ARC (1980) are broadly similar to the MLC allowances with a slightly higher requirement for energy and a slightly lower one for protein, for the ewe maintaining weight. It differs in proposing a slight fall in protein requirement when the ewe is utilizing reserves, because it is assumed that protein will become available for the synthesis of milk.

The emphasis in recent work on protein intake and the amounts and sources of protein that avoid degradation in the rumen has led to changes in the formulation of diets for the lactating ewe. Protein contents are being increased and protein sources with low rumen degradability such as fish meal, are being used in diets for early lambing flocks and for ewes at pasture, prior to the spring flush of grass growth. This trend has been particularly strong in Scotland but there is a need for some caution and more quantification of the exact extent of the response of the lactating ewe. For example, Vipond (1979) found an increase in lamb growth when a fish meal supplement was fed to Finnish Landrace cross ewes, but not when it was fed to Scottish Halfbred ewes.

As in the majority of systems, lactating ewes are at pasture with little or no availability of other nutrients, except possibly for minerals, it is very

Table 7.1 DAILY ALLOWANCES FOR ENERGY AND PROTEIN FOR 75 kg EWES WITH TWIN LAMBS

		From MLC (1981)[a]			*From ARC (1980)*[b]	
		ME (MJ)	*Protein* (g)		*ME* (MJ)	*Protein* (g)
			High degradability	*Low degradability*		
Last 2 weeks of pregnancy		18.6	255	215	19.6	160
First month of lactation (milk yield of 3 kg/day)						
Maintaining weight		31.5	380		32.0	360 (RDP 250)[c] (UDP 110)
Loss in condition score in 6 weeks	½	27.3	415	375	28.2[d]	335 (RDP 220) (UDP 115)
	1	23.4	455	390		

[a]For diet with ME concentration of 10 MJ/kg dry matter;
[b]For diet with ME:GE ratio (q) of 0.6
[c]RDP = Rumen-degradable protein; UDP = Undegraded protein
[d]Approximately equivalent to losing 0.5 condition score in six weeks

important to define the extent to which the lactating ewe can obtain its high requirements for nutrients from pasture. Unfortunately, the understanding of the interrelationships between intake, pasture characteristics and management decision is very poor. Experiments have produced isolated pieces of information and have not been sufficient in number to adequately quantify responses, which can be applied more generally.

Few studies give a clear idea of the patterns of intake of herbage by lactating ewes when they are not restricted by low herbage availability or quality. In housed ewes voluntary intake of high quality diets rises rapidly in the first and second weeks of lactation and then more slowly to a peak 6–8 weeks after lambing (Foot and Tissier, 1978). A similar pattern of intake with a peak in the fifth and sixth weeks was found by Maxwell *et al.* (1979) in ewes with twins gaining weight on very high quality pasture (*Figure 7.2*). Even in this experiment it is probable that low herbage availability restricted intake in the first and second weeks. When high herbage allowances of 100–116 g organic matter (OM)/kg liveweight/day were provided from the start of lactation, intakes were high initially and did not increase after the second or third week of lactation (Gibb and Treacher, 1978; Gibb, Treacher and Shanmugalingam, 1981).

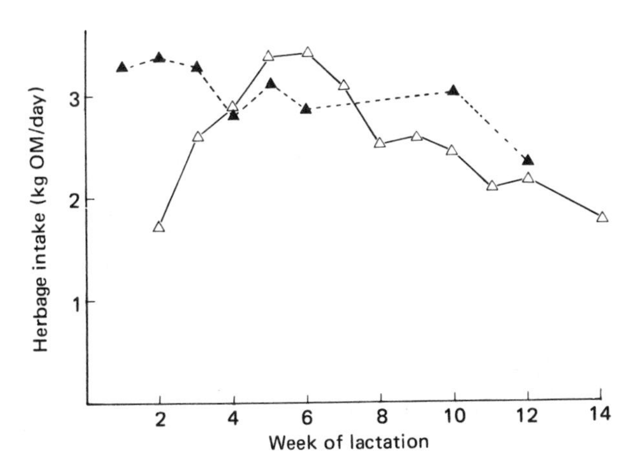

Figure 7.2 Patterns of intake of herbage organic matter (kg OM/day) by lactating ewes. △——△ Maxwell (1979); ewe weight 58 kg. ▲----▲ Gibb and Treacher (1978); ewe weight 75 kg

In spring there is usually a steady increase in herbage allowance as herbage growth increases; this gives a steady rise in intake by the ewe during the first few weeks of lactation. Later intake may be affected by reductions in herbage digestibility which may occur gradually or suddenly when grazing of regrowths is started. Both these points are illustrated in *Figure 7.3* taken from Gibb and Treacher (1980). Intake increased steadily from 1.69–2.95 kg OM/day in week 7 as daily herbage allowance increased from 27–56 g/kg liveweight and then fell to 1.55 kg in week 9 when herbage availability and OM digestibility both fell. The ewes lost weight rapidly (mean loss of 417 g/day in the first month) but maintained milk production at approximately 3 kg/day, not far below the potential for the breed. The

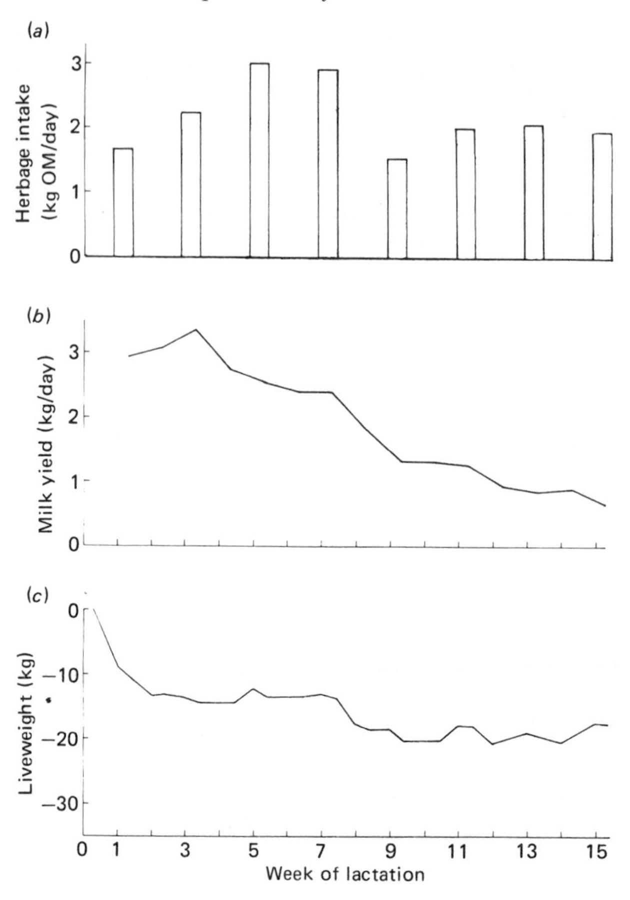

Figure 7.3 Patterns of (a) herbage intake (kg OM/day), (b) milk yield (kg/day) and (c) liveweight (kg) for ewes suckling twin lambs. From Gibb and Treacher (1980)

high lamb growth rate (270 g/day) corresponds reasonably with the estimated milk production.

This situation, in which ewes lose large amounts of weight but achieve high milk yields, is not uncommon and suggests that the protein requirements of the ewe are being supplied reasonably adequately. Unfortunately this aspect has not been considered in most papers on herbage intake by lactating ewes, and as they have not reported protein contents of the herbage eaten, calculations of protein intake etc. must be very speculative.

If it is assumed that a ewe losing weight rapidly must have an intake of crude protein of 400 g/day to sustain a milk production of 3 kg/day, then protein to energy ratios in the herbage of 22 and 16 g CP/MJ of ME will be necessary to provide the protein requirements at the levels of intake in weeks 1 and 3 in the results reported by Gibb and Treacher (1980). This would require protein contents of approximately 245 and 180 g/kg of herbage dry matter. Data for cut herbage (e.g. Corrall, Lavender and Terry, 1979) indicate that these levels are quite possible in April, but direct information as to the protein levels in herbage selected by sheep grazing in early spring is needed. Later in the season unpublished data from Hurley indicates that crude protein levels in grasses can fall to 80–100 g/kg DM even on swards receiving regular applications of nitrogenous fertilizer. Thus milk production may occasionally be limited by low protein content of grass rather than by low intake. In the study of Gibb and Treacher (1978), milk yields of ewes on the high allowances of herbage were well below the potential for the breed; this may have been due to the low protein content of herbage which declined from 131 to 94 g/kg DM between weeks 1 and 3 in samples collected with oesophageal fistulated sheep.

Thus, in most cases it appears that grass in early spring will contain sufficient protein to supply the total protein requirements of the lactating ewe, even when intake of herbage is restricted. But, as the rumen degradability of protein in fresh grass is high, 0.8–0.9 (Hvelpund and Moller, 1982; and by recalculation from Beever *et al.*, 1978) production of milk may be reduced by the low amounts of undegraded dietary protein (UDP) available from grazed herbage. A response to supplements differing in protein content and degradability was found by Penning and Treacher (1981) in ewes offered *ad libitum* freshly cut ryegrass with 162 g CP/kg DM and protein:energy ratio of 14.1 g CP/MJ of ME. Supplements of barley alone, or barley together with either soya bean meal, soya bean and fish meal, or fish meal were offered to give 10 MJ of ME daily with 40 g of crude protein for the barley supplement and 190 g for the three protein supplements. There was no response to the barley alone but milk yield was increased by 12, 24 and 25% above the unsupplemented control by the soya, soya with fish meal, and fish meal supplements respectively. Similar responses occurred when the barley and fish meal supplements were fed at the same levels to ewes grazing at two herbage allowances (Orr *et al.*, 1981). Lamb growth rates in the first six weeks were increased by 1% and 12% by the barley and fish meal supplements in comparison to the unsupplemented treatments at the low herbage allowance and 11 and 19% at the high allowance.

In view of the small responses to cereal supplements except at very high stocking rates (Young, Newton and Orr, 1980) further studies of the response by lactating ewes at pasture to high protein supplements may be valuable for application in systems where the aim is to achieve both high lamb growth rates and high stocking rates.

References

ALEXANDER, G. and DAVIES, H.L. (1959). The relationship of milk production to the number of lambs born or suckled. *Australian Journal of Agricultural Research* **10**, 720–724

ALLDEN, W.G. and SCOTT YOUNG, R. (1964). The summer nutrition of weaner sheep: herbage intake following period of different nutrition. *Australian Journal of Agricultural Research* **15**, 989–1000

ARC (1980). *The Nutrient Requirements of Ruminant Livestock*. Slough, Commonwealth Agricultural Bureaux

ARNOLD, G.W. and BIRRELL, M.A. (1977). Food intake and grazing behaviour of sheep varying in body condition. *Animal Production* **24**, 343–353

BARNICOAT, C.R., LOGAN, A.G. and GRANT, A.I. (1949). Milk secretion studies with New Zealand Romney sheep. Part II. Milk yields of ewes and factors affecting them. *Journal of Agricultural Science, Cambridge* **39**, 47–55

BARNICOAT, C.R., MURRAY, P.F., ROBERTS, E.M. and WILSON, G.S. (1957). Milk secretion studies with New Zealand Romney ewes. Parts V and VI. Experimental yield and composition of ewe's milk in relation to growth of the lamb. *Journal of Agricultural Science, Cambridge* **48**, 9–18

BEEVER, D.E., TERRY, R.A., CAMMELL, S.B. and WALLACE, A.S. (1978). The digestion of spring and autumn harvested perennial ryegrass by sheep. *Journal of Agricultural Science, Cambridge* **90**, 463–470

BINES, J.A., SUZUKI, S. and BALCH, C.C. (1969). The quantitative significance of long term regulation of food intake in the cow. *British Journal of Nutrition* **23**, 695–704

BOCCARD, R. (1963). Etude de la production de la viande chez les ovins. VII. Note sur les relations entre l'indice de consummation et la croissance de l'agneau. *Annales de Zootechnie* **12**, 227–230

BONSMA, F.W. (1939). Factors influencing the growth and development of lambs, with special reference to cross-breeding of Merino sheep for fat-lamb production in South Africa. II. Milk production of sheep and its effect on the growth of lambs. Faculty of Agriculture Publication No. 48, Pretoria

BOYAZOGLU, J.G. (1963). Aspects quantitatif de la production laitiere des brebis. *Annales de Zootechnie* **12**, 237–296

BURRIS, M.J. and BAUGUS, C.A. (1955). Milk consumption and growth of suckling lambs. *Journal of Animal Science* **14**, 186–191

BUTTERWORTH, M.H. and BLORE, T.W.D. (1969). The lactation of Persian Blackhead ewes and the growth of lambs. *Journal of Agricultural Science, Cambridge* **73**, 133–137

COOMBE, J.B., WARDROP, I.D. and TRIBE, D.E. (1960). A study of milk production of the grazing ewe, with emphasis on the experimental technique employed. *Journal of Agricultural Science, Cambridge* **54**, 353–359

CORRALL, A.J., LAVENDER, R.H. and TERRY, C.P. (1979). Grass species and varieties—seasonal patterns of production and relationship between yield, quality and date of first harvest. Grassland Research Institute Technical Report No. 26

COWAN, R.T., ROBINSON, J.J. and McDONALD, I. (1982). A note on the effects of body fatness and level of food intake on the rate of fat loss in lactating ewes. *Animal Production* **34**, 355–357

COWAN, R.T., ROBINSON, J.J., GREENHALGH, J.F.D. and McHATTIE, I. (1979). Body composition changes in lactating ewes estimated by serial slaughter and deuterium dilution. *Animal Production* **29**, 81–90

COWAN, R.T., ROBINSON, J.J., McDONALD, I. and SMART, R. (1980). Effects of body fatness at lambing and diet in lactation on body tissue loss, feed intake and milk yield of ewes in early lactation. *Journal of Agricultural Science, Cambridge* **95**, 497–514

COWAN, R.T., ROBINSON, J.J., McHATTIE, I. and PENNIE, K. (1981). Effects of protein concentration in the diet on milk yield, change in body composition and the efficiency of utilization of body tissue for milk production in ewes. *Animal Production* **33**, 111–120

DELOUIS, C. (1981). Les parametres physiologiques de la formation et du functionement de la mamelle. *6e Journée de la Recherche Ovine et Caprine*, pp. 5–34

DELOUIS, C., DJIANE, J., HONDEBINE, L.M. and TERQUI, M. (1980). Relation between hormones and mammary gland function. *Journal of Dairy Science* **63**, 1492–1513

DONEY, J.M., PEART, J.N., SMITH, W.F. and LOUDA, F. (1979). A consideration of the techniques for estimation of milk yield by suckled sheep and a comparison of estimates obtained by two methods in relation to the effect of breed, level of production and stage of lactation. *Journal of Agricultural Science, Cambridge* **92**, 123–132

FLAMANT, J-C. and CASU, S. (1978). Breed differences in milk production potential and genetic improvement of milk production. In *Milk Production in the Ewe* (J.G. Boyazoglu and T.T. Treacher, Eds.), pp. 1–20. EAAP Publication 23

FLAMANT, J-C. and LABUSSIERE, J. (1972). Premieres observations sur les aptitudes des brebis de race Romanov. *Annales de Zootechnie* **21**, 375–384

FOOT, J.Z. and TISSIER, M. (1978). Voluntary intake of feed by lactating ewes. In *Milk Production in the Ewe* (J.G. Boyazoglu and T.T. Treacher, Eds.), pp. 66–72. EAAP Publication No. 23

GALL, C. (1975). Milk production from sheep and goats. *World Animal Review*, No. 13, pp. 1–8

GARDNER, R.W. and HOGUE, D.E. (1964). Effects of energy intake and number of lambs suckled on milk yield, milk composition and energetic efficiency of lactating ewes. *Journal of Animal Science* **23**, 935–942

GARDNER, R.W. and HOGUE, D.E. (1966). Milk production, milk composition and energetic efficiency of Hampshire and Corriedale ewes fed to maintain body weight. *Journal of Animal Science* **25**, 789–795

GIBB, M.J. and TREACHER, T.T. (1978). The effect of herbage allowance on herbage intake and performance of ewes and their twin lambs grazing perennial ryegrass. *Journal of Agricultural Science, Cambridge* **90**, 139–147

GIBB, M.J. and TREACHER, T.T. (1980). The effect of ewe body condition at lambing on the performance of ewes and their lambs at pasture. *Journal of Agricultural Science, Cambridge* **95**, 631–640

GIBB, M.J. and TREACHER, T.T. (1982). The effect of body condition and nutrition during late pregnancy in the performance of grazing ewes during lactation. *Animal Production* **34**, 123–129

GIBB, M.J., TREACHER, T.T. and SHANMUGALINGAM, S. (1981). Herbage intake and performance of grazing ewes and of their lambs when weaned at 6, 8, 10 or 14 weeks of age. *Animal Production* **33**, 223–232

GONZALEZ, J.S., ROBINSON, J.J., McHATTIE, I. and FRASER, C. (1982). The effect in ewes of source and level of dietary protein on milk yield and the relationship between the intestinal supply of non-ammonia nitrogen and the production of milk protein. *Animal Production* **34**, 31–40

GUYER, P.Q. and DYER, A.J. (1954). Study of factors affecting sheep production. *Missouri Agricultural Experimental Station Research Bulletin*, No. 558

HODGE, R.W. (1966a). The apparent digestibility of ewe's milk and dried pasture by young lambs. *Australian Journal of Experimental Agriculture and Animal Husbandry* **6**, 139–144

HODGE, R.W. (1966b). The relative pasture intake of grazing lambs at two levels of milk intake. *Australian Journal of Experimental Agriculture and Animal Husbandry* **6**, 314–316

HUNTER, G.L. (1957). The maternal influence on size in sheep. *Journal of Agricultural Science, Cambridge* **48**, 36–60

HVELPUND, T. and MOLLER, P.D. (1982). Nitrogen metabolism in the rumen of cows fed rations of sugar beet, ammonia treated straw, grass or grass silage. In *Proceedings of the Third Symposium on Protein Metabolism and Nutrition*, pp. 414–419. EAAP Publication No. 27

JOYCE, J.P. and RATTRAY, P.V. (1970). The intake and utilization of milk and grass by lambs. *Proceedings of the New Zealand Society for Animal Production* **30**, 94–105

KELLY, P. (1979). Hypomagnesaemia in sheep: a review. *ADAS Quarterly Review* No. 34, 151–166

KOVNEREV, I.P. (1974). Biological reserves of Romanov sheep. *Ovtsevodstro* **11**, 29–30 (*Animal Breeding Abstracts* **43**, 2944)

LABUSSIERE, J. (1981). Aspects physiologiques et anatomiques de l'aptitude à la traite des brebis. *6ᵉ Journeés de la Recherche Ovine et Caprine*, pp. 74–90

LABUSSIERE, J. and PETREQUIN, P. (1969). Relations entre l'aptitude à la traite des brebis et la perte de production laitiere constatee au moment du sevrage. *Annales de Zootechnie* **18**, 5–15

LANGLANDS, J.P. (1968). The feed intake of grazing sheep differing in age, breed, previous nutrition and live-weight. *Journal of Agricultural Science, Cambridge* **71**, 167–172

LANGLANDS, J.P. (1972). Growth and herbage consumption of grazing Merino and Border Leicester lambs reared by their mothers or fostered by ewes of the other breed. *Animal Production* **14**, 317–322

LANGLANDS, J.P. (1973). Milk and herbage intakes by grazing lambs born to Merino ewes and sired by Merino, Border Leicester, Corriedale, Dorset Horn and Southdown rams. *Animal Production* **16**, 285–291

LANGLANDS, J.P. (1977). The intake and production of lactating Merino ewes and their lambs grazed at different stocking rates. *Australian Journal of Agricultural Research* **28**, 133–142

LANGLANDS, J.P. and BENNETT, I.L. (1973). Stocking intensity and pastoral production. II. Herbage intake of Merino sheep grazed at different stocking rates. *Journal of Agricultural Science, Cambridge* **81**, 205–209

LANGLANDS, J.P. and DONALD, G.E. (1975). The intakes and growth rates of grazing Border Leicester × Merino lambs weaned at 21, 49 and 77 days. *Animal Production* **21**, 175–181

LOUDA, F. and DONEY, J.M. (1976). Persistency of lactation in the Improved Valachian breed of sheep. *Journal of Agricultural Science, Cambridge* **87**, 455–457

McCANCE, I. and ALEXANDER, G. (1959). The onset of lactation in the Merino ewe and its modification by nutritional factors. *Australian Journal of Agricultural Research* **10**, 699–719

MASON, I.L. and DASSAT, P. (1954). Milk, meat and wool production in the Langhe sheep of Italy. *Zeitschrift fur Tierzuchtung und Zuchtungsbiologie* **62**, 97–234

MAXWELL, T.J., DONEY, T.P., MILNE, J.A., PEART, J.N., RUSSEL, A.J.F., SIBBALD, A.R. and MacDONALD, D. (1979). The effect of rearing type and prepartum nutrition on the intake and performance of lactating Greyface ewes at pasture. *Journal of Agricultural Science, Cambridge* **92**, 165–174

MLC (1981). *Feeding the Ewe*. Bletchley, Meat and Livestock Commission

MOORE, R.W. (1962). Comparison of two techniques for the estimation of the milk intake of lambs at pasture. *Proceedings of the Australian Society for Animal Production* **4**, 66–68

MOORE, R.W. (1966). Genetic factors affecting the milk intake of lambs. *Australian Journal of Agricultural Research* **17**, 191–199

MORAG, M., RAZ, A. and EYAL, E. (1970). Mother–offspring relationships in Awassi sheep. IV. The effect of weaning at birth or after 15 weeks, on lactational performance in the dairy ewe. *Journal of Agricultural Science, Cambridge* **75**, 183–187

NIKOLAEV, A.I. and MAGOMEDEV, I.M. (1976). Milk production of Romanov sheep. *Zhivotnovadstvo* **3**, 38–40 (*Animal Breeding Abstracts* **43**, 3704)

ORR, R.J., PENNING, P.D., TREACHER, T.T. and NEWTON, J.E. (1981). Annual Report, Grassland Research Institute, pp.94–95

ØRSKOV, E.R. (1977). Nitrogen digestion and utilization by young and lactating ruminants. *World Review of Nutrition and Dietetics* **26**, 225–257

ØRSKOV, E.R., HUGHES-JONES, M. and McDONALD, J. (1980). Degradability of protein supplements and utilization of undegraded protein by high-producing dairy cows. In *Recent Advances in Animal Nutrition* (W. Haresign, Ed.), pp. 85–88. London, Butterworths

OWEN, J.B. (1957). A study of the lactation and growth of hill sheep in their native environment and under lowland conditions. *Journal of Agricultural Science, Cambridge* **48**, 387–411

PEART, J.N. (1967). The effect of different levels of nutrition during late pregnancy on the subsequent milk production of Blackface ewes and on the growth of their lambs. *Journal of Agricultural Science, Cambridge* **68**, 365–371

PEART, J.N. (1968). Some effects of live weight and body condition on the milk production of Blackface ewes. *Journal of Agricultural Science, Cambridge* **70**, 331–338

PEART, J.N., DONEY, J.M. and MacDONALD, A.J. (1975). The influence of lamb genotype on the milk production of Blackface ewes. *Journal of Agricultural Science, Cambridge* **84**, 313–316

PEART, J.N., EDWARDS, R.A. and DONALDSON, E. (1972). The yield and composition of the milk of Finnish Landrace × Blackface ewes. I. Ewes and lambs maintained indoors. *Journal of Agricultural Science, Cambridge* **79**, 303–313

PEART, J.N., EDWARDS, R.A. and DONALDSON, E. (1975). The yield and composition of the milk of Finnish Landrace × Blackface ewes. II. Ewes and lambs grazed on pasture. *Journal of Agricultural Science, Cambridge* **85**, 315–324

PENNING, P.D. and GIBB, M.J. (1979). The effect of milk intake on the intake of cut and grazed herbage by lambs. *Animal Production* **29**, 53–67

PENNING, P.D. and TREACHER, T.T. (1981). Effect of protein supplements on performance of ewes offered cut fresh ryegrass. *Animal Production* **23**, 374–375 (Abstract)

PENNING, P.D., BRADFIELD, P.G.E. and TREACHER, T.T. (1971). A note on the performance of artificially reared lambs fed cold milk substitute from birth to slaughter. *Animal Production* **13**, 365–368

PENNING, P.D., PENNING, I.M. and TREACHER, T.T. (1978). The effects of heat treatment and protein quantity on digestibility and utilization of milk substitute by lambs. *Journal of Agricultural Science, Cambridge* **90**, 221–228

RICORDEAU, G. and DENAMUR, R. (1962). Production laitiere des brebis Prealpes du Sud pendant les phases d'allaitement, de sevrage et de traite. *Annales de Zootechnie* **11**, 5–38

ROBINSON, J.J. (1978). Response of the lactating ewe to variation in energy and protein intake. In *Milk Production in the Ewe* (J.G. Boyazoglu and T.T. Treacher, Eds.), pp. 53–65. EAAP publication No. 23

ROBINSON, J.J. (1980). Energy requirements of ewes during late pregnancy and early lactation. *Veterinary Record* **106**, 282–284

ROBINSON, J.J. and FORBES, T.J. (1970). Studies on protein utilization by ewes during lactation. *Animal Production* **12**, 601–610

ROBINSON, J.J., FOSTER, W.H. and FORBES, T.J. (1969). The estimation of milk yield of a ewe from body weight data on the suckling lamb. *Journal of Agricultural Science, Cambridge* **72**, 103–107

ROBINSON, J.J., FRASER, C., GILL, J.C. and McHATTIE, I. (1974). The effect of dietary crude protein concentration and time of weaning on milk production and body-weight change in the ewe. *Animal Production* **19**, 331–339

ROBINSON, J.J., McHATTIE, I., CALDERON CORTES, J.F. and THOMPSON, J.L. (1979). Further studies on the response of lactating ewes to dietary protein. *Animal Production* **29**, 257–269

ROY, J.H.B., BALCH, C.C., MILLER, E.L., ØRSKOV, E.R. and SMITH, R.H. (1977). Calculation of the N requirement for ruminants from nitrogen metabolism studies. In *Protein Metabolism and Nutrition* (S. Tamininga, Ed.), pp. 126–129. EAAP Publication 22

SLEN, S.B., CLARK, R.D. and HIRONAKA, R. (1963). A comparison of milk production and its relation to lambs growth in five breeds of sheep. *Canadian Journal of Animal Science* **43**, 16–21

SPEDDING, C.R.W., BROWN, T.H. and LARGE, R.V. (1963). The effect of milk intake on nematode infestation of the lamb. *Proceedings of the Nutrition Society* **22**, 32–41

STARKE, J.S. (1953). Studies on the inheritance of milk production in sheep. *South African Journal of Science* **49**, 245–254

STERN, D., ADLER, J.H., TAGARI, H. and EYAL, E. (1978). Responses of dairy ewes before and after parturition to different nutritional regimes during pregnancy. II. Energy intake, body-weight changes during lactation and milk production. *Annales de Zootechnie* **27**, 335–346

THOMSON, W. and THOMSON, A.M. (1953). Effect of diet on milk yield of the ewe and growth of her lambs. *British Journal of Nutrition* **7**, 263–274

TREACHER. T.T. (1970a). Apparatus and milking techniques used in lactation studies in sheep. *Journal of Dairy Research* **37**, 289–295

TREACHER, T.T. (1970b). Effects of nutrition in late pregnancy on subsequent milk production in ewes. *Animal Production* **12**, 23–36

TREACHER, T.T. (1978). The effects on milk production of the number of lambs suckled and age, parity and size of ewe. In *Milk Production in the Ewe* (J.G. Boyazoglu and T.T. Treacher, Eds.), pp. 31–40. EAAP Publication No. 23

VIPOND, J.E. (1979). Effect of clipping and diet on intake and performance of housed pregnant and lactating ewes. *Animal Production* 28, 451–452 (Abstract)

WALLACE, L.R. (1948). Growth of lambs before and after birth in relation to the level of nutrition. *Journal of Agricultural Science, Cambridge* **38**, 93–153

WILLIAMS, C.M.J., GEYTENBECK, R.E. and ALLDEN, W.G. (1976). Relationships between pasture availability, milk supply, lamb intake and growth. *Proceedings of the Australian Society for Animal Production* **11**, 333–336

YOUNG, B.A. and CORBETT, J.L. (1972). Maintenance energy requirement of grazing sheep in relation to herbage availability. II. Observations on grazing intake. *Australian Journal of Agricultural Research* **23**, 77–85

YOUNG, N.E., NEWTON, J.E. and ORR, R.J. (1980). The effect of cereal supplementation during early lactation on the performance and intake of ewes grazing perennial ryegrass at three stocking rates. *Grass and Forage Science* **35**, 197–202

8

NUTRITION OF LAMBS FROM BIRTH TO SLAUGHTER

E.R. ØRSKOV
Rowett Research Institute, Bucksburn, Aberdeen, UK

Since lambs can be fed in many different ways, the time taken to reach a certain slaughter weight or carcass weight can vary from little more than two months to more than a year. Variations in the time taken to reach slaughter weight are mainly determined by the feed availability at different times of the year and the price paid for a lamb carcass, and is not due to ignorance of lamb nutrition. Differences in time taken to reach slaughter weight have little or no effect on the final chemical composition of the carcass (Andrews and Ørskov, 1970) even if the carcasses are of identical weight.

It is intended here to discuss the nutrition of the lamb in terms of the feeding options available to the producer at different stages in the life of the lamb. Sometimes the feeding systems are enforced on the producer due to shortage or abundance of feeds, and sometimes they are deliberately imposed to ensure a particular carcass weight at a given time.

Nutritional options at birth

At birth there are only two options available to the producer: the lambs will either have to suck from their dams, or they will have to be reared artificially on liquid milk substitutes. Whichever choice is made it is essential to ensure that the lambs have received adequate colostrum as soon as possible after birth. If lambs are weak or if the ewe has inadequate colostrum, cow colostrum can be given at about 50 ml/kg liveweight of the lambs (Robinson and Ørskov, 1976).

REARING ON THE EWE

For almost all sheep enterprises where the production of meat is the primary purpose, the rearing of lambs on the ewe is the commonest and, in general, the most economical method. Only in cases of multiple births (three or more lambs), death of the ewe, sickness, or where rapid re-breeding is required are lambs reared artificially. Little attention needs to be given to the nutrition of sucking lambs but care must be taken to ensure that the quality and quantity of food for the ewe are adequate. The nutrition of the ewe may also be adjusted to prepare the lambs for early weaning onto solid food. Early weaning may be required in breeding systems where suckling leads to a reduction in body condition of the ewe which in turn may cause a very significant reduction in conception rate and ovulation rate (Hunter, 1968). However, these factors are discussed in detail in Chapters 7 and 24.

ARTIFICIAL REARING

Management

Artificial rearing on milk substitutes is of particular interest when the ewe's milk is a saleable product, as in many Mediterranean countries. Whatever the reason that lambs are reared artificially, it is important that they are taken from their dams within about 24 hours of birth and taught to drink from the sucking device or trough. Delaying the time of separation only results in increasing difficulties in imprinting the lambs on the substitute for the mother's teat. Inadequate imprinting leads to an inadequate functioning of the oesophageal groove which conducts the milk directly to the abomasum (Ørskov, 1975). If milk enters the rudimentary rumen it can cause fermentation, which in turn reduces the efficiency of utilization of milk.

Composition of milk substitute

The milk replacer must be liquid to ensure that the oesophageal groove functions adequately, but the composition of the milk substitute can be varied. It has been shown by Penning (1975) that the lactose:fat ratio in milk substitutes for lambs may vary quite widely. The source of fat may also vary provided that the fat is adequately homogenized. It has also been demonstrated that the source of carbohydrate can be varied, milk lactose being replaced by glucose or partially hydrolysed starch (Soliman *et al.*, 1979). Even the source of protein can be varied although difficulties with clotting have occurred where proteins other than those of milk have been used. Fish protein hydrolysate used in combination with frequent feeding

Table 8.1 THE EFFECT OF PROTEIN, FAT AND CARBOHYDRATE SOURCE ON PERFORMANCE OF ARTIFICIALLY REARED LAMBS

Protein source	*Fat source*	*Carbohydrate source*	*Liveweight gain* (g/day days 0–33)	*Food conversion efficiency* (kg dry matter/ kg gain)
Casein	Butter	Lactose	177	1.02
Casein	Butter	Starch	183	1.06
Casein	Lard	Starch	154	1.08
Fish protein	Lard	Starch[a]	170	1.00

[a]Partially hydrolysed.
From Soliman, Ørskov and Mackie (1979)

of the lambs (Soliman, Ørskov and Mackie, 1979) has been found to be a suitable protein. The results from one experiment by Soliman, Ørskov and Mackie (1979) demonstrated that lamb milk substitute can be based totally on non-milk constituents, in that milk substitutes based on starch, lard and fish protein hydrolysates gave results similar to milk substitutes based on casein, lactose and butterfat (*Table 8.1*).

Early weaning onto solid food

Early weaning of lambs onto solid food alone is attractive in several types of sheep production systems. For lambs which are artificially reared, milk substitutes are invariably more expensive than solid food, and the provision of milk generally requires a greater attention to management. Early weaning to solid feed is also useful in frequent breeding systems to ensure a high conception and ovulation rate in the ewes (Robinson *et al.*, 1974). Sometimes it may be advisable to early wean one of each pair of twins from hill ewes where inadequate nutrition of the ewes may limit their milk yield. There may also be occasions where early weaning could be of interest if the economic value of the carcasses from ewes in the spring was high. For out-of-season lamb production it may be economical to wean the lambs early rather than feed the ewes for high milk yield on expensive supplements.

MANAGEMENT OF LAMBS DESTINED FOR EARLY WEANING

For whatever reason that early weaning onto solid food is chosen, the most important physiological factor determining successful early weaning is the state of rumen development. Solid food will enter the rumen and the energy-yielding nutrients produced therein will be almost entirely volatile fatty acids rather than glucose which is the product of digestion of milk or milk substitutes. The state of rumen development is influenced by the intake of solid feed, because volatile fatty acids stimulate rumen development (Flatt, Warner and Loosli, 1958). The intake of solid feed can be affected by the management before weaning, the most important aspect of which, apart from availability of solid feed, is the amount of liquid milk offered (Davies and Owen, 1967). However, during the first 2–3 weeks after birth lambs have little inclination to consume solid feed regardless of the intake of liquid feed.

Restricting the access to liquid feed is relatively easy in systems of artificial rearing; it is much more difficult where the lambs are sucking the ewes. Restricting the feed intake of the ewes in order to decrease their milk yield may have the effect of increasing weight loss of the ewe, which in intensive breeding systems is undesirable. Restriction of protein allowance is one possible way of decreasing milk yield while maintaining high intakes of energy-yielding nutrients (Robinson *et al.*, 1974). The feed intake is generally not recorded for individual lambs, and as a result age at weaning is taken as the most useful guide to time of weaning. At the Rowett Institute a system of weaning at 28 days of age has been adopted provided that the lambs have had access to palatable creep feed and that the milk intake has been restricted by some means. If the lambs are weaned in groups then the youngest lamb must be at least four weeks of age. Nevertheless, lambs weaned by this method are likely to suffer a check in growth (Ørskov, Fraser and Gill, 1973), although they rapidly recover. There may be differences among breeds in the time at which the lambs will consume solid feed since Hogue, Brooks and Rotter (1980) have indicated successful early weaning of lambs and kids at 2–3 weeks of age.

NUTRITION OF EARLY-WEANED LAMBS

Although it is possible to wean lambs at four weeks of age, it does not mean that the rumen has reached mature proportions. The small volume of the rumen restricts the choice of feed that the lambs should be given. In order to compensate for the small rumen the feed must be of a type which allows a rapid turnover in the rumen and therefore the feeds must not only be highly digestible but rapidly degraded.

Andrews, Kay and Ørskov (1969) showed that oats were unsuitable for lambs due to the accumulation of oat husks in the rumen which could only partly be alleviated by grinding. However, if the oat husks are made more digestible by treating the whole grain with sodium hydroxide the accumulation of oat husks in the rumen can be prevented (Ørskov *et al.*, 1981). The rapid rate of breakdown of lucerne in the rumen enabled Jagusch, Clark and Jay (1970) to successfully wean lambs onto clover and lucerne pasture at an early age.

FATTENING OF EARLY-WEANED LAMBS

When the early-weaned lamb is about 8–10 weeks old there are more options open as far as its nutrition is concerned. However, if the lambs have been fed intensively for four weeks they are normally continued on this feeding system mainly because they can reach slaughter weight at 12–14 weeks of age.

The rapid rate of fermentation of cereals in the rumen makes fattening based on cereals a relatively easy process. The problem of the proportion of propionic acid in the rumen increasing to a level which exceeds the metabolic capacity of the liver of lambs and giving rise to odd- and branched-chain fatty acids has been discussed elsewhere (Ørskov, Fraser and Gordon, 1974; Garton, 1975), and will not be further discussed here, except to state that the production of soft fat and the resulting adverse effect on carcass quality is alleviated by giving unprocessed grain. The use of whole rather than processed grain is now generally accepted in practice as it is cheaper and also prevents rumenitis. In *Table 8.2* data are given on the performance of early-weaned lambs fed on whole or processed grain (Ørskov, Fraser and Gordon, 1974).

As mentioned earlier lambs can be fattened on forage-based diets but the forage must be highly digestible and rapidly degraded. Jagusch, Clark and Jay (1970) showed that early-weaned lambs could be fattened by

Table 8.2 EFFECT OF PROCESSING OF DIFFERENT CEREALS IN LAMB DIETS ON LIVEWEIGHT GAIN, FOOD CONVERSION EFFICIENCY, RUMEN pH AND PROPORTIONS OF ACETIC AND PROPIONIC ACID

Cereal	*Form*	*Liveweight gain* (g/day)	*Food conversion efficiency*	*Rumen pH*	*Molar proportions of*	
					Acetic acid	*Propionic acid*
Barley	Whole loose	347	2.8	6.4	52.5	30.1
Barley	Ground pelleted	340	2.8	5.4	45.0	45.3
Maize	Whole loose	346	2.6	6.1	47.2	38.7
Maize	Ground pelleted	345	2.5	5.2	41.3	43.2
Oats	Whole loose	238	3.3	6.7	65.0	18.6
Oats	Ground pelleted	251	3.1	6.1	53.2	37.5
Wheat	Whole loose	323	2.6	5.9	53.2	32.2
Wheat	Ground pelleted	303	3.0	5.0	34.2	42.6
S.E.M.		15	0.1	0.14	2.4	3.2

grazing on lush lucerne pastures, and Greenhalgh *et al.* (1976) showed that ground dried grass could successfully replace cereals in diets for early-weaned lambs.

FATTENING OF LAMBS WHICH ARE SUCKING THE EWE

If the ewes are grazing lush pastures, and in particular, suckling single lambs, the lambs will normally reach slaughter weight at 2–5 months of age. The limited fermentation capacity of the rumen in the earlier stages of growth in such lambs is compensated for by milk by-passing the rumen; the large amounts of protein required during the early rapid phase of growth are supplied by the milk entering the abomasum directly.

FATTENING OF LATE-WEANED LAMBS

When lambs are grazing with ewes in hill and upland areas or at high stocking rates they will normally not reach slaughter weight with the ewes, particularly if two lambs are being suckled. Within such systems they are normally weaned at 4–5 months of age and may be fattened on lowland grazing or root crops, to reach slaughter weight at 5–8 months of age.

In other sheep production systems the lambs are left to graze sparse pastures immediately after weaning, during which time they will only maintain or even lose weight before being fattened later to attain a higher market price in the Spring. Late fattening of this type depends on the availability of concentrate supplements or access to root crops.

Protein and energy requirements for different systems of fattening

In *Table 8.3* the energy requirement of lambs on different systems of fattening are given. It is assumed that the maintenance energy requirement is 440 kJ/kg $W^{0.75}$. It is further assumed that the metabolizability of the diet

Table 8.3 INCREASE IN BODY ENERGY CONTENT FROM 10–40 kg EMPTY BODY WEIGHT (MEAN OF CASTRATES AND FEMALES) AND THE METABOLIZABLE ENERGY (ME) REQUIRED FOR MAINTENANCE AND FATTENING FOR LAMBS SLAUGHTERED AT DIFFERENT AGES

	Age at slaughter (days)		
	100	*200*	*300*
Increase in body energy (MJ) from 10–40 kg empty body weight	483	490	497
ME for maintenance (MJ)	393	928	1473
Estimated metabolizability	0.65	0.55	0.55
Estimated efficiency for fattening, k_f	0.513	0.435	0.435
ME for fattening (MJ)	941	1126	1142
Total ME required (MJ)	1334	2054	2615

for lambs slaughtered at 100 days of age was 0.65 and the metabolizability of the diet for late-weaned and fattened lambs was 0.55. The efficiency of utilization for growth (k_f) was assumed to be 0.513 and 0.435 for the two diets respectively (ARC, 1980). The energy content of the empty body weight was calculated from the data summarized by ARC (1980) and taken as the average for castrated male and female lambs. The liveweight at slaughter was assumed to be 45 kg and empty body weight 40 kg, with a carcass weight of 20 kg. The values are calculated from an initial 10 kg empty body weight, at which time it was assumed that the lambs were 28 days of age. The small differences in body energy content between the different ages at slaughter is due to the weight of the wool, also calculated according to ARC (1980). It can be seen from these calculations that the metabolizable energy (ME) is about twice as much for slaughter at 300 days than for slaughter at 100 days.

In *Table 8.4* a similar calculation has been made for protein. The wool protein and the body gain of protein from 10–40 kg empty body weight were calculated from ARC (1980) assuming that the lambs were a mixture of castrated males and females.

The protein requirement for tissue maintenance has been calculated from the more recent observations on excretion of nitrogen using nitrogen-free infusates (Ørskov, MacLeod and Grubb, 1980), and a mean value of 350 mg nitrogen/kg $W^{0.75}$ is taken. This is considerably greater than the values calculated by ARC (1980). However, the values used by ARC

Table 8.4 INTAKE OF PROTEIN REQUIRED FOR DIFFERENT RATES OF GAIN AND THE CONTRIBUTION OF MICROBIAL PROTEIN (10–40 kg EMPTY BODY WEIGHT) TOWARDS THIS REQUIREMENT

	Age at slaughter (days)		
	100	*200*	*300*
Wool (g protein)	616	916	1216
Requirement of protein for tissue maintenance (g)	1953	4665	7378
Requirement of protein for body gain (g)	4080	4080	4080
Total net requirement of protein (g)	6649	9661	12674
Microbial crude protein produced (g)	10418	16042	20423
Net amino acids available from microbial nitrogen (g)	5625	8662	11028
Net amino acids required/MJ	5.0	4.7	4.8

(1980) were based on endogenous urinary nitrogen (EUN) derived by extrapolating to zero nitrogen intake from a series of intakes of protein whereas the more recent ability to feed ruminants by infusion of nutrients has made it possible to estimate the true maintenance protein requirement. The total net requirement expressed as g net amino acids is calculated and compared with the average production of microbial protein calculated from the ME input (ARC, 1980). The net availability of microbial amino acids has been calculated from the values derived by Storm and Ørskov (1982). By infusion of isolated rumen microorganisms into the abomasum of lambs the utilization of microbial nitrogen was found to be 0.54.

The adequacy of microbial protein to meet the requirement has also been calculated. As expected, the greatest inadequacy occurs with rapidly growing lambs. What is perhaps surprising is that the requirement is slightly greater relative to ME for the slowly growing lambs than for the lambs with medium growth rates. This is due to the higher requirement for tissue maintenance. For a maintenance energy requirement of 440 kJ/kg $W^{0.75}$, a production of 1.25 g microbial nitrogen/MJ of ME (ARC, 1980), and with a utilization efficiency of 0.54, the amount of net microbial amino acid nitrogen available for maintenance is 297 mg/kg $W^{0.75}$ or 1856 mg net amino acids. The requirement was 350 mg nitrogen/kg $W^{0.75}$ or 2188 mg net amino acids. Therefore, about 85% of maintenance protein requirement is supplied by microbial protein.

EFFECT OF DISTRIBUTION OF ENERGY DURING THE FATTENING PERIOD ON PROTEIN REQUIREMENT

Having ascertained the amount of protein required, assuming an average growth rate throughout the fattening period, it is of interest to look at the effects of differences in growth rates during the fattening period with particular attention to the late weaning systems.

Lambs fattened early at 100 days of age

It is obvious that if the lambs are to achieve a liveweight increase from 12.5 kg to 45 kg in 72 days they must achieve an average growth rate of 450 g/day. While such average growth rates are indeed possible (Ørskov *et al.*, 1971) the nutrition must ensure maximum growth at all times. Although the protein requirement per unit of feed energy decreases with increasing weight (Andrews and Ørskov, 1970), a change in dietary crude protein concentration is probably not advisable as the lambs are able to compensate to a large extent after periods of deficiency (Ørskov *et al.*, 1976). Andrews and Ørskov (1970) showed that while growth rate of lambs slaughtered at 27.5 kg liveweight was greater for those given diets containing 17.5% rather than 15.0% of crude protein, there were no differences in growth rate for lambs receiving 17.5% or 15.0% protein when they were slaughtered at 40 kg liveweight. However, lambs fattened with the ewes will obtain less protein per unit of energy as their liveweight increases, milk yield decreases and the intake of grass by the lambs increases.

Lambs weaned late and fattened early

In the example given in *Table 8.5* it is assumed that the lambs were weaned at 120 days and fattened during the following 2–3 months. In the UK this would be the case with lambs born in April, weaned in September and subsequently fattened during October/November on lowland pastures and root crops. It is assumed that the lambs reached 35 kg liveweight before weaning thus achieving a growth rate of about 240 g/day, followed by a growth rate of 125 g/day from weaning to slaughter.

Table 8.5 EFFECT OF CHANGES IN LIVEWEIGHT GAIN ON PROTEIN REQUIREMENTS FROM 10–40 kg EMPTY BODY WEIGHT (LAMBS SLAUGHTERED AT 200 DAYS)

	Age at slaughter (days)		*Total*
	28–120	*120–200*	
Liveweight (kg)	12.5–35	33–45	12.5–45
Wool (g protein)	490	426	916
Protein retained in body (g)	2795	1285	4080
Protein requirement for tissue maintenance (g)	2133	2783	4916
Total net protein requirement (g)	5418	4494	9912
ME required for maintenance (MJ)	421	560	981
ME required for production (MJ)	648	478	1126
Total ME requirement	1069	1038	2107
Net requirement of amino acids (g/MJ)	5.1	4.3	4.7

In this particular example it is also assumed that there was no period of maintenance feeding but rather a steady rate of gain until the slaughter weight was reached. The calculations indicate that the energy required is slightly greater than that given in *Table 8.3* due to the fact that low growth rates are incurred at heavier weights thus increasing the total maintenance energy required. If it is further assumed that the protein requirement was fully met during the time the lambs were sucking the ewes, then it can be seen that the protein requirement is low for the lambs during the fattening stage and very close to the microbial contribution. The microbial contribution of net amino acids was calculated to be 4.2 g/MJ of ME using the values given earlier. In other words, there is probably no need for protein supplementation. If diets contain less nitrogen than that required by the microorganisms, a non-protein source of nitrogen (e.g. urea) can be used.

Lambs weaned late and fattened late

In the example summarized in *Table 8.6* it was assumed, as for *Table 8.5*, that the lambs were weaned at 120 days but that after weaning they were given a maintenance energy diet for 4–5 months and then fattened for a final 50 days to achieve slaughter weight. In the first period growth rates of 240 g/day, and in the final period growth rates of 200 g/day are assumed. As for the example in *Table 8.5*, the requirements of energy and protein for maintenance are greater than those given in *Tables 8.3* and *8.4*. This is

Table 8.6 EFFECT OF CHANGES IN LIVEWEIGHT GAIN ON PROTEIN REQUIREMENTS FROM 10–40 kg EMPTY BODY WEIGHT (LAMBS SLAUGHTERED AT 300 DAYS)

	Age at slaughter (days)			*Total*
	28–120	*120–250*	*250–300*	
Live weight (kg)	12.5–35	35	35–45	12.5–45
Wool (g protein)	411	581	224	1216
Protein retained in body (g)	2795	−1202	2487	4080
Protein requirement for tissue maintenance (g)	2133	4095	1890	8118
Total net protein requirement	5339	3474	4601	13414
ME required for maintenance (MJ)	421	824	380	1625
ME required for production (MJ)	648	0	494	1142
Total ME requirement	1069	824	874	2767
Net amino acid requirement (g/MJ)	5.0	4.2	5.3	4.8

again due to the fact that the period of maintenance energy feeding occurred at a liveweight greater than the average used in *Table 8.3*.

Using the values for tissue maintenance derived by Ørskov and MacLeod (1982), the protein requirement during the final fattening period appears to be greater than during the early growth period, and requires further discussion.

It must be recognized that the reason for the higher protein requirement during the final stage of growth (*Table 8.6*) is due entirely to the higher maintenance requirement, as has been estimated with lambs maintained on intragastric nutrition. However, there are many recent observations which provide support for this phenomenon. In order to test the validity of the ARC protein requirement, Ørskov and Grubb (1978) carried out an experiment with lambs weighing 40 kg. Their protein requirement was estimated to be low and so there should have been no difference in growth rate between lambs receiving supplements of urea or fish meal with a barley based diet. In fact, the lambs receiving the fish meal gained 452 g/day during the first two weeks while the urea-supplemented lambs gained 250 g/day. After this compensatory phase, rates of gain became similar.

As distinct from other experiments where similar rates of gain for lambs given urea or fish meal had been noted at 40 kg liveweight, the lambs used for this trial had previously experienced a period of undernutrition when they had been given diets based on straw. Results with these lambs confirmed the calculations made in *Table 8.6*; the 4–5 months of maintenance energy feeding had altered their protein status and they showed immediate compensatory growth similar to lambs having been given diets low in protein (Ørskov *et al.*, 1976). Similar observations have been made by Kempton and Leng (1979); they showed that lambs given a high roughage diet increased liveweight gain from 55–112 g/day when the protein supplement was protected from degradation in the rumen by treating with formalin. Hovell and Ørskov (1981) showed that groups of lambs given diets of NaOH-treated straw increased liveweight gain from 37–108 g/day when they were given a supplement of fish meal. They also showed that the response to fish meal was better if the lambs had been given the straw diet for 70 days before fish meal was given.

It will be of interest to explore further how rapidly the protein can be repleted during a period of compensatory growth after a period of low-protein nutrition. Such observations may contribute to an explanation of the mechanisms of compensatory growth following a period of restricted feeding.

References

ANDREWS, R.P. and ØRSKOV, E.R. (1970). The nutrition of the early weaned lamb. 1. The influence of protein concentration and feeding level on rate of gain in body weight. *Journal of Agricultural Science, Cambridge* **75**, 11–18

ANDREWS, R.P., KAY, M. and ØRSKOV, E.R. (1969). The effect of different dietary energy concentrations on the voluntary intake and growth of intensively fed lambs. *Animal Production* **11**, 173–187

ARC (1980). *The Nutrient Requirements of Ruminant Livestock.* Slough, Commonwealth Agricultural Bureau

DAVIES, D.A.R. and OWEN, J.B. (1967). Some factors affecting performance in the liquid feeding period. *Animal Production* **9**, 501–510

FLATT, W.P., WARNER, R.C. and LOOSLI, J.K. (1958). Influence of purified materials on the development of the ruminant stomach. *Journal of Dairy Science* **41**, 1593–1600

GARTON, G.A. (1975). The occurrence and origin of branched-chain fatty acids in bacterial, avian and mammalian lipids. *Annual Report of Rowett Institute* **31**, 124–135

GREENHALGH, J.F.D., ØRSKOV, E.R. and FRASER, C. (1976). Pelleted herbages for intensive animal production. *Animal Production* **22**, 148–149

HOGUE, D.E., BROOKS, D. and ROTTER, M. (1980). Artificial rearing and early weaning of kids. *Proceedings of Cornell Nutrition Conference*, pp. 69–70

HOVELL, F.D. and ØRSKOV, E.R. (1981). The response to protein of lambs on a low plane of nutrition. *Animal Production* **32**, 374 (Abstract)

HUNTER, G.L. (1968). Increasing the frequency of pregnancy in sheep. Some factors affecting rebreeding during the post partum period. *Animal Breeding Abstracts* **36**, 347–378, 533–553

JAGUSCH, K.T., CLARK, V.R. and JAY, N.P. (1970). Lamb production from animals weaned at 3–5 weeks of age on to lucerne. *New Zealand Journal of Agricultural Research* **13**, 808–814

KEMPTON, T.J. and LENG, R.A. (1979). Protein nutrition of growing lambs. The effect on nitrogen digestion of supplementing a low protein cellulosic diet with either urea, casein or formaldehyde-treated casein. *British Journal of Nutrition* **42**, 289–302

ØRSKOV, E.R. (1975). Physiological conditioning in ruminants and its practical implications. *World Animal Review* **16**, 31–36

ØRSKOV, E.R. and GRUBB, D.A. (1978). Validation of new systems for protein evaluation in ruminants. *Journal of Agricultural Science, Cambridge* **91**, 483–486

ØRSKOV, E.R. and MacLEOD, N.A. (1982). The determination of the minimal nitrogen excretion in steers and dairy cows and its physiological and practical implications. *British Journal of Nutrition* **47**, 625–636

ØRSKOV, E.R., FRASER, C. and GILL, J.C. (1973). A note on the effect of time of weaning and weight at slaughter on feed utilization of intensively fed lambs. *Animal Production* **16**, 311–314

ØRSKOV, E.R., FRASER, C. and GORDON, J.G. (1974). Effect of processing of cereals on rumen fermentation, digestibility, rumination time and firmness of subcutaneous fat. *British Journal of Nutrition* **32**, 59–69

ØRSKOV, E.R., MacLEOD, N.A. and GRUBB, D.A. (1980). New concepts of basal N metabolism in ruminants. *Proceedings of the 3rd Symposium on Protein Metabolism and Nutrition, Braunschweigh, W. Germany*, EAAP Publication No. 27

ØRSKOV, E.R., GILL, J.C., FRASER, C. and CORSE, E.L. (1971). The effect on intensive production systems of type of cereal and time of weaning on the performance of lambs. *Animal Production* **13**, 485–492

ØRSKOV, E.R., MacDEARMID, A., GRUBB, D.A. and INNES, G.M. (1981). Utilization of alkali-treated grain. Alkali-treated grain in complete diets for steers and lambs. *Animal Food and Science Technology* **6**, 273–283

ØRSKOV, E.R., McDONALD, I., GRUBB, D.A. and PENNIE, K. (1976). Effect on growth rate, food utilization and body composition of changing from a low to a high protein diet. *Journal of Agricultural Science, Cambridge* **86**, 411–423

PENNING, I.M. (1975). Nutrition of the liquid-fed lamb. PhD Thesis, University of Reading

ROBINSON, J.J. and ØRSKOV, E.R. (1976). An integrated approach to improving the biological efficiency of sheep meat production. *World Review of Animal Production* **11**, 63–76

ROBINSON, J.J., FRASER, C., GILL, J.C. and McHATTIE, I. (1974). The effect of dietary crude protein concentrations and time of weaning on milk production and body weight change in the ewe. *Animal Production* **19**, 331–339

SOLIMAN, H.S., ØRSKOV, E.R. and MACKIE, I. (1979). Utilization of fish protein in milk substitutes for lambs. *Journal of Agricultural Science, Cambridge* **93**, 37–46

SOLIMAN, H.S., ØRSKOV, E.R., ATKINSON, T. and SMART, R.I. (1979). Utilization of partially hydrolysed starch in milk replacers by new born lambs. *Journal of Agricultural Science, Cambridge* **92**, 343–349

STORM, E. and ØRSKOV, E.R. (1982). Biological value and digestibility in the small intestine of rumen microbial protein in lambs. *Proceedings of Nutrition Society* **41**, 78A

9

MEETING THE MINERAL REQUIREMENTS OF SHEEP

N.F. SUTTLE
Moredun Research Institute, Edinburgh, UK

The objective of this chapter is to devise an approach to the mineral nutrition of sheep whereby deficiencies are prevented at least cost to the farmer. This practical objective will require the derivation of mineral allowances for sheep which will ensure that the requirements of the majority of individuals in any environment will be met. This will in turn require a detailed appraisal of the genetic and environmental factors which can cause requirements to vary. Finally a strategy will be devised for meeting deficits between the recommended allowances and what forage-based diets in the UK are likely to provide.

Average mineral requirements

Average mineral requirements are commonly formulated by factorial models of the type used by the Agricultural Research Council (ARC, 1980):

$$\text{Dietary requirement} = (E + G + P + L)/A$$

where E = the inevitable loss of the element from the body in faeces and urine; G = daily retention of the element at the specified rate and stage of growth; P = daily retention of the element in the foetus and adnexa; L = daily secretion of the element in milk; and A = the coefficient of absorption i.e. the amount of a mineral supplied in the diet that enters the body from the gut divided by the amount ingested.

The model embraces a replacement principle i.e. that it makes nutritional sense for a diet to provide sufficient absorbed mineral to replace that lost endogenously and that secreted during lactation or deposited during all forms of growth (foetal, wool, etc.). All components of the model should be qualified in the way that E is qualified since they may be used to regulate metabolism and therefore vary with intake. An animal may react to a high mineral intake by decreasing A or increasing E, G, P or L and it would be nutritional nonsense to regard the replacement of such increased 'losses' as obligatory. In their recently published mineral requirements, ARC (1980) acknowledged this only with respect to their calculations of calcium and phosphorus requirements. By using high values for the efficiency of calcium absorption and low values for the endogenous loss of phosphorus from experiments where intakes were low, they decreased requirements by 30–40% compared with those derived by a previous working party from similar data (ARC, 1965).

Of the various requirements which ARC (1980) calculated, only those for calcium and phosphorus can be strictly regarded as minimum dietary requirements although there is no evidence at present that requirements for the other major minerals would be significantly reduced by allowing for intake effects. Requirements for the two trace elements which were calculated by a factorial model, copper and zinc, would however be reduced. A recent study has indicated that sheep are capable of absorbing zinc far more efficiently from natural diets low in zinc than the ARC (1980) suspected (coefficient 0.8 v. 0.2–0.3; Suttle, Lloyd Davies and Field, 1982). Studies with cattle (Suttle, 1978; Simpson, Mills and McDonald, 1981) suggest that the copper content of growing tissue is positively related to copper intake and that the minimum net requirement for growth may be 50% less than the common value used for sheep and cattle by ARC (1980).

Sources of variation in minimum requirements

The predominant source of variation in requirements is the divisor A (the coefficient of absorption) in the model. There is no evidence that genotype or environment influences the minimum values for G, P or L. Although the faecal endogenous component of E for phosphorus will later be shown to vary, this may merely reflect the efficiency with which the secreted element is reabsorbed, i.e. a process closely related to absorption. What then are the principal factors known to influence A and hence the mineral requirements?

INDIVIDUAL AND GENETIC DIFFERENCES BETWEEN ANIMALS

Differences in mineral absorption between individuals of the same breed under identical experimental conditions can be striking. Two to four-fold variations have been reported for magnesium (Field, McCallum and Butler, 1958; Field, 1960), copper (Suttle, 1974) and phosphorus (Field, 1981) and they would almost certainly be associated with similar variations in requirement.

Individual differences in mineral metabolism are partly heritable and have been detected as differences between breeds. With copper, the differences are believed to arise from differences in A which may be three times as high in North Ronaldsay as in Blackface ewes (Wiener *et al.*, 1978) and twice as high in Texel × Blackface as in pure Blackface lambs (Woolliams *et al.*, 1983). Published differences in magnesium metabolism between breeds of sheep are confined to plasma magnesium concentrations (Wiener and Field, 1969) but in cattle two-fold differences between pairs of identical twins in their magnesium absorption have been reported (Field and Suttle, 1979).

The message is clear: breeds, like individuals, can differ substantially in their mineral requirements and for both we have only just begun to appreciate the extent of the differences.

DIFFERENCES BETWEEN DIETS

Although it is generally assumed that the composition of a diet will influence the efficiency with which minerals are absorbed from it, there are few clear examples in the literature. Increases in dietary potassium to the upper limit of the normal range for herbage increase the magnesium requirement by about 50% (Suttle and Field, 1969), probably by reducing the efficiency of magnesium absorption by a similar margin (Suttle and Field, 1967; Field and Suttle, 1979). The influence of dietary molybdenum on the absorption of copper was recognized by ARC (1980) who allowed for a four-fold variation in copper requirement as molybdenum and sulphur concentrations varied over the normal ranges for herbage. Recently, large dietary supplements of iron (800 mg/kg dry matter) have been reported to halve the efficiency of copper absorption in sheep (Suttle, Abrahams and Thornton, 1982) and zinc and cadmium may have similar effects (Bremner, Young and Mills, 1976; Bremner and Campbell, 1978).

Again, we are only just beginning to appreciate the importance of dietary composition in determining requirements. We do not know if the effects of the various antagonists of copper absorption are additive. We do not know if some foodstuffs contain poorly available calcium and phosphorus or if the calcium:phosphorus ratio in the diet really influences the absorption of one or both elements.

INTERACTIONS BETWEEN ANIMALS AND DIETS

The extent to which individual and genetic variation is dependent on dietary composition has received little attention although, in the case of copper, individual variation appears to be lower in sheep given cut fresh herbage than in those fed semi-purified diets (Suttle and Price, 1976). There is also evidence that the differences in copper metabolism between breeds can be reduced on diets high in molybdenum (Woolliams *et al.*, 1981). On the other hand, studies with monozygotic twin cattle have shown that the equally large genetic and potassium effects on magnesium absorption are independent i.e. additive (Field and Suttle, 1979). Thus the least efficient twins on a high potassium diet absorbed only one-seventh of the magnesium that the most efficient twins on a low potassium diet absorbed.

DISEASE

Nutrient requirements are invariably calculated for healthy animals. It is, however, normal for grazing sheep to carry populations of intestinal parasites and Wilson and Field (1983) have found that subclinical infestations of *T. colubriformis* can significantly reduce the absorption of dietary phosphorus and the reabsorption of secreted phosphorus. The combined effect was to increase phosphorus requirements by 30% and there was evidence that the phosphorus concentration in the basal diet, while adequate for uninfected lambs, was inadequate for infected lambs. It is quite possible that other parasites affect absorption and that minerals other than phosphorus will be affected.

LEVEL OF PRODUCTION

It is widely believed that mineral requirements are increased when high levels of production are attained; this is not necessarily so. Most minerals are used as 'catalysts' rather than building blocks. As yields increase, the provision of more food to meet increased needs for energy and protein will often provide more than the required increment in mineral needs. Only when production is increased by feeding a more digestible diet will mineral requirements in concentration terms increase (*Table 9.1*).

Table 9.1 THE EFFECTS OF LEVEL OF PRODUCTION AND DIET DIGESTIBILITY ON MINERAL REQUIREMENTS OF LACTATING EWES. THE FIGURES DEMONSTRATE THAT FEEDING A DIET OF HIGHER DIGESTIBILITY HAS A GREATER EFFECT ON MINERAL REQUIREMENTS THAN INCREASES IN PRODUCTIVITY *PER SE*.

Milk yield (kg/day)	*Diet* q *value*	*ARC (1980) requirement* (/kg dry matter)[a]					
		Ca (g)	*P* (g)	*Mg* (g)	*Na* (g)	*Zn* (mg)	*Cu* (mg)
1	0.6	2.8	2.6	1.6	1.7	31	5.9
3	0.6	3.0	2.9	1.5	1.2	32	7.1
3	0.7	3.7	3.6	1.8	1.5	39	8.7

[a]Values for ewes weighing 75 kg: q values reflect the metabolizable energy content of the diet.

The derivation of mineral allowances for sheep

The task of devising mineral allowances for sheep by taking minimum mineral requirements and allowing for known sources of variation is

Table 9.2 DIETARY ALLOWANCES OF MINERALS FOR SHEEP CONTRASTED WITH THE AVERAGE MINERAL COMPOSITION OF FORAGES IN THE UK

Mineral		*Recommended allowance*[a]			*Forage composition*[b]		*Need to supplement*
		Growth[c]	*Pregnancy*[c]	*Lactation*[c]	*Mean*[c]	*(Range)*[c]	
Ca		2.9–4.5	1.6–3.0	2.8–3.0	5.9	(3–10)	0
P		2.1–3.8	2.4–3.3	3.9–4.4	3.9	(1.5–4.5)	+
Mg		1.2–2.1	1.5–2.0	2.1–2.5	1.7	(1.0–3.0)	++
Na		0.8–1.5	1.5–2.2	1.1–2.7	2.3	(1.0–6.0)	0
Zn		8–19	12–14	14–21	51	(20–60)	0
Cu	Mo<1	8–10	9–11	7–8			+
	Mo>3	17–21	19–23	14–17	9	(2–15)	+++
Co		0.08–0.11			0.13	(0.05–0.30)	+
I	Winter	0.40			0.23	(0.1–0.4)	+++
	Summer	0.15					0
Se		0.05			0.07	(0.02–0.15)	++
Fe		30			–	(70–500)[d]	0
Mn		25			102	(25–250)	0

[a]Based on ARC (1980) requirements for Ca, Na, Co, I, Se, Fe and Mn: ARC values for P and Mg increased by 50%: new factorial requirements calculated for Zn and Cu—see text for details.
[b]From ADAS (1975).
[c]Major mineral (Ca, P, Mg, Na) concentrations in g/kg DM, others in mg/kg DM.
[d]Suttle, N.F. (unpublished data).

fraught with difficulties. The ARC (1980) made just one attempt to do this. For magnesium, they used an alternative value for A (0.17) which was lower than the average (0.29) and generated allowances for magnesium which were 70% higher than their minimum requirements. They were conscious of the belief that magnesium might be poorly absorbed from spring grass. The average value for magnesium in sheep given grass was, however, 0.22±0.06, indicating a narrower safety margin than was intended and making no allowance for individual or genetic variation.

The mineral allowances advocated in *Table 9.2* were derived from the ARC (1980) requirements in various ways.

MAJOR MINERALS

For magnesium, an absorption coefficient of 0.11 was used to allow for dietary and individual extremes, giving values 1.5 times the ARC requirement. For phosphorus, an absorption coefficient of 0.40, the minimum individual value reported by Field (1981), was used to give allowances 50% above the ARC (1980) recommendations. In the absence of known sources of variation in requirements for calcium and sodium, the uncorrected ARC (1980) values have been used.

TRACE MINERALS

New factorial requirements for copper were calculated using a new value for G of 0.45 mg/kg liveweight gain (Suttle, 1978; Simpson, Mills and McDonald, 1981) and assuming absorption coefficients of 0.03 for animals predominantly consuming grass in summer. An additional estimate was made using an absorption coefficient of 0.015 which would be appropriate for autumn pasture or improved pastures of slightly increased molybdenum content (>3 mg molybdenum/kg dry matter) (Suttle, 1981a). Since the all important values for A_{Cu} were obtained from a breed, the Scottish Blackface, which absorbs copper less efficiently than any so far examined, the requirements are assumed to be satisfactory as allowances for most other breeds.

New requirements were also calculated for zinc using values of 0.11 mg/kg liveweight for E and 0.60 for A after Suttle, Lloyd Davies and Field (1982) and no additional allowances were made.

The allowances for trace elements other than copper and zinc are those given by ARC (1980) since the manner in which they were calculated makes them allowances rather than requirements. Their approach was generally to survey field outbreaks of deficiencies for dietary concentrations below which clinical or biochemical signs of deficiency were likely to develop. These outbreaks are likely to have occurred when the combination of environmental and genetic conditions was maximizing the requirement and thus a safety factor was fortuitously incorporated. A further safety factor is provided by the fact that many trace element requirements are highest for the young, rapidly growing animal (iron: Suttle, 1979; iodine: Statham and Bray, 1975). A dietary concentration which meets the requirement at the most vulnerable stage of development should be more than adequate at other times.

An assessment of priorities

In order to assess where the emphases should be placed in meeting mineral requirements the average and ranges of mineral analyses in forage samples submitted to ADAS are set against the allowances in *Table 9.2*. When fresh or conserved pasture provides the bulk of the dry matter intake in the UK, deficiencies of calcium, sodium and potassium should rarely develop. The situation is less satisfactory for phosphorus and magnesium, and the principal time of risk is early lactation. However supplementary feeding would normally be practised then and the extent to which the supplementary foods provide minerals becomes relevant. Cereals are rich in phosphorus and contain magnesium which is relatively well absorbed, and both vegetable and animal protein sources are rich in all major minerals. Supplementation to meet the additional energy and protein requirements during lactation should automatically improve the magnesium and phosphorus status of the diet. Leafy and root brassica crops generally contain less calcium and phosphorus than pasture (Cornforth *et al.*, 1978) but are rarely fed to sheep during lactation in the UK.

As far as trace elements are concerned, there are some marked contrasts. Whereas normal pastures barely contain sufficient copper to meet requirements they invariably contain sufficient zinc, manganese and iron. With iodine there is a marked seasonal effect, forages being adequate in summer but providing less than the requirement in winter. The element which is least well provided appears to be selenium. The mean forage value for selenium is slightly above, and the lowest value 60% below, the allowance. The situation for cobalt is marginally better than that for selenium although a number of samples appear to contain only 50% of the recommended dietary allowance for this element.

Meeting the priorities for mineral supplementation

CURRENT PRACTICE

The principal means by which sheep farmers attempt to meet the mineral requirements of their flocks is through the use of dietary supplements. It has been estimated that some 9000 tons of mineral supplements are used annually for free access and for mixing into feeds at a cost of some £2.4m to the industry. At least 25 different brands are available and they vary widely in their composition (North of Scotland College of Agriculture, 1979). The mean and ranges of mineral concentrations in a group of 13 mineral supplements which had one factor in common, namely that they were rich in magnesium, is given in *Table 9.3*. These minerals were selected because they would be used primarily in spring for the specific purpose of avoiding hypomagnesaemic tetany in ewes. They should certainly be effective in this context because a daily intake of only 20 g of the supplement would meet the entire allowance of magnesium for a 75 kg ewe yielding 2.5 kg milk/day. In other respects, however, there is little evidence that the composition of such supplements is based on nutritional principles. If, for example, one estimates the corresponding intakes of other minerals needed to meet the ARC (1980) requirements, one finds that cobalt and iodine are grossly over-provided (*Table 9.3*). While calcium and phosphorus are required in roughly equal concentrations and phosphorus is the

Table 9.3 MINERAL COMPOSITION OF THIRTEEN MAGNESIUM-RICH SUPPLEMENTS[(a)] FOR SHEEP AND THEIR CONTRIBUTION TO A MIXED DIET WHEN INCLUDED AT 2.5%

Major minerals				*Trace minerals*		
	Mean	*S.D.* (g/kg)	*Average provision*		*Mean S.D.* (mg/kg)	*Average provision*
Ca	63	45	1.6	Co	84±58	2.1
P	26	18	0.7	Mn	1660±1332	41
Mg	214	104	5.4	Zn	750±729	19
				I	136±73	3.4
Na	77	41	1.9	Fe	5238±4233	131

[(a)]From North of Scotland College of Agriculture (1979).

more likely to be limiting in pasture (*Table 9.2*), on average the supplements provide 2.4 times more calcium than phosphorus. Of the trace elements, the most abundant is iron and this is the one least likely to be deficient.

The primary factors which appear to determine the composition of mineral supplements are the need to provide a cheap and harmless matrix, the sales appeal of a coloured product and the low toxicity of certain elements. Thus calcium carbonate and the iron oxides are popular constituents and cobalt and iodine are included in generous proportions.

In addition to the problem of imbalanced composition, dietary mineral supplements suffer another big disadvantage—the imprecision with which they are fed. All means of supplementing the mineral intakes of sheep involve free access to the supplement and thus intakes will vary, particularly with the mineral in loose or block form. It has been demonstrated that significant proportions of a flock (mean 19%, range 5–67%) may consume none while at the other extreme 15% may consume excessive amounts of a mineral-containing block (0.4–1.4 kg/day; Ducker *et al.*, 1981). The proportion of non-consumers was particularly high at low stocking rates and in younger sheep. Low intakes of copper-containing cattle minerals (600 mg copper/kg) by individual ewes given free access to it probably accounted for the persistence of hypocupraemia in 25–35% of those sampled in two hill flocks (Suttle, N.F., unpublished data). Even when the mineral is given with other food supplements, intakes are likely to vary. Foot and Russell (1973) reported coefficients of variation from 13.3–35.8% for individual food consumption in group-fed ewes in outdoor pens. Variability was greatest when intake was rapid, i.e. when pelleted food was given and the differences in energy (and also mineral) intakes were 35–80%.

In allowing for the reluctance of some individuals to consume free or compounded minerals, one can only overfeed the avid consumer with a mixture which is already more than adequate. If the ewe stored the excess minerals provided then there might be some justification for overfeeding minerals to some animals for part of the year to provide a reserve for times when supplements were not fed. With the exception of copper and possibly cobalt (as vitamin B_{12}) however, the ewe will not store a current excess of mineral but will rather excrete that excess via the faeces, urine or milk.

The overfeeding of minerals, on the other hand, might in some instances be positively harmful: take iron for example. It has been known for some time that high iron intakes (approximately 1000 mg/kg dry matter) reduce the storage of copper in the liver of sheep (Abdellatif, 1968) and cattle (Campbell *et al.*, 1974). Recent studies indicate that a supplement of 800 mg iron/kg dry matter halves the efficiency with which sheep absorb copper (Suttle, Abrahams and Thornton, 1982). It is highly probable that the use of most mineral supplements will carry iron concentrations in the whole diet above this level. During autumn and winter, soil contamination of pastures can increase their iron content to 200–1000 mg/kg dry matter (Suttle, N.F., unpublished data) and the ingestion of 50 g/day of the average magnesium-rich mineral would add a further 260 mg iron/kg dry matter to the diet. It is ironic that the manufacturer is forbidden by law to add copper to his product, since copper is likely to be present in inadequate amounts in the pastures being grazed (*Table 9.2*) and is susceptible to antagonism from other permitted constituents (zinc as well as iron; Bremner, Young and Mills, 1976).

Risks associated with overfeeding of other minerals may occur although they have yet to be proven. For example, an increasing proportion of dietary cobalt is incorporated into analogues of vitamin B_{12} rather than the true vitamin as intakes increase (Elliot, 1980). It is not known to what extent these analogues are absorbed and might impair vitamin B_{12} function but impairment of another B vitamin, thiamine, by analogues generated in the gut (Edwin *et al.*, 1979) provides a notable precedent.

It is known that the underfeeding of calcium during late pregnancy in the bovine can reduce the incidence of milk fever. The mechanism is believed to involve the enhanced activity of the parathyroid gland which increases the efficiency of calcium absorption so that the animal is primed to react to the increased demands of calcium at the onset of lactation. It would be surprising if the same adaptation did not occur in sheep on low calcium intakes. Although the ovine differs from the bovine in showing no marked increase in calcium requirement at parturition (*Table 9.4*), it may be less

Table 9.4 CALCIUM REQUIREMENTS OF SHEEP AND COWS IN RELATION TO STAGE OF PREGNANCY AND LACTATION. NOTE THAT THE ONSET OF LACTATION IS NOT ASSOCIATED WITH AN INCREASE IN REQUIREMENTS IN THE EWE AS IT IS IN THE COW

	ARC (1980) calcium requirements (g/kg dry matter)		
	Early pregnancy	*Late pregnancy*	*Lactation*
Bovine[(a)]	3.1	3.2	3.4
Ovine[(b)]	1.7	3.6	3.0

[(a)]Friesian, 600 kg liveweight; 42 kg calf; milk yield 30 kg/day.
[(b)]Ewe 75 kg liveweight; twin foetuses; milk yield 3 kg/day.

able to cope with a sudden reduction in calcium intake (e.g. during snow cover, difficult lambing, handling and transport) if a previous excessive calcium intake had lulled the parathyroid gland into an inactive state. An impaired rate of bone resorption might decrease the supply of magnesium and phosphorus as well as calcium from the skeleton at these critical times.

A further argument which might be made in favour of overfeeding minerals is that allowances such as those in *Table 9.2* and the requirements from which they were derived (ARC, 1980) assume generous levels of food intake which may not always be found in practice. It would, however, be pointless to correct a deficit in mineral nutrition without correcting general undernutrition and it is not therefore a valid reason for overfeeding minerals. There are thus grounds for suggesting that the current practice results in the overfeeding of some minerals and that this is not only wasteful but may indeed be harmful to the ewe.

AN ALTERNATIVE STRATEGY

Encouraged by the difficulties experienced in inducing deficiencies of both major and minor elements on diets containing far less calcium, phosphorus, magnesium, copper and zinc than the current recommendation (*Table 9.2*), it is tempting to suggest a radical alternative to continuous multi-element supplementation. Its central theme would be the optimal use of minerals inherent in the diet and it would be supported by specific supplementation in instances of proven need. Priority would be given to those elements which forage analysis data suggest are the more likely to be deficient (*Table 9.2*).

Magnesium

In the field, clinical magnesium deficiency is associated primarily with the grazing of the first flush of potassium-rich grass in spring. Experiments with lactating ewes have shown only a mild, transient fall in plasma magnesium when intakes were suddenly reduced from 3–1.5 g/day (Suttle and Field, 1969) but when potassium intakes were concurrently increased, hypomagnesaemic tetany developed. The use of fertilizers with minimal potassium contents, the concurrent dusting of strips of pasture with calcined magnesite (Young, 1981) and the gradual introduction of new grass into the diet could each make cost-effective contributions to the control of the disease. Particular care should be given to the older twin-bearing ewes in a flock which might be most at risk.

Where isolated cases of deficiency occur, serious consideration should be given to culling affected ewes which did not possess other outstanding performance attributes. There is ample evidence that hypomagnesaemia is, in part, a repeatable and inheritable characteristic (Wiener and Field, 1969; Field and Suttle, 1979) and it should therefore be amenable to some degree to control by selection.

Phosphorus

It has been argued that nutritional deficiencies of phosphorus might be rare in the UK. Exceptions might arise if the basal diet consisted largely of poor quality roughage or if parasitic infections restricted the absorption of phosphorus and increased maintenance requirements (Wilson and Field, 1983). Even so, there are alternatives to the use of specific mineral supplements. The phosphorus content of home-mixed feeds or ensiled products can be increased considerably by the inclusion of poultry waste which is rich in absorbable phosphorus (Field, Munro and Suttle, 1977). Where good quality forage is available and good control over parasitic infection is maintained, it is questionable whether specific phosphorus supplementation is merited in the majority of flocks, particularly since the pool of phosphorus in the skeleton can be drawn upon during temporary deficiencies.

Calcium

Calcium deficiency is even less likely to occur than phosphorus deficiency, the principal risk being when heavy reliance is placed on homegrown cereals as a winter feed supplement. The situation may be further aggravated if vitamin D status is low. Smith and Wright (1981) have found that levels of 25-OHD in plasma can fall to very low levels in some ewes between December and April in Scotland. While the critical levels for an adverse effect on calcium absorption are not known, vitamin D supplementation by the oral or parenteral route in late pregnancy may well be advised. Supplementation with calcium *per se*, however, may be unnecessary, particularly since the ewe is able to mobilize large amounts of calcium during undernutrition and replenish the skeleton when lactation wanes (Field, Suttle and Gunn, 1968).

Selenium

The comparisons made in *Table 9.2* suggest that selenium is the trace element most likely to be deficient in UK forages. This assessment is given support by the low blood selenium values which are found in many grazing ewes at pasture (Anderson, Berrett and Patterson, 1979). While the significance of low blood selenium (or the closely correlated activity of glutathione peroxidase) values in clinical and production terms has yet to be fully evaluated, they indicate an enhanced risk of deficiency. An inseparable factor which contributes to the risk of disorder is the vitamin E status of the diet. The two nutrients, selenium and vitamin E, perform closely related functions and sometimes have mutually sparing properties (ARC, 1980). Pasture is usually rich in vitamin E but losses occur during conservation. It has recently been found that vitamin E is also destroyed during the storage of moist grain, a process which is believed to contribute to the increased incidence of acute myopathy in cattle (Allen *et al.*, 1974). Straw and turnips are also relatively poor in vitamin E. The vitamin E status of sheep is therefore likely to be minimal after winter feeding on home-produced feeds and nutritional myopathy has been detected when in-wintered hoggs were sent for slaughter. Vitamin E supplementation of the diet of ewes in late pregnancy can confer protection from myopathy on her offspring.

The case for routinely increasing intakes of selenium via supplements to the diet or pasture fertilization has yet to be made convincingly. Nutritional myopathy (WMD) is not a common disease in the national flock despite a generally low blood selenium status. Furthermore, selenium is a highly toxic element and the risks of toxicity may equal the risk of lost production. Small weight gains have been reported in selenium-treated lambs in some trials (31/76 farms in Scotland; Blaxter, 1963) but the overall effect can be non-significant even in flocks of low blood selenium status (Paynter, Anderson and McDonald, 1979). Prompt parenteral treatment with selenium and vitamin E following clinical signs and/or biochemical evidence of muscle damage in the form of elevated serum creatine phosphokinase (CPK) values (>1000 iu/litre) may be the best course of action to take.

Copper

The delicate balance between intake and requirements for copper (*Table 9.2*) is confirmed by the 40% incidence of hypocupraemia in ovine blood samples submitted for analysis by the Scottish Veterinary Investigation Service in May and June of 1981 (Suttle, N.F., unpublished data). The presence of low concentrations of copper in the bloodstream (<0.6 mg/litre) provides unequivocal evidence that the diet is providing less copper than the sheep requires, but equivocal evidence that health and productivity might be impaired. Fear of deficiency and its often irreversible manifestation as swayback is, however, sufficient to initiate the parenteral administration of copper in pregnancy. As many as one in four of the ewes in the national flock may be treated at an annual cost of £1.4m but the ultimate necessity for copper administration on such a scale cannot be ascertained.

Copper deficiency has also been manifested recently as growth retardation in lambs on improved hill pastures (Whitelaw *et al.*, 1979). Again the delicate balance between deficiency and adequacy is apparent in that the triggering factor was believed to be the small increments in pasture molybdenum (2 mg/kg dry matter) and sulphur (2 g/kg dry matter) which together lowered the efficiency of copper absorption from the improved pasture. It should be noted that the lambs remained severely hypocupraemic (plasma copper <0.2 mg/litre) from a few weeks of age and the frequency with which deficiency of that severity occurs elsewhere is entirely unknown.

The reasons why the balance between dietary provision and requirement is so delicate for copper have become apparent in studies of the absorbability of copper from natural foodstuffs (*Table 9.5*). There is almost a ten-fold variation in the mean efficiency with which Scottish Blackface ewes absorb copper from autumn pasture (1.2%) and from leafy brassicas (13.2%). The benefits of high copper absorbability in crops such as cereals, roots and the

Table 9.5 MEAN TOTAL AND ABSORBABLE COPPER CONCENTRATIONS IN CROPS AND FORAGES FED TO SHEEP IN SCOTLAND

	Total copper[a] (mg/kg)	*Absorption*[b] *coefficient*	*Absorbable copper* (mg/kg)
Summer pasture	6.5	0.025	0.16
Autumn pasture	8.5	0.012	0.10
Silage	8.2	0.049	0.40
Hay	5.5	0.073	0.40
Roots	2.5	0.068	0.17
Kale/rape	3.9	0.132	0.51
Cereals	4.3	0.091	0.39

[a]Mean of samples submitted to Scottish agricultural colleges.
[b]From Suttle (1981a).

leafy brassicas are partially offset by low copper concentrations. Nevertheless, as sole constituents of the diet they would just meet the demands for copper by the twin-bearing ewe in late pregnancy. Autumn pasture, on the other hand, is grossly inadequate and the situation is little improved for the lactating ewe on summer pasture, giving no opportunity to establish a liver copper reserve. It is not known what happens to copper absorption when the diet is a mixture of foods but it seems clear that the low incidence of swayback after severe winters may be due in large measure to the increased reliance placed upon supplementary foods such as hay and barley of high copper absorbability.

Bearing in mind that trends towards pasture improvement and the increased use of home-grown feeds are likely to aggravate an already delicate situation, the copper status of sheep in parts of the UK should be improved. The choice of reliable methods is, however, limited. Fertilization of pastures is unreliable and inefficient because uptake from the soil into the plant and then from the plant into the animal is poor. Legislation prohibits the commercial supplementation route. Direct treatment of the animal by injection is widely practised but it has disadvantages (Suttle, 1981b) and mass medication can hardly be accepted as a long-term answer. While various innovations such as cupric oxide needles (Whitelaw *et al.*, 1980; Dewey, 1979) and soluble glasses (Moore, Hall and Sansom, 1982) are likely to improve the range and effectiveness of direct treatment methods, two further approaches suggest themselves. First, the direct application of copper to hay in the bale to achieve copper concentrations of 20 mg/kg dry matter would provide sufficient absorbable copper to substantially boost the copper status of the ewe during winter and thus prevent swayback. Where growth retardation affects the lamb, it may not be possible to meet its requirements via the ewe (Whitelaw *et al.*, 1980) but a single supplementary dose of cupric oxide needles is effective.

The second option would take advantage of the growing knowledge that the ability of a sheep to absorb copper is determined in part by its genetic makeup. Differences between the pure Scottish Blackface and Welsh Mountain breeds in their susceptibility to swayback has been known for some time (Wiener and Field, 1969) and have been attributed to varying efficiencies of copper absorption (Wiener *et al.*, 1978). More recently, Woolliams *et al.* (1983) have found that Texel-X lambs absorb 13% of the copper from a concentrate, East Friesland-X 8.3% and the pure breed

(Blackface) only 6.4%. It would be surprising if summer grass did not meet the requirement of pure breeds such as the Texel and winter feeds provide them with an excess of copper. While commercial considerations may preclude the use of crossing specifically to improve copper status, variations within breeds are also partly heritable and selection of ram lambs within breeds on the basis of plasma copper concentrations has caused values to diverge in 'high' and 'low' lines by 0.4 mg/litre over four generations (Wiener *et al.*, unpublished data). The culling of ewes whose offspring develop swayback should also lead to a slow improvement in the copper status of the flock.

Cobalt

The seriousness with which the lesser risk of cobalt or B_{12} deficiency should be treated is hard to assess. Clinical deficiency is not easy to define because symptoms are non-specific and take the form of general unthriftiness. Confidence in serum B_{12} as a diagnostic aid has declined following observations that values often fall to exceedingly low levels (<0.2 ng/litre) before clinical symptoms develop. A further factor of importance is the extent of soil ingestion. Soil contains far more cobalt than herbage and analytical chemists will commonly reject high cobalt values in herbage samples that are also high in iron and titanium. Sheep are less discriminating and it has long been held that the ingestion of soil could significantly improve the B_{12} status of the sheep (Healy, 1967). Cobalt from soil can be incorporated into B_{12} by continuous cultures of rumen microorganisms although the extent is not a function of either the total or acetate-extractable cobalt contents of the soil (McDonald and Suttle, unpublished data). Analyses of carefully washed herbage samples will therefore tend to underestimate the dietary supply of biologically available cobalt.

The importance of cobalt deficiency would be made clearer by reports of cobalt responsiveness in controlled studies accompanied by data for the cobalt or B_{12} status of soil, pasture and animal. In thus establishing the need for cobalt, injection of the pre-formed vitamin is perhaps a more reliable way of boosting B_{12} status than the cobalt pellet and data on the urinary or plasma metabolites, methylmalonic and formiminoglutamic acid further facilitate the diagnosis of functional B_{12} deficiency (Millar and Lorentz, 1979). As with other deficiencies with non-specific symptoms (copper and occasionally selenium in the growing lamb), a response to supplementation is the surest indication that intake has failed to meet requirement. Subsequently, the routine prevention of a proven deficiency would be best achieved by fertilization of the pasture (2–3 kg hydrated cobalt sulphate/hectare) or the administration of cobalt bullets.

Iodine

There would appear to be some justification for a closer examination of the iodine status of ewes during late pregnancy. The newborn lamb is particularly susceptible to goitre and may develop histologically abnormal thyroids in the absence of visibly enlarged glands (Statham and Bray, 1975). Subclinical iodine deficiency could reduce the viability of the newborn lamb. Protein-bound iodine in plasma is no longer regarded as a reliable guide to iodine status and the specific assay of tri-iodothyronine or thyroid stimulating hormone is advocated. If a deficiency is suspected, parenteral administration of iodized poppy-seed soil provides an alternative to the conventional dietary methods of supplementation.

The use of goitrogenic crops such as the leafy and root brassicas may increase iodine requirement by impairing its uptake by the thyroid gland. However, their quantitative effects on requirement are unknown (ARC, 1980) and their effects on production uncertain (Barry *et al.*, 1981). There are other goitrogenic foodstuffs (e.g. rape seed meal) which may impair the iodination of tyrosine within the gland and the effect cannot be countered by supplementary iodine; their use in foodstuffs in late pregnancy must be viewed with caution.

Conclusions

It would appear that there is a need for a careful reappraisal of the extent to which mineral deficiencies constrain animal health and productivity and of the methods used to avoid such constraints. The routine mass-medication or mass-supplementation of the majority of animals is a costly, needless and possibly harmful strategy for meeting the mineral requirements of sheep. The encouragement of specific supplementation in situations of proven deficiency would clarify the true scale of the problem. Furthermore, it would assist in assessing the effort which should be put into long term preventative measures such as breeding programmes for forages and crops with readily utilized mineral components and for animals with minimal mineral requirements.

References

ABDELLATIF, A.M.M. (1968). Conditioned hypocuprosis: some effects of diet on copper storage in ruminants. *Verslagen van Landbouwkundige Onderzoekingen* **709**, 43–67

ADAS (1975). *The Important Mineral Elements in Animal Nutrition and their Optimum Concentrations in Forages.* Advisory Paper No. 16

ALLEN, W.M., PARR, M.R., BRADLEY, R., SWANNACK, K., BARTON, C.R.Q. and TYLER, R. (1974). Loss of Vitamin E in stored cereals in relation to a myopathy in yearling cattle. *Veterinary Record* **94**, 373–375

ANDERSON, P.H., BERRETT, S. and PATTERSON, D.S.P. (1979). The biological selenium status of livestock in Britain as indicated by sheep erythrocyte glutathione peroxidase activity. *Veterinary Record* **104**, 235–238

ARC (1965). *The Nutrient Requirements of Livestock. No. 2 Ruminants.* 1st Edition. pp. 39–55. London, HMSO

ARC (1980). *The Nutrient Requirements of Livestock No. 2.* 2nd Edition. pp. 183–267. Slough, Commonwealth Agricultural Bureaux

BARRY, T.N., REID, T.C., MILLAR, K.R. and SADLER, W.A. (1981). Nutritional evaluation of kale (*Brassica oleracea*) diets. 2. Copper deficiency, thyroid function and selenium status in young cattle and sheep fed kale for prolonged periods. *Journal of Agricultural Science, Cambridge* **96**, 269–282

BLAXTER, K.L. (1963). The effect of selenium administration on the growth and health of sheep on Scottish farms. *British Journal of Nutrition* **17**, 105–115

BREMNER, I. and CAMPBELL, J.K. (1978). Effect of copper and zinc status on susceptibility to cadmium intoxication. *Environmental Health Perspectives* **25**, 125–128

BREMNER, I., YOUNG, B.W. and MILLS, C.F. (1976). Protective effect of zinc supplementation against copper toxicosis in sheep. *British Journal of Nutrition* **36**, 551–561

CAMPBELL, A.G., COUP, M.R., BISHOP, W.H. and WRIGHT, D.E. (1974). Effect of elevated iron intake on the copper status of grazing cattle. *New Zealand Journal of Agricultural Research* **17**, 393–399

CORNFORTH, I.S., STEPHEN, R.C., BARRY, T.N. and BAIRD, G.A. (1978). Mineral content of swedes, turnips and kale. *New Zealand Journal of Experimental Agriculture* **6**, 151–156

DEWEY, D.W. (1979). An effective method for the administration of trace amounts of copper to ruminants. *Search* **8**, 326–327

DUCKER, M.J., KENDALL, P.T., HEMINGWAY, R.G. and McCLELLAND, T.H. (1981). An evaluation of feedblocks as a means of providing supplementary nutrients to ewes grazing upland/hill pastures. *Animal Production* **33**, 51–58

EDWIN, E.E., MARKSON, L.M., SHREEVE, J., JACKMAN, R. and CARROLL, P.J. (1979). Diagnostic aspects of cerebrocortical necrosis. *Veterinary Record* **104**, 4–8

ELLIOT, J.M. (1980). Propionate metabolism and vitamin B_{12}. In *Proceedings of the 5th International Symposium on Ruminant Physiology* (Y. Ruckebusch and P. Thivend, Eds.), pp. 485–504. Lancaster, MTP Press

FIELD, A.C. (1960). The absorption and excretion of magnesium in the ruminant. In *Proceedings of the British Veterinary Association Conference on Hypomagnesaemia, London*, pp. 1–8

FIELD, A.C. (1981). Some thoughts on dietary requirements of macro-elements for ruminants. *Proceedings of the Nutrition Society* **40**, 267–272

FIELD, A.C. and SUTTLE, N.F. (1979). Effect of high potassium and low magnesium intakes on the mineral metabolism of monozygotic twin cows. *Journal of Comparative Pathology* **89**, 431–439

FIELD, A.C., McCALLUM, J.W. and BUTLER, E.J. (1958). Studies on magnesium in ruminant nutrition. 1. Balance experiments on sheep with herbage from fields associated with lactation tetany and from control pastures. *British Journal of Nutrition* **12**, 433–446

FIELD, A.C., MUNRO, A.C. and SUTTLE, N.F. (1977). Dried poultry manure as a source of phosphorus for sheep. *Journal of Agricultural Science* **89**, 599–604

FIELD, A.C., SUTTLE, N.F. and GUNN, R.G. (1968). Seasonal changes in the body composition and mineral content of the body of hill ewes. *Journal of Agricultural Science, Cambridge* **71**, 303–310

FOOT, J.Z. and RUSSELL, A.J.F. (1973). Some nutritional implications of group feeding hill sheep. *Animal Production* **16**, 293–302

HEALY, W.B. (1967). Ingestion of soil by sheep. *Proceedings of the New Zealand Society of Animal Production* **27**, 109–120

MILLAR, K.R. and LORENTZ, P.P. (1979). Urinary methyl malonic acid as an indicator of the vitamin B_{12} status of grazing sheep. *New Zealand Veterinary Journal* **27**, 90–92

MOORE, P.R., HALL, G.A. and SANSOM, B.F. (1982). The use of copper-containing controlled release glasses for systemic supplementation of sheep with copper. *Proceedings of The Nutrition Society* **41**, 84A

NORTH OF SCOTLAND COLLEGE OF AGRICULTURE (1979). Mineral–vitamin supplements for cattle and sheep. Information Note No. 144

PAYNTER, D.I., ANDERSON, J.W. and McDONALD, J.W. (1979). Glutathione peroxidase and selenium in sheep. II. The relationship between glutathione peroxidase and selenium-responsive unthriftiness in Merino lambs. *Australian Journal of Agricultural Research* **30**, 703–709

SIMPSON, A.M., MILLS, C.F. and McDONALD, I. (1981). Tissue copper retention or loss in young growing cattle. In *Proceedings of the Fourth International Symposium on Trace Element Metabolism in Man and Animals, Perth* (J.McC. Howell, J.M. Gawthorne and C.L. White, Eds.), pp. 133–136. Canberra, Australian Academy of Science

SMITH, B.S.W. and WRIGHT, H. (1981). Seasonal variation in serum 25-hydroxyvitamin D concentrations in sheep. *Veterinary Record* **109**, 139–141

STATHAM, M. and BRAY, A.C. (1975). Congenital goitre in sheep in Southern Tasmania. *Australian Journal of Agricultural Research* **26**, 751–768

SUTTLE, N.F. (1974). A technique for measuring the biological availability of copper to sheep using hypocupraemic ewes. *British Journal of Nutrition* **32**, 395–405

SUTTLE, N.F. (1978). Determining the copper requirements of cattle by means of an intravenous repletion technique. In *Proceedings 3rd International Symposium on Trace Element Metabolism in Man and Animals* (M. Kirchgessner, Ed.), pp. 473–480. Arbeitskreis Tierernharung Weihenstephan

SUTTLE, N.F. (1979). Copper, iron, manganese and zinc concentrations in the carcasses of lambs and calves and the relationships to trace element requirements for growth. *British Journal of Nutrition* **42**, 89–96

SUTTLE, N.F. (1981a). Predicting the effects of molybdenum and sulphur concentrations on the absorbability of copper in grass and forage crops to ruminants. In *Proceedings of the Fourth International Symposium on Trace Element Metabolism in Man and Animals, Perth* (J.McC. Howell, J.M. Gawthorne and C.L. White, Eds.), pp. 545–548. Canberra, Australian Academy of Science

SUTTLE, N.F. (1981b). Comparison between parenterally administered copper complexes of their ability to alleviate hypocupraemia in sheep and cattle. *Veterinary Record* **109**, 304–307

SUTTLE, N.F. and FIELD, A.C. (1967). Studies on Mg in ruminant nutrition. 8. Effect of K and water intakes on the metabolism of Mg, Ca, Na, K and P in sheep. *British Journal of Nutrition* **21**, 819–831

SUTTLE, N.F. and FIELD, A.C. (1969). Studies of Mg in ruminant nutrition. 9.

Effect of K and Mg intakes on development of hypomagnesaemia in sheep. *British Journal of Nutrition* **23**, 81–90

SUTTLE, N.F. and PRICE, J. (1976). The potential toxicity of copper-rich animal excreta to sheep. *Animal Production* **23**, 233–241

SUTTLE, N.F., ABRAHAMS, P.W. and THORNTON, I. (1982). The importance of soil type and dietary sulphur in the impairment of copper absorption in sheep which ingest soil. *Proceedings of the Nutrition Society* **41**, 83A

SUTTLE, N.F., LLOYD DAVIES, H. and FIELD, A.C. (1982). A model for zinc metabolism in sheep given a diet of hay. *British Journal of Nutrition* **47**, 105–112

WHITELAW, A., ARMSTRONG, R.H., EVANS, C.C. and FAWCETT, A.R. (1979). A study of the effect of copper deficiency in Scottish Blackface lambs on improved hill pasture. *Veterinary Record* **104**, 455–460

WHITELAW, A., ARMSTRONG, R.H., EVANS, C.C., FAWCETT, A.R., RUSSELL, A.J.F. and SUTTLE, N.F. (1980). Effects of oral administration of copper oxide needles to hypocupraemic sheep. *Veterinary Record* **107**, 87–88

WIENER, G. and FIELD, A.C. (1969). The concentration of minerals in the blood of genetically diverse groups of sheep. V. Concentrations of copper, calcium, phosphorus, magnesium, potassium and sodium in the blood of lambs and ewes. *Journal of Agricultural Science, Cambridge* **76**, 513–520

WIENER, G., SUTTLE, N.F., HERBERT, J.G., FIELD, A.C. and WOOLLIAMS, J.A. (1978). Breed differences in copper metabolism in the sheep. *Journal of Agricultural Science, Cambridge* **91**, 433–441

WILSON, W.D. and FIELD, A.C. (1983). Absorption and secretion of calcium and phosphorus in the alimentary tract of lambs infected with daily doses of *Trichostrongylus colubriformis* or *Ostertagia circumcincta* larvae. *Journal of Agricultural Science, Cambridge* (in press)

WOOLLIAMS, J.A., SUTTLE, N.F., WIENER, G. and FIELD, A.C. (1981). Genetic and dietary factors in Cu accumulation by sheep. In *Proceedings of the Fourth International Symposium on Trace Element Metabolism in Man and Animals, Perth* (J.McC. Howell, J.M. Gawthorne and C.L. White, Eds.), pp. 137–140. Canberra, Australian Academy of Science

WOOLLIAMS, J.A., SUTTLE, N.F., WIENER, G., FIELD, A.C. and WOOLLIAMS, CAROL (1983). The effect of breed of sire on the accumulation of copper in lambs with particular reference to copper toxicity. *Animal Production* (in press)

YOUNG, P.W. (1981). Hypomagnesaemia in beef cattle. In *Proceedings of the Sheep and Beef Cattle Society of the New Zealand Veterinary Association's Eleventh Seminar, Massey University, Palmerston North, N.Z.*, pp. 62–69

III

Forage Growth and Utilization

10

FACTORS AFFECTING THE GROWTH AND UTILIZATION OF SOWN GRASSLANDS FOR SHEEP PRODUCTION

T.J. MAXWELL
Hill Farming Research Organisation, Bush Estate, Penicuik, Midlothian, UK

Grassland management is concerned with the production and utilization of grass and is considered here in relation to the management of systems of sheep production in which grass is the main food source. A review of some of the more important factors which have to be taken into account in making decisions with respect to grassland management is given and is used as a background to develop some ideas as to how decision making can become more objective.

The aim in managing grass for animal production is frequently to maximize output/unit area, but beyond a certain point this can only be done at the expense of individual animal performance. In making decisions about how much grass to grow and how it should be utilized, farmers attempt to achieve an acceptable optimum between output/unit area and output/animal. The optimum they choose will be set by the level of economic performance that they regard as being acceptable.

It is not possible to have absolute control over systems of animal production which include substantial periods of grazing, not least because of variable climatic effects; the grazing manager must be able to continuously respond to changing conditions throughout a grazing cycle to achieve his desired targets of performance.

The environment (climate and soil) and the species and varieties of plants which are considered suitable for that environment, the number of animals and the area of land devoted to them, provide the framework within which grassland management decisions have to be made. These decisions can be summarized under two headings:

1. those concerning the manipulation of grass production; and
2. those concerning the conversion of herbage into animal product (i.e. utilization).

Manipulation of grass production is essentially a matter of increasing the rate, and extending the period, of herbage growth. To understand how this might be achieved requires a consideration of the morphology and physiology of the grass plant and the ways in which it adapts to defoliation by the grazing animal. Positive manipulation of grass growth rate is achieved by fertilizer application, particularly nitrogenous fertilizer, and information about the response to fertilizer nitrogen is necessary if any measure of control of herbage production is to be achieved. The presence of clover in a sward will have a significant effect on its nitrogen economy; therefore an understanding of the factors which affect its ability to compete with grass in a sward has to be considered.

Utilization of the sward requires the presence of animals and decisions about the number of animals that graze a given area at any given point in the grazing cycle. The factors that affect the amount of herbage eaten by the animal are clearly important here and though there is an increasing amount of information becoming available, the farmer requires a simple objective parameter on which to base decisions. There is good reason to believe that herbage mass can be used for this purpose.

Environment

On the whole, grass is well suited to the climatic environment of the UK, particularly on the west coasts of England, Wales and south-west Scotland. This is because of the relatively even distribution of rainfall and the moderate range of temperatures experienced during the spring, summer and autumn. In some areas the lack of rainfall in the summer can limit growth quite severely, especially in the south and east of England. Rainfall and soil water-holding capacity together have a substantial effect on the level of annual grass production (Morrison, Jackson and Sparrow, 1980).

Where moisture is not limiting, the initiation, rate and cessation of growth is related to temperature which varies with altitude. Growth of grass begins when the temperature is in excess of 5.5 °C and the number of days when the temperature is above this limit gives the potential length of the growing period of an area.

The coastal areas of Devon and Cornwall, south-west Wales and Anglesey have a mean above 300 potential growing days, while the south-western and western lowlands of Britain have around 250 growing days. The hills and mountain areas of England, Wales and Scotland have around 200 growing days (Holmes, 1980). Differences between years in the date at which temperature reaches 5.5 °C in the spring and falls below that level in the autumn can cause a variation in the length of the growing season of some 20–30 days.

The annual variation in the temperature and rainfall, and of the incoming radiation from the sun, are largely responsible for the seasonal pattern of growth and for setting the upper limit to annual grass production. These factors are largely outwith the control of management. The exception would be lack of rainfall or poor water-holding capacity which might be alleviated by irrigation, but on the basis of a recent study by Doyle (1981) the indications are that for the foreseeable future only a small percentage of the total grassland area in the UK can be irrigated profitably.

The efficiency with which grass grows is much influenced by its physiological status. In the spring when it is in its reproductive phase, it more effectively exploits its environment than later in the year when it becomes vegetative. Where nutrient supply is non-limiting, growth rates of S24 and S23 ryegrasses, for example, reach 150 kg dry matter (DM)/hectare/day in the spring, prior to heading, whereas in the late summer and/or early autumn they are of the order of 40–60 kg DM/hectare/day (Corrall and

Fenlon, 1978). The calculated potential yields from theoretical models and extrapolation from field experiments estimate potential harvestable levels of production of between 27 and 30 tonnes DM/hectare (Cooper, 1968; Leafe, 1978). Leafe (1978) concludes, however, that a realistic target for even the best farmers using modern technology might be nearer 20 tonnes DM/hectare. This compares with a national average grass crop yield, excluding rough grazings, which is thought to be 6 tonnes DM/hectare (Robson, 1981).

The yield of the grass crop represents the sum of the yields of individual plants and it is the morphological and physiological characteristics of the individual plant which determines the way in which the grass crop is managed to achieve specific rates of growth and levels of production.

Morphology and physiology of the grass plant

It is useful to think of the tiller, composed of a tubular structure made up of the leaf sheaths surmounted by the leaf blades, as the fundamental unit of the grass crop. Grass growth then can be regarded as the regular production of new leaves from the tiller and the production of secondary tillers from its base. Generally the tiller is composed of three to four leaves. As a new leaf emerges, senescence begins at the leaf tip of the oldest leaf, transfer of nutrients to other parts of the tiller takes place and the oldest leaf dies. The length of life of a leaf in the summer is about four weeks and it is twice as long in the winter.

The disposition of leaves on a tiller affect its ability to capture sunlight. Even though swards may have a similar total area of leaf relative to the ground from which they grow (i.e. their leaf area index, LAI), those which have leaves which are long and erect will allow better light penetration, and therefore more light is absorbed by the canopy; as a consequence there will be a greater capacity for growth than in a prostrate-leaved crop. However, the leaves of an intensively, continuously grazed crop are subject to frequent defoliation; only short leaves or partially defoliated leaves make up such a crop and their total area is relatively small. In these circumstances it is better that the leaves are prostrate so that the maximum potential light interception is achieved. Thus the morphological characteristics of a crop relative to the purpose for which the grass is grown will determine its growth potential (Wilson, 1981).

The rate of tillering, i.e. the production of secondary tillers, is much influenced by the light reaching the tiller base (Langer, 1963); it also differs between grass species. Tiller development is stimulated by defoliation, though the number of new tillers initiated, and the rate at which they develop, is likely to be strongly affected by sward conditions both before and after defoliation. Grant *et al.* (1981) found that the rate at which tiller populations diverged in swards maintained at different levels of herbage mass was much lower before the summer solstice when light conditions were steadily improving, than after the solstice when they were steadily declining, though all the swards were tillering actively. The loss of tillers is increased by both severe shading and severe defoliation.

The production of a reproductive stem and inflorescence follows from a sequence of events, starting, for most grasses, with vernalization in the winter (i.e. exposure to low temperature), the formation in the spring of a juvenile inflorescence on the meristematic apex (growth point), which is enclosed at the base of the tiller by leaf sheaths (the pseudostem), and concludes with the elongation of the reproductive stem which carries the inflorescence upwards to emerge from the sheath of the uppermost leaf in early summer. The time at which ear emergence occurs is used as a basis for describing grass varieties since it is associated with earliness of growth and is closely related to the period when significant changes in the digestibility of the grass plant takes place. Within a species, early heading is also associated with a less densely tillering plant and, since persistency and sward longevity depend on tillering capacity, the later heading varieties which have greater tillering capacity are more suitable for long duration leys.

Thus it is clear also that the morphological and physiological characteristics of different species and varieties are important in determining their suitability for particular environments, the duration of time they are required in leys and their persistence under systems of grazing management.

GRASS VARIETIES

In England and Wales the National Institute of Agricultural Botany, and in Scotland the Scottish Agricultural Colleges, issue each year classifications and lists of recommended species and varieties of grass and clover (e.g. NIAB 1981/82 and SAC 1981/82). It is not the purpose of this chapter to review what they conclude but rather to draw attention to some of the plant characteristics which may be important in relation to sheep production systems based on the utilization of grasslands.

Much emphasis is placed on the annual yield of a variety in the recommended lists along with earliness of growth, digestibility of the crop at a stage of growth suitable for cutting for conservation, persistence and winter hardiness. Some concern has been expressed about the fact that such recommendations are made for the most part on the basis of cutting trials, though these are designed in some cases to simulate grazing management. In reviewing evidence from a number of experiments Hodgson (1981) concluded that comparative measurements of herbage production in cutting trials are unlikely to be a reliable guide to performance under grazing conditions, particularly amongst genotypes which vary in growth habit, and that more attention must be paid to measurements under grazing management. A selection of the more promising varieties are tested in the presence of the grazing animal in secondary trials (Aldrich, 1977), but measurements are limited to the effect of presence of the animal on plant productivity and persistence with no information on herbage intake and digested nutrients. Nevertheless, the ability of a plant to persist under grazing management is an attribute which must be regarded as being vitally important in providing for stable systems of production.

Over the last 30 years there has been a progressive improvement in the persistence of newly bred varieties. Wright (1981) has estimated that there has been about a 70% increase in persistence in terms of ground cover of sown species of the recent top cultivars of grasses. Even so the percentage yield gain/year for grasses over the last 30 years has been only 0.32% on first year grass and 0.43% on second year grass (Wright, 1981). In fact, modern varieties which have been bred for high potential yield have improved little in the last 40 years since the introduction of S24 and S23 perennial ryegrasses (Robson, 1981). In the short term substantial improvement in yield by genetic means seems unlikely. Much of the progress that has been made in increasing the yield of cereals has been achieved by selecting plants which partition nutrients away from straw into the grain. Robson (1981) observes that for grasses there is no organ analogous to the cereal grain into which an increased fraction of the above-ground dry matter can be profitably diverted. Except for an amount necessary to support regrowth all the above-ground dry matter of the grass plant can be utilized. There is little scope for selecting genotypes which divert more assimilates towards shoots since the grass crop rarely partitions more than 15%, and during stem elongation less than 5% of assimilate to roots (Parsons and Robson, 1981). A more promising approach is to consider ways of increasing assimilate production or reducing subsequent losses, i.e. improving the productivity of the plant.

In the knowledge that there is ample scope for increasing yield of herbage by fertilizer nitrogen in most sheep production systems, and bearing in mind the physiological difficulties of improving yield by genetic means, the choice of variety for animal grazing systems ought to be more concerned with distribution of yield and the nutritional value of the ingested material.

The distribution of yield in meeting the nutritional needs of the grazing ewe or lamb is clearly important in systems of production, and there is obvious scope for choosing varieties which effectively extend the grazing season and to some extent the patterns of herbage supply. However, the prospect of extending the grazing season of a species by selection within that species is not great (Green and Anslow, 1972), although the differences of 10–14 days which currently exist between the early perennial ryegrasses (NIAB, 1981) are of considerable practical significance.

The nutritional value of herbage throughout the year is not only of importance in its own right, but because it is also one of the main determinants of how much the grazing animal will eat. However, the measurement of the digestibility and voluntary intake of cut herbage fed to animals indoors cannot take account of some of the non-nutritional factors which are likely to influence the intake of the animal; e.g. it is quite likely that the spatial distribution of material will influence the accessibility of the more nutritious components of the sward. The relative prostrateness or erectness of a genotype is therefore a characteristic which may have a profound effect on the ability of an animal to prehend a diet of sufficient quality and quantity to sustain adequate levels of animal performance.

The types of plant required for grazed pasture are those that have characteristics which, on the one hand, contribute to herbage production and sward longevity under various systems of grazing management and, on the other, contribute to high intake potential and efficient utilization (Hodgson, 1981). There are varieties which fulfil some of these requirements, but it is doubtful on the basis of present evidence whether there are any which fulfil them all. In any case, a farmer would have some difficulty in finding information on new plant material on which he could base reliable estimates of relative levels of animal production under grazing conditions (Hodgson, 1981). Nevertheless, while it may be true that present opportunities and selection criteria for breeding plants particularly suited to grazing are limited, it would be misleading to suggest that existing recommended varieties seriously constrain levels of output from systems of sheep production.

In practice the choice of variety is determined for the short-term ley by the need to have early spring growth, plants which will tiller readily and achieve maximum ground cover as quickly as possible; Italian ryegrass varieties are used successfully in such leys. Earliness of growth and tillering capacity are equally important for the longer leys and permanent grasslands, but persistence and winter hardiness must also feature strongly in the choice of species and variety for these areas.

The success of making full use of the species and varieties of grass and clover which are currently available will be dependent upon providing adequate levels of plant nutrients and this is a matter about which management can be quite positive.

NITROGEN AND GRASS GROWTH AND PRODUCTION

Of the soil nutrients that influence the growth and total annual production of grass, nitrogen is the most important though, of course, the potential benefits of mineral nitrogen applications will only be realized if other elements, in particular phosphorus and potassium, are not limiting. It has been demonstrated in many experiments and confirmed in practice, that the application of nitrogen fertilizer is a ready means of generating greater output of animal product/unit area; up to certain limits it has also been shown to be profitable. Though there is a positive linear relationship between the level of nitrogen application and stocking rate among lowland and upland sheep farms surveyed by MLC (1978), it is clear that there is great variability in the stocking rate for any given nitrogen level. Some of this variation will undoubtedly be due to differences in grazing management and levels of utilization but it also arises because of differences in soil type and climate, and their effects on soil moisture-holding capacity and rainfall (Morrison, Jackson and Sparrow, 1980). The presence of clover, and the rate of breakdown of organic nitrogen to the pool of inorganic nitrogen will also influence the level of response to nitrogen fertilizer application, and therefore the ability to support a given number of animals/unit area of pasture.

All soils require additions of inorganic nitrogen to achieve economic levels of production from sown grassland unless it also contains a legume. The supply of inorganic or mineral nitrogen in the soil supporting a grazed sward comes from three sources, mineral nitrogen fertilizer application, return of nitrogen in the urine and faeces of animals, and the breakdown of organic nitrogen, i.e. mineralization of organic nitrogen. The major source of organic nitrogen is decomposing plant material. As decomposition takes place, mineralization takes place and nitrogen enters the inorganic pool of nitrogen which then is available to the growing plant. The rate of decomposition is dependent upon the nitrogen content of the material in relation to its carbon content; the lower the carbon:nitrogen ratio the more readily will nitrogen enter the inorganic pool. The micro-organisms which break down highly structured organic matter such as woody stems and straws and highly lignified plants with a high carbon:nitrogen ratio, will actually draw upon the pool of inorganic nitrogen before decomposition of this material can take place. This is one of the reasons why peaty soils tend to be low in mineral nitrogen. Peaty soils are common in the hills and upland areas of the UK and where pasture improvement takes place on such marginal soils substantial initial amounts of nitrogen have to be applied to establish grass swards successfully. It is also the reason why 'raw' straw-based farmyard manure may well reduce the available inorganic nitrogen in the soil until it reaches a more advanced stage of decomposition.

At any time, and depending on the particular soil and the nature of its organic matter content, the available pool of inorganic nitrogen can vary enormously. Morrison, Jackson and Sparrow (1980) measured the yield of nitrogen in herbage from unfertilized control plots and used this as an index of available soil nitrogen. The mean values for individual sites ranged from 11–157 kg nitrogen/hectare. It is therefore not surprising that response to nitrogen varies between sites (Davies and Munro, 1974; Richards, 1977). The study of Morrison, Jackson and Sparrow (1980) nevertheless indicated that response between years at the same site was reasonably consistent and characteristic for particular combinations of weather and soil.

This same study, which was carried out in the absence of the grazing animal but covered a wide range of sites, indicated a linear response to nitrogen applications up to 300 kg nitrogen/hectare and that the slope of the relationship was directly related to the annual amount of nitrogen which could be applied before the response fell below 10 kg DM/kg nitrogen. This point on the response relationship was arbitrarily chosen as the optimum amount of nitrogen to apply, and the mean value of nitrogen input at which it was reached for all sites in this study was 388 kg nitrogen (with a range of 260–530 kg). Over a 168-day period the apparent nitrogen requirement to achieve an 'optimum' response can be calculated to be in the range 1.7–3.0 kg nitrogen/day. The lower value is similar to that calculated by Brockman (1966) for the response to nitrogen at North Wyke. The data suggest that to achieve optimum response a general recommendation to give applications corresponding to about 2.5 kg nitrogen/hectare/day of growth would be appropriate. However, since sheep farmers are unlikely to be operating systems of management which would effectively utilize the amount of grass produced at levels of nitrogen above 300 kg/hectare, adequate levels of grass production could be sustained, with a marginal improvement in the efficiency of use of fertilizer nitrogen, on the basis of total annual applications less than 300 kg/hectare corresponding to 1.8–2.0 kg nitrogen/hectare/day of growth.

Under grazing the supply of nitrogen to the sward is increased by dung and urine, and it has been estimated that the apparent recovery of applied nitrogen under grazing would be 94% (Richards, 1977). However, grazing animals also cause sward damage through treading and urine scorch and this tends to reduce yields at higher levels of nitrogen application, but up to 300 kg nitrogen/hectare/year grazing can increase the dry matter yield from a sward (Richards, 1977) as a consequence of recycled nitrogen.

The time of lambing, stock numbers, area of land allocated to a system of production and winter forage requirement all affect the pattern through time, of the relationship between feed provision and feed demand of the grazing animal. It is clear from systems studies (Maxwell, 1978), for example, that during the period of conservation the balance between feed provision for the winter and lamb growth and ewe liveweight recovery during the early summer will be much influenced by the time of lambing and stocking rate. The ability to manipulate herbage growth by adjusting both the level and timing of fertilizer application with respect to meeting the needs of the grazing animal in this context is important. Similarly, the pattern of herbage supply in the late summer and autumn will have a significant impact on the reproductive performance of the ewe, thus judicious applications of nitrogen in the middle and late summer have the potential of influencing the total output of the system to a significant extent by providing adequate herbage prior to and during mating.

To maximize growth and production in early spring to meet both the nutrient requirements of the lactating ewe and provide a surplus of pasture which can be conserved for winter feed, and even though the grass plant is at its most physiologically efficient stage of growth at this time, there appears to be little benefit from applying more than 80 kg nitrogen/hectare in the spring (Brockman, 1966; Morrison, Jackson and Sparrow, 1980). Indeed for annual levels of application of nitrogen approaching 300 kg nitrogen/hectare, the evidence would suggest that higher appplications in mid-season would give a better distribution of production for grazing systems without reducing annual yield. Presumably the response reported to increased use of nitrogen in mid-season is associated with a less efficient use of nitrogen by the plant at this time. Present evidence suggests that for grazing systems spring application should be between 60–80 kg nitrogen (or 30–40% of total) depending on the total annual application of nitrogen used. Mid-season applications probably ought to be in the region of 50% of the total, leaving about 20% for a late summer application. However, these guidelines may require some modification if clover is expected to make a significant contribution to the nitrogen economy of the sward.

Clover and its contribution to the nitrogen economy of the grass sward

Clover has a direct and beneficial effect on the nutrition of the grazing sheep. The growth rate of lambs is demonstrably greater on white clover as compared to grass and organic matter intakes are greater at similar digestibilities (Thompson, 1977). Clover has a particularly important role in the grazing situation when requirements for protein are high, as in early lactation and during the phase of growth in lambs when they become increasingly dependent upon herbage as milk supply declines.

The main benefit to the sward of an association between grass and clover is related to the fixation of nitrogen by symbiotic rhizobia allowing the clover to become independent of soil nitrogen, and in time this fixed nitrogen may become available to the grass.

Much of the grassland acreage in the UK is made up of grass/clover swards. The clover in such swards has an inherent ability, under normal conditions of climate and management, to supply sufficient nitrogen to sustain an annual herbage production of 5000–6000 kg DM/hectare (Brockman and Wolton, 1962). It is difficult, however, to define precisely what the 'normal conditions of management' ought to be to ensure the continued presence of clover in the sward. The presence of clover in hill and upland areas is particularly important. The rates of organic matter decomposition and nutrient mineralization are slow, which leads to a situation where the rooting zone of upland soils contains high total amounts of nitrogen (10 tonnes/hectare) and phosphorus (2 tonnes/hectare) but little of which is available for uptake by plants (Newbould, 1982). The inclusion of clover in the newly sown grasslands of these areas is regarded as the key to sustained pasture production and to economically viable systems.

The establishment of white clover in a sward requires that soil pH must be increased to about 5.2 and that phosphorus and potassium are not limiting. Normally for new reseeds 40–60 kg phosphorus/hectare and 60–100 kg potassium/hectare are essential. On peat soils a balance between phosphorus and potassium must also be maintained; much of the potassium recycled in the urine of sheep grazing such areas is apparently lost from the system unlike the case for mineral soils (Floate, Rangeley and Bolton, 1981). Rhizobia are usually present in previously cultivated soils but may be absent in the undisturbed acid soils of the hills and uplands, in which case inoculation may be necessary for satisfactory establishment.

The ability of clover to remain in a grass sward is dependent on its ability to compete with grass. In the early stages of establishment, and particularly before nodulation has taken place, clover will compete with grass for mineral nitrogen. It may continue to do so, but as nodulation and nitrogen fixation take place the transfer of mineral nitrogen to the soil pool will occur. Under most conditions nitrogen transfer occurs through cycling, by grazing animals and the decay of legume herbage and roots and/or nodules. The amount of nitrogen transferred from legumes to associated grasses varies with season, species of legume, species of grass, proportion of legume in the sward, age of sward and type of management.

Generally those grasses which are high yielding as pure swards are also high yielding when grown with clover (e.g. cocksfoot), but when conditions favour grass growth they are likely to shade out the clover plant which leads to the decay of its above- and below-ground parts and the release of nitrogen which further enhances the growth of grass (Cowling and Lockyer, 1965; 1967). Perennial ryegrass, however, has been found to be highly compatible with clover (Chestnutt and Lowe, 1970). Brougham (1959), for example, has shown that the annual growth curves of white clover and short rotation ryegrass in a mixture are complementary. During midsummer and early autumn, when light and temperature are near optimal for white clover growth, white clover is able to spread through the swards by stoloniferous growth. The greater growth rates of ryegrass under low temperatures and light regimes of late autumn, winter and spring, which are suboptimal for white clover, result in ryegrass dominance during this period. Ryegrass dominance is also partially explained by the balance between nitrogen fixation by the clover, subsequent decay of clover and release of nitrogen which is then taken up by ryegrass leading to increased dominance during these periods. The contribution to yield by ryegrass and clover therefore occurs at different times of the year, resulting in a more balanced dry matter production throughout the year. Brougham (1959) suggests that about 30% clover content on average throughout the year results in an equilibrium being obtained. In the UK, experience suggests that, in general, levels of less than 10% are common.

In summarizing differences in foliage architecture Haynes (1981) concluded that legumes are generally more prone to be shaded by competitors than are grasses and compete poorly for light because legumes generally possess horizontally-inclined leaves, while in grasses light is distributed more evenly throughout the canopy because of their more upright leaves.

In grazed swards the relative competition for light between the two species is likely to be modified by defoliation. There is no strong evidence for selective grazing of clover. Milne *et al.* (1982) showed that although the proportion of white clover in the diet was generally greater than in the sward, the proportion of white clover in the grazed horizon of the sward explained much more of the variation in the proportion of white clover in the diet. A high proportion of clover leaves tend to be in the upper horizons of the sward whereas grass leaves tend to be more evenly distributed throughout the horizons. As grazing takes place through the upper horizons a higher proportion of clover leaf compared to grass leaf is removed, which puts the clover plant at a relative disadvantage in capturing sunlight. To some extent the severity of the effect will depend on the species of clover used and whether it is an erect or prostrate type (Brougham, 1959; Harkness, Hunt and Frame, 1970), and also the relative vigour of the grass with which it is growing. Though the clover plant replaces leaves relatively quickly, it is likely that the frequency of defoliation will have a significant impact on its ability to compete and, indeed, survive in swards that are continuously grazed at a high level of utilization. Such an analysis suggests that periods of rest from grazing are required to achieve a stable balance between grass and clover in a sward.

Another factor which has an important bearing on the maintenance of clover in the sward is the need to maintain the phosphorus level in the soil. Clover competes poorly for phosphorus, and though its symbiotic association between its roots and mycorrhizal fungi may improve its ability to obtain phosphorus and thereby improve its nitrogen fixing ability, the continued application of phosphorus to the soil is likely to be an important feature in the maintenance of useful grass/clover swards.

Clover is also affected by the level of mineral nitrogen in the soil. High levels of applied nitrogen reduce the number of effective nitrogen fixing nodules on its roots. High levels of applied nitrogen also enhance the competitive ability of grass because, invariably, grass will take up about 90% of the mineral nitrogen applied and therefore, unless defoliated differentially relative to clover, will become dominant. If defoliated by an animal much of this nitrogen will be subsequently returned to the soil. Though in practice it has been assumed that clover will not survive with fertilizer applications above 100–150 kg nitrogen/hectare (Brockman,

1966), it would seem possible that if appropriate defoliation or grazing regimes were adopted clover may still have a role in grass/clover swards where fertilizer applications are greater than those currently regarded as being acceptable. With reasonable levels of utilization, and the fact that current average use of nitrogen on grassland is around 100 kg nitrogen/hectare (Morrison, Jackson and Sparrow, 1980), there is clearly ample scope for making use of the potential nitrogen contribution from clover.

UTILIZATION: FACTORS AFFECTING INTAKE

The nutrient value of the ingested herbage, described in terms of its digestibility, directly affects the amount of herbage eaten. Though there is evidence to show positive linear responses in intake to increased herbage digestibility in grazing cattle there is no direct evidence for sheep (Hodgson, 1977). The evidence for this effect in sheep is based on cut herbages of known digestibility fed indoors.

There are an increasing number of experiments which report data supporting an asymptotic relationship between herbage intake and herbage mass though there is some variation in the herbage mass below which herbage intake is depressed. Hodgson (1981) suggests this may reflect in part the effect of associated changes in the nutritive value of the herbage consumed; it is quite possible, for example, that there is a progressive decline in herbage digestibility with increasing mass which tends to limit the response to variation in mass alone. Other factors, such as height and bulk density of the sward, will also influence herbage intake (Hodgson, 1981), and it is apparent that there are specific morphological characteristics of the plant which appear to have some influence on what the sheep will eat. It has been suggested (Hodgson, 1981) that a probable explanation for the effect of sward height on intake is that grazing sheep seldom penetrate into the horizons containing pseudostem or dead material, and that there is a direct relationship between the surface height of the sward and the depth of the grazed horizon (Barthram, 1980).

The bulk density of the sward, its height and the spatial distribution of the preferred plant parts, will all influence the ability of the animal to prehend its diet. Bite size is a fundamental unit of measure of intake and it is reasonable to assume that in a stable set of sward conditions it will vary very little. There is evidence to suggest that under conditions of low herbage mass the number of bites/unit time increases as bite size is reduced. In order to compensate for a smaller bite size, animals will also increase their grazing time. At low herbage mass these compensatory mechanisms fail to be adequate in maintaining high levels of intake.

UTILIZATION: FACTORS AFFECTING HERBAGE GROWTH

The adaptability of the sward under grazing is a function of its capacity to produce daughter tillers and thereby increase the potential growth points of the sward. Tiller numbers on continuously stocked ryegrass/clover swards tend to increase progressively with reductions in herbage mass down to levels of 700 kg DM/hectare, but fall sharply with further reductions in herbage mass (Bircham and Hodgson, 1981). There is an inverse relationship between tiller population density and tiller weight (Hodgson *et al.,* 1981). Under continuous sheep grazing maximal tiller populations, which normally occur immediately after the conclusion of the reproductive phase of growth, can be as high as 40 000–60 000 tillers/m^2 (Bircham, 1981). Where management of swards allows a frequent change and a wide amplitude in herbage mass, tiller populations are continuously switching between release of inhibition on dominant tillering buds at the base of the tiller and tiller loss by self-thinning; seasonal effects are confounded with effects of management (Grant *et al.*, 1981). However, King, Lamb and McGregor (1979) showed that in frequently and closely defoliated swards tiller population increased, leaf angle declined and a greater degree of light penetration occurred than in laxly defoliated swards. Furthermore, because new tillers produce new leaves and because ageing leaf tissue is removed by defoliation, the proportion of new and young leaf tissue in the canopy is relatively high.

Parsons and Leafe (1981) have shown that the photosynthesis of expanding leaves and young fully expanded leaves is three to four times greater than old leaves, and some 76% of the total photosynthetic uptake of the canopy of an intensively grazed sward is contributed by the expanding leaves and fully expanded leaves, even though these leaf categories represent only 40% of the leaf and sheath area. In a laxly grazed sward there is tissue loss due to death and decay and a considerable proportion of the photosynthetic uptake is inevitably lost. This does not occur to anything like the same extent in more intensively grazed swards because tissue is harvested before it reaches this stage. Also, the energy maintenance requirement of an intensively grazed sward maintained at low herbage mass is less. Theoretically, at least, efficient utilization by the grazing animal of the energy captured by the sward requires that leaves should not reach the stage of senescence since animals are likely to consume only small amounts of such material, which therefore inevitably dies and decays.

It would appear from the above data, and that of Bircham (1981) (see *Figure 10.2*), that there is much scope for manipulating swards over a range from, say, 1000–2000 kg DM without substantially affecting the growth of these swards.

UTILIZATION: ANIMAL PRODUCTION AND MANAGEMENT

In a recent series of studies the levels of animal performance that can be obtained from swards managed under a system of continuous stocking within defined limits of herbage mass have been investigated, with the overall objective of defining annual herbage mass profiles for systems of upland sheep production.

In early lactation the potential intake of the lactating ewe is about 50% greater than the non-lactating ewe (Maxwell *et al.*, 1979) and therefore the demand for herbage in this period of lactation is high. In early spring before substantial growth of grass has taken place, herbage supply will invariably limit herbage intake and the production of milk is either maintained from a supplementary feed source or by the utilization of body reserves (Milne, Maxwell and Souter, 1981). Even when supplementation of the diet has taken place using protein of high undegradability, the influence of herbage mass on lamb growth is still significant (Milne *et al.*, 1982). It is concluded that a herbage mass in excess of 750 kg DM/hectare is essential in early lactation in order to obtain lamb growth rates in excess of 300 g/day for twin lambs.

It is clear also that ewes are extremely sensitive to changes in herbage mass during this period. For example, when herbage mass reached a level of 2000 kg DM under continuous grazing and thereafter grazing pressure was increased to a level which required ewes to graze the sward down to 1500 kg DM, there was a substantial reduction in the organic matter digestibility of ingested material by ewes, and a consequent reduction of 150 g/day in the growth rate of lambs. Both digestibility and growth rate increased again as grazing pressure was decreased and the sward was held at a stable herbage mass of 1500 kg DM (Milne, Maxwell and Souter, 1981).

To achieve a herbage mass of 750 kg DM/hectare early in the season, which is subsequently at least maintained, it is necessary to apply adequate levels of nitrogenous fertilizer when 100 mm soil temperatures reach 5.5 °C. Responses can be expected from applications up to 80–90 kg nitrogen/hectare. Stocking of the grazed area should be delayed until there is a clear indication that swards are growing adequately. An indication of this is given by measuring the accumulated degrees by which the daily 100 mm soil temperature exceeds 5.5 °C. A value of over 20 °C will be necessary before substantial growth can be assumed.

In practice, it may not be possible to avoid stocking areas before these conditions are met due to the lack of flexibility that occurs on some farms at this time of year. However, if herbage mass is reduced to levels below 500 kg DM/hectare the growth of the sward will be seriously impaired. This will delay the time at which areas can be closed for conservation and extend the period over which supplementary feeding has to be given. Stocking rate then has to be maintained at a low level requiring a greater allocation of land for grazing in early spring.

During the first half of lactation, and for upland ewes more specifically during mid-lactation, pastures will be at a stage when seed head emergence is taking place and when there is a need to consider also the closing of areas for conservation. The decision to close areas off from grazing will be influenced by a desire to prevent a decline in the digestibility of the ingested herbage. To do this effectively requires that the grass plant be defoliated with a frequency and intensity which effectively removes the stem and inflorescence at an early stage in their development. In practice this generally means that swards have to be maintained at below 1800 kg DM/hectare. Provided that herbage mass remains below this level, ingested organic matter digestibility can be maintained at a level above 80% well into the summer (Maxwell *et al.*, 1979; Milne, Maxwell and Souter, 1981). One way to achieve these conditions is to remove an area from grazing and use the released area for conservation.

The extent of the area closed for conservation will be influenced also by the amount of winter forage required and the need to sustain lamb growth during the period of conservation on the remaining grazed area. Ultimately, the decision is an economic one, relative to the seasonal price changes of the weaned and/or 'finished' lambs and the cost of 'bought in' winter feed for ewes. To make this assessment it is necessary to know something of the performance of stock during this period in terms of lamb growth and ewe liveweight recovery. Information about the latter is required since the economic calculation is not only concerned with lamb growth and date of 'finish' but also with the number of lambs 'finished'. The number of lambs 'finished' will be a function of the lambs born which is strongly related to the liveweight of the ewe at mating, this in turn being influenced by the rate of liveweight recovery in the summer. *Figure 10.1* shows the relationship between lamb growth and ewe liveweight change with herbage

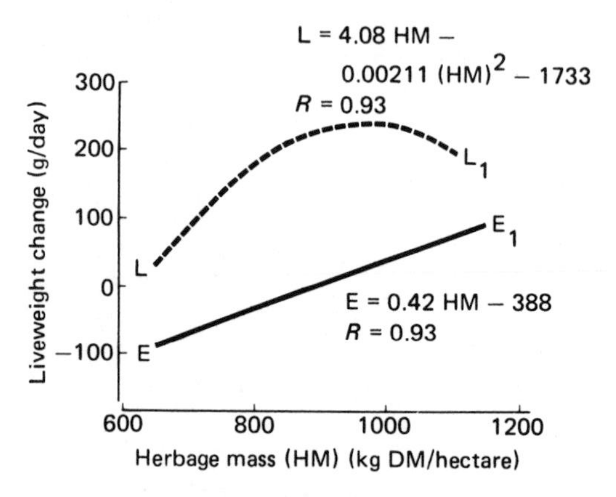

Figure 10.1 The liveweight change (E_1) of Greyface ewes (——) and their lambs (L_1; ----) during mid-lactation (6–12 weeks) related to herbage mass (HM). From T.J. Maxwell and R.D.M. Agnew (unpublished)

mass during the mid-lactation period of upland Greyface ewes (T.J. Maxwell and R.D.M. Agnew, unpublished). Similar relationships were obtained by Bircham (1981) over a similar range of herbage mass, though ewe liveweight change in his experiment reached an asymptote at 1000 kg/hectare. The differences in response of ewe liveweight change may be due to differences between the condition of the ewes in the two experiments.

The amount of winter forage required by the upland ewe is usually in the region of 80–100 kg. A hay crop will yield about 6.0 t/hectare in an upland environment with modest amounts of nitrogen. The area of pasture required to provide this feed can be calculated and estimates of animal performance, on the basis of the herbage mass remaining on the grazed areas, can be made from the above relationships (*Figure 10.1*). By using objective methods of management an optimum balance can be obtained between animal performance on the one hand and the provision of winter forage on the other.

It is also possible to be objective about the effect that stock manipulation and closure for conservation is likely to have on sward productivity. The growth of continuously grazed swards has been measured by Bircham

(1981) and related to herbage mass. *Figure 10.2* shows the relationship between net production and senescence over a range of herbage mass and it is clear from these relationships, for example, that a herbage mass of around 1250 kg DM/hectare is the level at which maximum net herbage production is achieved. In these experiments it was also the level at which maximum output of lamb/unit area was obtained.

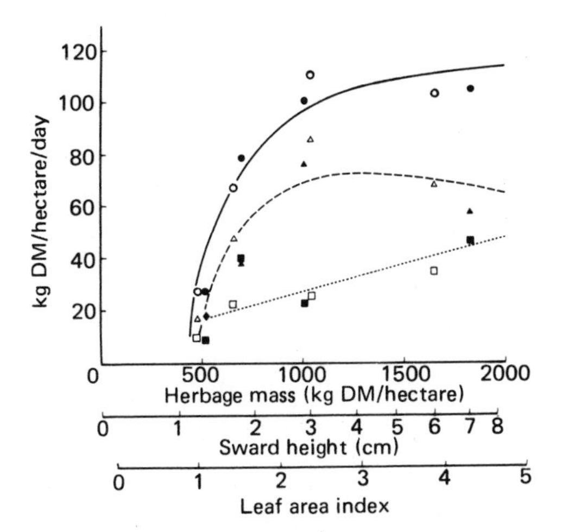

Figure 10.2 Relationships between herbage growth (——), net growth (----), senescence (········) and herbage mass in continuously grazed swards. From Bircham (1981)

At present the evidence from experiments on sown upland pastures suggests that, in the last phase of lactation, lamb growth and ewe liveweight change is more closely associated with herbage quality than herbage mass. More precise information is required for this period, a period of the grazing cycle which is likely to be important in determining the recovery of the ewe to a desired weight and level of body condition at mating.

Nevertheless, information so far suggests that a herbage mass in excess of 1600 kg DM/hectare has to be maintained if ewes are to achieve a satisfactory liveweight by the time of mating. However, excessive amounts of herbage (i.e. more than 2500 kg DM) will almost certainly incur a nutritional penalty as animals progressively graze through the sward and encounter dead and decaying herbage. To maintain herbage mass below these levels it will be necessary to increase grazing pressure in the late summer, particularly after weaning, which may in itself provide an option for taking a second cut for conservation.

The period prior to and during mating is one in which the growth of grass is slowing down considerably and has often ceased before mating is complete. The herbage intake of the ewes during this period will not only be affected by herbage mass and herbage digestibility, but also by the level of their body fat reserves (see Chapter 5). Present studies are investigating the effect of saving of pasture for the immediate pre-mating and mating period on herbage intake and reproductive performance.

As an approach to the practical implementation of more objective grazing management with respect to animal performance, the manipulation of swards either by stock movement or by closure of areas of grazing to achieve specific herbage masses, seems to be a worthwhile objective. Recently, New Zealand farmers have been strongly encouraged to use this procedure on the basis of data relating animal performance in their environment to herbage allowance as well as herbage mass (New Zealand Farmer, December 1981). The approach there is determined on the basis of residual herbage mass in the context of both rotational and set stocking systems, though the former is preferred.

Experiments which have attempted to compare rotational and set stocking methods of grazing management have rarely been conclusive and it has not always been possible to explain the reasons that lie behind the results. So far as sheep are concerned it is unlikely that substantial advantages arise from rotational grazing. Other methods of grazing management have been investigated with the aim of improving the nutrient intake of lambs from pasture, e.g. forward creep grazing and, though some have been successful in improving animal performance, they have not received widespread adoption. However, while rotational grazing may not necessarily have any inherent advantages over continuous stocking, it does allow feed resources to accumulate on a sward so that a grazing manager has an objective basis on which to plan his next phase of grazing in relation to the area which is currently being grazed. In using rotational grazing methods, however, the interval between defoliations should be no greater than three leaf appearance intervals because normally the oldest of the leaves after this time interval will begin to die. It follows that since leaf appearance interval varies with season, the interval between the successive grazing of paddocks in a rotation should vary also in a similar manner.

At present it is not possible to give a comprehensive recommendation with respect to the herbage masses that ought to be maintained throughout the production cycle but it is possible to begin to provide a guide to management decisions for specific phases of the grazing cycle. Herbage mass is a concept which farmers must have used in a qualitative fashion since they first attempted to manage pastures in relation to animals. To make more progress the concept must be put on a *quantitative* basis. Though direct measurements of herbage mass may be outwith the scope of many farmers, the measurement of the height of the sward with a ruler can be used as an index of herbage mass, but since herbage mass is also dependent upon the density of the sward this must also be taken into account. Though this will have to be done subjectively, experience will improve consistency.

Given the conditions whereby it is possible to have much greater control of grassland utilization and animal performance, it is also possible to achieve an appropriate balance between animal performance and sward productivity. This provides a basis on which to assess more precisely the stocking rate that should be used relative to the chosen level of nitrogen fertilizer application and given level of pasture production. Favourable economic results will be achieved with increasing certainty by improving the precision of decision making with respect to grassland management.

References

ALDRICH, D.T.A. (1977). Factors affecting the performance of grass and legume cultivars in the U.K. In *Proceedings of International Meeting on Animal Production from Temperate Grassland* (B. Gilesnan, Ed.), pp. 20–22. Irish Grassland and Animal Production Association, Dublin

BARTHRAM, G.T. (1980). Sward structure and the depth of the grazed horizon. *Grass and Forage Science* **36**, 130–131 (Abstract)

BIRCHAM, J.S. (1981). Herbage growth and utilisation under continuous stocking management. PhD Thesis, Edinburgh University

BIRCHAM, J.S. and HODGSON, J. (1981). The dynamics of herbage growth and senescence in a mixed-species temperate sward continuously grazed by sheep. *XIIIth International Grassland Conference, Lexington, USA* (in press).

BROCKMAN, J.S. (1966). The growth rate of grass as influenced by fertiliser nitrogen and stage of defoliation. In *Proceedings of the Xth International Grassland Congress, Helsinki, Finland*, pp. 234–240

BROCKMAN, J.S. and WOLTON, K.M. (1962). The use of nitrogen on grass/white clover swards. *Journal of the British Grassland Society* **18**, 7–13

BROUGHAM, R.W. (1959). The effects of frequency and intensity of grazing on the productivity of a pasture of short-rotation ryegrass and red and white clover. *New Zealand Journal of Agricultural Research* **2**, 1232–1248

CHESTNUTT, D.M.B. and LOWE, J. (1970). Agronomy of white clover/grass swards. In *White Clover Research* (J. Lowe, Ed.), pp. 191–213. British Grassland Society Occasional Symposium No. 6

COOPER, J.P. (1968). Energy and nutrient conversion in a simulated sward. *Report of the Welsh Plant Breeding Station 1967*, 10–11

CORRALL, A.J. and FENLON, J.S. (1978). A comparative method for describing the seasonal distribution of production from grasses. *Journal of Agricultural Science, Cambridge* **91**, 61–67

COWLING, D.W. and LOCKYER, D.R. (1965). A comparison of the reaction of different grass species to fertiliser nitrogen and to growth in association with white clover. I. Yield of dry matter. *Journal of the British Grassland Society* **20**, 197–204

COWLING, D.W. and LOCKYER, D.R. (1967). A comparison of the reaction of different grass species to fertiliser nitrogen and to growth in association with white clover. II. Yield of nitrogen. *Journal of the British Grassland Society* **22**, 53–61

DAVIES, D.A. and MUNRO, J.M.M. (1974). Potential pasture production in the uplands of Wales. 4. Nitrogen response from sown and natural pastures. *Journal of the British Grassland Society* **29**, 149–158

DOYLE, C.J. (1981). Economics of irrigating grassland in the United Kingdom. *Grass and Forage Science* **36**, 297–306

FLOATE, M.J.S., RANGELEY, A. and BOLTON, G.R. (1981). An investigation of problems of sward improvement on deep peats with special reference to potassium responses and interactions with lime and phosphorus. *Grass and Forage Science* **36**, 81–90

GRANT, S.A., KING, J., BARTHRAM, G.T. and TORVELL, L. (1981). Responses of tiller populations to variation in grazing management on continuously stocked swards as affected by time of year. In *Plant Physiology and Herbage Production* (C.E. Wright, Ed.), pp. 81–84. British Grassland Society Occasional Symposium No. 13

GREEN, J.O. and ANSLOW, R.A. (1972). Comparative yields of herbage species. In *Grasses and Legumes in British Agriculture* (C.R.W. Spedding and F.C. Diekmahns, Eds.), pp. 438–444. Slough, Commonwealth Agricultural Bureaux

HARKNESS, R.D., HUNT, I.V. and FRAME, J. (1970). The effect of variety and companion grass on the productivity of white clover. In *White Clover Research* (J. Lowe, Ed.), pp. 175–186. British Grassland Society Occasional Symposium No.6

HAYNES, R.J. (1981). Competitive aspects of the grass–legume association. *Advances in Agronomy* **22**, 227–261

HODGSON, J. (1977). Factors limiting herbage intake by the grazing animal. In *Proceedings of International Meeting on Animal Production from Temperate Grassland* (B. Gilesnan, Ed.), pp. 70–75. Dublin, Irish Grassland and Animal Production Association

HODGSON, J. (1980). Testing and improvement of pasture species. In *Grazing Animals* (F.H.W. Morley, Ed.), pp. 309–317. Amsterdam, Elsevier Scientific

HODGSON, J. (1981). Influence of sward characteristics on diet selection and herbage intake by the grazing animal. In *Proceedings of International Symposium on Nutritional Limits to Animal Production from Pastures* (J.B. Hacker, Ed.), pp. 153–166. Slough, Commonwealth Agricultural Bureaux

HODGSON, J., BIRCHAM, J.S., GRANT, S.A. and KING, J. (1981). The influence of cutting and grazing management on herbage growth and utilisation. In *Plant Physiology and Herbage Production* (C.E. Wright, Ed.), pp. 51–62. British Grassland Society Occasional Symposium No. 13

HOLMES, W. (1980). *Grass: its Production and Utilisation*. London, Blackwell Scientific (for British Grassland Society)

KING, J., LAMB, W.I.C. and McGREGOR, M.T. (1979). Regrowth of ryegrass swards subject to different cutting regimes and stocking densities. *Grass and Forage Science* **34**, 107–118

LANGER, R.M.M. (1963). Tillering in herbage grasses. *Herbage Abstracts* **33**, 141–148

LEAFE, E.L. (1978). Physiological, environmental and management factors of importance to maximise yield of the grass crop. In *Maximising Yields of Crops*, pp. 37–49. London, HMSO

MAXWELL, T.J. (1978). Management decisions in grazing systems. In *Sheep on Lowland Grass,* pp. 21–27. Summer Meeting of the British Society of Animal Production

MAXWELL, T.J., DONEY, J.M., MILNE, J.A., PEART, J.N., RUSSEL, A.J.F., SIBBALD, A.R. and MACDONALD, D. (1979). The effect of weaning type and prepartum nutrition on the intake and performance of lactating Greyface ewes at pasture. *Journal of Agricultural Science, Cambridge* **92**, 165–174

MEAT AND LIVESTOCK COMMISSION (1978). *Sheep Improvement Services.*

Data Summaries on Upland and Lowland Sheep Production. MLC Sheep Improvement Service Data Sheet 6

MILNE, J.A., MAXWELL, T.J. and SOUTER, W. (1981). Effect of supplementary feeding and herbage mass on the intake and performance of grazing ewes in early lactation. *Animal Production* **32**, 185–195

MILNE, J.A., HODGSON, J., THOMPSON, R., SOUTER, W.G. and BARTHRAM, G.T. (1982). The diet ingested by sheep grazing swards differing in white clover and perennial ryegrass content. *Grass and Forage Science* **37**, 209

MORRISON, J., JACKSON, M.V. and SPARROW, P.E. (1980). Response of perennial ryegrass to fertiliser nitrogen in relation to climate and soil. *Grassland Research Institute Technical Report 27*

NATIONAL INSTITUTE OF AGRICULTURAL BOTANY (1981). *Recommended Varieties of Grasses*

NEW ZEALAND FARMER (December 1981). Supplement *Grazing Management*

NEWBOULD, P. (1982). Biological nitrogen fixation in upland and marginal area of the U.K. *Philsophical Transactions of the Royal Society, London* **B296,** 405–417

PARSONS, A.J. and LEAFE, E.L. (1981). Photosynthesis and carbon balance of a grazed sward. In *Plant Physiology and Herbage Production* (C.E. Wright, Ed.), pp. 69–72. British Grassland Society Occasional Symposium No. 13

PARSONS, A.J. and ROBSON, M.J. (1981). Seasonal changes in the physiology of S24 perennial ryegrass (*Lolium perenne* L). III. Partition of assimilates between root and shoot during the transition from negative to reproductive growth. *Annals of Botany* **45**, 733–744

RICHARDS, I.R. (1977). Influence of soil and sward characteristics on the response to nitrogen. In *Proceedings of International Meeting on Animal Production from Temperate Grassland* (B. Gilesnan, Ed.), pp. 45–49. Dublin, Irish Grassland and Animal Production Association

ROBSON, M.J. (1981). Potential production—what is it and can we increase it? In *Plant Physiology and Herbage Production* (C.E. Wright, Ed.), pp. 5–18. British Grassland Society Occasional Symposium No. 13

SCOTTISH AGRICULTURAL COLLEGES (1982). *Classification of Grass and Clover Varieties for Scotland 1981–82*, Publication 78

THOMPSON, D.J. (1977). The role of legumes in improving the quality of forage diets. In *Proceedings of International Meeting on Animal Production from Temperate Grassland* (B. Gilesnan, Ed.), pp. 131–135. Dublin, Irish Grassland and Animal Production Association

WILSON, D. (1981). The role of physiology in breeding herbage cultivars adapted to their environment. In *Plant Physiology and Herbage Production.* (C.E. Wright, Ed.), pp. 95–108. British Grassland Society Occasional Symposium No. 13

WRIGHT, C.E. (1981). Introductory remarks. In *Plant Physiology and Herbage Production* (C.E. Wright, Ed.), pp. 1–4. British Grassland Society Occasional Symposium No. 13

11

GRASSLAND MANAGEMENT FOR THE LOWLAND EWE FLOCK

W. RUTTER
East of Scotland College of Agriculture, St. Boswells, Melrose, Roxburghshire

Successful grazing management of ewes and lambs depends on the provision of clean grazing, the timeous application of nitrogen and a willingness to manipulate stocking rate so that the sward is kept in a vegetative state and herbage allowance is maintained within acceptable limits. The grazing system itself must be sufficiently simple to be implemented by people with little experience or knowledge of grassland management, sufficiently flexible to cope with our unpredictable weather and sufficiently conservative to maintain the sward over its expected life. If such a system can be implemented by following simple and objective management guidelines, future refinements can be made with experience.

Stocking rate and gross margin

There are two ways of increasing stocking rate; one is by keeping the same number of sheep on less ground, the other is by keeping more sheep on the same ground and there is a profound difference between the two. The former obviously requires that the area of ground which has been released will be used for more productive purposes but it needs no additional capital for the sheep enterprise. With a basic gross margin of £40/ewe before forage costs have been assumed and if lamb growth rates and nitrogen usage/ewe remain constant, doubling the stocking rate from 10 ewes/hectare to 20 ewes/hectare during summer grazing, gives at 1982 prices an increase of 80% in gross margin/hectare (*Table 11.1*). This increase is only 80% because a constant area of silage, in this case 2.5 hectares/100 ewes, is associated with each stocking rate. The second method, that of increasing ewe numbers on the same area, involves additional capital for the extra ewes and if interest is charged on their purchase price, doubling the stocking rate during summer grazing gives an increase in gross margin/hectare of only 56% (*Table 11.2*).

These are maximum increases and they will be lower if supplementary feeding after lambing or requirements of nitrogen/ewe and phosphate and potash/hectare increase with stocking rate. They are also calculated without any clear knowledge about the interrelationships between stocking rate, grazing pressure or herbage allowance, herbage intake and lamb growth other than that any relationship is likely to be curvilinear rather

Table 11.1 EFFECT OF INCREASING STOCKING RATE BY REDUCING LAND AREA FOR 100 EWES

Ewes/hectare during summer grazing	*Grazing +*[a] *silage/ 100 ewes* (ha)	*Grazing +*[b] *silage variable costs* (£)	*Gross margin*[c]	
			/100 ewes (£)	*/hectare* (£)
10	10 + 2.5	1270	2730	218
12	8.35 + 2.5	1200	2800	258
14	7.15 + 2.5	1144	2856	296
16	6.25 + 2.5	1106	2894	331
18	5.55 + 2.5	1072	2928	364
20	5 + 2.5	1050	2950	393

[a]Silage requirements based on 2.5 hectares/100 ewes
[b]Silage variable costs/hectare including aftermath fertilizer, are taken as £140. Grazing variable costs/hectare are calculated as: 12 kg nitrogen/ewe @ 40p plus 50 kg P_2O_5 @ 40p plus 50 kg K_2O @ 18p plus £15 share of grass seeds. In summary, grazing costs/hectare are £4.80 × no. of ewes + £44
[c]Based on gross margin before forage costs of £40/ewe for all stocking rates (see text)

Table 11.2 EFFECT OF INCREASING STOCKING RATE BY INCREASING EWE NUMBERS ON 12.5 HECTARES

Ewes/hectare during summer grazing	*Total number of ewes (and hectares for grazing + silage)*	*Grazing +*[a] *silage variable costs* (£)	*Gross margin*[b]	
			/100 ewes (£)	*/hectare* (£)
10	100(10 + 2.5)	1270	2730	218
12	115(9.6 + 2.9)	1381	2669	246
14	130(9.3 + 3.2)	1480	2631	274
16	143(8.9 + 3.6)	1581	2594	297
18	155(8.6 + 3.9)	1664	2572	319
20	167(8.3 + 4.2)	1750	2551	340

[a]Calculated as in *Table 11.1*
[b]Based on a gross margin before forage of £40/ewe for the first 100 ewes and £30/ewe for the additional ewes. This reduction represents an interest charge of £10 on the capital cost of additional ewes (e.g. £63 @ 16%)

than linear (Hodgson, 1975). Over the whole range of stocking rates it appears at first sight that the increase in gross margin can comfortably accommodate any significant reductions in lamb growth. But in any particular situation the effect on animal performance of a small increase in stocking rate is of more concern, particularly if the relationship is curvilinear, and the tabulated figures indicate that no reduction in lamb growth can be tolerated. Even with the more profitable method of intensifying, that of grazing the same number of ewes on less area (*Table 11.1*), as stocking rate/grazing hectare is increased from, for example, 16 ewes with say 29.3 lambs to 18 ewes with 33 lambs, these 33 lambs need only reduce in value by £1/head to erode the expected increase in gross margin of £33/hectare. In the first year this reduction would be acceptable if it was more than compensated by whatever enterprise utilized the hectares which were released; valuable experience would have been gained on the problems of intensification but thereafter there would be opportunity to readjust ewe numbers and the principles outlined in *Table 11.2* would then apply. The expected increase in gross margin of £22/hectare would be eroded by a decrease in lamb value of £0.7/head and at current prices this is equivalent to 0.8 kg liveweight or less than 10 g/day over a 12 week grazing period. Such small increases in stocking rate can be associated with higher reductions in lamb growth (Young and Newton, 1975; Nolans *et al.*, 1977) and there may be further penalties associated with overstocking in terms of higher parasite levels and their costlier control, disrupted pattern of lamb sale and a reduction in subsequent ewe condition and performance (Macleod, 1975; Gibb and Treacher, 1978). These are in addition to the earlier problems which can arise when there is a lengthy period between lambing and the start of active herbage growth; more costly supplementary feeding or a higher mortality rate could themselves erode the expected increase in gross margin.

Traditional stocking rates

It is important from a financial aspect that stocking rate is not increased to the point where individual lamb performance is likely to be impaired and this is presumably one reason why stocking rates of ewes and lambs at least in the south east of Scotland (e.g. eight ewes with approximately 14 lambs/grazing hectare) have traditionally been low in relation to stocking rates with cattle or to the crop-growing potential of the land. Other possible reasons were identified by Runcie (1981) at a recent British Grassland Society symposium where he stated that 'the promotions of more intensive systems of sheep production which produce more poor quality lambs, as has happened in the past, have been singularly ineffective in spite of possibly more financial return. The maintenance and improvement of lamb quality so that good shepherds can be proud of the stock which they produce for sales has an important bearing on the effectiveness of the advisory or adoption process'.

Attempts through advisory work to achieve better grass utilization by increased stocking rates and high applications of nitrogen fertilizer have proved disappointing in terms of individual lamb performance and lead to the unfounded conclusion by many farmers that high nitrogen use and good lamb growth are incompatible; unfounded because it is difficult to distinguish between the effect of nitrogen and the detrimental effect of increased levels of gastrointestinal parasites.

Mixed grazing of cattle and sheep is known to reduce the overall levels of gastrointestinal parasites which are infective to lambs and any benefit in growth rate for either species is likely to be greatest when this species is the minor component of the mixture (Brelin, 1979). But both Baxter (1959) and Black (1960) have shown that it is not effective as a means of eradicating and preventing disease from an already present *Nematodirus* spp. contamination and, moreover, mixed grazing is not always feasible, either because of disparate numbers of sheep and cattle or because of an inability to provide adequate grass for the cattle in the early part of the grazing season when they are turned out to join the sheep on an already overgrazed sward.

It has long been established that the simplest and most effective way of controlling gastrointestinal parasites in lambs is to graze ewes and lambs on *clean* pasture, i.e. pasture which has not carried lambs or young sheep during the previous twelve months (Black, 1960). In this way the life cycle of most of the important worm parasites including *Nematodirus* spp. is broken and the primary source of worm infection for lambs is reduced. Furthermore, Gibson (1965) stressed that if ewes receive an anthelmintic dose before they enter their *clean* fields, their post-parturient worm egg output and thus the secondary source of worm infection for their lambs is also reduced.

The East College clean grazing system

The provision of *clean* grazing is easily achieved, for example, by alternating sheep grazing areas and cattle grazing areas each year and this is the basis of an integrated beef and sheep clean grazing system which has operated since 1973 on the Bush Estate of the Edinburgh School of Agriculture. It was designed with no primary comparative treatments and with sufficient size and flexibility to be representative of a mixed farming enterprise so that it could demonstrate to visiting farmers and students that clean grazing is an effective way of reducing the risk of gastrointestinal parasites and of allowing the intensification of sheep without any reduction in lamb performance.

The system has been described in detail by Rutter (1975) and Rutter *et al.* (1977), but in summary it involved the integration of 150 Scotch Halfbred ewes lambing to a Suffolk sire in late March, 33 autumn-calving beef cows and hay conservation on approximately 25.5 hectares of long-term pasture. The grass was divided into three areas of similar size (see *Figure 11.1* on p.213) to allow a rotation from year to year of cattle, sheep and hay so that ewes and lambs were grazed on relatively *clean* grass each year.

All grazing was on a set-stocked basis and lambs and calves grazed the hay aftermath after weaning, but the critical feature was that lambs and young sheep were denied access to the cattle grazing fields for the whole year (late April to late April) because this was set aside as next year's clean grazing for sheep. But in late autumn, during the later stages of mating and after the cows and young calves were housed, it was often necessary to graze the ewes over the cattle grazing fields, i.e. next year's clean sheep grazing, in which case they were first dosed with an anthelmintic to avoid pasture contamination. All the available pasture, including last year's sheep fields, was used at lambing by ewes and lambs but at least the majority of sheep were denied access to their clean grazing fields for as long as possible on the assumption that the more grass that was available (at least 8 cm in height) when full grazing started, the better would be its continued carrying capacity without further supplementary feeding. Full grazing normally started in late April when ewes and any lambs over four weeks of age were dosed on entering their clean fields. There was no further routine dosing until weaning; lambs were sold as they reached an estimated carcass weight of 19 kg and at weaning in late July all remaining lambs were dosed and moved to aftermath.

Results for the first five years are summarized in *Table 11.3* and show that the management principles adopted allowed good growth rate of lambs at relatively high summer stocking rates of 17.5 ewes with 29 lambs/grazing hectare, with an average nitrogen usage on the grazing of 211 kg/hectare. Average liveweight and age of lambs at sale were 41.0 kg and 137 days respectively and liveweight gains from birth to sale of 275 g/day compare favourably with those recorded by Barton (1975) and Speedy *et al.* (1977) on private farms in the surrounding area where stocking rates with similar breeds of ewes and lambs were only 40% of those in the East College System.

Table 11.3 GROWTH RATE OF LAMBS IN THE EAST COLLEGE CLEAN GRAZING SYSTEM

	Summer grazing period				
Variables	*1973*	*1974*	*1975*	*1976*	*1977*
Number of ewes	150	150	150	150	150
Lambing date	16 March	4 April	9 April	24 March	26 March
Rainfall in May, June and July (mm)	215	142	119	134	181
Lamb growth to mid-June					
Number of lambs	263	246	247	248	239
Liveweight (kg)	33.3	31.6	26.8	30.8	27.2
Age (days)	94	81	62	77	73
Daily gain (g)	303	332	357	337	312
Lamb growth to sale					
Number of lambs	262	246	233	248	239
Liveweight (kg)	40.1	41.8	40.8	44.0	42.0
Age (days)	135	131	128	144	147
Daily gain (g)	266	284	282	272	254

The incidence of worm eggs in dung samples from ewes and lambs was negligible and worm larvae counts from pasture sampling were very low with peaks of 95 larvae/kg of grass compared with an expected peak in July of over 1000 under continuous sheep grazing conditions (R.J. Thomas, personal communication).

The five-year period embraced notable variations in climate, particularly a low mean temperature during May 1975 and abnormally dry summers in the three middle years. Unfortunately no objective measurements of herbage mass were made but variations in herbage availability and thus in grazing pressure, between fields within years and between years, were tolerated without adjustments to stocking rate or the use of supplementary feeding and without any noticeable effect on lamb performance. But such data on herbage mass would have been incomplete without simultaneous measurements of intake. It may be that in some cases an apparent high herbage mass is a reflection of low intakes or vice versa and current work at the Hill Farming Research Organisation will hopefully add to our limited knowledge on the complex relationships between herbage mass, digestible organic matter intakes and lamb growth rates.

Practical clean grazing systems

On any particular farm it is necessary to identify acceptable lower and upper limits of stocking rate for both cattle and sheep. For example, with appropriate fertilizer applications, a 10 hectare field of productive grass might be grazed successfully with either 120–180 ewes with lambs or 40–60 store cattle of approximately 300 kg or 27–40 cows with calves. Thus, in the extreme, the ratio of sheep to store cattle could vary from 2:1 to 4.5:1 yet they could still be regarded as requiring similar areas for summer grazing. These ratios, however, do not equate in terms of area required for conservation of winter feed, at least on farms where home-conserved grass provides the majority of winter roughage (*Table 11.4*). In general, the

Table 11.4 SILAGE REQUIREMENTS (AT 25% DRY MATTER) OF SHEEP, STORE CATTLE OR SUCKLER COWS

Type of stock	*Duration of winter* (days)	*Daily intake* (kg/head)	*Total intake* (tonnes)	*Hectares required at silage yields/hectare*		
				17	*20*	*25*
100 ewes	130	4	50	3	2.5	2
33 store cattle	160	19	100	6	5	4
22 suckler cows with calves	180	25	100	6	5	4

winter roughage requirement for ewes is only half that of cattle, primarily because they have a longer grazing season and because concentrates may provide a higher proportion of their intake.

A ewe flock normally requires, or at least can make good use of, a larger area of aftermath than that provided by its own conservation needs of, say, 2.5 hectares/100 ewes. With a weaning percentage of 170 and if half the lambs are sold by weaning, the remaining 85 lambs/100 ewes would need up to 3.5 hectares of aftermath. This is another reason why the integration of sheep and cattle is so beneficial; the cattle need a higher proportion of the total conserved product but sheep need a higher proportion of the total conservation area for grazing during the rest of the year. With these factors in mind, it is relatively simple to design integrated grazing systems which suit different types of grassland.

ALTERNATIVE ROTATIONS ON PERMANENT OR LONG-TERM GRASS

1. The two grazing areas for cattle and sheep are alternated annually. Any conservation (hay or silage) is taken from a separate area and the aftermath is grazed by any stock (*Figure 11.1(a)*).
2. The two grazing areas for cattle and sheep are alternated annually and conservation (of fixed or variable area) is taken from within each grazing area. The aftermath in the cattle area is *not* grazed by weaned lambs since this may be part of next year's clean sheep grazing (*Figure 11.1(b)*).
3. Three areas of similar size each follow the rotation of cattle, sheep, conservation etc. as in the East College system. The aftermath is grazed by any stock (*Figure 11.1(c)*).

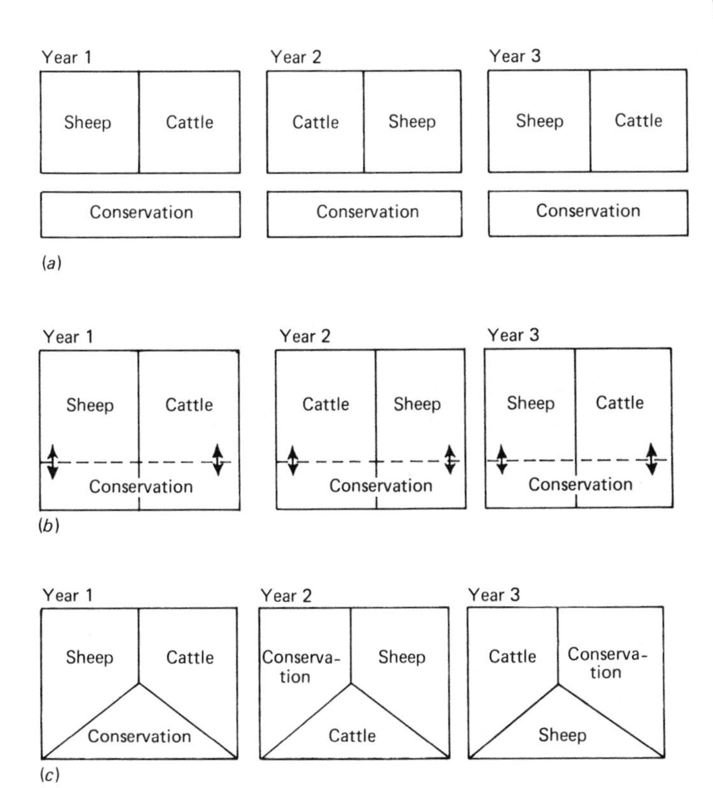

Figure 11.1 Three alternative rotations on permanent or long-term grass leys, integrating cattle, sheep and conservation into a clean grazing system

ALTERNATIVE ROTATION ON SHORT-TERM GRASS

These cover different lengths of ley and accommodate varying proportions of sheep and cattle, but note that if sheep follow conservation (con.) it is necessary to keep weaned lambs off the aftermath (*Table 11.5*).

Table 11.5 ALTERNATIVE ROTATIONS ON SHORT-TERM LEYS

Alternatives	*Autumn of reseeding year*	*Spring and Summer*			
		1st year grass	*2nd year grass*	*3rd year grass*	*4th year grass*
1	ewes	sheep	con.	cattle	sheep
2	ewes	sheep	cattle	sheep	con.
3	lambs	cattle	sheep	con.[a]	sheep
4	ewes	sheep	con.	cattle	–
5	ewes	sheep	cattle	con.	–
6	lambs	cattle	sheep	con.[a]	–
7	ewes	sheep	con.	–	–

[a]If sheep follow conservation do not use weaned lambs

FORWARD PLANNING

These previous alternatives give a broad guideline as to how the farm resources can be organized to accommodate a clean grazing system and there is no reason to assume that the cattle will gain any less benefit than the sheep, but once a system or combination of systems has been established, any forward planning can be as follows:

1. Each year, just before *weaning*, identify next year's clean grazing, i.e. fields which are currently being used for either cattle, or hay and silage, or undersown cereals or direct reseeding. Keep *weaned* lambs and young sheep off these fields and graze them instead with cattle or dosed ewes up to and including tupping (mating).
2. After *lambing*, dose the ewes and any lambs over four weeks of age immediately *before* they enter their clean grazing. Allocate the remaining grass to cattle and conservation as convenient, ensuring that young weaned cattle receive priority for clean grazing.

It is important to dose the ewes immediately before they go into their clean grazing fields so that the build-up of worms during the grazing season and the carryover of worms to the following spring is minimized. These same fields can then be grazed safely by ewes and lambs during next lambing time until ewes and any lambs over four weeks of age are again dosed and moved to their next clean grazing fields.

Conversely, failure to dose before grazing may lead to a cumulative build-up of worms to the extent that the system finally breaks down.

Grazing management

If the foregoing recommendations on dosing are followed there should be no need for further routine dosing of lambs at least until normal weaning. If lambs show symptoms of worm infestation they should for expediency be dosed but this should be preceded by dung sampling and any other tests which a veterinary surgeon and soil scientist might recommend to identify any other cause. One other aspect of lamb health which is often overlooked is that of water; troughs should be designed so that lambs can get out as well as in and they should also be sited so that any prematurely orphaned lamb can obtain water.

In general, no more than 150 ewes with lambs should be grazed in one group, particularly during the first weeks after lambing when mismothering can occur or when there is a lack of experience or skill in handling large numbers. This is one reason why set stocking may be preferred to rotational grazing apart from the fact that it avoids the need to erect fences and provide water in each paddock. Moreover, rotational grazing is only likely to show a benefit at relatively high stocking rates and anyone contemplating adopting it should only do so after having made a success of set stocking. The following recommendations are made with regard to set stocking but anyone wishing further details about management within a rotational grazing system should obtain a copy of *Farmer's Booklet No. 1* from the Grassland Research Institute at Hurley, Berks.

FERTILIZER APPLICATIONS

In the absence of more accurate local knowledge of nitrogen requirements an objective method is to base it on 12 kg/ewe, half of which (6 kg) should be applied in March and a quarter (3 kg) each in mid-May and mid-June. Thus at 16 ewes/hectare, these three dressings would be 96 kg, 48 kg and 48 kg respectively, but the need for any further applications will depend mainly on the contribution being made by clover, pattern of lamb sale and expected time of weaning. Most advisory services now predict the optimum date for the first application based on some aspects of soil temperature, but the need for the second application in mid-May (or even earlier in the more southerly areas) is not so obvious. This is often the period of most active growth, and in the College system, this application was made regardless of herbage availability on the assumption that it would forestall an otherwise inevitable grass shortage in June.

Professional advice should be sought to investigate any suspected trace element deficiency, and phosphate and potash application must be made with regard to maintaining sward quality and long-term soil fertility. Time of phosphate and potash applications is of less importance although there is a common if not well substantiated belief that potash applied in early spring may induce magnesium deficiency. It should be appreciated that nitrogen in the form of ammonium nitrate or ammonium sulphate has an acidifying effect on the soil and high applications will significantly increase lime requirements.

START OF FULL GRAZING

During the first weeks after lambing there is an inherent and understandable temptation to spread ewes and lambs over all the available grass, usually at the expense of silage or hay yield or early turnout of cattle. It is good husbandry to retain ewes and young lambs in small groups to prevent mis-mothering but competition from other enterprises may be such that full stocking needs to be achieved sooner than desirable. Freedom from grazing during the previous winter and timeous nitrogen application will encourage earlier herbage availability and, if full supplementary feeding is continued for as long as necessary, there may be no initial effect on milk yields or lamb growth rate (Gibb and Treacher, 1978). Data presented by Gibb and Treacher (1978) suggest that when herbage availability is unrestricted, daily organic matter intake of a ewe with twins reaches a maximum of up to 40 g/kg liveweight at about the second week of lactation and declines thereafter. This decline in the ewe's intake will be more or less countered by the increasing herbage intake of her lambs and in broad generalization, if herbage availability is unrestricted, daily organic matter intake of each family unit will remain more or less constant at, say, 2.8 kg/day. Thus a hectare stocked with 18 ewes and twins would need to provide 50 kg organic matter (56 kg dry matter) daily if maximum intakes are to be achieved. Data from the West of Scotland College of Agriculture (I.A. Dickson and J. Frame, personal communication) indicate that such yields are only likely to be attained with highly fertilized productive swards and it must be appreciated that overgrazing in the early stages and the poaching effect of tractors and trailers during supplementary feeding might have long-term consequences, particularly on young grass, in terms of sward recovery and ewe and lamb performance. These are among the factors to consider when deciding maximum stocking rates.

MANIPULATING GRAZING PRESSURE

With clean grazing and late March lambing a proportion at least of single lambs should be saleable in early June. This is a major advantage of mixing singles and twins; sale of lambs can help reduce stocking rate and thus grazing pressure as herbage growth slows down. However, earlier in the season any need to adjust stocking rate can be achieved by one of the following methods.

1. Set stock in a number of fields. Each field is stocked initially at less than its potential and thereafter, between-field variations in grazing pressure are reduced by moving a proportion of ewes and lambs, provided that families are not split during moving. If overall undergrazing occurs, cattle can be brought in or the sheep can be grouped to release a field for cattle or conservation. This method could obviously provide a lead-in to a future rotational grazing system.
2. Set stock on only a proportion (say 80%) of the grazing area at a rate where undergrazing is unlikely. Reserve the remaining 20% as a buffer area or 'safety valve' so that it can be used for conservation if it is not needed for grazing.

Any decision to conserve the buffer area or 'safety valve' is unlikely to be regretted; aftermath is soon available after silage removal, and wet weather which delays the removal of hay may simultaneously encourage growth on the grazing area. Conversely, during periods of overgrazing, a more difficult decision to make is when to move the sheep to their 'safety valve'. Low herbage allowance may be compensated to some extent by higher digestibility and, if warm moist weather conditions are imminent, it is surprising how closely cropped the grass can become without any noticeable effect on lamb growth. But there is obviously a stage where reduced intakes are likely to have a permanent effect on ewe and lamb performance; experienced shepherds may suggest that changes in animal behaviour are a good indication of this, but in their absence any decision to move the flock will depend not only on current herbage mass but also on expected future growth. It is unlikely that anyone managing ewes and lambs at grass will understand the physiological processes which affect sward recovery after a period of severe defoliation; suffice it to say that recovery will be slower during cold dry periods. In the absence of grazing, recovery is often quicker than expected and if the sward is to be maintained in a vegetative state, it is important that herbage mass does not become too great before the sheep are returned. An absence of between three and five days is all that may be required but an alternative could be to retain the sheep in the 'safety valve' and to use part or all of their previous grazing area as a new 'safety valve'.

If there is no 'safety valve' other than a field which is intended for next year's clean sheep grazing, there is no alternative, apart from supplementary feeding, but to relax the rules of clean grazing. This is obviously preferable to the alternative of a permanent reduction in flock performance and the later in the season that such a move occurs, the more important it is to dose both ewes and lambs before moving so that the risk of overwintered worm larvae is minimized. But a need to move the flock could indicate an over-ambitious stocking rate and it might be prudent to reduce it in future years.

Conclusion

No direct reference to liveweight recovery of ewes is made; in the absence of contrary evidence it is assumed that this has a close positive correlation both with lamb growth rate and the rate at which lambs can be marketed before normal weaning. Space limitations do not allow discussion on seeds mixtures or the merits of clover as an alternative nitrogen source. Farms under little or no financial pressure may be able to afford the luxury of lower stocking rates under a high clover/low nitrogen system and on some fields this system may be obligatory because of terrain. But whichever system is chosen it is vital that increases in stocking rate are made gradually and with the buffering capacity of an area set aside as a 'safety valve'.

The long-term effect of a successful grazing system should be gradual improvement in sward quality or at least the maintenance of an already adequate sward. It should provide a better basis on which to make sensible decisions about reseeding and seeds mixtures and it should give a better understanding of the interrelationships between herbage growth and animal response so that if economic circumstances dictate, a successful transition can be made to systems which require greater reliance on clover or other legumes as a source of nitrogen.

References

BARTON, G.M. (1975). The effect of variation in physical production factors on the profitability of upland sheep production. *East of Scotland College of Agriculture Report to the Meat and Livestock Commission*, pp. 45–46

BAXTER, J.T. (1959). Mixed grazing and *Nematodirus* disease of lambs. *Veterinary Record* **71**, 820–823

BLACK, W.J.M. (1960). Control of *Nematodirus* disease by grassland management. *Proceedings of 8th International Grassland Congress*, pp. 723–726

BRELIN, B. (1979). Mixed grazing with sheep and cattle compared with single grazing. *Swedish Journal of Agricultural Research* **9**, 113–120

GIBB, M.J. and TREACHER, T.T. (1978). The effect of herbage allowance on herbage intake and performance of ewes and their two lambs grazing perennial ryegrass. *Journal of Agricultural Science, Cambridge* **90**, 139–147

GIBSON, T.E. (1965). Helminthiasis in sheep. *Veterinary Record* **77**, 1034–1041

HODGSON, J. (1975). The influence of grazing pressure and stocking rate on herbage intake and animal performance. In *Pasture Utilization by the Grazing Animal* (J. Hodgson and D.K. Jackson, Eds.), pp. 93–103. Occasional Symposium No. 8 of the British Grassland Society

MACLEOD, J. (1975). Systems of grazing management for lowland sheep. In *Pasture Utilization by the Grazing Animal* (J. Hodgson and D.K. Jackson, Eds.), pp. 129–134. Occasional Symposium No.8 of the British Grassland Society

NOLANS, T., FLANAGAN, S.P., GRENNAN, E. and O'TOOLE, M.A. (1977). Potential of Irish grassland for sheep production. In *Proceedings of International Meeting on Animal Production from Temperate Grassland* (B. Gilsenan, Ed.), pp. 79–87. The Irish Grassland and Animal Production Association, An Foras Talúntais, Belclare, Galway

RUNCIE, K.V. (1981). The promotion and adoption of new ideas in hill and upland farming. In *The Effective Use of Forage and Animal Resources in the Hills and Uplands* (J. Frame, Ed.), pp. 105–110. Occasional Symposium No. 12 of the British Grassland Society

RUTTER, W. (1975). *Sheep from Grass*. Bulletin 13. East of Scotland College of Agriculture

RUTTER, W., BLACK, W.J.M., CARSON, I.S., FITZSIMONS, J. and SWIFT, G. (1977). An integrated sheep and beef system—its development and extension. *Proceedings European Association for Animal Production, 28th Annual Meeting, Brussels*, pp. 1–5

SPEEDY, A.W., MACKENZIE, C.G., USHER, H.J. and MCKELVIE, D. (1977). The intensification of upland sheep flocks. *Animal Production* **24**, 159 (Abstract)

YOUNG, N.E. and NEWTON, J.E. (1975). A comparison between rotational grazing and set stocking with ewes and lambs at three stocking rates. *Animal Production* **21**, 303–311

12

MEETING THE FEED REQUIREMENTS OF THE HILL EWE

A.J.F. RUSSEL
Hill Farming Research Organisation, Bush Estate, Penicuik, Midlothian, UK

The feed requirements of the hill ewe for satisfactory reproduction, to ensure an acceptable performance at lambing time, and for the production of sufficient milk to promote high lamb growth rates, have been presented in previous chapters. This chapter outlines briefly the extent to which these requirements are likely to be met within a traditional system of hill sheep management, considers some of the consequences to production arising from the annual cycle of nutrition which that system affords, and gives an example of the effects on production of a different system of management which seeks to provide improved nutrition at certain critical stages in the annual cycle.

The traditional system of hill sheep management

In the traditional system of hill sheep management ewes are set-stocked on hill pastures throughout the year and in many instances never have access to improved pasture at any time. The stock are often territorial and little if any control is exercised over their grazing, although in some areas they are shepherded to the lower ground in the morning and driven to the higher ground later in the day. They are dependent on grazed herbage at all times, but it is becoming common practice to supplement this with some form of concentrate feeding in the weeks prior to lambing. Hay is usually given only as storm feeding when the pastures are covered by snow and grazing is difficult or impossible.

Stocking rates are generally low, varying from about 0.75–4.0 hectare/ewe, and are determined by the 'winter carrying capacity'. This is the number of ewes which experience has shown can be overwintered commensurate with an acceptable lambing performance and the minimum of supplementary feeding. Levels of production are low in comparison with other forms of animal production from pastoral resources, typical outputs being of the order of 20 kg weaned lamb/ewe or 15 kg lamb/hectare, with large variations around these values.

NUTRITION IN THE TRADITIONAL SYSTEM

The quality of herbage available for grazing on hill pastures, and thus the quantity of nutrients ingested, are a function of the interaction between the

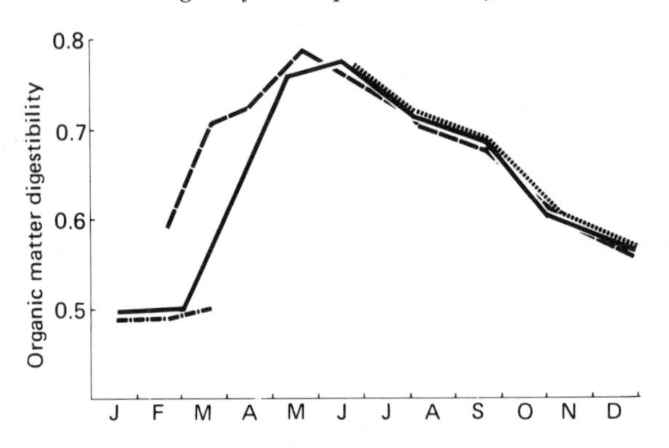

Figure 12.1 Seasonal changes in the quality of herbage ingested by free-grazing sheep on a grassy hill pasture. ---- 1961; —— 1962; ······ 1963; -·-·- 1964

year-round stocking of the pastures and the highly seasonal nature of pasture growth. Seasonal changes in the quality of herbage ingested by wether sheep, set-stocked on a predominantly grassy hill pasture, were studied by Eadie (1967). These results, illustrated in *Figure 12.1*, show that organic matter digestibility (OMD) is at a maximum of about 0.75 in May and June when the pasture is growing most actively. At this time, however, only a very small proportion of the current season's growth is consumed, and the major part of the production matures and deteriorates in quality. Although sheep graze selectively, some of the mature material is inevitably consumed along with the newer growth and dilutes the quality of the intake. Thus by July the OMD of the intake has fallen to about 0.70, and continues to decline throughout the remainder of the pasture growing season. From October the sheep are entirely dependent on mature and senescing material, and the quality of the intake declines even more rapidly, reaching around 0.50 to 0.48 in January, as the sheep continue to graze selectively and the quality of the material available deteriorates further. In this part of the cycle variation between years is relatively small, but the rapid increase in pasture quality following the initiation of the new season's growth is dependent on soil temperature and can vary markedly from one year to another.

The pattern of energy intake over the year is broadly similar to that of ingested pasture quality but, because of the relationship between voluntary intake and diet quality (Hodgson, 1977) the amplitude of the cycle is very much greater. The information presented in *Figure 12.2* on seasonal changes in the energy intake of hill sheep grazing different plant communities has been calculated from estimates of ingested herbage quality (Eadie, 1967; Eadie and Black, 1968; Milne, Bagley and Grant, 1979). The conventional value used for the maintenance requirements of a 50 kg ewe is around 6.8 MJ ME/day, but for ewes on hill grazing the figure is likely to be considerably greater although actual values have not been determined. On even the better hill pastures (the *Agrostis–Festuca* communities) energy intakes are likely to range from sub-maintenance from November to March, with minimum values of less than 5 MJ ME/day, to maximum values of more than 20 MJ ME/day in lactating ewes in June. On poorer

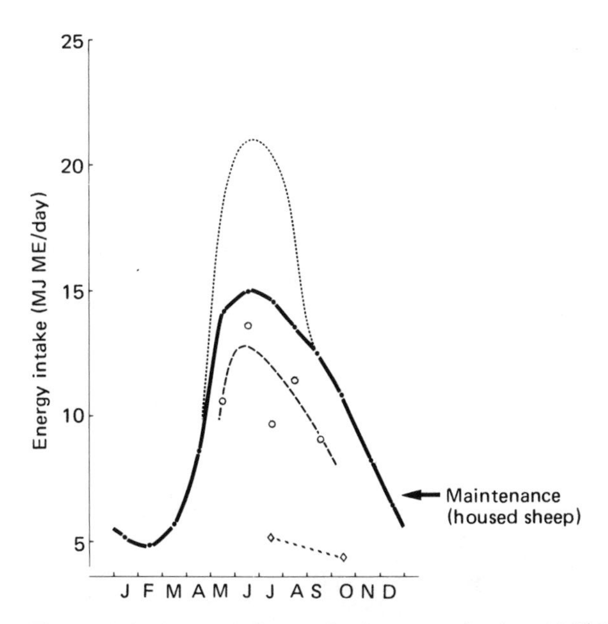

Figure 12.2 Seasonal changes in the energy intake of hill sheep grazing different plant communities. —— *Agrostis–Festuca* (dry sheep); ······ *Agrostis–Festuca* (lactating sheep); ---- *Nardus*; ◇----◇ *Calluna*

plant communities, such as those containing *Nardus* or *Calluna* (heather), energy intakes will be lower. Even the better values appear low in relation to normally accepted standards, but it has been shown (Eadie, 1970) that increases in the low stocking rates without other changes in management would depress the minimum energy intakes in winter to still lower and quite unacceptable levels. Thus the traditional system of hill sheep management perpetuates a vicious cycle in which the low stocking rates required to ensure an acceptable minimum level of winter nutrition lead to a level of undergrazing during the growing season which lowers diet quality and nutrient intake.

SEASONAL CHANGES IN EWE LIVEWEIGHT AND BODY COMPOSITION

The characteristic annual patterns of ingested pasture quality (*Figure 12.1*) and energy intake (*Figure 12.2*) are determined to a large extent by the year-round, set-stocked system of grazing, but at the same time have a pronounced effect on the stock themselves, resulting in an equally characteristic annual cycle of ewe liveweight change (*Figure 12.3*). Typically, ewe liveweights are at a maximum in late October, and begin to decline before mating in late November and December. The extent of the apparent loss in weight over winter is variable, but values of 8–10 kg between October and February are not uncommon. By the end of February the weight of the products of conception is probably about 3 kg, but in calculating net maternal weight losses this is offset by a reduced weight of gastrointestinal contents. There is normally a small increase in liveweight over the final weeks of pregnancy as a result of foetal growth. Substantial losses of liveweight are not necessarily of great biological significance in themselves. What is important in the case of the hill ewe is the relatively low levels of fatness over which these changes occur. The typical hill ewe at peak condition in the autumn carries only some 6 kg fat (about 12–14% of liveweight), and by the end of pregnancy more than half of this modest reserve has generally been utilized. It has also been shown that some 20% of the ewe's body protein and minerals are catabolized during pregnancy (Russel, Gunn and Doney, 1968; Russel, Foot and McFarlane, 1982).

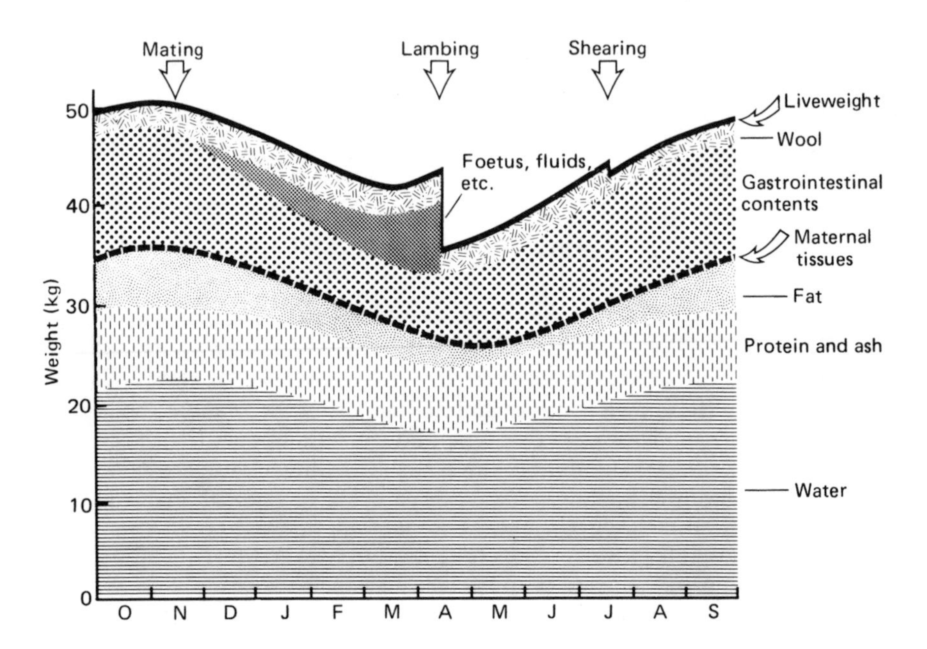

Figure 12.3 Seasonal changes in the liveweight and body composition of ewes in a typical hill environment

Parturition does not signify the end of the period of dependence on body reserves. Liveweight generally increases fairly quickly following lambing, but this is due principally to a rapid expansion of voluntary intake which greatly increases gut fill. The early weeks of lactation are the period of highest nutrient requirement, and during this time body reserves are generally further depleted with amounts of body fat frequently decreasing to less than 1 kg or 2% of liveweight.

The main recovery of liveweight and replenishment of body reserves of fat and protein take place in the later stages of lactation and after weaning in August. As described earlier, the quality of ingested herbage has by that time declined substantially and this places a major constraint on the extent of the gain in liveweight and maternal tissues which can be achieved before the next period of submaintenance nutrition.

Effects of nutrition on the components of production

In Chapter 5, Gunn has presented evidence showing that hill ewes are capable of ovulation rates of around 200%, provided that they are in good body condition at mating (a condition score of 3–3.5 as described by Russel, Doney and Gunn, 1969). However, the annual cycle of nutrition afforded by the traditional system of management is such that most ewes are in only moderate body condition (condition score 2–2.5) at their maximum in late October, and are generally losing weight and condition prior to and during mating in late November–December (*Figure 12.3*). Energy intakes during November range typically from 6–9 MJ ME/day, depending on the nature of the vegetation, whereas an intake of around 15 MJ ME/day would be required to bring about the improvement in body condition needed to ensure a condition score of 3 in a 50 kg ewe at that time. The continuing low level of nutrition post-mating, frequently linked with adverse climatic conditions, is also likely to lead to a high incidence of embryonic mortality (Gunn and Doney, 1975; Doney, Smith and Gunn, 1976).

Throughout the second and third months of pregnancy energy intakes are invariably lower than the maintenance requirements quoted for housed ewes, and considerably below the recommended allowance of 9–10 MJ ME/day for ewes in less than condition score 3.5 at mating (Meat and Livestock Commission, 1981). Recent evidence (Russel *et al.*, 1981; Russel, unpublished) indicates that the combination of relatively poor body condition and low energy intake at this time will reduce lamb birthweight, irrespective of the level of nutrition in late pregnancy. Although supplementary feeding in late pregnancy is now practised much more widely than was formerly the case, intakes of both energy and protein are substantially lower than those advocated in Chapter 6 by Robinson. It has been estimated that the consequent undernourishment, even where supplementary feeding is used, is likely to reduce the birthweight of twin lambs by 25%, thus severely prejudicing their chances of survival (Russel, Doney and Reid, 1967).

The hill ewe has a considerable potential for high levels of milk production, as has the hill lamb for growth. Peak levels of milk production of 3 kg/day in twin-suckling ewes and lamb growth rates of 350–400 g/day have been recorded on many occasions (e.g. Peart, 1967; HFRO, 1979a). In traditionally managed flocks, however, peak levels of milk production are typically 1.0–1.2 kg/day, and lamb growth rates of around 230 g/day to marking generally decline to less than 160 g/day from marking to weaning (Armstrong and Eadie, 1973). The differences between potential and actual levels of production are wholly explicable in terms of nutrition. Peak energy intakes of lactating ewes grazing *Agrostis–Festuca* pastures are in excess of 20 MJ ME/day for only a short period (*Figure 12.2*) while those grazing communities in which *Nardus* or *Calluna* predominate are unlikely to approach that figure at any stage, and fall far short of the allowances of around 19 and 26 MJ ME/day required to produce reasonable levels of milk production in single- and twin-suckling ewes respectively (Treacher, Chapter 7).

From the above brief summary, and from work in HFRO and elsewhere reviewed by Russel (1971; 1978a), Lucas (1975) and HFRO (1979a), it is clear that nutrition is the main factor limiting production from hill sheep. In the majority of instances energy is the first limiting nutrient. Hill pasture species are not noted for their high protein levels but in most cases the nitrogen content (MacRae, Campbell and Eadie, 1975) is greater than the equivalent of 10 g crude protein (CP)/MJ ME and therefore not low in relation to energy. MacRae *et al.* (1979) have shown, for example, that a typical *Agrostis–Festuca* diet of 0.545 OMD and 14.3 g nitrogen/kg (ME = 7.6 MJ/kg dry matter (DM) and CP = 89.3 g/kg) would supply excess nitrogen relative to the energy available in the rumen. One exception to this generalization is where heather forms a major component of the diet. Milne (1974) found that in heather varying in OMD from 0.427–0.568, and in CP content from 79–101 g/kg DM, the apparent nitrogen digestibility ranged from only 0.267–0.422. The supplementation of heather diets with small quantities of nitrogen has been shown to result in marked increases in both voluntary intake and digestibility (Milne, Christie and Russel, 1979).

Requirements for improved performance

In the traditional free-grazing system of management where the annual stocking rate is, and must be, determined by the winter carrying capacity, the level of nutrition during the winter has been regarded as the principal limiting factor, and for a long time there was difficulty in looking beyond this. The work on the effects of nutrition on the components of performance has, however, demonstrated that hill ewes have a high potential production which is dependent for its realization on improved nutrition at times of year other than the winter (Russel, 1971).

The first requirement for an improvement in per ewe production is for improved nutrition in the 6–8 weeks prior to mating. This is essential if the ewes are to achieve the body condition required for higher ovulation and conception rates, but it also serves to provide the ewe with greater body reserves with which to meet the inevitable low levels of nutrition over the major part of the winter.

Given increased conception rates and higher levels of body condition, the second requirement is for improved nutrition in late pregnancy to ensure satisfactory birthweights, particularly of twin lambs, and adequate amounts of colostrum at parturition.

The third requirement is for improved nutrition as early as possible in the spring and throughout the summer months to enable the ewe to achieve the levels of milk production necessary for the satisfactory growth of twin lambs, and to ensure that these lambs maintain high growth rates as they become more dependent on herbage intake and less dependent on milk (Armstrong and Eadie, 1977). Thus the emphasis changes from a preoccupation with winter nutrition to a realization of the importance to production of nutrition at every other stage of the annual cycle.

The early work of Hunter (1962) and later studies by Eadie and Black (1968) were of central importance in seeking means of achieving these prescribed levels of improved nutrition in the hill environment. It was shown, for example, that controlled grazing of *Agrostis–Festuca* areas could reduce the amount of dead herbage in the sward from around 2000 to 600 kg DM/hectare. This not only had the effect of increasing the quality of herbage ingested by between 4 and 10 units with consequent increases in nutrient intake, but also subsequently resulted in a doubling of the efficiency of utilization (i.e. the amount harvested by grazing). This demonstrated the possibility of substantially improving nutrient intake during the spring, summer and autumn months by a system of controlled grazing of the better indigenous hill pasture communities. It was also evident at this stage that the necessary improvement in late pregnancy nutrition would require to be provided by purchased feedstuffs, particularly if the system of grazing control denied stock access to the better hill areas in the winter months (Eadie, 1971).

Improved nutrition from pasture

A detailed consideration of the techniques of pasture improvement applicable to the various soil types and plant communities of hill land is outwith the scope of this chapter. In summary, however, the first and most important requirement is for fencing to provide the means of grazing control necessary in all situations. On the better soils, such as the freely-draining brown earths supporting high grade species-rich *Agrostis –Festuca* grassland, a worthwhile degree of improvement can be achieved by grazing control alone. Subsequent improvements of such areas include successively the addition of lime, phosphate and, where necessary, potassium to improve the existing vegetation, followed by the introduction of white clover (*Trifolium repens*). In some cases it may be necessary to consider the alteration of the existing plant species by the use of selective herbicides, e.g. the eradication of bracken (*Pteridium aquilinum*) by asulam. Improvements beyond these stages require the complete destruction of the existing vegetation either by the use of herbicides or by cultivation, or both, and the replacement of the sward with sown species in which perennial ryegrass (*Lolium perenne*) and white clover predominate. On poorer soils supporting less productive plant species little worthwhile improvement in herbage quality or quantity can be achieved through grazing control alone, and it is necessary to begin the process of improvement at one of the later stages described above.

The particular problem of blanket bog situations merits mention. In addition to the extreme acidity and difficulties of the *Tricophorum –Eriophorum–Calluna* communities encountered on blanket bog are added problems of drainage and access. A system of mosaic reseeds has been developed in which the more accessible and drier parts, comprising perhaps 30% of an enclosed area, are sown with ryegrass and white clover following treatment with lime and phosphate. The whole of the enclosed area is regarded as a production paddock and used as described below.

For a fuller discussion of the establishment and maintenance of improved pasture on hill land the reader is referred to recent reviews by HFRO (1979b) and Newbould (1981).

The two-pasture system

THEORY

The appraisal of the results of the studies on the effects of nutrition on the components of production (Russel, 1971) and of the early studies on grazing ecology (Eadie, 1971) indicated the possibility of developing a new system of hill sheep management radically different to that practised traditionally. The central feature of the new system was the integrated use of two distinctly different pasture types, hence the origin of the term 'two-pasture system'. In this system the improved pastures or 'production areas', which are always enclosed and have been upgraded by one of the

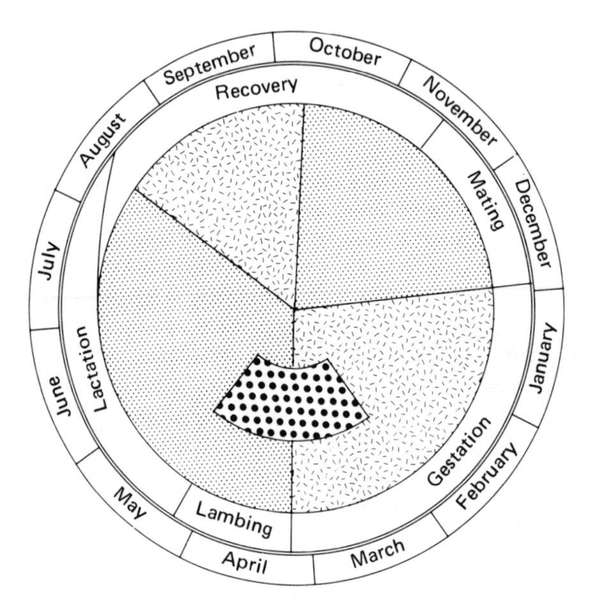

Figure 12.4 Stock management in the two-pasture system. ▨ improved pasture; ▨ hill grazings; ▨ supplementary feeding

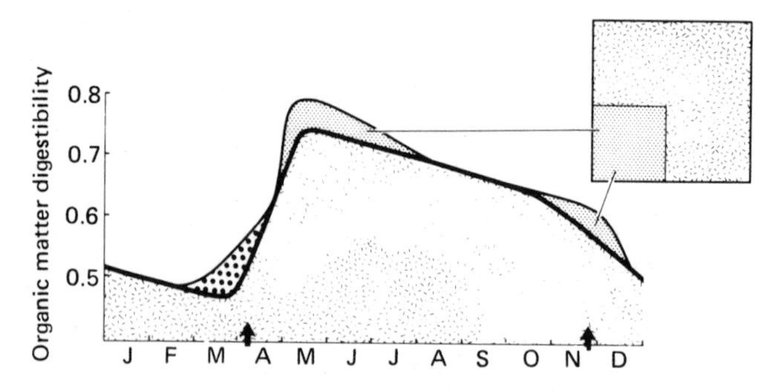

Figure 12.5 Seasonal changes in diet quality of ewes in the two-pasture system of management. ▨ Hill grazings; ▨ improved pasture; ▨ purchased concentrates

methods described above, are grazed by productive stock at those times when the largest production responses to improved nutrition can be achieved, i.e. during lactation and again in the period before and during mating (*Figure 12.4*). The unimproved hill grazings serve the equally important purposes of supporting the ewe stock at those times when high levels of nutrient intake are not essential and of carrying the replacement ewe stock and any non-producing adult animals. The effect on diet quality of the strategic use of improved pasture and purchased supplements is shown in *Figure 12.5*.

The integrated use of the improved pastures and hill grazings throughout the year is illustrated in *Figure 12.4*. The ewe stock are wintered on the hill grazings, receive their late pregnancy supplementary feeding there, and are put on to improved pasture immediately after lambing. Ewes with lambs remain on the improved areas throughout lactation and return to the hill grazings at weaning. This allows the improved pasture a rest period in which to accumulate an adequate supply of high quality herbage to provide the necessary nutritional boost prior to and during mating. The length of time the autumn-saved pasture will support the ewe stock in the latter part of the year is dependent to a large extent on weather conditions and particularly on the severity of frosts and the timing of the first heavy snowfalls. Experience has shown that ewes can generally be kept on these areas with advantage until mid to late December, i.e. until at least the first cycle of mating has been completed.

Where pasture improvement is sought by means of only grazing control, the initial grazing demands the removal of a considerable proportion of the fund of accumulated dead herbage. Although this must incur some degree of nutritional penalty to the animals used, this need not be unduly severe if sufficient stock are available to eat off the dead material quickly. Where possible, dry sheep and cattle should be used. Once the required degree of improvement has been achieved pasture quality can be maintained relatively easily if a sufficient grazing pressure can be sustained. Again, a short period of stocking with dry sheep and cattle immediately before the pasture is rested in August can be beneficial.

One of the central points on which the two-pasture system is based is increased herbage utilization, not only on the improved areas but also on the unimproved hill grazings. If in the first instance stock numbers remain unchanged and the area of improved pasture created is such as to satisfy the needs of all breeding ewes, the grazing pressure on the unimproved hill, stocked for most of the summer with only juvenile flock replacements and any dry ewes, will be much reduced. This in turn will reduce the degree of utilization of the hill herbage, indicating a potential for increasing stock numbers if the area of improved pasture can be extended correspondingly. While most of the argument up to this point has been directed to the improvement of individual animal production, a major part of the increased output resulting from the practice of the two-pasture system stems, as will be shown later, from increased stock numbers. It must also be made clear that a major proportion of the increased output can be attributed to the greater utilization of the unimproved hill grazings as well as to the increased production and utilization of the improved pasture.

The objective of the use of improved pasture in the pre-mating and mating period is to increase conception rate, and the needs of the greater number of ewes carrying twin foetuses must be recognized in the policy of supplementary feeding in late pregnancy. In the Hill Farming Research Organisation's development programme set up to test the two-pasture system, inputs of supplements have been determined objectively on the basis of assessments of the severity of undernourishment using plasma ketone or 3-hydroxybutyrate concentrations as described by Russel (1978b). This has allowed the same standards to be applied between years and between flocks, and although the approach is unlikely to have application in commercial enterprises, it has enabled patterns of weight change, which can be used as an indication of the adequacy of nutrition, to be defined.

In the two-pasture system the input of supplementary feeding during late pregnancy must increase and do so at a greater rate than the increase in ewe numbers. The areas selected for pasture improvement are generally the better areas which, under a free-grazing system, were probably also the more heavily grazed areas. Thus, in the winter months the increased number of ewes has to be carried out not only on a smaller area than formerly, but also on the poorer parts of the unit. There is therefore a need, as the system becomes more intensive, to begin supplementation at an earlier stage and to provide it in amounts which will cater for the needs of the greater number of twin-bearing ewes.

The two-pasture system has been tested by HFRO in the varied soil, vegetation and climatic environments of three of its research stations. These projects have served to test rigorously the concept of the new system on a reasonable scale, and to establish from a management viewpoint that the system could be operated satisfactorily. These same projects have also provided the means of evaluating the system by permitting measurement of responses in total output and in individual animal performance to specific changes in inputs over a range of stocking rates. They have provided biological data whereby the relative contribution to total output of each of the components of this system might be assessed, and have allowed the analysis in economic terms of relationships between inputs and outputs.

This approach to obtaining an understanding of biological and economic responses has meant that it has been necessary to drive the systems beyond limits which would be reasonable in normal commercial enterprises, e.g. to increase stock numbers to and beyond the point at which output/animal is depressed before applying further inputs. An example of this is given below and illustrated diagrammatically in *Figure 12.7*.

The projects in which the two-pasture system has been tested and evaluated have also proved invaluable as a means of identifying, and indeed of highlighting, areas requiring further research, and as such have stimulated a two-way flow between research and development.

IN PRACTICE

The Hairney Law–Auchope project, on the Organisation's Sourhope Research Station in the Cheviot Hills of south east Scotland, was the first example of the two-pasture system to be established. It was initiated in

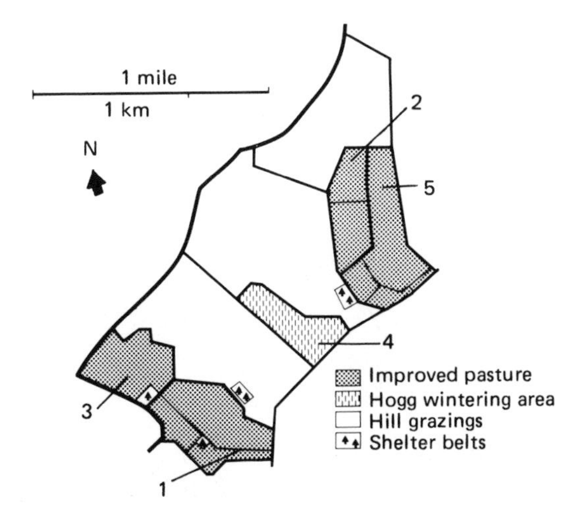

Figure 12.6 The Hairney Law–Auchope project at Sourhope Research Station

1968 on a hirsel of some 283 hectares, rising from 250 to 490 m and stocked with 387 ewes of both Hill North Country Cheviot and South Country Cheviot breeds. The stock also included 25 spring-calving cross Hereford cows which still graze on the unit from May until early winter each year. Most of the area carries a predominantly *Nardus* and *Molinia* dominant grass heath vegetation with about 93 hectares of *Agrostis–Festuca* communities, much of it covered with bracken. A map of the project area is shown in *Figure 12.6.*

Phase I, 1968–1973

The initial objective was to create the production paddocks or areas of improved pasture as inexpensively as possible. Capital inputs were restricted at that stage to fencing production paddocks 1 and 2, each of about 20 hectares, and to the retention of additional ewe lambs to increase stock numbers. In the production areas, and subsequently in paddocks 3, 4 and 5 which were enclosed in 1969, 1970 and 1972 respectively, pasture improvement was achieved by the use of only grazing control, using a heavy stocking of both sheep and cattle in the initial grazing to remove most of the accumulated dead herbage. Paddock 4 serves a dual purpose, being used as a permanent wintering area for the ewe hoggs each year, and as a production area for lactating ewes in the early and mid-summer months.

Ewe numbers and production data are presented in *Table 12.1.* These show that during Phase I ewe numbers were increased substantially, and that despite this the mean pre-mating liveweight (November) also increased progressively. This was accompanied by improvements in weaning percentage and mean weight of lamb weaned, which together resulted in a considerable increase in total weight of lamb produced and total wool weight. In percentage terms, ewe numbers increased by 48% while total output of lamb and wool rose by 79% and 80% respectively.

Table 12.1 FLOCK PRODUCTION DATA IN THE HAIRNEY LAW–AUCHOPE PROJECT

	Base data	*Phase I*					*Phase II*						
	5 yr. av.	*1969*	*1970*	*1971*	*1972*	*1973*	*1974*	*1975*	*1976*	*1977*	*1978*	*1979*	*1980*
Stock numbers	387	398	451	518	528	573	600	601	620	621	623	622	631
Pre-mating body weight (kg)		50.9	53.2	56.0	57.7	59.3	53.7	55.7	58.0	53.6	57.7	59.1	58.1
Weaning percentage	90.6	84.7	86.7	102.5	104.7	99.5	91.5	102.7	108.5	106.9	105.1	113.3	118.2
Mean lamb weaning weight (kg)	22.6	23.1	23.5	26.7	25.4	24.9	26.1	26.0	26.6	26.5	25.2	25.5	26.1
Total weight lamb weaned (kg)	7924	7786	9188	14177	14046	14193	14329	16042	17902	17596	16506	17978	19471
Weight lamb weaned/ ewe mated (kg)	20.5	19.6	20.4	27.4	26.6	24.8	23.9	26.7	28.9	28.3	26.5	28.9	30.9
Total weight wool (kg)	869	850	1017	1253	1369	1561	1454	1535	1543	1503	1523	1601	1887

Table 12.2 GROSS CAPITAL INPUTS[a] TO THE HAIRNEY LAW–AUCHOPE PROJECT (£)

	Phase I	*Phase II*							*Total capital invested*	
	1968–1973	*1974*	*1975*	*1976*	*1977*	*1978*	*1979*	*1980*	*Prevailing costs*	*At 1980 costs*
Fencing	1377	—	—	—	196	157	—	455	2185	6194
Incorporation of two lambing paddocks fully into resource	—	1800	—	—	—	—	—	—	1800	3394
Lime and phosphate	—	1496	—	—	921	946	1279	—	4642	7946
Bracken spraying	—	288	—	376	—	—	196	—	860	1573
Reseeding	—	—	—	—	164	895	332	683	2074	2183
Overseeding	—	—	673	—	—	—	15	—	688	1112
Fertilizer	—	—	—	160	—	—	139	50	349	450
Weed control	—	—	—	—	—	—	67	27	94	101
	1377	3584	673	536	1281	1998	2028	1215	12692	22953

[a] All gross inputs subject to 50% grant

Table 12.3 GROSS MARGINS OF THE HAIRNEY LAW–AUCHOPE PROJECT (£)

	Base data 5 yr. av.	*1969*	*1970*	*1971*	*1972*	*1973*	*1974*	*1975*	*1976*	*1977*	*1978*	*1979*	*1980*
(a) Prevailing costs and prices													
Ewe numbers	387	398	451	518	528	573	600	601	620	621	623	622	631
Income													
Lambs	1387	1235	1329	2197	3071	4287	3576	4495	7668	9398	9691	10060	11545
Wool	430	421	470	670	720	852	801	1012	1223	1653	1698	1846	1957
Cast ewes	326	229	176	533	391	1170	744	963	1661	2359	2300	1887	2187
Gimmers (surplus)									586				
Subsidy	476	490	555	829	871	946	1800	2163	2232	2236	2243	2550	3470
TOTAL	2619	2375	2530	4229	5053	7255	6921	8633	13370	15646	15932	16343	19159
Expenditure													
Feed	195	498	787	490	722	893	1492	1477	2297	4616	3638	5523	4866
Grazing	—	33	33	92	121	119	—	—	—	—	—	—	—
Fertilizer	—	—	—	—	—	—	22	50	48	140	97	227	506
Other costs	301	398	460	508	491	636	798	968	1178	1310	1788	2034	1880
TOTAL	496	929	1280	1090	1334	1648	2312	2495	3523	6066	5523	7784	7252
Flock gross margin	2123	1446	1250	3139	3719	5607	4609	6138	9847	9580	10409	8559	11907
Gross margin/ewe	5.5	3.6	2.7	6.0	7.0	9.8	7.7	10.2	15.9	15.4	16.7	13.8	18.9
(b) 1980 costs and prices													
Flock gross margin	6813	4582	4648	9486	8927	9903	10059	11227	12405	9521	9546	8240	11907
Gross margin/ewe	17.6	11.5	10.3	18.3	16.9	17.3	16.8	18.7	20.0	15.3	15.3	13.3	18.9

At that stage (1973) the total investment (excluding the value of the additional stock retained) was in fencing which had cost £1377 (*Table 12.2*) and which on a commercial enterprise would have been eligible for a 50% grant. The flock gross margin had, in the meanwhile, increased by 164% (*Table 12.3*) at the then prevailing costs and prices. The cash flow at that stage was positive with a cumulative balance (excluding the valuation of the extra stock) of £1395.

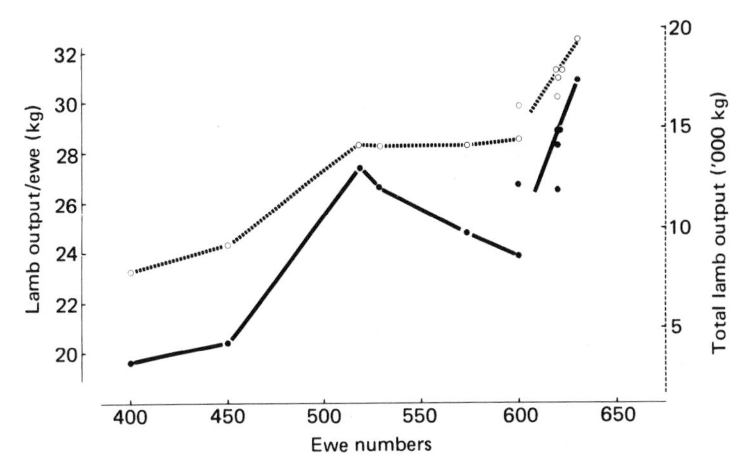

Figure 12.7 The relationship of lamb output/ewe (———) and in total (··········) to ewe numbers in the Hairney Law–Auchope project

By 1973, however, it was apparent that output/ewe had declined as a result of reductions in weaning percentage and in mean lamb weaning weight (*Table 12.1*) although total lamb output had been maintained as a consequence of increased stock numbers (*Figure 12.7*). These responses, supported by a detailed examination of patterns of ewe liveweight change throughout the year, indicated the need for more improved pasture.

Phase II, 1974–1980

In 1974 a programme designed to upgrade the improved pasture in paddocks 1 and 2 was initiated. Some 33 hectares received applications of ground magnesium limestone and ground mineral phosphate, and about 10 hectares of this were sprayed to eradicate bracken, 'spike-bar' rotovated and sown with a grass/clover seeds mixture. In addition, two small areas which had previously been used only for lambing were brought into the project on a year-round basis. The capital investment of £11 315 during Phase II (*Table 12.2*) has been wholly financed from the cumulative balance arising from earlier investments.

These investments in pasture improvement reversed the decline in individual animal performance. The data in *Table 12.1* show that the pre-mating liveweights again began to increase and were accompanied, as would be expected, by a substantial response in weaning percentage. Mean lamb weaning weight remained relatively constant at a very satisfactory

Table 12.4 CASH FLOW ANALYSIS OF THE MARGINAL INVESTMENT (£) IN THE HAIRNEY LAW–AUCHOPE PROJECT

	1969	*1970*	*1971*	*1972*	*1973*	*1974*	*1975*	*1976*	*1977*	*1978*	*1979*	*1980*
(a) Prevailing costs and prices												
Cumulative balance	−1187	−2138	−1528	−593	1395	966	3694	7980	10584	13312	14679	19515
Valuation extra stock												7898
(b) 1980 costs and prices												
Cumulative balance	−4210	−7647	−5792	−4057	−1279	2279	2892	8378	10426	12268	13021	17912
Valuation extra stock												7898

level, but the higher weaning percentage from a further increased number of ewes resulted in a very substantial increase in total weight of lamb produced. Wool production also increased substantially, in both the total and per ewe terms.

Thus, by 1980 the output of lamb from the project had increased from 7924 kg to 19 471 kg, an increase of 146%. This had been achieved by increasing ewe numbers (by 63%) and by a very substantial improvement in the weight of lamb weaned/ewe (51%). Increased wool production (122%) also contributed to the financial viability of the project.

Financial data

The effects in financial terms of the increased output resulting from the adoption of the two-pasture system of management are reflected in the flock gross margins shown in *Table 12.3* and in the cash flow analysis in *Table 12.4*. The increased flock gross margin at 1980 costs and prices is less impressive than that at the prevailing figures, reflecting the greater upward movement in costs than in returns over the period. Nonetheless, the increase of 75% in flock gross margin at fixed costs and prices is substantial. This figure has, however, to be considered in relation to the investments made and to the consequent cash flow situation during the life of the project.

The cash flow assessed on either prevailing or 1980 costs and prices must be considered to be satisfactory with cumulative balances of some £19 500 and £18 000 respectively. The policy of an initial low level of investment allowed capital to be generated within the system so that by the fifth year a positive cumulative balance had been achieved. The later higher levels of investment were totally funded from this balance and thus there has been no requirement to return to overdraft financing.

The true measure of the cash worth of the flock's performance must include the value of the additional stock (some 250 ewes and 60 ewe hoggs). Adding this to the cumulative balances gives a total figure in 1980 of almost £27 500. These figures refer only to the changes brought about by the adoption of the two-pasture system of management. The actual balances from the unit are greater since in real terms they also include the income deriving from the original 387 ewes producing 7924 kg of lambs.

A fuller account of the earlier years of the project has been reported by Armstrong, Eadie and Maxwell (1978) and a more detailed consideration of the development and testing of the two-pasture system has been presented by HFRO (1979c).

Conclusions

The annual cycle of nutrition afforded by the traditional system of hill sheep management undoubtedly limits production. It cannot, however, be claimed that it fails to meet, on an annual basis, the ewe's nutrient requirements for the low level of production it supports; the ewes do, after all, return to approximately the same weight at the end of each cycle. The potential production of hill ewes is clearly considerably greater than can normally be expressed in a truly free-grazing, set-stocked system. The recognition of the responses in production possible from the strategic improvement of nutrition at certain critical times, and of the means whereby this might be achieved in the hill environment, has led to the development of a new system of management in which the potential production of the hill ewe is more closely approached. The two-pasture system has been shown to operate successfully in a range of varied environments, and although details of management and of strategies of land improvement may vary from one farm to another, the principles have now been well tested and found worthy of adoption on a large scale.

Acknowledgements

The contributions of many colleagues, not all appearing in the list of references, are most gratefully acknowledged. Particular thanks are due to Dr R.H. Armstrong, Officer-in-Charge, Sourhope Research Station, who has been responsible with J. Eadie and Dr T.J. Maxwell for the project outlined above, and to A.R. Sibbald who analysed and interpreted the economic data.

References

ARMSTRONG, R.H. and EADIE, J. (1973). Some aspects of the growth of hill lambs. *HFRO 6th Report, 1971–73*, 57–68

ARMSTRONG, R.H. and EADIE, J. (1977). The growth of hill lambs on herbage diets. *Journal of Agricultural Science, Cambridge* **88**, 683–692

ARMSTRONG, R.H., EADIE, J. and MAXWELL, T.J. (1978). The development and assessment of a modified hill sheep production system at Sourhope in the Cheviot hills (1968–1976). *HFRO 7th Report, 1974–77*, 79–101

DONEY, J.M., SMITH, W.F. and GUNN, R.G. (1976). Effects of post-mating environmental stress or administration of ACTH on early embryonic loss in sheep. *Journal of Agricultural Science, Cambridge* **87**, 133–136

EADIE, J. (1967). The nutrition of grazing hill sheep: utilisation of hill pastures. *HFRO 4th Report, 1964–67*, 38–45

EADIE, J. (1970). Sheep production and pastoral resources. In *Animal Populations in Relation to their Food Resources* (A.C. Watson, Ed.), pp. 7–24. British Ecological Society Symposium No. 10. Oxford, Blackwell

EADIE, J. (1971). Hill pastoral resources and sheep production. *Proceedings of the Nutrition Society* **30**, 204–210

EADIE, J. and BLACK, J.S. (1968). Herbage utilisation of hill pastures. In *Hill-land Productivity* (I.V. Hunt, Ed.), pp. 191–195. British Grassland Society Occasional Symposium No. 4

GUNN, R.G. and DONEY, J.M. (1975). The interaction of nutrition and body condition at mating on ovulation rate and early embryonic mortality in Scottish Blackface ewes. *Journal of Agricultural Science, Cambridge* **85**, 465–470

HILL FARMING RESEARCH ORGANISATION (1979a). Hill sheep production and nutrition. In *Science and Hill Farming*, pp. 41–76. Penicuik, Hill Farming Research Organisation

HILL FARMING RESEARCH ORGANISATION (1979b). Soils and vegetation of the hills and their limitations. In *Science and Hill Farming*, pp. 9–21. Penicuik, Hill Farming Research Organisation

HILL FARMING RESEARCH ORGANISATION (1979c). Systems research. In *Science and Hill Farming*, pp. 102–135. Penicuik, Hill Farming Research Organisation

HODGSON, J. (1977). Factors limiting herbage intake by the grazing animal. In *Proceedings of the International Meeting on Animal Production from Temperate Grassland* (B. Gilesnan, Ed.), pp. 70–75. Dublin, Irish Grassland and Animal Production Association and An Foras Taluntais

HODGSON, J. and GRANT, S.A. (1981). Grazing animals and forage resources in the hills and uplands. In *The Effective Use of Forage and Animal Resources in the Hills and Uplands* (J. Frame, Ed.), pp. 41–57. British Grassland Society Occasional Symposium No. 12

HUNTER, R.F. (1962). Hill sheep and their pasture: a study of sheep grazing in south-east Scotland. *Journal of Ecology* **50**, 651–680

LUCAS, I.A.M. (1975). The contribution of science to the improvement of hill sheep production. *Proceedings of the British Society of Animal Production* **4**, 45–67

MACRAE, J.C., CAMPBELL, D.R. and EADIE, J. (1975). Changes in the biochemical composition of herbage upon freezing and thawing. *Journal of Agricultural Science, Cambridge* **84**, 125–131

MACRAE, J.C., MILNE, J.A., WILSON, S. and SPENCE, A.M. (1979). Nitrogen digestion in sheep given poor-quality indigenous hill herbages. *British Journal of Nutrition* **42**, 525–534

MEAT AND LIVESTOCK COMMISSION (1981). *Feeding the Ewe*, 2nd Edition. Bletchley, Meat and Livestock Commission

MILNE, J.A. (1974). The effects of season and age of stand on the nutritive value of heather (*Calluna vulgaris*, L. Hull) to sheep. *Journal of Agricultural Science, Cambridge* **83**, 281–288

MILNE, J.A., BAGLEY, L. and GRANT, S.A. (1979). Effects of season and level of grazing on the utilisation of heather by sheep. 2. Diet selection and intake. *Grass and Forage Science* **34**, 45–53

MILNE, J.A., CHRISTIE, A. and RUSSEL, A.J.F. (1979). The effects of nitrogen and energy supplementation on the voluntary intake and digestion of heather by sheep. *Journal of Agricultural Science, Cambridge* **92**, 635–643

NEWBOULD, P. (1981). The potential of indigenous plant resources. In *The Effective Use of Forage and Animal Resources in the Hills and Uplands* (J. Frame, Ed.), pp. 1–15. British Grassland Society Occasional Symposium No. 12

PEART, J.N. (1967). The effect of different levels of nutrition during late pregnancy on the subsequent milk production of Blackface ewes and on the growth of their lambs. *Journal of Agricultural Science, Cambridge* **68**, 365–371

RUSSEL, A.J.F. (1971). Relationships between energy intake and productivity in hill sheep. *Proceedings of the Nutrition Society* **30**, 197–204

RUSSEL, A.J.F. (1978a). The relative contributions of nutrition and genetics to improvements in the efficiency of sheep production. *Agricultural Progress* **58**, 92–97

RUSSEL, A.J.F. (1978b). The use of measurement of energy status in pregnant ewes. In *The Use of Blood Metabolites in Animal Production*, pp. 31–40. British Society of Animal Production Occasional Publication No. 1

RUSSEL, A.J.F., DONEY, J.M. and GUNN, R.G. (1969). Subjective assessment of body fat in live sheep. *Journal of Agricultural Science, Cambridge* **72**, 451–454

RUSSEL, A.J.F., DONEY, J.M. and REID, R.L. (1967). The use of biochemical parameters in controlling nutritional state in pregnant ewes, and the effect of undernourishment during pregnancy on lamb birth weight. *Journal of Agricultural Science, Cambridge* **68**, 351–358

RUSSEL, A.J.F., FOOT, J.Z. and MCFARLANE, D.N. (1982). The use of tritiated water in the estimation of body composition in grazing ewes. In *Studies of Water Metabolism of Herbivores using Tritiated Water*. Vienna, International Atomic Energy Agency

RUSSEL, A.J.F., GUNN, R.G. and DONEY, J.M. (1968). Components of weight loss in pregnant hill ewes during winter. *Animal Production* **10**, 43–51

RUSSEL, A.J.F., FOOT, J.Z., WHITE, I.R. and DAVIES, G.J. (1981). The effect of weight at mating and of nutrition during mid-pregnancy on the birth weight of lambs from primiparous ewes. *Journal of Agricultural Science, Cambridge* **97**, 723–729

13

THE USE OF FORAGE CROPS FOR STORE LAMB FATTENING

S. FITZGERALD
Agricultural Institute, Western Research Centre, Belclare, Co. Galway, Eire

During the Middle Ages and well into the 17th century the traditional system of arable farming consisted of the arable three-field system i.e. winter corn, spring corn, followed by fallow. This system of farming was widely practised in Britain and throughout Europe, with little emphasis on the production of livestock, which had to be fed either indoors or grazed under commonage on pasture and stubbles. Due to the very meagre sources of winter feed e.g. straw or poor quality hay, most livestock, other than those kept for breeding, were slaughtered in the autumn (Fussel, 1966).

The introduction of turnips and clover into Britain from the continent in the late 17th century together with improved husbandry techniques and the gradual enclosure of land, resulted in the development of the 'Norfolk four course rotation', consisting of wheat, turnips, barley and clover/grass. This new system of farming brought about a considerable improvement in agricultural production both in Britain and abroad and virtually transformed farming from a meagre subsistence type of activity in the Middle Ages to a much more progressive state in the 18th and 19th centuries and was able to meet the growing demand for food from the increasing population following the Industrial Revolution.

The inclusion of turnips and clover in the crop rotation had many benefits which helped to increase crop yields and produce better livestock. Inter-row cultivations in turnips in place of the fallow helped to control weeds and cereal diseases; the feeding of roots and clover hay to livestock improved both the quality and quantity of winter feed available for livestock; the dung from such stock and the nitrogen fixation from the clover break helped to build up soil fertility for the subsequent crops; and more stock could be fed over the winter and brought to heavier weights before slaughter. Greater selection of breeding stock was therefore possible on the basis of their ability to produce milk or to fatten, and provided the opportunity for the development of much better breeds of cattle and sheep.

The increased demand for meat and other animal products in the late 18th and 19th centuries resulted in a considerable expansion in livestock, including sheep. This led to the era of the 'golden hoof' in the traditional tillage areas of Britain, where sheep were folded on turnips and other brassica crops in the autumn and winter and grazed on short-term grass/clover leys in summer. It can, therefore, be stated that the humble turnip and other brassica crops which followed its introduction played a substantial role, not only in the expansion of sheep in Britain and the development of improved sheep breeds, but in agriculture generally.

Decline in the role of roots and forage crops in agriculture

During the last century and the earlier part of this century root crops and green forage crops constituted a major part of the winter feed available for livestock in Ireland and in Britain, to supplement the poor quality roughages available. However, there has been a considerable decline in the acreage of these crops grown for livestock feeding in both countries since the start of this century and this trend has accelerated in the last 20 years (*Table 13.1*).

In Ireland the acreage of root crops has declined by 76% since 1901 or by 68% since 1950. While there was some increase in the acreage of mangels up to 1940, this has since declined considerably. The acreage of green forage crops declined by 50% up to 1950, but increased to previous levels by 1960, mainly due to an increase in kale for feeding to dairy cows. However, the acreage of green forage crops has declined rapidly since then (–62%). If however, the acreage of sugar beet tops, which has increased since 1940, is included with green forage crops, then the total acreage of green forage crops available has increased since 1901 (+67%) and has remained reasonably stable over the last 30 years. Against a background of decline in the total acreage of tillage crops from 16 to 10% of the total area under crops and pasture since 1940, the proportion of tillage land devoted to roots and green forage crops in Ireland has declined from 20% to 8% since 1901, or, if sugar beet tops are included, from 20% to 15%, with the greatest reduction occurring in root crops.

In the UK the overall acreage of fodder roots has declined by 88% since 1908 or by 74% since 1950. The growing of mangels, despite some recovery in the 1950s, has practically ceased. In the case of green forage crops, there was a considerable increase in the acreage of rape and particularly of kale in the 1940–1960 period, but these crops have since declined in popularity. While the proportion of land under tillage crops in the UK increased from 32–41% between 1908 and 1950 and has remained at that level since, the proportion of tillage devoted to root and forage crops has declined from 21% to 3%, or from 21% to 7% if the acreage of sugar beet tops is included. While potentially available as a feed for livestock, a large proportion of beet tops are, however, ploughed in as a green crop in the intensive tillage areas of Britain.

The acreage of green forage crops is, however, likely to be considerably underestimated since most green forage crops are grown as catch crops in July/August and are not, therefore, included in the census of crop acreages which is recorded in June. Some regional differences in the distribution and relative importance of roots and forage crops within the UK also exist. Swedes are mainly grown in north east England and Scotland, while rape is quite popular in Wales and Central Scotland and kale in south west

Table 13.1 TRENDS IN THE ACREAGE OF FORAGE CROPS IN IRELAND AND BRITAIN

	1901	1940	1950	1960	1970	1976	% Change 1901–76	% Change 1950–76	Source
			('000 hectares)						
S. Ireland									
Turnips and swedes	88	61	52	42	38	21	−76	−60	Statistical abstracts of Ireland (1977)
Mangels	29	38	32	27	12	7	−76	−78	
Sugar beet (tops)	–	26	24	28	26	35		+46	
Green crops	27	13	14	26	14	10	−63	−29	
All forage crops (minus S. beet)	144	111	97	96	65	37	−74	−62	
Forage crops/tillage crops (%)	(20.3)	(14.9)	(13.6)	(14.1)	(12.3)	(7.8)			
Forage crops + S. beet tops	144	137	122	122	91	72	−50	−41	
Forage crops + S. beet/tillage crops (%)	(20.3)	(18.4)	(17.2)	(18.1)	(17.2)	(15.2)			
	1908	*1939*	*1950*	*1960*	*1970*	*1979*	*1908–79*	*1950–79*	
UK (GB + NI)									
Turnips and swedes	628	294	242	184	100	87	−86	−64	Agricultural Statistics, UK (various issues)
Mangels	173	88	112	54	10	6	−97	−95	
Sugar beet (tops)	–	140	174	176	187	214	–	+23	
Green crops	70	69	166	205	92	60	−14	−64	
All forage crops (minus s. beet)	871	451	520	442	202	153	−82	−71	
Forage crops/tillage crops (%)	(20.8)	(12.6)	(10.1)	(9.8)	(4.1)	(3.1)			
Forage crops + S. beet tops	871	591	694	619	389	367	−58	−47	
Forage crops + S. beet/tillage crops (%)	(20.8)	(16.6)	(13.5)	(13.7)	(8.0)	(7.4)			

England (Willey, 1971; Morrison, 1971). In Ireland roots and green forage crops are mainly grown in the tillage areas of Leinster and East Munster.

What has brought about this drastic reduction in what was once a major source of winter feed for stock? A number of factors have been involved:

1. High labour requirements in growing root crops, as conventionally grown, combined with a drastic decline in the farm work force;
2. Increased mechanization of grass conservation has made grass conservation more attractive than forage crops as a source of winter feed;
3. Continuous cereal growing in the intensive tillage areas, to the exclusion of sheep and cattle on many such farms;
4. Modern developments e.g. artificial fertilizers, herbicides and fungicides, have made the role of root crops as cleaning crops and soil fertility builders obsolete;
5. Increased importations of cheap frozen lamb from New Zealand to Britain considerably reduced the profitability of fattening native store lambs on arable crops, both in Britain and Ireland;
6. The growing of root crops and forage crops became associated with a more traditional and conservative type of farming compared with more 'modern' farming practices e.g. conserving grass as silage, barn dried hay or even as dried grass and the feeding of livestock on high concentrate diets;
7. The availability of relatively cheap home grown or imported cereals were much more convenient to feed to livestock, particularly as supplements to roughage based diets, than were forage crops;
8. Forage crops, because of their high moisture content and bulky nature, became synonymous with low quality feed in the minds of many, including several agricultural scientists and advisors, suitable only as a maintenance type of feed, similar to roughages.

The expression of the composition and feeding value of feeds on a fresh weight basis, as in MAFF (1975) rather than on a dry matter basis as is now done in Technical Bulletin No. 33 (1975), perhaps lent credence to the erroneous philosophy that the feeding value of roots and green crops was low, simply because animals required 5–10 times as much weight of such feeds as cereals to provide the same amount of nutrients. It was also believed by many that livestock could not physically consume sufficient of such bulky feeds to meet their requirements for high levels of production and if they were fed on high levels of such crops that all sorts of complications could arise (digestive disorders, scouring, poisoning, lambing difficulties and prolapsed vaginas). While some problems undoubtedly were associated with some forage crops, the fact that livestock farmers had, for centuries, been feeding livestock on such crops at high levels and getting good levels of production, and were well aware of any problems which were likely to arise, was unfortunately ignored.

There has, however, been a reappraisal of the value of root and forage crops as sources of feed for livestock and renewed interest among farmers and others in such crops over the last decade, despite the continued overall decline in the acreages of such crops. There are some indications that the decline in acreage of fodder crops is slowing down. This change in attitude has come about for several reasons:

1. Rapidly rising costs of cereals and grass conservation;
2. Wide variation in the quality and preservation of conserved roughages, depending on stage of growth when cut, weather conditions, etc.;
3. Roots and forage crops are relatively cheap sources of feed with a consistently high feeding value;
4. High yields of fodder can be obtained from root crops and reasonably good yields from rapidly growing forage crops when sown as catch crops in mid-summer or early autumn;
5. Increased mechanization of sowing, cultivating and harvesting of such crops has considerably reduced their labour requirements;
6. Easy systems of feeding involving little labour e.g. grazing of forage crops and some roots *in situ*, or mechanized feeding of roots indoors.

Types of forage crops and their uses

Forage crops are grown as specific crops to provide a source of feed to livestock at a time when other sources of feed (e.g. grass or conserved roughages) are either unavailable or of inadequate quality to sustain a good level of animal performance. They are generally grown as once-off crops which are either grazed or harvested and must be renewed each year. They are therefore generally associated with tillage farms, though not exclusively so. Since a wide range of such crops exist, a sequence of crops can be chosen which will provide a source of high quality feed right through from early autumn (August/September) to the following spring (March/April).

Forage crops can be classified on the basis of their type and main use into two broad categories namely (a) green forage crops and (b) fodder root crops and cabbages.

GREEN FORAGE CROPS

These crops are mostly leafy crops, mainly belonging to the Brassica family, and are usually grown as catch crops for utilization in autumn or early winter. They are mainly grazed *in situ* since the yield obtained would not justify the cost of harvesting. The various types of green forage crops available and their general suitability for lambs are set out in *Table 13.2*. In general rape has been the most commonly grown crop for fattening lambs, while kale has been fed mainly to cattle. In recent years Dutch or stubble turnips have become quite popular, due to their rapid growth, for autumn grazing but are less frost resistant than rape and the bulbs can become rather spongy and unpalatable late in the season. Some new hybrid turnips are somewhat leafier, have better root anchorage than Dutch turnips, have the ability to regrow if leniently grazed, and are as winter hardy as rape. Fodder radish and mustard grow very rapidly but become fibrous at the onset of flowering and are of limited use, mainly in early autumn. The hybrid *Raphano brassica* and some late flowering types of radish e.g. Hailstone are more suitable than the early flowering types. Other green feeds (grazing rye, winter cereals, ryegrasses e.g. westerwolds and Italian ryegrass) are grown in autumn and are generally grazed in spring by early lambing ewes, but could be used for lamb fattening in early spring.

Table 13.2 TYPES OF GREEN FORAGE CROPS AVAILABLE AS SOURCES OF FEED FOR STORE LAMBS

Green forage crops	*Type*	*Normally sown*	*Normally grazed*	*Common varieties*	*Comments*
Rape (*Brassica napus*)		June–Aug	Oct–Jan	Emerald, Lair, Canard, Nevin	Main forage crop for lambs, relatively winter hardy
Kale (*Brassica oleracea*)	(a) Autumn	May–June	Sept–Nov	Marrow stem varieties, Merlin, Proteor	
	(b) Winter	July–Aug	Nov–Jan	Thousand Head (Dwarf), Maris Kestrel, Canson	Leafy, more suitable for lambs in early winter
Dutch turnips (Stubble turnips)	(a) Bulbous	May–June	Sept–Nov	Debra, Vobra, Marco Civasto	Rapid growth, for early autumn use, high wastage of bulbs
	(b) Leafy	July–Aug	Nov–Jan	Record, Taronda, Ponda, Jobe	Late growers, alternative to rape less winter hardy
Hybrid turnips (Dutch turnips × cabbages)		June–Aug	Oct–Dec	Appin, Perko, Tyfon, Ballater	Leafy, can regrow, winter hardy
Fodder radish (*Raphanus sativus*)	(a) Early flowering	July–Aug	Sept–Oct	Rapide, Siletta	Rapid growth, matures fast, limited use
	(b) Late flowering	July–Aug	Sept–Oct	Slobolt, Raifort, Champetre, Hailstone	Slower to mature, more suitable for autumn grazing
Raphano brassica (Kale × radish hybrid)		July–Aug	Sept–Nov	Radicole	Slower to mature, more suitable for autumn grazing
Mustard (*Brassica alba*)		June–July	Aug–Sept		Matures early, limited use
Grazing cereals (Rye, winter cereals)		Aug–Oct	Feb–Apr	Rye, Rheidol, lavasz-patonai, winter barley, oats or wheat	Of limited use for lambs, used mainly for early lambing ewes in spring
Ryegrass (short-term)		Aug–Sept	Feb–Apr	Westerwolds ryegrass, Italian ryegrass	

When sown as catch crops usually from June to August green forage crops can be sown either (a) in tillage farms after harvesting an early main crop e.g. winter cereals, early potatoes, peas; (b) into grassland, following a cut of hay or silage, particularly where such land is due to be ploughed for tillage or reseeding; (c) on reclaimed land either as a pioneer crop before reseeding (Davies, 1978), as a first crop after tilling for 1–2 years, or as a cover crop, undersown with grass seeds. Some green forage crops can also be sown earlier in the year (April/May) if required for summer/early autumn grazing (July/September) in areas susceptible to drought and where grass is likely to be scarce. They can then be followed with winter cereals or the land can be reseeded to pasture.

The methods of sowing adopted vary widely and include conventional ploughing and tilling, minimal cultivations, direct drilling into cereal stubbles, grassland or virgin soil, or over-sowing into cereals before harvesting. Each method of sowing has its own merits in a particular situation. The emphasis in tillage farms is usually on speed of sowing, quick establishment and conservation of moisture, hence methods such as direct drilling or over-sowing are usually preferred. However, the results with each method can be variable, being very dependent on soil moisture or rainfall subsequent to sowing.

FODDER ROOT CROPS AND CABBAGES

These crops are grown as main crops, usually between two cereal crops but can be grown in ley ground, and occupy the land for a full growing season. Consequently high yields are needed to justify their cost of production and the devotion of land to such crops in preference to cereals or other cash crops. Cabbage is included in this category, even though it is a leafy crop, since it is grown as a main crop. Such crops are used for feeding over the late autumn or winter months either grazed *in situ* or harvested, stored and fed either indoors or outdoors at pasture. The main types of root and cabbages and their suitability for fattening lambs are set out in *Table 13.3*.

The soft white fleshed turnips are of limited use since they do not keep very well. While the yellow fleshed turnips are much hardier they are generally much poorer yielders than swedes. The softer swedes are generally more suitable for lambs than the harder types which can result in teeth losses if roots are grazed *in situ* or fed in whole form indoors. Mangels and low dry matter (DM) fodder beets are quite soft and suitable for feeding to lambs in whole form or grazed *in situ* in frost-free areas but the medium or high DM types of fodder beet are generally too hard and must be harvested and chopped if fed to lambs. Cabbages are generally grazed *in situ*, mainly in autumn and early winter, although the hardier winter types can be grazed up to early spring. Sugar beet tops are included in this category, but they should be regarded as a green forage crop since they consist mainly of leaf.

Composition and nutritive value of green forage and root crops

How green forage crops and root crops compare in terms of their main chemical constituents and nutritive value (ME concentration or digestibility) with other sources of feed on a dry matter basis (Technical Bulletin No. 33, 1975) is set out in *Table 13.4*.

Table 13.3 FODDER ROOT CROPS AND CABBAGES AVAILABLE FOR FEEDING TO LAMBS

Root crops	*Type*	*Sown*	*Used*	*Common varieties*	*Comments*
Turnips (*Brassica rapa*)	(a) White fleshed (7–8% DM)	May/June	Oct/Nov	Green Globe, Mammoth Purple Top	Soft, poor keepers, limited use
	(b) Yellow fleshed (8–9% DM)	May/June	Nov–Mar	Aberdeen Green Top, the Bruce, the Wallace	Hardy, good keepers, suitable for grazing late
Swedes (*B. rutabaga*) (*B. napo-brassica*)	(a) Soft types (8–9% DM)	May/June	Nov–Mar	Merrick, Doon Major, Marian, Best of All, Criffel	Suitable for winter grazing, partly susceptible to frost
	(b) Hard types (9–11% DM)	May/June	Nov–Mar	Bangholm, Wilhelmsburger, Ruta Otofte, Magres	Good keepers, rather hard for lambs, unless chopped
Mangels and fodder beet (*Beta vulgaris*)	(a) Mangels (10–11% DM)	April	Dec–Apr	Yellow Globe, Winter Gold	Very soft, susceptible to frost, stores well, fed whole
	(b) Low DM fodder beet (11–14% DM)	April	Dec–Apr	Peramono, Monara, Capax, Brigadier	Soft, susceptible to frost versatile for feeding
	(c) Medium DM fodder beet (15–17% DM)	April	Dec–Apr	Monoval, Kyros, Trivert	Fairly hard, stores well, suitable only for feeding indoors, fed chopped or whole
	(d) High DM (18–20% DM)	April	Dec–Apr	Monorosa, Monoblanc, Solano, Red Otofte	Very hard, stores well, must be chopped and clean
Beet tops		April	Oct–Dec	Fodder beet and sugar beet varieties	Mainly leaves, not frost hardy, equivalent to a green forage crop
Cabbages	(a) Autumn	April/May	Oct–Dec	Octema, Novema, Brunswick	Although green crops, cabbages are grown as main crops
	(b) Winter	April/May	Jan–Mar	Drumhead, January King, Celtic, Savoy types	High yielders, frost hardy

Table 13.4 CHEMICAL COMPOSITION AND NUTRITIVE VALUE OF GREEN FORAGE AND ROOT CROPS RELATIVE TO BARLEY, GRASS OR CONSERVED ROUGHAGES

Crop	*DM* (g/kg)	*Crude protein* (g/kg DM)	*Crude fibre* (g/kg DM)	*Ash* (g/kg DM)	*ME* (MJ/kg DM)	*DOMD* (%)	*Barley equivalent* (ME basis)
Barley	860	108**	53*	26*	13.7****	86	100
Green crops (rape, kale, cabbage)	140–160	136–200***	160–250***	93–107***	9.5–11.1**	59–71	69–81
Leaves of roots (turnips, mangels, sugar beet)	110–160	125–218***	100–145**	182–212****	9.0–9.9*	57–62	66–72
Brassica roots (turnips and swedes)	90–120	108–122**	100–111**	58–78**	11.2–12.8***	72–82	82–93
Beet roots (mangels, fodder beet, sugar beet)	110–230	48–91*	48–64*	30–69*	12.5–13.7****	79–87	91–100
Conserved roughages							
Hay (very poor–good)	850	85–132**	291–366****	69–85**	7.0–10.1*	47–67	51–74
Silage (poor–good)	200	160–170***	300–380****	80–100***	7.6–10.2*	52–67	55–74
Grass pasture (3–4 weeks growth)	200	175–265****	130–225***	90–105***	11.2–12.1***	72–75	82–88

Rating: ****very high level, ***high level, **low to moderate level, *very low level.

Green crops

These crops have a fairly low DM content, a high crude protein content, a moderate to high fibre and ash content and have a moderate metabolizable energy (ME) content and DOMD value, approximately 70–80% of that in barley, but are generally higher than that of conserved roughages and somewhat lower than grazed pasture. The values, however, include the stem which is often not fully eaten by stock, hence the nutritive value of the edible leafy and upper stem is likely to be higher.

Leaves of roots

These are by-products of harvested roots and are available for grazing *in situ*, ensiling or in combination with roots when the entire crop is grazed. They have a generally low DM content, are relatively high in crude protein, have a moderate fibre content but are very high in ash and consequently have a relatively low ME concentration and D value equivalent to 66–77% of barley, and are similar to good quality roughages.

Brassica roots

These have a very low DM content, a moderate level of protein, fibre and ash and have a relatively high ME concentration and D value equivalent to 82–93% of barley, and are similar to grazed pasture but superior to even good quality conserved roughages.

Beet roots

These crops have a wide range in DM content, are quite low in crude protein, fibre and ash, are mainly carbohydrate in composition and therefore have a very high ME concentration and D value, almost equivalent to barley (90–100%) and are superior to either grazed pasture or conserved roughages in energy but are deficient in protein and some minerals.

It is evident from this summary that green forage crops and the leaves of roots are relatively rich in protein and in most minerals but are somewhat lacking in energy, although they are at least comparable in feeding value to good quality conserved roughage. Root crops on the other hand are rich sources of energy, almost equivalent to barley and superior to conserved roughages, although mangels and fodder beet are deficient in protein and minerals. When these deficiencies are corrected, both forage and root crops are very useful sources of feed which can be used to either replace or supplement other sources of feed.

Evaluation of green forage and root crops as sources of feed to fattening lambs

The chemical composition and nutritive value of forage and root crops as set out in *Table 13.4* would suggest that lamb performance on such crops should be quite good, at least as good if not better than on autumn pasture or on conserved roughages, though possibly less than that obtainable on a high concentrate diet based on cereals such as barley. However, many other factors have to be taken into consideration when assessing the value of such crops for fattening lambs. These can be broadly grouped as follows:

1. Factors affecting lamb performance—nutritive value, feed intake, teeth losses, health problems, type and condition of store lambs;
2. Deficiencies in forage crops in terms of protein, energy, fibre, or minerals and the responses in lamb growth rate to supplementation of such crops;
3. Factors affecting lamb production/hectare—crop yield, utilization, feeding method, feeding capacity, feed conversion efficiency;
4. Economic aspects—the ability of forage crops to finish lambs at a reasonably low cost compared with conserved roughages or cereals, to leave margins which can compare favourably with those obtainable from other uses of such feeds or from alternative enterprises.

The more important of these factors are dealt with in detail for both green forage crops and fodder root crops in the remainder of this chapter.

Green forage crops

COMPARATIVE YIELDS

When assessing the various green forage crops available for fattening lambs, the first criterion to be considered is their comparative yields. A considerable amount of such data is available from national variety trials carried out in Ireland and Britain and from other forage crop or lamb

Table 13.5 COMPARATIVE YIELDS OF GREEN FORAGE CROPS (IRISH AND BRITISH VARIETY TRIALS)

(Recommended varieties only)	*Rape*	*Autumn kale*	*Winter kale*	*Dutch turnips*	*Hybrid turnips*	*Fodder radish*
Ireland[a]						
No. of varieties	5	5	5	4	1	
Average DM yield (t/hectare)	6.4	6.0	5.8	4.3	3.9	
Yield relative to rape	100	95	91	67	61	
% Leaf (range)	49–60	40–50	42–54	32–52	62	
Britain[b]						
No. of varieties	9	7	5	11	2	2
Average DM yield (t/hectare)	4.6	8.3	5.6	3.0	2.9	3.7
Yield relative to rape	100	180	122	65	63	80
DOMD (%)	70	68	70			66

[a]From Department of Agriculture, Dublin (1979)
[b]From NIAB (1980a)

production trials. The comparative yields for recommended varieties both in Ireland (Department of Agriculture, 1979) and in Britain (NIAB, 1980a) are set out in *Table 13.5*.

In Ireland both autumn and winter kales produced somewhat less feed than rape while the yields of Dutch turnips and hybrid turnips were considerably less. In Britain, on the other hand, the yield of kale, particularly autumn kale, was considerably higher than that of rape, while the yields of Dutch turnips and hybrid turnips, and to a lesser extent that of fodder radish, were lower than that of rape. A wide range in leafiness between types and varieties exist, particularly in the case of Dutch turnips.

The comparative yields of green forage crops from various other agronomic production or feeding trials are listed in *Table 13.6*. Yields

Table 13.6 YIELD OF GREEN FORAGE CROPS FROM VARIOUS EXPERIMENTS

		DM yield (tonnes/hectare)					
Source	*No. of trials*	*Rape*	*Kale*	*Dutch turnips*	*Hybrid turnips*	*Fodder radish*	Raphano brassica
Black (1967)	1	3.3	3.5			3.5	
Boyd and Dickson (1966)	4	3.4				4.0	
Paterson *et al.* (1977)	2	5.2		6.9			
Harper and Compton (1980)							
1976: 30/5–23/7	1	4.1	5.6	2.4	2.2	–	
1977: 31/5–24/6	1	6.2	6.4	5.0	5.4		7.2
Speedy *et al.* (1980)	2	3.6		6.4			
Sheldrick *et al.* (1981)	3	3.1		3.4	3.4		
ADAS (1976)	1	2.9		4.6	4.3		3.7
Thomas (1976)	1	2.0		3.1	2.4		
Thomas (1974)	1						
Relative yield		(100)		(165)		(93)	
Yield relative to rape							
No. of comparisons			(3)	(12)	(7)	(6)	(2)
Weighted average		100	115	131	106	112	122
(Range)			(103–107)	(59–178)	(54–158)	(93–118)	(116–128)

varied widely both between and within the comparisons made, due to the wide range of conditions and sowing dates encountered. Mean yields relative to rape, and weighted on the basis of the number of comparisons made, are also shown. These figures are broadly in line with those quoted by Kilkenny (1976) for forage crops grown for use in summer or autumn/winter, although he quoted much higher yields for Dutch turnips grown for summer use. The yields of both Dutch turnips and their hybrids were noticeably much more variable than that of rape. For example, Speedy *et al.* (1980) obtained much higher yields for Dutch turnips than for rape in two experiments. Others have also obtained better yields for Dutch turnips than for rape (Paterson, Dickson and Berlyn, 1977; ADAS, 1976; Thomas, 1974, 1976), although lower yields of Dutch turnips have also been recorded (Harper and Compton, 1980). The yields of hybrid turnips appear to be less than that of Dutch turnips and rather similar to, but much more variable than, that of rape. However, hybrid turnips appear to have other advantages which may compensate for their lower yield in that they are generally leafier than Dutch varieties, are more winter hardy, have better root anchorage which could result in less soiling and wastage when grazed, and they have the ability to regrow following a lenient grazing.

There would appear to be some divergence between the results of national variety trials and many of the other trials outlined in *Table 13.6*, particularly in relation to the difference in yield between Dutch turnips and rape. With any type of forage crop there are also some differences between varieties in yield (±10%) and in leafiness. In general the leafier types are preferred for grazing with sheep, since there is less wastage, and they are probably more nutritious than stemmy or bulbous types which can become unpalatable or deteriorate in quality as they mature.

The yields of green forage crops are notoriously variable and many factors other than the type or variety have a large influence on the yield produced. The most important of these factors are sowing date, soil moisture, soil fertility, geographic location and method of sowing. Yields ranging from 8.5 tonnes DM/hectare for kale sown in late May down to virtual failure (0.05 tonnes DM/hectare) when sown in late August have been obtained by Harper and Compton (1980), with similar wide ranges in yield for rape (4.5→0.06 tonnes DM/hectare) and stubble turnips (2.8→0.5 tonnes DM/hectare). Speedy *et al.* (1981) also obtained a large reduction in the yield of Dutch turnips sown from mid-June to late August (10→1 tonnes DM/hectare). Thomas (1974) found that the yield of green forage crops was reduced by 35–45% by delaying sowing from mid-August to early September, though the rate of decline in yield due to delayed sowing was less with rape (–10%) up to late August than for Dutch turnips and fodder radish (–22%).

UTILIZATION OF GREEN FORAGE CROPS

The amount of forage dry matter actually consumed by sheep, while generally of the order of 60–70% can be very variable, ranging from as low as 20% up to almost complete utilization (95%) in some cases (*Table 13.7*). Utilization depends on many factors including type of crop, its leafiness or stage of maturity, yield, stocking density, method of grazing, weather and soil conditions. ADAS trials summarized by Bastiman and Slade (1978) indicated that wastage can be as high as 70–80% under wet conditions in areas subject to high rainfall compared with 40–60% in Wales and 20–40% in the drier regions of the east Midlands, Yorks/Lancs and south-east England. On the other hand, they quoted that wastage can be quite low (4–11%) under dry conditions, even in western areas. In Ireland, utilization of forage crops ranged from 48–75% under conditions of high rainfall, depending on the crop, yield and weather conditions. High wastage tends to be associated with high yields (Greenall, 1958), due to a high proportion of inedible stem or root in addition to loss of leaf, although not always so, (Speedy *et al.*, 1980; Fitzgerald, 1969, 1970, 1971), poor anchorage of some varieties of stubble turnips particularly in mineral soils (Bastiman and

Table 13.7 UTILIZATION OF GREEN FORAGE CROPS WHEN GRAZED BY LAMBS

Source	*No. of trials*	*Range or mean value*					
		Rape	*Kale*	*Dutch turnips*	*Hybrid turnips*	*Fodder radish*	Raphano brassica
Greenall (1958)	1	52–78 (65)					
Black (1967)	1	71	75			70	
Fitzgerald (1969; 1970; 1971)	3	49–61 (53)					
Craig (1970)	1	60–89 (75)					
ADAS (1976)	1	87		68	70–75		85
Paterson *et al.* (1977)	2	25–75 (44)		19–74 (46)			
Sheldrick and Young (1977)	1	59			45–56 (51)	59	
Speedy *et al.* (1980)	2	43–73 (58)		59–92 (76)			
Bastiman and Slade (1978)							
Liscombe EHF		30–38	30–75	20–70			
High Mowthorpe EHF				77–89			
South West Region				38–96			
Wales		60–100		40–95			
Yorks/Lancs/East Midlands		80–83		60–68			
Weighted average (% of DM)		62	64	64	64	64	(85)
(Range)		(25–100)	(30–75)	(19–96)	(45–75)	(59–70)	

Slade, 1978) and a low stocking density (Paterson, Dickson and Berlyn, 1977). It is desirable to obtain a reasonably good degree of utilization of the crop (70–80%) provided that it is compatible with good lamb performance, which may not always be the case.

FEEDING CAPACITY OF GREEN FORAGE CROPS

The feeding capacity of a forage crop as measured in terms of lamb grazing days/hectare (LGD/hectare) is a function of its yield, degree of utilization and the amount of forage eaten/lamb/day. These factors are not independent of each other since intake will depend to some extent on yield or maturity of the crop, especially in the case of Dutch turnips and fodder radish, and on the intensity of grazing. Intake is also likely to be low as lambs adapt to the crop but will increase over time. Given the large variation in yield and utilization of forage crops, the feeding capacity of such crops is also quite variable, with a two- to three-fold difference from the limited amount of data available e.g. 1200–2800 LGD/hectare on rape and 1370–4800 LGD/hectare on Dutch turnips (Black, 1967; Fitzgerald, 1969, 1970, 1971; Appleton, 1969; Craig, 1970; Thomas, 1974; Speedy *et al.*, 1980). In general the feeding capacity of stubble turnips is greater and in some cases double that of rape (Speedy *et al.*, 1980), reflecting the higher yields of turnips in such trials, while that of kale and fodder radish is rather similar to rape (Black, 1967). The feeding capacity of fodder radish may in some cases be greater than that of rape due to the reduced intake of fibrous material if not fully grazed before the onset of flowering.

LAMB PERFORMANCE ON GREEN FORAGE CROPS

Lamb performance on green forage crops, measured mainly in terms of daily liveweight gain (DLWG), has been rather variable on different forage crops, as set out in *Table 13.8*, despite their general similarity in terms of composition and nutritive value.

In general, the performance of lambs on Dutch turnips was similar to, but much more variable than on rape, while the limited number of comparisons made between rape and kale or the hybrid, *Raphano brassica*, would indicate that lamb performance on these crops is also fairly similar to that obtained on rape. However, lamb gains on fodder radish and on the hybrid turnip, Appin, were considerably less than on rape. It is noteworthy that lamb performance on autumn pasture was also less than on rape.

There were, however, some wide divergences in the relative performance of lambs on different forage crops in some of these comparisons. Black (1967) and Appleton (1969) found that lambs fed on kale gained less weight than those fed on rape. However, Ewer and Sinclair (1952) found that lambs performed equally well on kale and rape and better than on short rotation ryegrass. In the case of Dutch turnips, Paterson, Dickson and Berlyn (1977) obtained poorer gains with Blackface lambs compared with those fed at similar forage allowances on rape, when fed either alone or supplemented with cereals. A lower level of performance on stubble turnips (Ponda) than on rape was also recorded in ADAS trials (1976). On the other hand, Thomas (1973, 1974) found that lamb gains on rape and stubble turnips were generally similar. Speedy *et al.* (1980) obtained much better lamb gains on stubble turnips than on rape, both in the case of Suffolk-X lambs and Blackface lambs, while lamb gains on grass aftermath were poorer. Poorer lamb gains were obtained on fodder radish than on rape both in the case of lowland lambs (Black, 1967) and Blackface lambs (Boyd and Dickson, 1966), although there was some variation between trials in the latter case. The poorer lamb gains on fodder radish was partly due to the onset of flowering, which reduced the quality of the feed towards the end of the trial. Prior to that the performance was similar to that obtained with kale but less than on rape (Black, 1967). Fodder radish would, therefore, appear to be suitable only as a quick growing crop for use in early autumn and grazed over a relatively short period before the onset of flowering. On the other hand the hybrid *Raphano brassica*, which is late maturing, would appear to be equivalent to rape in terms of lamb gains and yield and is a promising crop. The hybrid turnip, however, particularly Appin which has been the variety most studied, has given disappointing results compared with rape and Dutch turnips (ADAS, 1976; Young *et al.*, 1981) in the limited number of grazing trials carried out to date.

Table 13.8 AVERAGE LIVEWEIGHT GAIN (g/day) OF LAMBS ON VARIOUS GREEN FORAGE CROPS

Source	*No. of trials*	*Type of lambs*	*Rape*	*Kale*	*Dutch turnips*	*Hybrid turnips*	*Fodder radish*	Raphano brassica	*Grass*	*SE*
Ewer and Sinclair (1952)	2	Corridale-X, South Down (25 kg)	177	182					133	
Boyd and Dickson (1966)	2	Blackface (25–27 kg)	65				45			
Black (1967)	1	Suffolk-X (32 kg)	152	108			91			±6.3
ADAS (1976)	1	N.A.	147		109	53		130		
Paterson *et al.* (1977)	(2)	Blackface								
1975 forage only	1	(28 kg)	83		74					±5.5
1976 forage + cereal	1	(28 kg)	142		98					±5.1
Speedy *et al.* (1980)	1	Suffolk-X	147		218				106	±7.8
		Blackface	111		145				80	±7.8
Young *et al.* (1981)	1	Dorset Down × Masham (30 kg)	101			61				±10.2
Sharman *et al.* (1981)	1	N.A.	135					141		
Thomas (1973)	1	Suffolk-X	130		127					
Thomas (1974)	1	Suffolk-X	167		156					
Range in DLWG (g/day)			65–177	108–182	74–218	53–61	45–91	130–141	80–133	
Gain relative to rape										
(No. of comparisons)				(3)	(7)	(2)	(3)	(2)	(4)	
Weighted average			100	92	102	48	66	96	74	
(Range)				(71–103)	(74–148)	(36–60)	(60–69)	(88–104)	(72–75)	

Under dry conditions in Australia, rape produced similar lamb gains to kale, turnips or millet when grazed by weaned lambs during a dry period but the production of lamb meat/hectare was highest on rape (Tribe, Boniwell and Aitken, 1960).

LAMB PRODUCTION/HECTARE FROM GREEN FORAGE CROPS

Production of lamb meat/hectare from forage crops is a function of lamb performance and feeding capacity. From the limited amount of data available, liveweight gain/hectare on rape ranged from 194–350 kg/hectare on rape alone (Black, 1967; Appleton, 1969; Paterson, Dickson and Berlyn, 1977) up to 563 kg/hectare when a cereal supplement was fed with rape (Paterson, Dickson and Berlyn, 1977). In terms of carcass gain/hectare, the level of output ranged from 111–155 kg/hectare on rape alone, and up to 335 kg/hectare on rape plus cereals (Black, 1967; Paterson, Dickson and Berlyn, 1977). Production/hectare was much less both from kale and fodder radish than from rape (Black, 1967), but was generally better from stubble turnips albeit with a range from similar levels (Paterson, Dickson and Berlyn, 1977; Thomas, 1974) up to much higher levels of output (Speedy *et al.*, 1980). In terms of carcass gain/hectare, Paterson, Dickson and Berlyn (1977) obtained a 28% better output, despite a lower level of lamb gains, on stubble turnips compared with rape, due to the higher yield and feeding capacity of the crop. While these levels of production may not be spectacular compared with roots or other crops, e.g. cereals, they are nevertheless quite respectable considering that such crops are grown only in the latter half of the growing season, and would not be achieved on grass alone in late autumn or on conserved roughage without a considerable input of supplementary concentrate feed.

EFFECT OF STOCKING DENSITY OR FORAGE ALLOWANCE ON UTILIZATION AND LAMB PERFORMANCE

The amount of forage made available/lamb on a daily basis will affect both intake, and hence lamb performance, and crop utilization. To determine the optimum forage allowance as a percentage of lamb bodyweight (W), Paterson, Dickson and Berlyn (1977) grazed Blackface lambs on both the rape and stubble turnips at three forage allowances, 4.5, 6.75 and 9.0% W in 1975 and 4.5, 6.0 and 7.5% W in 1976 along with a cereal supplement of 150–350 g/day to provide approximately 10, 15 and 20% of their total intake at the low, medium and high allowances, respectively. Forage DM utilized was quite low at the high allowance (21–54%) and increased as the forage allowance was reduced (43–74%) and was much better in 1976 than in 1975, despite a higher yield of forage and feeding a barley supplement. While lamb gains were depressed at the low allowance, this was more than offset by the higher utilization and increased feeding capacity in the case of the rape crop, resulting in a much higher output of both liveweight gain/hectare and carcass gain/hectare compared with lambs stocked at the high allowance, particularly in 1976. In the case of stubble turnips, reducing the feeding allowance below 6.0% W also reduced lamb gains. However, this was not sufficiently offset by the increased feeding capacity and utilization of the crop, and consequently increasing the grazing pressure on stubble turnips resulted only in a marginal increase in production.

Young *et al.* (1981) also found that a reduction in the forage DM allowance from 8 or 6% W to 4% W reduced lamb growth rate when lambs were fed either on a radish/rape sequence or on a hybrid turnip, Appin. However, they were unable to detect any difference in organic matter intake between the forage allowances or the types of forage used. Craig (1970) also found that increasing the stocking rate on rape from 50 to 60 lambs/hectare with Blackface lambs, which represented a reduction in the forage allowance from approximately 8 to 6% W over the 55-day grazing period, reduced feed intake and increased the feeding capacity of the crop without affecting lamb performance, and increased output of liveweight gain/hectare by 22%.

It would appear from these results that a grazing pressure which will provide a forage DM allowance of about 6% W is needed to obtain a satisfactory level of lamb performance on forage crops. However, a higher grazing pressure or a lower allowance of 4.5% W can be tolerated on a crop such as rape, particularly if a cereal supplement is fed with the crop to ensure that all lambs are finished when the crop is grazed off. While a high degree of wastage will occur at higher allowances (6% W in the case of stubble turnips or other forage crops), the remaining residue can be grazed off with animals whose requirements are less critical e.g. ewes in mid pregnancy or ewe lamb replacements.

METHOD OF GRAZING FORAGE CROPS

Forage crops can be grazed by means of set-stocking, break-feeding, or by strip-grazing or folding, using an electric mesh fence to provide a new allocation of forage every 1–2 days. Both break-feeding and strip-feeding require fencing and is, therefore, more labour demanding than set-stocking but provides some degree of control over feed allocation. Set-stocking demands careful matching of the number of lambs to be fed in relation to the amount of forage available to ensure that all lambs are finished and ready for sale by the time that the crop is grazed off and to ensure a reasonable degree of utilization.

In terms of lamb performance, Ewer and Sinclair (1952) in several trials found little or no advantage in break-feeding rape (one week breaks) compared with set-stocking. In one trial they found that daily strip-grazing of rape produced poorer lamb gains than set-stocking, but found no difference between strip-grazing and set-stocking of sweet blue lupins. Slade (1977) also found that two day strip-grazing of rape and kale, particularly in the case of soft and hardy turnips, reduced lamb performance compared with block grazing using one week breaks. The feeding capacity of the crops was reduced by about 10% when strip-grazed, with no improvement in utilization.

It is likely that strip-grazing has to be very carefully managed to ensure that lambs have an adequate amount of forage available to them without at the same time leading to excess wastage, otherwise lamb performance may suffer. While a break-feeding system may not have any advantage over set-stocking in terms of lamb performance, some form of break-feeding e.g. 1–2 weeks may be desirable from a management point of view to allow the feed to be apportioned as required.

EFFECT OF SUPPLEMENTING FORAGE CROPS WITH CEREALS ON LAMB PERFORMANCE

Since the ME concentration of green forage crops, particularly rape, appears to be rather low (Technical Bulletin No. 33, 1975) and considering the rather modest lamb gains obtained on forage crops in some experiments, the effect of supplementing rape with cereals has been investigated. In one trial, Fitzgerald (1969) obtained a response to supplementing Suffolk-X lambs with 225 g rolled barley/lamb/day both in terms of liveweight gain (+39%) and carcass gain (+31%). In a subsequent trial (Fitzgerald, 1970), the response to the same level of supplementing barley in Suffolk-X and Galway store lambs was much less (+10%) and no response was obtained with Blackface mountain lambs (Fitzgerald and O'Toole, 1970). However, feeding a high level of rolled barley (450 g/day) actually depressed lamb performance compared with rape alone (Fitzgerald, 1969, 1970). This was due to complete substitution of barley for rape, and possibly also reduced the digestibility of rape. Craig (1970), on the other hand, obtained a high response (+77%) in the performance of hill lambs grazing on rape when supplemented with 225 g/day of a barley/oats mixture during the first 34 days, but no response was evident during a second period of 21 days with the remaining unfinished lambs. Intake of rape, as estimated from pre- and post-grazing clips, was, however, much higher (+44%) for the lambs given 225 g concentrates/lamb/day than for those grazing rape alone and may account for much of the apparent response to cereal supplementation. Craig (1970) also found no further response when feeding a higher level of cereal (450 g/day). Ewer and Sinclair (1952) could find no benefit in any of four trials in supplementing rape with 225 g of oats or barley.

In general, therefore, the response to supplementing forage crops such as rape with cereal supplements is variable and at best is limited to a lower level of supplement (225 g/day) for up to five weeks grazing on the crop. The generally poor overall response to supplementing rape with cereals would suggest that the intake and nutritive value of forage crops such as rape is quite good, even though the level of performance obtained on such crops, at least in terms of liveweight gain, is often not very spectacular. However, because of the substitution effect of rape for cereal which was of the order of 0.9–1.1 kg rape DM/kg barley DM fed (Fitzgerald, 1969, 1970) the feeding capacity of the crop was increased by 15–33% at the low and high level of cereal fed, respectively. This resulted in an increase in production of 33–43% in terms of liveweight gain/hectare and 30–34% in terms of carcass gain/hectare by supplementing rape with 225 g barley/lamb/day. Consequently, cereal supplementation of rape at a low level may be economically worthwhile, even in situations where no worthwhile growth response is obtained, since the extra feeding capacity of the crop will enable more lambs to be finished/hectare of forage crops grown. This could be important in areas where the acreage of forage crops which can be grown is limited, particularly in relation to the number of lambs to be fattened, as in hill areas where a high degree of utilization is required but without adversely affecting lamb performance.

SUPPLEMENTATION OF FORAGE CROPS WITH ROUGHAGE OR PASTURE

It has generally been observed that when lambs are first introduced to a forage crop such as rape they tend to graze any pasture around the headlands first and appear to find rape unpalatable and only gradually begin to eat it when roughage is no longer available. During this time they tend to either gain very little weight or may even lose weight, but they recover rapidly when fully accustomed to the rape. It was therefore felt that the provision of a runback onto pasture at night or the feeding of roughage such as hay to lambs confined to rape would help to get over this phase without losing weight and could thereby benefit overall lamb performance. The results of such an experiment are set out in *Table 13.9* (Fitzgerald, 1971). The lambs consisted of Suffolk-X stores (37 kg liveweight) and were serially slaughtered after 28, 56 and 91 days on the treatments. While the lambs given a runback overnight to pasture at 25 lambs/hectare or fed hay *ad libitum* had a somewhat better, though not a significantly greater, overall rate of liveweight gain than those on rape alone, this was mainly due to a lower reduction in gut contents, particularly over the first four weeks or so. There was, however, no overall benefit in providing either pasture or feeding hay with rape in terms of fasted weight gain, empty body weight gain, carcass gain or offal gain. Lamb performance on the rape diets was, however, much better than on autumn pasture fed alone, on which lambs did little more than maintain body weight with very little increase in carcass gain. It is also apparent that liveweight gain considerably underestimated the true rate of gain (empty body weight gain) due to the large reduction in gut contents, which was of the order of 8.8 kg (24% of liveweight) for lambs coming off grass in autumn compared with 4.8–5.8 kg (11.6–14.0% of liveweight) for lambs finished on the rape diets. Ewer and Sinclair (1952) also found no advantage in terms of liveweight gain to either providing a runback onto pasture for lambs grazing on rape in several comparisons made and, in fact, lamb performance was reduced by about 20% by providing access to pasture. They also obtained no benefit to feeding a roughage supplement with rape or to including Italian ryegrass in a mixture with rape compared with rape alone.

While the provision of a pasture runback to lambs fed on rape had no overall benefit in terms of lamb performance, it did, however, have a sparing effect on the rape crop by reducing the intake of rape and consequently increasing the feeding capacity of the crop from 2060–2980 LGD/hectare whereas the feeding of hay to lambs confined to rape made little difference (Fitzgerald, 1971). Because of this, lamb gains/hectare were increased substantially in the case of liveweight gain (154 v. 246 kg/

Table 13.9 EFFECT OF ROUGHAGE SUPPLEMENTS AND DATE OF SLAUGHTER ON THE PERFORMANCE OF LAMBS FED ON RAPE

		Diets				*Slaughter group (rape diets only)*			*S.E.* (g/day) (n = 21)
		Rape (control)	*Rape + pasture*	*Rape + hay*	*Autumn pasture*	S_1 *(28 days)*	S_2 *(56 days)*	S_3 *(91 days)*	
Liveweight gain	(kg)	4.4	4.8	4.7	1.0	0.3	4.9	8.6	
	(g/day)	49	72	73	6	12	87	94	10.3
Fasted weight gain	(kg)	5.6	5.3	5.3	2.3	1.5	5.2	9.5	
	(g/day)	78	85	85	42	52	93	104	11.3
Gut fill loss	(kg)	–4.0	–3.0	–3.2	–2.0	–3.1	–3.1	–4.1	
	(g/day)	–89	–59	–62	–49	–111	–55	–44	5.1
Empty body weight gain	(kg)	8.4	7.9	7.9	2.9	3.5	8.0	12.7	
	(g/day)	138	132	135	56	123	142	140	9.8
Carcass gain	(kg)	3.9	3.6	3.5	0.6	0.7	3.2	7.1	
	(g/day)	57	50	52	6	25	56	77	6.6
Offal gain	(kg)	4.5	4.3	4.4	2.3	2.8	4.8	5.6	
	(g/day)	81	82	85	50	100	85	62	5.1
Weight gain ratios									
Carcass gain/liveweight gain		0.89	0.74	0.75	0.62	2.15	0.65	0.82	
Carcass gain/fasted weight gain		0.69	0.68	0.66	0.25	0.49	0.61	0.74	
Carcass gain/empty body weight gain		0.46	0.46	0.44	0.20	0.21	0.40	0.56	

From Fitzgerald (1971)

hectare) and to a lesser extent in the case of empty body weight gain/hectare (296 v. 401 kg/hectare) and carcass gain (137 v. 182 kg/hectare) by providing a pasture runback, whereas the feeding of hay had little or no effect on lamb gains/hectare. However, consumption of hay was considerably reduced after four weeks on rape. Consequently the provision of a runback could be important, not in terms of lamb performance, but in finishing a greater number of lambs from a limited acreage of rape, and may also help to keep lambs cleaner in areas of high rainfall. However, grass production in the runback area will be delayed in the following spring.

VARIATIONS IN THE RATE OF LIVEWEIGHT GAIN AND ITS COMPONENTS IN LAMBS FED ON FORAGE CROPS

Lamb performance on forage crops such as rape, as measured in terms of liveweight gain, can often appear to be quite poor and varies considerably from one experiment to another (*Table 13.8*) and even within an experiment (*Table 13.9*) depending on the length of time lambs graze the crop. This is partly due to the fact that lambs gain very little or may even lose weight over the first 2–3 weeks (Fitzgerald, 1969, 1970, 1971). Consequently the overall rate of gain will depend on how long lambs are fed on the crop. For example, in one experiment (Fitzgerald, 1970) lambs slaughtered at 26 days (S_1) had a low rate of liveweight gain (41 g/day DLWG) and a somewhat better rate of carcass gain (45 g/day DCG), whereas lighter lambs slaughtered subsequently at 54 or 67 days (S_2 or S_3) had a longer period of recovery and much higher rates of both liveweight gain (105–108 g/day DLWG) and carcass gain (73–74 g/day DCG). However, when averaged over all slaughter groups the overall rate of lamb gain was only 85 g/day DLWG or 64 g/day DCG. Thus the ratio of carcass gain to liveweight gain averaged 0.75 but ranged from 1.10 (S_1) to 0.68 (S_2 and S_3), depending on the slaughter date. Similar high ratios of carcass to liveweight gain were obtained for lambs slaughtered at 32 or 55 days (0.63–0.70) in a previous experiment (Fitzgerald, 1969). However, a much lower ratio of carcass gain to liveweight gain (0.43–0.49) has also been obtained in an experiment where much better lamb liveweight gains though not carcass gains have been recorded (Black, 1967).

These results would suggest that liveweight gain is not a very reliable indicator of lamb performance on forage crops, particularly during the early stages. This is illustrated clearly in *Table 13.9*. Lamb liveweight gains over the first 28 days averaged across the three rape-based diets was quite low (12 g/day DLWG) compared with that of lambs slaughtered at 56 or 91 days (81 and 99 g/day DLWG). Fasting the lambs for 24 hours prior to weighing indicated a better rate of gain over the first 28 days but had much less effect on lambs slaughtered later. However, when the large effect of gut fill changes was removed, empty body weight gains at 28 days were quite good and were little less than the overall rate of empty body weight gain of lambs slaughtered 4–9 weeks later. Despite the relatively constant rate of empty body weight gain throughout the feeding period, the rate of carcass gain was quite low initially but increased with time. Most of the gain in the early stages consisted of offal which generally declined with time as more of the gain was deposited in the carcass.

Despite the lack of any overall effect of feeding roughage, some differences were apparent in terms of lamb liveweight gains up to 28 days. During the first 28 days, the lambs fed on rape alone had apparently lost weight (–44 g/day DLWG), whereas those given access to pasture (+30 g/day DLWG) or fed hay with rape (+49 g/day DLWG) had gained some weight. The differences in terms of fasted weight gain between those fed on rape alone and those given roughage with rape were somewhat smaller and not significant. However, when the large effect of gut fill changes were removed, it became clear that the lambs fed on rape alone had actually gained in empty body weight and carcass gain over 28 days to a similar extent as those given pasture or hay with rape. Thus the apparent advantage in terms of liveweight gain of feeding roughage with rape over the initial stages was illusionary rather than real and merely reflected differences in the degree of gut fill loss rather than any difference in true body weight gains. From 28 days onwards there were no apparent differences either in the rate of liveweight gain or of any of its components between those fed on rape alone and those given access to roughage with rape. However, both the intake of hay and the availability of pasture declined rapidly, particularly over the last five weeks, and made little contribution to the lambs' diet once they were fully accustomed to the rape diet. Using liveweight gain as a predictor of carcass gain (by expressing lamb carcass gains as a proportion of liveweight gain) would indicate that a very high proportion of the overall liveweight gain on rape-based diets consisted of carcass gain (0.74–0.89) as was observed in other experiments (Fitzgerald, 1969, 1970). This ratio was somewhat less when expressed as a proportion of fasted weight gain (0.66–0.69) and particularly when expressed as a proportion of empty body weight gain (0.44–0.46). However, none of these ratios are very constant and are very much influenced by slaughter date as indicated in *Table 13.9*.

It is not then perhaps too surprising that such a large variation should occur in lamb performance on forage crops, as indicated by liveweight gains in *Table 13.8* from the various experiments reported, given such wide differences in terms of days on trial, type and condition of lambs, gut fill content, etc. Unusually large increases in lamb weight gains have also been recorded on forage crops from time to time which can overestimate their true rate of gain. This is usually associated with contamination of the fleece with soil or moisture when grazing under wet conditions, which is pretty frequent in areas subject to high rainfall. Consequently in any assessment of lamb performance on forage crops, liveweight gain is at best only a crude indicator of lamb performance and is very unreliable for short term studies. Fasting of lambs prior to weighing, while reducing the effect of gut fill changes at least in the early stages, is not very practical in many situations where lambs have to be weighed in the field and could adversely affect their performance if done frequently. A reasonable length of time (at least six weeks) should be allowed to enable lambs to attain a satisfactory rate of gain, particularly carcass gain, on forage crops. In many experiments lambs are drafted for slaughter at a particular weight or condition. Consequently, given the wide variation in weight and condition in most batches of store lambs, some lambs will be drafted for slaughter at an early stage and will consequently reduce the overall rate of gain on the crop. Ultimately, lamb performance on forage crops should be measured in terms of empty body weight gain and its components and assessed on the basis of output of saleable meat/hectare of feed. Unfortunately, this has rarely been done when assessing forage crops.

HEALTH PROBLEMS ASSOCIATED WITH GREEN FORAGE CROPS AND THEIR EFFECT ON LAMB PERFORMANCE

Some problems have been encountered when feeding green forage crops to livestock which can adversely affect their health and performance.

Haemolytic anaemia (kale poisoning)

This problem is mainly associated with the feeding of large amounts of kale to cattle and is characterized by a rapid rise in Heinze–Ehrlich bodies in the blood which are excreted and result in a fall in blood haemoglobin levels (Smith and Greenhalgh, 1977). The main constituent involved which is present in kale and other brassicas has been identified as S-methyl cysteine sulphoxide (SMCO). It is broken down in the rumen to dimethyl disulphide, which is the toxic agent. SMCO levels can vary between brassicas, being quite high in some types of kales e.g. Maris Kestrel and in the secondary growth of leaves and flowers compared with older leaves. SMCO levels also increase as plants mature, from 4–14 g/kg DM in Maris Kestrel kale from August to January (Smith and Greenhalgh, 1977) and from 4–10 g/kg DM in rape (Lair) and radicole from October to late December (Sharman, Lawson and Whitelaw, 1981). Consequently the toxic effects are often associated with the onset of frost. SMCO levels in the roots of swedes decline with maturity but that of the leaves increase. Daily intakes of 15–20 g SMCO/100 kg liveweight are considered to be necessary to produce acute anaemia in cattle but intakes of 10 g/kg DM, while not manifesting itself clinically, could adversely affect animal performance.

Anaemia appears to be less of a problem with lambs even though they are often fed exclusively on forage crops, probably because they are seldom grazed on kale which appears to be the major culprit. However, sub-clinical or a low grade anaemia may occur which could adversely affect lamb performance. Sharman, Lawson and Whitelaw (1981) found that lambs introduced abruptly to rape (Lair) or radicole developed a slight anaemia but then recovered to normal again despite an increase in SMCO levels in the crops. There was no mortality and lamb growth rates were quite good, producing well finished carcasses with no apparent adverse effects. Young *et al.* (1981) found that the SMCO level in rape (6.2–7.7 g/kg DM) was higher than in Appin turnips (3.9–5.4 g/kg DM) which was also higher than in fodder radish (3.7–4.5 g/kg DM) grazed in November. After one week's grazing on the crops anaemia changes became evident, with a drop in haemoglobin, red blood cell counts and packed cell volume.

While some recovery in these parameters was evident in the case of lambs grazing on Appin after 5 weeks, this did not occur with rape. Despite the higher level of SMCO and more prolonged anaemia on rape, lamb performance was better on the radish/rape sequence than on Appin. There is no evidence from these limited results, therefore, that the level of SMCO normally found in forage crops grazed by lambs has any serious adverse effect on lamb performance. However, the effect could be serious in certain circumstances, e.g. where lambs are grazed on kale or other crops, particularly following secondary growth or at a flowering stage.

Goitrogens

Brassica crops contain goitrogenic substances such as thiocyanates which inhibit the uptake of iodine by the thyroid gland, but this can be overcome by iodine supplementation. Thiocyanates are found mainly in the roots and leaves of brassicas (Greenhalgh, 1971). A second substance called goitrin interferes with the synthesis of thyroid hormone and cannot be overcome by iodine supplementation, but is generally less of a problem since it occurs mainly in the seeds of brassica crops. While the feeding of green brassica crops such as kale to pregnant ewes has produced goitre and thus high mortality in newborn lambs (Sinclair and Andrews, 1958) this has been overcome by supplementing the ewes with iodine (KIO_3). In the case of fattening lambs, Russel (1967) found that lambs grazing on rape and apparently suffering from a mild degree of goitre showed no response in growth rate when supplemented with iodine, despite a reduction in the size of the thyroid glands. Young *et al.* (1981) found that the thiocyanate level in rape was greater than on Appin turnips, or on fodder radish, yet lamb gains were better on a radish/rape sequence than on Appin turnips. Sharman, Lawson and Whitelaw (1981) could also find no relationship between the weight of thyroid glands and the growth rate of the lambs grazing on forage crops. These results would suggest that lambs fed on green forage crops can overcome any goitrogenic activity in such crops by enlarging the thyroids without adversely affecting their performance.

Nitrate poisoning

Very high levels of nitrate in brassica crops (3% of DM in the roots and up to 5% of DM in the leaves) have been reported. These are much higher than the 1.5–2.0% level which is normally regarded as the minimum toxic level, and has been implicated in nitrate poisoning of cattle fed on turnip tops and roots (Greenhalgh, 1971). The high levels of soluble carbohydrate present in brassica crops, particularly in the roots, may render the high levels of nitrate less toxic. Levels of nitrate are likely to be higher in young rapidly growing crops, particularly in crops grown in soils with a high nitrogen content or given heavy applications of nitrogen fertilizer, than on more mature plants and in freshly harvested roots rather than in roots stored for some time (Toosey, 1972). There is no strong evidence available, however, to suggest that such high levels of nitrate in brassica or root crops have any adverse effect on lamb performance, particularly where such crops are grazed later in the year or following a period of storage.

Copper levels

Sharman, Lawson and Whitelaw (1981) found that plasma copper levels in lambs grazing kale or radicole dropped rapidly during October/November, resulting in very low plasma and liver copper levels in lambs slaughtered in December. They indicated that the levels of copper, molybdenum and sulphur present in these crops were such that, if found in grass, they would severely inhibit the availability of copper to the lambs. Nevertheless, they found no correlation between liver copper levels and lamb growth rates to slaughter.

On the basis of these results, therefore, neither the level of SMCO, goitrogenic substances, nitrates, nor minor elements such as copper, molybdenum or sulphur present in forage crops can be definitely associated with having a very serious adverse effect on the health or performance of grazing lambs. However, any of these factors could have a serious effect in specific situations, which to date have not been adequately defined or documented.

Fodder root crops and cabbages

Fodder root crops and cabbages are grown as main crops to produce a high yield of high quality fodder for feeding to livestock over the winter months (November to April). They fall into the three main categories of brassica roots, beet-type roots and cabbages, as outlined in *Table 13.3.*

COMPARATIVE YIELDS OF ROOT CROPS AND CABBAGES

These crops are generally much higher yielding than green forage crops, mainly because they are sown earlier in the growing season. Sizeable differences, however, exist between these crops, particularly between the brassica root crops and the beet-type crops as indicated by the data in *Tables 13.10* and *13.11*. In Britain the recommended varieties of swedes yielded 5.3–6.0 tonnes root DM/hectare compared with 10.7–13.0 tonnes root DM/hectare for mangels and fodder beet (NIAB, 1980b). In Ireland somewhat higher yields of roots were recorded, 7.2–7.5 tonnes root DM/hectare for swedes compared with 12.4–15.0 tonnes root DM/hectare for the fodder beets (Department of Agriculture, Dublin, 1979). The relative yield of tops also appeared to be higher on medium and high DM fodder beets than on swedes but that of mangels and low DM fodder beet were somewhat lower. Within each category of root the high DM types appear to produce a higher yield than the low DM types. The proportion of fodder beet roots which is underground appears to increase with DM content from 40–70%.

Table 13.10 COMPARATIVE YIELDS OF FODDER ROOT CROPS (NATIONAL VARIETY TRIALS IN BRITAIN AND IRELAND)

Recommended varieties only	*Swedes*		*Mangels*	*Fodder beet*		
	Low DM (7–9% DM)	*High DM (9–11% DM)*	*(10–11% DM)*	*Low DM (11–13% DM)*	*Medium DM (14–17% DM)*	*High DM (18–20% DM)*
Britain[a]						
No. of varieties	(3)	(9)	(3)	(1)	(4)	(2)
Average root yield (tonnes DM/hectare)	5.3	6.0	10.7	11.5	12.2	13.0
Yield relative to low DM swedes	100	113	200	215	228	243
Relative size of tops (0–9)	4.7	5.0	3.7	5	6.8	6.0
% of root under ground	–	–	34–38	37	48–54	65–69
Ireland[b]						
No. of varieties	(4)	(5)		(1)	(3)	(3)
Average root yield (tonnes DM/hectare)	7.3	7.5	–	12.4	14.6	15.0
Yield relative to low DM swedes	100	103		171	201	207
Relative yield of tops (1–9)	6.5	5.8		4	7.7	8.0
% of root under ground				40	50–65	70–75

[a]From NIAB (1980b)
[b]From Department of Agriculture, Dublin (1979).

Table 13.11 YIELDS OF ROOT CROPS FROM VARIOUS TRIALS IN IRELAND

Source	*Years on trials*	*Turnips*	*Swedes*		*Mangels*	*Fodder beet*			*Sugar beet*
		(7–8% DM)	*Low DM (9–10%)*	*High DM (10–11%)*	*(10–11%)*	*Low DM (11–14%)*	*Medium DM (15–17%)*	*High DM (17–20%)*	*(23% DM)*
Feeley (1964a,b)	2–3		(15)[a]	(5)	(6)	(4)	(4)	(5)	(1)
Root yields (tonnes DM/hectare)			7.1	7.7	7.9	8.8	8.9	8.9	8.5
(rel. to swedes)			(100)	(109)	(112)	(125)	(126)	(127)	(120)
Leonard (1970)	3		(3)	(2)					
Root yield (tonnes DM/hectare)			7.2	8.0					
Roots + tops (tonnes DM/hectare)			8.1	8.8					
Thomas (1974)	2					(2)		(2)	
Root yields (tonnes DM/hectare)						11.6		12.2	
Roots + tops (tonnes DM/hectare)						15.3		17.6	
Storey and Barry (1979)	1–4								
Root yields (tonnes DM/hectare)						11.5	11.8	11.7	
Roots + tops (tonnes DM/hectare)						15.0	15.4	16.3	
Fitzgerald (1975; 1976; 1977a)	1–3	(2)	(3)	(2)	(2)	(3)	(2)	(1)	(1)
Root yield (tonnes DM/hectare)		4.8	7.1	7.7	12.1	11.9	13.7	12.7	10.2
(rel. to swedes)		(66)	(100)	(107)	(176)	(170)	(200)	(187)	(136)
Roots + tops (tonnes DM/hectare)		7.2	9.8	10.0	14.1	14.8	17.2	18.0	13.7
(rel. to swedes)		(70)	(100)	(97)	(147)	(151)	(179)	(176)	(132)

[a] No. of varieties

Results of other experiments carried out in Ireland (*Table 13.11*) indicate yields of swedes in the range of 7–8 tonnes DM/hectare for roots or 8–10 tonnes DM/hectare, including the tops and do not appear to have changed much over the last 20 years. The harder, high DM swedes produced about 10% more root DM than the softer types. The yield of turnips was lower (–33%) than that for swedes. In the early sixties, the yield of mangels and fodder beet was only 12–27% greater than that of softer swedes (Feeley, 1964a,b) and there was no difference in yield between the low and high DM fodder beets. More recent trials, however, have shown that much greater yields of mangels and fodder beet have been achieved (Thomas, 1974; Fitzgerald, 1975, 1976, 1977; Storey and Barry, 1979). This is probably due to improvements in crop husbandry and fertilizer use, and the introduction of more high yielding monogerm varieties of fodder beet, although some of the older varieties have produced yields just as high as the newer varieties (Storey and Barry, 1979). Relative to low DM swedes, the yields of mangels and fodder beet roots were 70–100% greater and the total DM yields, inclusive of tops, were 50–80% greater (Fitzgerald, 1975, 1976, 1977a). Medium and high DM varieties of fodder beet yielded marginally more total DM than low DM types (3–15%), while that of mangels was about 10% lower than low DM fodder beet. However, there are far greater differences in yield between varieties than between DM types (Feeley, 1964b; Storey and Barry, 1979).

Results of comparisons made between brassica roots and cabbages in Britain and New Zealand (*Table 13.12*) show that yields of swedes were

Table 13.12 YIELD OF BRASSICA CROPS FROM BRITISH AND NEW ZEALAND TRIALS

Source	*No. of trials* (years)	*Turnips* (tonnes DM/ hectare)	*Swedes* (tonnes DM/ hectare)	*Cabbage* (tonnes DM/ hectare)
Bastiman (1977)				
(High Mowthorpe EHF)	1	4.1	5.9	6.9
Barry and Drew (1978)	1	5.2	6.6	
Speedy *et al.* (1981)	2	8.9	11.1	7.8
Rutherford and Dover (1981b)				
Nov/Dec	4	6.2	8.5	7.5
Jan/Feb			8.8	5.8
Kilkenny (1976)				
Typical yields		6.0	7.2	12.7
(Range)		(4.4–7.9)	(5.3–8.9)	(9.1–14.6)
Yield rel. to swedes				
(No. of comparisons)		(8)		(11)
Weighted average		75	100	79
(Range)		(69–80)		(66–117)

more variable than those recorded in Ireland (Bastiman, 1977; Barry and Drew, 1978; Speedy *et al.*, 1980; Rutherford and Dover, 1981b). Yields of turnips were also variable and about 20–30% lower than that of swedes, while the yields of cabbage were generally lower (12–35%) than that of swedes, with the exception of one comparison at High Mowthorpe (+17%; Bastiman, 1977). In New Zealand kale considerably out-yielded turnips and swedes in one comparison (Barry and Drew, 1978), and in Australia turnips produced similar but rather poor yields (2.5–2.7 tonnes DM/hectare) to rape and kale under dry conditions (Tribe, Boniwell and Aitken, 1960). Kilkenny (1976) quoted far higher DM yields for cabbage compared with turnips or swedes than indicated from the results of these trials.

It would appear from these results that the yields of turnips and cabbages are generally less than those of swedes while those of mangels and fodder beets are considerably greater. However, many factors can influence yield besides the type or variety of root crop chosen, the most important of these being sowing date, location, weather conditions, soil fertility, weed control and spacing. Sowing of fodder beet later than 4th May (Storey and Barry, 1979) or of swedes later than 3rd June (Leonard, 1970) depressed yields in Ireland, but there was little difference in yield for crops sown earlier than these dates. Differences in plant spacing, resulting in a wide range in plant population/hectare, had little effect on the yield of swedes within the range of 50000–160000 plants/hectare (Leonard, 1970) or of fodder beet within the range of 50000–130000 plants/hectare (Storey and Barry, 1979), but obviously had a considerable effect on root size. A much narrower range of 60000–80000 plants/hectare is desirable to produce reasonably uniform sized roots to facilitate either harvesting or a good degree of utilization if grazed *in situ*, since a greater proportion of smaller roots lie underground (Leonard, 1970).

UTILIZATION OF ROOT CROPS AND CABBAGES

Utilization of root crops grazed *in situ* by store lambs was within the range of 80–92%, assuming complete utilization of leaves, in the case of swedes, mangels and fodder beets (Fitzgerald, 1977). The lower value was obtained on medium DM fodder beet, due to the high proportion of the root which lay underground. Somewhat lower values have been recorded for lambs grazing on different varieties of swedes by Leonard (1970). Speedy *et al.* (1980) obtained equally high values on turnips and swedes, with a slightly higher degree of utilization on cabbages. However, in ADAS trials much more variable utilization rates have been obtained, from as low as 30% up to 90% on turnips and swedes, depending on soil and weather conditions (Bastiman and Slade, 1978). The lower values were generally found in the wetter regions, but under dry conditions high values (75–95%) have also been obtained in these areas. Higher utilization rates have been recorded for cabbages (67–99%) than for swedes (39–90%) or turnips (60–70%) at High Mowthorpe EHF. Soil contamination of roots under wet conditions would appear to be the main reason for rejection of roots by lambs. However, subsequent rain will help to clean such roots for further grazing. A high degree of utilization (70–90%) is essential with root crops, due to the high cost of growing them.

In the case of harvested roots, an estimate of 90% utilization for sheep fed indoors was found by Fitzgerald (1975, 1977a) although field losses of the order of 15% for swedes and fodder beet during harvesting have been quoted (Bastiman and Slade, 1978). Losses during storage of 13–17% in swedes (McNaughton and Thow, 1972) and up to 50% when harvested in wet conditions compared with 8–20% in fodder beet have been recorded (Bastiman and Slade, 1978). Large losses due to soft rot in harvested swedes and turnips have been found by Fitzgerald (1976, 1977). Thus, losses during harvesting and storage of roots can be considerably greater than for roots grazed *in situ*. However, severe frost can also result in high losses of roots left in the field, particularly in the case of fodder beet (Fitzgerald, 1979).

FEEDING CAPACITY OF ROOT CROPS AND CABBAGES

The feeding capacity of root crops, as measured in terms of lamb grazing days (LGD)/hectare is a function of yield, utilization, feed intake and the length of the feeding period. Other factors such as weather conditions and teeth losses which can affect intake or utilization will also have an influence on the feeding capacity of the crop.

On the rather opimistic assumption that 90% of the harvested roots fed indoors are utilized, the feeding capacity (LGD/hectare) for lowland store lambs (40 kg), averaged over three trials (Fitzgerald, 1975, 1976, 1977), ranged from 6800 LGD/hectare on turnips, 10 000 LGD/hectare on high DM swedes, 11 000 LGD/hectare on mangels, 16 000 LGD/hectare on medium DM fodder beet, up to an exceptionally high value of 20 500 LGD/hectare on sugar beet when all were fed whole. Relative to soft swedes the feeding capacity of whole roots was 23% lower on turnips, 20% higher on high DM swedes and mangels and 60–70% higher on fodder beet. Feed intake was considerably higher on pulped roots, and consequently the feeding capacity of roots was much reduced (4750–6900 LGD/hectare on turnips and swedes, 8620–12 370 LGD/hectare on mangels and fodder beet and up to 14 170 LGD/hectare on pulped sugar beet). These values are, however, considerably greater than the feeding capacity of cereal crops (3460 LGD/hectare on a barley-based concentrate, assuming a yield of 5 tonnes barley/hectare). Relative to pulped low DM swedes, the feeding capacity of pulped turnips was again 23% lower, that of high DM swedes was somewhat higher (+14%), while mangels and low DM fodder beet was 60% greater and that of medium to high DM fodder beets was over twice as great (+109–125%), compared with a 40% lower feeding capacity on barley-based concentrates. In the case of lighter hill lambs (26 kg) the feeding capacity of root crops, based on one year's results (Fitzgerald, 1977), was 40–50% higher than for lowland lambs due to a lower intake of roots, and ranged from 12 800–24 400 LGD/hectare on whole swedes, mangels and low to medium DM fodder beet, compared with 7600–15 550 LGD/hectare on pulped roots and 4500 LGD/hectare on barley-based concentrates. The differences in feeding capacity between swedes, mangels, low and medium DM fodder beets for hill lambs followed a similar pattern to that obtained for lowland lambs.

Where roots were grazed *in situ* without any protein supplementation but including the tops (Fitzgerald, 1977c), the feeding capacity of roots ranged from 7400 LGD/hectare on swedes up to 14 000 LGD/hectare on medium DM fodder beet for lowland lambs and from 11 400–20 700 LGD/

hectare for Blackface mountain lambs. The feeding capacity on grazed roots was 10–15% less than estimated for lambs fed on whole roots indoors, due to a greater estimated intake of roots and a somewhat lower utilization. However, a much lower feeding capacity was obtained on low DM fodder beet (8700 LGD/hectare) with lowland lambs in a second grazing experiment (Fitzgerald, 1978), due to a much higher estimated level of intake than in the previous trial (Fitzgerald, 1977c).

In grazing trials in Britain the feeding capacity of turnips ranged from 3200–4600 LGD/hectare, which was less than half that obtained on swedes while that for cabbages was intermediate at 4470–6690 LGD/hectare and paralleled their relative differences in yield (Bastiman, 1977; Bastiman and Slade, 1978; Speedy *et al.*, 1980; Rutherford and Dover, 1981a).

It is clear from these results that the higher yielding fodder beets can feed considerably more lambs/hectare than swedes, whether fed indoors or grazed *in situ*, while the feeding capacities of cabbages and turnips are less than that of swedes and considerably less than that of fodder beet and mangels.

FEED INTAKE OF LAMBS FED ON ROOT DIETS

In evaluating a wide range of root crops in terms of their suitability for lamb fattening, the intake of such roots was recorded when fed as the basic diet in whole or pulped form to lowland lambs (40 kg) and supplemented with a protein/mineral/vitamin source, which constituted 5–18% of total DM intake, depending on the protein content of the root crop, to provide about 120 g crude protein/kg DM (Fitzgerald, 1975, 1976, 1977a). The results for these experiments, averaged over the three years, are presented in *Table 13.13*.

Some root types were not included every year and there was some year to year variation in intake, particularly in the case of high DM fodder beet. Large differences in intake between the various root types were recorded when fed in whole form, ranging from as low as 473 g/day DMI on sugar beet to 690–760 g/day DMI on turnips and swedes and 713–1032 g/day DMI on fodder beets and mangels. Relative to low DM swedes, the intake of whole roots was slightly less for turnips and high DM swedes (–9%), considerably less for sugar beet (–42%), similar for high DM fodder beet, though much more variable (–30 to +20%), and was somewhat higher for low to medium DM fodder beets (+9 to +20%) and particularly for mangels (+42%).

Pulping of such roots increased intake on all types by 27–51%, with intakes ranging from 983–1310 g/day DMI, with the exception of pulped sugar beet. The intake of pulped roots was much less for sugar beet (–36%) than low DM swedes, slightly less for turnips and high DM fodder beet (–10%), similar for high DM swedes, low to medium DM fodder beets and concentrates (–5 to +4%) and somewhat higher on mangels (+10%).

In the case of hill lambs (26 kg) the pattern of intake on a more limited range of root crops was similar to that for lowland lambs (Fitzgerald, 1977b) but at about a 30% lower level of intake due to their lower body weight (26 v. 40 kg). Relative to low DM swedes, intake of whole roots was

Table 13.13 DRY MATTER INTAKE (DMI) OF STORE LAMBS FED ON WHOLE OR PULPED ROOTS INDOORS OR GRAZED *IN SITU*

Dry matter intake (DMI) (g/day)	*Soya/ min. fed*	*Turnips*	*Swedes*		*Mangels*	*Fodder beet*			*Sugar beet*	*Barley conc.*	*Source*
			Low DM	*High DM*		*Low DM*	*Med DM*	*High DM*			
Lowland stores (Galway: 40 kg)											
(No. of trials)		(2)	(3)	(2)	(2)	(3)	(2)	(2)	(1)	(2)	Fitzgerald (1975–1977a)
Whole roots	+	692	758	696	1032	821	866	713	473		
Pulped roots	+	993	1145	1052	1310	1189	1210	983	673	1226	Fitzgerald (1975–1977a)
Grazed roots	–		(1)		(1)	(1)	(1)				Fitzgerald (1977c)
(estimated)			1020		987	942	727				
Hill lambs (Blackface: 26 kg)											
Whole roots	+		530		738	666	562				Fitzgerald (1977b)
Pulped roots	+		859		946	935	861			938	
Grazed roots (estimated)	–		781		681	656	482				

26–39% greater for low DM fodder beet and mangels but was similar for medium DM fodder beet (+6%). Pulping of roots also considerably increased root intake by 28–62%.

Feed intake on a body weight basis (W) was of the order of 1.7–2.5% W for lowland lambs on whole roots compared with 2.5–3.3% W on pulped roots. These results indicate that lambs can consume just as much dry matter in the form of pulped roots, despite their bulky nature and high moisture content, as they can on an all concentrate diet. Bulkiness *per se* is not a limiting factor affecting root intake, due to the very high digestibility of root crops (Barry, Drew and Duncan, 1971; Fitzgerald, 1976). The lower intake of whole roots, particularly that of high DM types, would appear to be associated with the hardness of such roots, which is directly related to their DM content, and which lambs have difficulty in eating. The time spent eating and jaw fatigue are probably the main factors limiting intake of soft whole roots.

In the case of lambs grazing roots *in situ* without any protein/mineral supplements but including the tops (Fitzgerald, 1977c), the estimated intake from pre- and post-grazing yields was highest on swedes and declined with increasing root DM content for mangels and fodder beet (*Table 13.13*). Compared with whole roots fed indoors, the intake of grazed swedes was 36–47% greater for Galway and Blackface lambs respectively but was 9–16% lower than that of pulped swedes. Intake of grazed low DM fodder beet was, however, similar to that for whole roots fed indoors while that of grazed mangels (–9%) and medium DM fodder beet (–14%) was lower than for whole roots fed indoors. The higher intakes of swedes grazed *in situ* compared with whole swedes fed indoors could be due to a high proportion of leaf in the crop and poor storage in the harvested swedes which developed soft rot. Lack of protein may have limited intake of grazed mangels and fodder beet.

Rutherford and Dover (1981a) also obtained much greater intakes of chopped swedes when fed to Suffolk-X lambs (30–35 kg) either outdoors (890 g/day DMI) or indoors (760 g/day DMI) compared with folding roots *in situ* (390 g/day DMI). Despite feeding a higher level of supplement (0.28–0.36 kg DM) with the folded swedes compared with the chopped swedes (0.08–0.10 kg DM), total DM intake was 22–32% less and lamb performance was considerably less. This was attributed to a much higher incidence of teeth loss (1.0 v. 1.8 teeth lost/lamb) in lambs grazing swedes. The size of the chopped root particles and the cleanliness of the root may also affect lamb performance (Bastiman, 1977). Barry, Drew and Duncan (1971) in New Zealand also found that the intake of digestible energy was higher in lambs grazing on swedes and turnips than on mangels, fodder beet or kale and consequently such lambs gained more weight. However, no protein supplement was fed with the roots.

To summarize, higher intakes have been obtained with mangels and low to medium DM fodder beets when fed in whole form indoors than with turnips, swedes or high DM fodder beets. When pulped, intake of all roots is considerably improved but that of the low to medium DM fodder beets and mangels tends to remain somewhat higher than that of turnips, high DM swedes or high DM fodder beets and similar to that of an all concentrate barley-based diet. On the other hand, when roots are grazed *in situ*, intake of soft swedes appears to be greater than that of mangels, low DM fodder beet and particularly medium DM fodder beet. The latter type of root is not really suitable for grazing due to its hardness and a high proportion of the root being underground. However, more recent results with low DM fodder beet when supplemented with protein/minerals indicate that feed intake and lamb performance on fodder beet grazed *in situ* can be higher than that of lambs fed on whole roots indoors (Fitzgerald, 1979, 1981) and comparable to that obtained on swedes grazed *in situ*. Only low DM fodder beets which are sufficiently soft and are mostly above ground level are suitable for grazing *in situ*, but they are, however, much more susceptible to damage from severe frost than swedes.

LAMB PERFORMANCE ON ROOT CROPS AND CABBAGES

The performance of store lambs fed indoors on whole or pulped roots with protein supplementation or on roots grazed *in situ* without any supplement is shown in *Table 13.14* (Fitzgerald, 1975, 1976, 1977a,b,c). In general, lamb liveweight (DLWG) and carcass gains (DCG) for lowland store lambs fed on whole roots were much better on the softer mangels and low to medium DM fodder beets than on either turnips, swedes or high DM fodder beets, reflecting the differences obtained in intakes. Lamb performance between years was quite variable on the high DM fodder beet when fed whole (–19 to +123 g/day DLWG) and lambs lost weight on whole sugar beet due to its extreme hardness, high sugar content and consequent low intake. Pulping of roots increased lamb performance on all diets by 56–114 g/day DLWG or 23–49 g/day DCG, due to increased intake. The best lamb gains on pulped roots were obtained on low to medium DM fodder beets, which in terms of carcass gain was equivalent to about 80% of that obtained on the barley-based concentrate diet at similar levels of intake. Somewhat lower gains were obtained on pulped swedes, mangels and high DM fodder beet (–9 to 24%), while that on pulped sugar beet was much poorer due to the generally lower intakes of roots on those diets.

Lamb performance on grazed roots without any protein supplementation (Fitzgerald, 1977a) was generally better on swedes or mangels than on low DM fodder beet and particularly so on medium DM fodder beet, on which lambs only maintained weight. Lamb performance on grazed swedes was better than for lambs fed on whole swedes indoors, though not as good as for pulped swedes. However, the performance of lambs grazing on mangels and particularly on fodder beet was much worse than that of lambs fed on similar roots in whole form indoors, due partly to a lower intake and a higher maintenance requirement and to the lack of protein/mineral supplementation for grazing lambs, even though they had access to the leaves which are relatively high in protein and ash.

In the case of hill lambs, a similar trend in performance was observed as shown in *Table 13.15* (Fitzgerald, 1977b). Moderate lamb gains were obtained on mangels and low DM fodder beet when fed in whole form indoors, compared with poor gains on whole swedes and medium DM fodder beet. Performance was also considerably improved when roots were fed pulped (+73–122 g/day DLWG or +31–58 g/day DCG), with gains

Table 13.14 EFFECT OF TYPE OF ROOT AND METHOD OF FEEDING ON THE PERFORMANCE OF LOWLAND (GALWAY) STORE LAMBS

	Soya /min fed	*Turnips*	*Swedes*		*Mangels*	*Fodder beet*			*Sugar beet*	*Barley concs.*	*S.E.*	*Source*
			Low DM	*High DM*		*Low DM*	*Med DM*	*High DM*				
(No. of trials)		(2)	(3)	(2)	(2)	(3)	(2)	(2)	(1)	(2)		Fitzgerald (1975–1977a)
Liveweight gain (g/day)												
Whole roots	+	31	67	52	130	101	113	52	–45		(±12.0)	
Pulped roots	+	125	158	152	186	203	227	152	25	254	(±12.0)	
Grazed roots (1977)	–		94		94	57	15				±9.8	Fitzgerald (1977c)
Carcass gain (g/day)												
Whole roots	+	28	39	34	66	57	56	36	–3		(±5.9)	
Pulped roots	+	69	78	83	89	104	103	78	22	129	(±5.9)	
Grazed roots (1977)	–		52		37	23	1				±5.4	
Feed conversion efficiency (kg feed DM/kg carcass gain)												
Whole roots	+	24.1	21.8	20.7	15.8	14.5	17.1	28.9	–			
Pulped roots	+	14.9	14.5	12.6	14.8	11.5	11.9	12.3	31.6	9.8		
Grazed roots (1977) (estimated)	–		19.6		26.7	41.0	–					

Table 13.15 EFFECT OF ROOT TYPE AND METHOD OF FEEDING ON THE PERFORMANCE OF HILL (BLACKFACE) LAMBS

	Soya /min feed	*Swedes Low DM*	*Mangels*	*Fodder beet*		*Barley concs.*	*S.E.*
				Low DM	*Med DM*		
Liveweight gain (g/day)							
Whole roots	+	–2	65	65	18		±11.0
Pulped roots	+	109	138	155	140	171	±11.0
Grazed roots	–	17	16	12	–19		±11.0
Carcass gain (g/day)							
Whole roots	+	18	38	50	31		±5.1
Pulped roots	+	60	73	81	89	105	±5.1
Grazed roots	–	12	10	7	–5		±5.1
Feed conversion efficiency (kg feed DM/kg carcass gain)							
Whole roots	+	29.4	19.3	13.3	18.2		
Pulped roots	+	14.3	12.9	11.5	9.7	9.0	
Grazed roots (estimated)	–	65.1	68.1	93.7			

From Fitzgerald (1977b)

equivalent to 77–85% of that obtained on a concentrate diet with fodder beet. The performance of hill lambs was poor on all types of root grazed *in situ*.

In other studies Leonard (1970) recorded poor performance for Galway and Suffolk-X store lambs grazing on a hard swede (Bangholm) which was associated with a high incidence of teeth loss. Much better lamb gains with Blackface and Cheviot hill lambs were obtained when grazed on a softer swede (Broadlands) and on which teeth losses were minimal. However, similar lambs gained much less weight on green top turnips even though no teeth were lost. In New Zealand, Barry, Drew and Duncan (1971) obtained much better lamb gains on turnips and swedes grazed *in situ* than on mangels or fodder beet due to a lower intake of roots and possibly protein deficiency. In a subsequent study Barry and Drew (1978) obtained similar lamb gains when grazed on turnips, swedes and kale. In Britain, Rutherford and Dover (1981a) obtained much better lamb gains on cabbage than on swedes over four years (118 v. 64 g/day DLWG), despite feeding a higher level of concentrate supplement with swedes. The poorer performance obtained on swedes was partly due to a greater incidence of teeth loss. Speedy *et al.* (1980) also obtained better lamb performance with Suffolk-X lambs grazing on cabbage than on either swedes or turnips, but the performance of Blackface lambs was similar on all three crops.

It would appear from these results that lamb performance on turnips and swedes when grazed *in situ* was fairly similar in both UK and New Zealand studies but that better lamb gains have been obtained on cabbages in most comparisons. In Irish studies better lamb gains have been obtained when lambs were fed on soft swedes than on either hard swedes or turnips when grazed *in situ* and to a lesser extent when fed whole or pulped indoors. Much better lamb gains have, however, been obtained with mangels and low to medium DM fodder beets when fed indoors and supplemented with protein/minerals than in the case of swedes or turnips, particularly when fed in whole form and to a lesser extent when fed in pulped form, and were in the region of 80–85% of that obtained with a cereal-based diet. Grazing of fodder beet *in situ* did result in poorer lamb gains, in the absence of protein supplementation, than obtained on swedes. However, more recent trials have given similar levels of performance when supplemented with protein/minerals. High DM fodder beet has given rather variable and generally poorer results when fed in whole form than low DM fodder beet. Even when fed in pulped form, intake and lamb performance on high DM fodder beet is generally 20–25% lower than on low or medium DM fodder beet, possibly because of the higher sugar content of such roots and the extreme hardness of such roots which may be difficult to chew even when pulped. Very high DM fodder beets, despite their advantage in yield, are therefore less suitable for feeding to lambs than low or medium DM types.

TEETH LOSSES ASSOCIATED WITH FEEDING OF ROOTS

Teeth losses in the indoor feeding trials (Fitzgerald, 1975, 1976, 1977a) were generally low and, if anything, were negatively related to the hardness or DM content of the root when fed whole, but rather were positively related to root intake. For example, lowland lambs lost on average 0.2, 0.9, 0.0 and 0.4 teeth/lamb on high DM fodder beets, mangels, hard and soft swedes respectively. On pulped roots teeth losses were minimal and indeed were 0.25/lamb on concentrates. Most of these losses occurred towards the end of the feeding period and appeared to be associated more with a high intake of roots than with the hardness of the root and were not always associated with poor lamb performance, although in individual cases a high incidence of teeth loss could be associated with poor lamb gains. A higher incidence of teeth loss was found for lambs grazing swedes (1.5 teeth lost/lamb) but very few teeth (0–0.13 teeth/lamb) were lost by lambs grazing mangels and fodder beet.

Teeth losses in Blackface mountain lambs fed on roots indoors were, however, much more serious (Fitzgerald, 1977b) with lambs losing between 1.6 and 3.2 teeth on whole roots, particularly on fodder beet, whereas very few teeth were lost on pulped roots. On grazed roots teeth losses among Blackface lambs were much more variable, with the highest incidence occurring on swedes, but this was not directly reflected in lamb performance.

In a further trial comparing low DM fodder beet with swedes grazed *in situ* (Fitzgerald, 1979), teeth losses were again generally low and seemed to be associated with a high intake of whole roots rather than on the hardness of the root *per se*. However, high incidences of teeth loss have been reported for lambs grazing on high DM swedes in other studies (Leonard, 1970; Rutherford and Dover, 1981a) and for hill lambs fed indoors on whole roots or on grazed swedes (Fitzgerald, 1977b) and this was associated with poor lamb performance.

FEED CONVERSION EFFICIENCY ON ROOT-BASED DIETS

The feed conversion efficiency of root-based diets (kg feed DM/kg carcass gain) fed whole to lowland lambs when averaged over three years (Fitzgerald, 1975, 1976, 1977a), ranged from 28.9–14.5:1 (*Table 13.14*). Feed conversion efficiency was much better and less variable on pulped roots, due to the increased intake and better lamb performance, ranging from 11.5–14.9:1 compared with 9.8:1 for barley-based concentrates. Feed conversion efficiency on grazed roots, in the absence of protein supplementation, was very much poorer for mangels and fodder beet, but similar for swedes. Compared with concentrates, therefore, the feeding value of roots in terms of their ability to put on carcass weight depends on the type of root fed and the method of feeding. Relative to concentrates the feeding value of roots ranged from 40–47% for whole turnips and swedes, 57–67% for whole mangels, low and medium DM fodder beet, down to 34% for high DM fodder beet. For pulped roots the feeding value was much better, ranging from 65–77% on pulped turnips, swedes and mangels and 79–85% on pulped fodder beet, with a very low feeding value (31%) for pulped sugar beet due to its extremely low intake.

In the case of Blackface mountain lambs, the feed conversion efficiency to carcass gain (*Table 13.15*) was generally poor on whole roots, particularly swedes, medium DM fodder beet and mangels, but was quite good on the low DM fodder beet. Pulping of roots considerably improved feed conversion efficiency whereas very poor feed conversion efficiencies were obtained with hill lambs grazed on roots. The feeding value of pulped roots relative to concentrates when fed to hill lambs was of the order of 63–70% for swedes and mangels and 78–93% for fodder beet, compared with 31–68% on whole roots.

It is obvious from these results that when roots are fed in pulped or chopped form and properly balanced with protein/minerals and vitamins, they can be converted to lamb meat with an efficiency close to that obtained by feeding a cereal-based concentrate diet.

PRODUCTION OF LAMB MEAT/HECTARE ON ROOTS

Forage crops should be ultimately assessed on the basis of how much saleable product (lamb meat)/hectare can be produced from them compared with other feeds. Consequently, their use as a feed for finishing store lambs can be justified only if they can produce a relatively high level of meat output at a reasonably low cost.

Production of lamb meat/hectare is dependent on the yield of the crop, its utilization and the efficiency with which the feed is converted into meat. Since yield can vary from one type of root crop to another and from year to year, it is first useful to assess fodder root crops on the basis of the amount of meat produced/tonne of root DM (*Table 13.16*), apart from any additional protein/mineral or vitamin supplements required to balance such diets. On such a basis, the output of lamb meat/tonne root DM eaten for lowland stores lay in the range of 45–55 kg for whole turnips and swedes fed indoors, 69–80 kg for whole mangels, low and medium DM fodder beets and 57 kg on high DM fodder beets fed whole (Fitzgerald 1975, 1976, 1977). The level of output/tonne root DM when fed pulped increased to 75–86 kg for turnips, swedes and mangels and 97–103 kg for fodder beets, compared with 103 kg for cereals.

The production of carcass meat/hectare of roots, excluding the tops, when fed to store lambs is also set out in *Table 13.16*. These figures take into account differences between crops in yield and assume a 90% utilization. For whole roots production of lamb meat is in the order of 200–410 kg/hectare on turnips and swedes and 640–860 kg/hectare on mangels and fodder beet using lowland lambs. Much higher levels of output are possible with pulped roots and indeed all but turnips produced higher meat yields/hectare than barley.

Lamb meat production from Blackface mountain lambs (26 kg) followed a similar trend to that obtained with lowland lambs (Fitzgerald, 1979b; *Table 13.16*), with the highest levels being obtained on low DM fodder beet. Production/hectare with hill lambs was lower on whole swedes and mangels than for lowland lambs, but was as good and in some cases better on fodder beet, pulped roots and cereals. As with lowland lambs meat production/hectare was considerably improved by feeding roots in a pulped form, with the greatest increase being obtained on medium DM fodder beet.

Table 13.16 OUTPUT OF CARCASS GAIN FROM ROOT CROPS FED INDOORS COMPARED WITH A CEREAL-BASED DIET

Carcass gain	*Turnips*	*Swedes*		*Mangels*	*Fodder beet*			*Barley concs.*
		Low DM	*High DM*		*Low DM*	*Med DM*	*High DM*	
(kg/tonne root DM)								
(a) Lowland lambs								
Whole roots	45	53	55	69	80	76	57	
Pulped roots	76	76	86	75	100	103	97	103
(b) Hill lambs								
Whole roots		37		59	85	68		
Pulped roots		75		87	98	123		111
Typical DM yield (tonnes/hectare) (90% utilized)	5.0	7.0	7.5	12.0	12.0	12.5	12.5	4.5
Carcass gain (kg/ha)								
(a) Lowland lambs								
Whole roots	203	335	412	746	862	858	641	
Pulped roots	344	480	583	812	1078	1158	1096	466
(b) Hill lambs								
Whole roots		234		632	921	763		
Pulped roots		475		944	1061	1387		498

EFFECT OF SUPPLEMENTING ROOTS WITH PROTEIN

The low crude protein (CP) content of roots, particularly fodder beet and mangels (Fitzgerald, 1975, 1976), is not adequate for lambs and consequently a diet based on fodder beet requires additional protein. This was clearly established in a recent experiment (Fitzgerald, 1981) in which a level of 110 g CP/kg DM was found to be adequate with fodder beet containing 80 g CP/kg DM when fed in pulped form to either lowland store lambs (40 kg) consuming 1100–1200 g DM/day, or to Blackface lambs (27 kg) consuming 900 g DM/day. There was no further response in lamb performance to feeding a higher level of protein (130 g CP/kg DM) to lambs fed on pulped roots, but the performance of lambs fed on whole roots was increased substantially by feeding the higher level of protein, mainly due to a large increase (+43%) in total DM intake. Thus, a total intake of about 130–140 g CP/day was required by lowland lambs and about 100 g CP/day by hill lambs fed on fodder beet diets to maximize their performance; this can be supplied by supplementing fodder beet containing 80 g CP/kg DM with about 150 g of a protein-rich supplement containing 130 g soyabean meal (450 g CP/kg DM) and 20 g mineral/vitamin supplement. A higher level of supplement would be required for roots with a lower protein content while a lower level would be adequate with roots containing a higher level of protein. Less degradable sources of protein e.g. fishmeal or meat and bone meal may have an advantage over soyabean meal as protein supplements with fodder beet which contain a high proportion (30–40%) of the crude protein content in a non-protein nitrogen form.

SUPPLEMENTATION OF ROOTS FED WHOLE OR GRAZED *IN SITU* WITH PROTEIN AND ENERGY-RICH CONCENTRATES

The cost of root harvesting, storage and increased labour costs involved in feeding roots can add 30–50% (+£200–£300/hectare) to their cost of production. Housing accommodation is also required for lambs fed indoors and is now becoming very expensive, unless old buildings can be suitably adapted for fattening lambs. Secondly, although lamb performance is considerably improved by pulping roots, the feeding capacity of the root crop is considerably reduced due to higher intakes. It was therefore decided to investigate further the possibility of grazing fodder beet *in situ* compared with swedes, and to supplement such roots with barley in addition to protein, minerals and vitamins as an alternative to harvesting and feeding in pulped form to lambs indoors. A low DM fodder beet (Peramono) was either fed whole indoors or grazed *in situ* alongside swedes by both lowland (Galway and Suffolk-X) and hill lambs (Blackface). The roots were fed either alone (control) or supplemented with 150 g/day of a soyabean meal/mineral/vitamin mixture, in addition to either 0, 150 or 450 g whole barley/day and compared with pulped roots fed indoors and supplemented with 150 g/day of soyabean meal/mineral mix (Fitzgerald, 1979).

Due to severe frost damage to the fodder beet crop grazed *in situ*, the feeding period for both lowland (seven weeks) and hill lambs (nine weeks) was reduced by two weeks compared with those fed on whole fodder beet indoors or swedes grazed *in situ*. The feed intake and performance of the lambs are given in *Table 13.17*. Lamb performance on whole fodder beet alone indoors was quite poor, with lambs doing little better than maintaining weight. There was, however, quite a good response to feeding both the protein/mineral supplement and barley for both types of lamb, and in many instances produced lamb gains in excess of that obtained on pulped fodder beet. In contrast to the poor lamb gains obtained with whole fodder beet

Table 13.17 EFFECT OF SUPPLEMENTING ROOTS FED WHOLE INDOORS OR GRAZED *IN SITU* BY STORE LAMBS WITH PROTEIN/MINERALS AND BARLEY

Supplement fed	*Whole or grazed roots*				*Pulped roots*	*S.E.*
Soya/min (g/day)	0	150	150	150	150	
Barley (g/day)	0	0	150	450	0	
Lowland lambs						
DM intake (g/day)						
1. Fodder beet (whole)	651	749	738	485	974	
Total DMI	651	882	993	985	1107	
Carcass gain (g/day)						(n = 18)
1. Fodder beet, whole (9 wks)	18	57	70	89	74	6.7
2. Fodder beet, grazed (7 wks)	54	100	103	113	–	8.6
3. Swedes, grazed (9 wks)	61	72	101	116	69	6.8
Hill lambs						
DM intake (g/day)						
1. Fodder beet, whole	476	522	484	422	775	
Total DMI	476	655	740	923	908	
Carcass gain (g/day)						(n = 15)
1. Fodder beet, whole (11 wks)	5	31	56	83	64	10.3
2. Fodder beet, grazed (9 wks)	35	55	67	73	–	10.2
3. Swedes, grazed (11 wks)	42	52	68	74	67	9.3

fed indoors, a very good level of performance was obtained when it was grazed *in situ,* even without any supplement. This was probably due to the softening effect which the severe frost had on the grazed fodder beet. There was a very good response to feeding a protein/mineral supplement with fodder beet in both cases. While there was very little further improvement in lamb gains to feeding barley with fodder beet grazed *in situ* in the case of lowland lambs, a worthwhile improvement in the gains of hill lambs was obtained by feeding 150 g barley/day with the protein supplement, but the incremental response to feeding a higher level of barley (450 g/day) was much poorer. Lamb performance on swedes grazed alone, which suffered less damage from frost than fodder beet, was similar to that obtained on fodder beet grazed alone. There was a much poorer response to the protein/mineral supplement due to the higher protein content of swedes. There was, however, a good response to supplementing swedes with 150 g barley/day in addition to protein/minerals in both lowland and hill lambs, giving levels of performance similar to those obtained on fodder beet diets and as good or better than obtained on pulped swedes fed indoors. As with fodder beet there was little further response to feeding a higher level of barley with swedes.

It would appear from these results that good lamb performance can be obtained on fodder beet or swedes grazed *in situ* when suitably supplemented with protein/minerals plus a low level of barley (300 g/day) and will give as good or even better weight gain than obtained on pulped roots fed indoors. A higher level of barley supplementation would appear to be necessary to maximize lamb gains with whole fodder beet fed indoors, although a subsequent trial (Fitzgerald, 1981; unpublished data) indicates that a level of 300 g/day is adequate with medium or large sized roots.

It is, therefore, feasible on well-drained soils and in relatively frost-free areas to graze roots *in situ* and supplement them with a moderate level of concentrates to obtain a high level of lamb performance comparable to that obtained on pulped roots fed indoors. A substantial saving in feed and overhead costs can be realized in this way while maintaining a higher feeding capacity, due to the lower level of root intake compared with pulped roots. Fodder beet is, however, much more susceptible to damage from severe frost (–8-–10 °C) than swedes although it can tolerate milder frosts (–4-–5 °C) pretty well. Therefore, many inland areas with a high risk of severe frost would not be suitable for grazing fodder beet, whereas swedes would be relatively safe. However, the yield and the feeding capacity of swedes is less than that of grazed fodder beet.

The response to supplementing swedes grazed *in situ* with concentrates has not been very consistent. Leonard (1970) obtained a relatively good response to supplementing a hard swede (Bangholm) with 450 g concentrates/lamb/day in the case of Suffolk-X lambs, but not with Galway lambs, even though both breeds had performed badly on swedes alone and had lost a lot of teeth. He also obtained a good response to supplementing turnips (Green top) with 340 g concentrates/day both in the case of Cheviot lambs and, to a lesser extent, with Blackface lambs in a situation where lamb performance on turnips alone was quite poor despite little or no loss of teeth. The feeding of 340 g concentrates/day to similar types of hill lambs grazing soft swedes (Broadlands), and on which lambs had gained reasonably well when fed alone with little loss of teeth, had, however, little additional benefit. A similarly poor response was obtained by Fitzgerald and O'Toole (1970) with Blackface lambs grazing Broadland swedes and supplemented with 225 g concentrates/day, even though lamb weight gains on swedes alone were poor.

It would, therefore, appear that the response to supplementation of turnips or swedes grazed *in situ* will depend on many factors, namely the type of root, its hardness, the level of gain obtained on the root when fed alone, the incidence of teeth loss and to some extent the breed type and possibly the size and condition of the lambs at the start of the root feeding period.

References

ADAS (1976). *Sheep: Brassica species for Store Lambs.* ADAS Regional Agricultural Science Service, Annual Report, pp. 128–129

APPLETON, M. (1969). *Fodder Crops for Lamb Fattening.* Experimental Husbandry Farms and Horticulture Station's Progress Report, pp. 66–68

BARRY, T.N. and DREW, K.R. (1978). Responses of young sheep grazing swedes, turnips and kale to intraperitoneal and oral supplementation with DL-Methionine. *New Zealand Journal of Agricultural Research* **21**, 395–399

BARRY, T.N., DREW, K.R. and DUNCAN, S.J. (1971). The digestion, voluntary intake and utilisation of energy by Romney hoggets fed five winter forage crops. *New Zealand Journal of Agricultural Research* **14**, 835–846

BASTIMAN, B. (1977). ADAS work on the utilisation and use of fodder brassica. In *Brassica Fodder Crops Conference*, Scottish Plant Breeding Station, Edinburgh, pp. 123–132

BASTIMAN, B. and SLADE, C.F.R. (1978). The utilisation of forage crops—a review of recent work undertaken by ADAS. *ADAS Quarterly Review*, No. 21, pp. 1–11

BLACK, W.J.M. (1967). The use of some fodder crops for fattening weaned lambs. *An Foras Taluntais, Animal Production Research Report*, pp. 45–46

BOYD, A.C. and DICKSON, I.A. (1966). Fodder radish or rape? *Agriculture* **73**, 217–222

CRAIG, P. (1970). The influence of stocking rate and supplementary feeding on the liveweight performance of lambs grazing rape. B.Sc. Thesis, Edinburgh University

DAVIES, M.A. (1978). Land improvement by pioneer cropping with Dutch turnips and subsequent reseeding. Redesdale EHF Annual Report, 40–47

DEPARTMENT OF AGRICULTURE, DUBLIN (1979). *Fodder Root and Fodder Crops, Recommended List.* National Cereal Breeding and Seed Propagation Centre, Backweston, Leixlip, Co. Kildare

EWER, T.K. and SINCLAIR, D.P. (1952). Lamb fattening trials in the Canterbury Plains. *New Zealand Journal of Science and Technology* **33**, 1–30

FEELEY, D.P. (1964a). Swede variety trials, 1960–62. *Irish Journal of Agricultural Research* **3**, 71–82

FEELEY, D.P. (1964b). Comparison of mangel and fodder beet varieties, 1960–62. *Irish Journal of Agricultural Research* **3**, 189–200

FITZGERALD, S. (1969). Fattening store lambs on rape supplemented with barley. *An Foras Taluntais, Animal Production Research Report*, p. 38

FITZGERALD, S. (1970). Effect of cereal supplementation on the performance of store lambs fed on rape. *An Foras Taluntais, Animal Production Research Report*, p. 40

FITZGERALD, S. (1975). Evaluation of root crops for fattening store lambs. *An Foras Taluntais, Animal Production Research Report*, pp. 157–158

FITZGERALD, S. (1976). Evaluation of root crops for fattening store lambs. *An Foras Taluntais, Animal Production Research Report*, pp. 163–164

FITZGERALD, S. (1977a). Performance of Galway store lambs on roots fed indoors. *An Foras Taluntais, Animal Production Research Report*, pp. 161–162

FITZGERALD, S. (1977b). Performance of Blackface Mountain store lambs on roots fed indoors. *An Foras Taluntais, Animal Production Research Report*, pp. 162–163

FITZGERALD, S. (1977c). Performance of store lambs on roots grazed *in situ. An Foras Taluntais, Animal Production Research Report*, pp. 163–164

FITZGERALD, S. (1978). Performance of store lambs grazed on fodder beet *in situ. An Foras Taluntais, Animal Production Research Report*, pp. 170–171

FITZGERALD, S. (1979). Effect of supplementing roots with protein and barley on the performance of store lambs. *An Foras Taluntais, Animal Production Research Report*, pp. 89–91

FITZGERALD, S. (1981). Effect of supplementing fodder beet with protein on the performance of store lambs. British Society of Animal Production, Winter Meeting, Harrogate, Paper No. 63; *Animal Production* **32**, 373–374 (Abstract)

FITZGERALD, S. and O'TOOLE, M.A. (1970). Methods of fattening Blackface mountain lambs. *An Foras Taluntais, Animal Production Research Report*, pp. 36–38

FUSSEL, G.E. (1966). *Farming Technique from Pre-historic to Modern Times*, pp. 106–170. London, Pergamon Press

GREENALL, A.F. (1958). Studies of grazing of crop forages by sheep. II. Wastage of rape by grazing wethers. *New Zealand Journal of Agricultural Research* **1**, 683–693

GREENHALGH, J.F.D. (1971). Problems of animal disease. In *The Future of Brassica Fodder Crops*, pp. 56–64. Occasional Publication No. 2, Rowett Research Institute

HARPER, F. and COMPTON, I.J. (1980). Sowing date, harvest date and the yield of forage brassica crops. *Grass and Forage Science* **35**, 147–157

KILKENNY, J.B. (1976). Forage crops for lamb finishing. *27th Annual Meeting EAAP, Zurich*

LEONARD, T.F. (1970). Modern methods of swede production and traditional methods of swede utilisation. *Journal of Irish Grassland and Animal Production Association* **5**, 153–171

MAFF (1975). Bulletin No. 48, *Rations for Livestock* (1960) London, HMSO

McNAUGHTON, I.H. and THOW, R.F. (1972). Swedes and turnips. *Field Crop Abstracts* **25**, 1–12

MORRISON, D. (1971). The present position of Brassica fodder crops in Scotland. In *The Future of Brassica Crops*, pp. 9–13. Occasional Publication No. 2, Rowett Research Institute

NIAB (1980a). *Recommended Varieties of Green Fodder Crops.* Farmers Leaflet No. 2, Cambridge

NIAB (1980b). *Varieties of Fodder Root Crops.* Farmers Leaflet No. 6, Cambridge

PATERSON, W.G.W., DICKSON, I.A. and BERLYN, P. (1977). Evaluation of forage brassica crops via fattening lambs. In *Brassica Fodder Crops Conference*, pp. 82–88. Scottish Plant Breeding Station, Edinburgh

RUSSEL, A.J.F. (1967). A note on goitre in lambs grazing on rape (*Brassica rapus*). *Animal Production* **9**, 131–133

RUTHERFORD, R.N. and DOVER, P.A. (1981a). Alternative finishing systems for store lambs incorporating chopped swedes, silage and cereal or chopped swede supplements compared with traditionally folded swedes. British Society of Animal Production, Winter Meeting, Harrogate, Paper No. 94; *Animal Production* **32**, 384 (Abstract)

RUTHERFORD, R.N. and DOVER, P.A. (1981b). Cabbage folded *in situ* for finishing store lambs compared with traditional finishing on folded turnips and swedes. British Society of Animal Production, Winter Meeting, Harrogate, Paper No. 94A; *Animal Production* **32**, 384–385 (Abstract)

SHARMAN, G.A., LAWSON, W.J. and WHITELAW, A. (1981). Potential growth limiting factors in the Brassicae. British Society of Animal Production, Winter Meeting, Harrogate, Paper No. 9; *Animal Production* **32**, 383–384 (Abstract)

SHELDRICK, R.D. and YOUNG, N.E. (1977). The role of Brassica catch crops in lamb production systems. In *Brassica Fodder Crops Conference*, pp. 111–118. Scottish Plant Breeding Station, Edinburgh

SHELDRICK, R.D., FENLON, J.S. and LAVENDER, R.H. (1981). Variation in forage yield and quality of three catch crops grown in Southern England. *Grass and Forage Science* **36**, 179–187

SINCLAIR, D.P. and ANDREWS, E.D. (1958). Prevention of goitre in new born lambs from kale fed ewes. *New Zealand Veterinary Journal* **6**, 87–95

SLADE, C.F.R. (1977). The performance of lambs fattening on two different types of forage crop in late autumn. *Animal Production* **24**, 161 (Abstract)

SMITH, R.H. and GREENHALGH, J.F.D. (1977). Haemolytic toxin of the brassica and its practical implications. In *Brassica Fodder Crops Conference*, pp. 96 and 101. Scottish Plant Breeding Station, Edinburgh

SPEEDY, A.W., GILL, W.D., JOHNSTON, A.A. and FITZSIMONS, J. (1980). An objective comparison of different lamb finishing systems. British Society of Animal Production, Winter Meeting, Harrogate, Paper No. 78; *Animal Production* **30**, 482 (Abstract)

SPEEDY, A.W., ILLIUS, A.W., RODGER, J.B.A. and DICKSON, A.C. (1981). Factors affecting the output per hectare of lamb finishing systems based on forage and root crops grazed *in situ*. British Society of Animal Production, Winter Meeting, Harrogate, Paper No. 95; *Animal Production* **32**, 385 (Abstract)

STOREY, T.S. and BARRY, P. (1979). Yield and quality of fodder beet as affected by cultivar, sowing date and plant population. *Irish Journal of Agricultural Research* **18**, 263–277

TECHNICAL BULLETIN 33 (1975). *Energy Allowances and Feeding Systems for Ruminants.* London, HMSO

THOMAS, T.M. (1973). Rape and stubble turnip. *An Foras Taluntais, Animal Production Research Report*, p. 23

THOMAS, T.M. (1974). Autumn sown stubble turnips. *An Foras Taluntais, Animal Production Research Report*, pp. 17–18

THOMAS, T.M. (1975). Productivity and value of arable fodder crops. *Journal of Irish Grassland and Animal Producers Association* **10**, 83–89

THOMAS, T.M. (1976). Brassica catch crops. *An Foras Taluntais, Animal Production Research Report*, p. 13

TOOSEY, R.D. (1972). The brassica green forage and root crops. In *Profitable Fodder Cropping*, pp. 59–85. Suffolk, Farming Press Ltd

TRIBE, D.E., BONIWELL, B.A. and AITKEN, Y. (1960). Fodder crops for fattening lambs in seasons of low rainfall. *Empire Journal of Experimental Agriculture* **28**, 171–180

WILLEY, L.A. (1971). The present position of brassica fodder crops in

England and Wales. In *The Future of Brassica Fodder Crops*, pp. 3–8. Occasional Publication No. 2, Rowett Research Institute

YOUNG, N.E., AUSTIN, A.R., NEWTON, J.E. and ORR, J.R. (1981). The performance and intake of lambs grazing forage crops. British Society of Animal Production, Winter Meeting, Harrogate, Paper No. 93; *Animal Production* **32**, 384 (Abstract)

IV

Health

14

NEONATAL MORTALITY OF LAMBS AND ITS CAUSES

F.A. EALES, J. SMALL and J.S. GILMOUR
Moredun Research Institute, Edinburgh, UK

Neonatal mortality accounts for approximately 35% of all sheep losses and is thus a major factor in reducing the profitability of a sheep enterprise (Howe, 1976). An understanding of the causes of this problem and its prevention is thus essential for all involved in sheep farming including the farmer himself, the veterinary surgeon, the agricultural adviser and the research worker. In this chapter the current knowledge of this subject is reviewed and ways in which the individual farmer may approach this problem in his flock are suggested.

Numerical assessment

Some estimates of neonatal mortality including stillbirth are shown in *Table 14.1*. Considerable farm to farm variation is evident but there is no

Table 14.1 SOME ESTIMATIONS OF NEONATAL MORTALITY IN LAMBS, INCLUDING STILLBIRTH, EXPRESSED AS A PERCENTAGE OF ALL LAMBS BORN FROM LOWLAND, UPLAND AND HILL FLOCKS

Source	*Flock type*	*Mortality (%)*
MLC (1976)	Lowland	11
Harker (1977)	Lowland	9
Johnston (1977)	Lowland	13
Saunders (1977)	Lowland	13
Purvis *et al.* (1979)	Lowland	12
MLC (1976)	Upland	9
Speedy *et al.* (1977)	Upland	5 (3–29)
Purser and Young (1959)	Hill	12, 19
Bannatyne (1977)	Hill	21
Whitelaw (1976)	Hill	10, 16
Howe (1976)	National flock	15

clear relationship between the rate of mortality and the type of flock. Losses nationally appear to lie around 15%, ranging on individual farms from 5–40%.

The cost

The cost of neonatal mortality to the sheep industry has been reviewed by Howe (1976) and estimates, at 1981 prices, are based on his work. The total cost of lamb mortality to weaning is approximately £26m. Deaths in the first week of life account for 75% or more of this loss (Whitelaw, 1976) and therefore the cost of neonatal mortality is approximately £20m. The cost to the 'average' farmer with 500 ewes can be calculated to be approximately £1140. To some farmers the loss will be considerably less but to others it will be considerably more.

Causes of neonatal mortality

Most investigations into the causes of neonatal mortality employ diagnostic categories similar to those given below:

1. Starvation/exposure
2. Abortion/stillbirth
3. Infectious disease
4. Accidents, predators etc.
5. Congenital defects

These categories may be useful for diagnostic purposes but they have serious practical shortcomings in the context of preventing mortality in the individual sheep flock. Two categories in particular are of limited value. These are starvation/exposure, and abortion and stillbirth. In the next two sections these two causes of mortality are reviewed and revised diagnoses which would be of increased practical use to the sheep farmer are suggested.

STARVATION/EXPOSURE

The term starvation/exposure implies that the condition, which is in fact hypothermia, is caused by a combination of starvation and exposure. Recent work has shown that in most cases this is not so. There are two distinct causes of hypothermia (Eales *et al.*, 1982); the first cause is excessive heat loss from the wet newborn lamb aged up to four hours, and the second is depressed heat production due to starvation which can occur as early as six hours of age but more commonly between 12 and 48 hours. The close relationship between the cause of hypothermia and the age of lamb affected makes the farm diagnosis of the cause a simple matter.

A number of factors can predispose a lamb to hypothermia. These include nutrition during pregnancy, the condition of the ewe, the age of the ewe and the size of the litter. Susceptibility to hypothermia in a lamb is increased if it comes from a ewe in poor condition, if it comes from either a young or an old ewe or if it is a twin or triplet (Eales *et al.*, 1982). Thus the prevention of hypothermia in a flock requires two types of information from past lambings:

1. What was the relative importance of the two major causes? If excess

heat loss was the major problem more attention should be paid to shelter whereas if depressed heat production due to starvation was the major problem attention should be paid to the nutrition of both the ewes and the lambs.

2. Which of the three factors ewe nutrition, ewe age and litter size were significant determinants of susceptibility to the condition? This information will help to identify the area of flock management which requires revision.

Under the best management some lambs will still become hypothermic. This must be acknowledged and appropriate preparations made. The single most important aspect of treatment is early detection. The temperature of any lamb that appears at all weak should be taken, and if treatment is instigated at an early stage it is simple and very likely to be successful. This is of particular importance where hypothermia is caused by depressed heat production following starvation. New techniques of treatment comprising the administration of glucose for the reversal of hypoglycaemia in starved lambs, rewarming in air at 37–40 °C and careful after-care have recently been developed and effectively used on commercial farms (Eales, Small and Gilmour, 1982).

ABORTION AND STILLBIRTH

Abortion is the premature birth of a non-viable lamb. Stillbirth is the birth of a dead lamb. There is inevitably some overlap between these two terms but for the purposes of this section the term stillbirth will be restricted to describe the birth of a dead lamb of mature gestational age.

Abortion may be caused by infectious disease such as toxoplasmosis or enzootic abortion, or it may be the result of some stress such as mishandling. Abortion of infectious origin requires investigation by the veterinary laboratory. The various diseases involved, their diagnosis and prevention have been well described elsewhere (Linklater, 1979). Abortion caused by stress such as mishandling is an unusual event and should be easily avoided.

Abortion, though a serious problem when it occurs, is hopefully an occasional one, but stillbirth is a constant feature of lambing in all flocks and can account for up to 50% of all neonatal mortality. *Table 14.2* shows the reported incidence of stillbirth in a variety of type of flocks. Two factors emerge from these figures. First there is considerable flock to flock

Table 14.2 SOME ESTIMATES OF THE INCIDENCE OF STILLBIRTH (EXPRESSED AS A PERCENTAGE OF ALL LAMBS BORN)

Source	*Flock type*	*Stillbirth rate (%)*
MLC (1976)	Lowland	7
Saunders (1977)	Lowland	5
Purvis *et al.* (1979)	Lowland	5
MLC (1976)	Upland	7
Speedy *et al.* (1977)	Upland	1
Whitelaw (1976)	Hill	4, 7

Table 14.3 THE RATES OF STILLBIRTH IN FOUR CLOSELY OBSERVED FLOCKS (EXPRESSED AS A PERCENTAGE OF ALL LAMBS BORN)

Source	*Lambs born/100 ewes lambing*	*Stillbirth rate (%)*		
Barlow, R.M. (1982)[a]	197	3		
Eales, F.A. (1982)[b]	167	4		
Eales, F.A. and Small, J. (1982)[b]	157	3		
Robinson, J.J. (1982)[a]	227	5	Singles	: 4
			Twins	: 2
			Triplets	: 4
			Quads and quins	: 15

[a] Personal communications
[b] Unpublished data

variation in the rate of stillbirth and secondly the rate overall appears disturbingly high. *Table 14.3* shows the rates of stillbirths in four flocks which were intensively observed at lambing time. The incidence of stillbirth in these flocks appears to be lower than in the flocks shown in *Table 14.2*. Two factors probably account for this apparent discrepancy. First, death during birth was probably a rare event in the intensively observed flocks since help was always at hand, and secondly it would seem possible that some of the figures shown in *Table 14.2* include early postnatal deaths. If a lamb suffers acute hypoxia during birth it may die of hypothermia within one hour (Eales and Small, 1980). If the shepherd is not present at the time of birth and arrives more than one hour later he may well record the death as a stillbirth. Further analysis of the records from one of the flocks shown in *Table 14.3* demonstrates this point (*Table 14.4*). The true rate of stillbirth was only 3.2%, but this figure can be inflated to 8.9% by the inclusion of lambs which did die within an hour of birth and lambs which would have died without prompt active intervention by the shepherd.

A true stillbirth may be caused by either foetal death before the birth process has begun or by death during birth itself (parturient death). In the context of reducing losses from stillbirth in the future it is crucial to a flockmaster that he should know which cause or causes are important in his flock. A high incidence of foetal death would indicate that pre-lambing management requires more attention whereas a high proportion of parturient deaths due to dystocia would suggest that changes in management

Table 14.4 STILLBIRTHS, EARLY POSTNATAL DEATHS AND POTENTIAL EARLY POSTNATAL DEATHS IN A FLOCK OF 244 EWES WHICH PRODUCED 382 LAMBS

	No. lambs	*Percentage of all lambs born*
Stillbirth (foetal death)	8	2.1
Stillbirth (parturient death)	4	1.1
Postnatal death in first hour of life	5	1.3
Potential postnatal death in the first hour of life (prevented by shepherd intervention)	17	4.4
TOTAL	34	8.9

From Eales and Small (1982, unpublished data)

Table 14.5 SOME CRITERIA WHICH AID THE DIFFERENTIAL DIAGNOSIS OF AN APPARENT STILLBIRTH. AN ASSOCIATION BETWEEN AN OBSERVATION AND A DIAGNOSIS IS INDICATED BY (×)

	Stillbirth		*Early postnatal death*
	Foetal death	*Parturient death*	
Decomposed or mummified foetus	×		
Observed dead immediately after birth	×	×	
No expansion of lungs	×	×	
Ruptured liver		×	
Partial expansion of lungs		×	(×)
Swollen head or limbs		×	×
Excessive size		×	×
Full expansion of lungs			×
Shrivelled cord			×
Food in stomach			×

during lambing are required. It may not be possible to diagnose the exact cause of stillbirth in every single affected lamb but it should be possible to indicate on a flock basis which of the two causes of stillbirth requires attention. In *Table 14.5* it is shown how evidence readily gathered on the farm can be used to classify many stillbirths.

LOSSES DUE TO INFECTIOUS DISEASE

In general infectious disease acquired after birth is not considered to be a major source of newborn lamb losses (Houston and Maddox, 1974; Purvis *et al.*, 1979; Watt, 1980). In individual flocks, however, outbreaks of infectious disease such as enteritis can cause considerable problems and veterinary advice should be sought. The whole subject of infectious disease in newborn lambs has been discussed elsewhere and will not be discussed here (Donald, 1977; Watt, 1980). It may be useful, however, to give some general indications of how infectious disease can be prevented:

1. Careful attention to management throughout pregnancy is essential including clostridial and other vaccination.
2. A high standard of lambing management should be maintained. Special attention should be paid to hygiene. Lambing pens must be clean and bottles and stomach tubes etc. regularly sterilized.
3. Navels should be effectively dressed. This is of particular importance in the intensive outdoor situation.
4. All lambs should get colostrum as early as possible and must never go hungry (this will be effective in reducing the incidence of hypothermia also).
5. Any disease should be promptly treated under veterinary advice.
6. Infected animals should be isolated.

OTHER SOURCES OF NEONATAL MORTALITY

Other sources of loss include congenital abnormalities, accidents and losses due to predators. In some circumstances losses due to predators are probably overestimated since it is likely that many of these lambs were already weakened by some other condition such as hypothermia or infectious disease. In some flocks congenital deficiencies of minerals such as copper can lead to heavy losses. Detailed veterinary investigation is required for an accurate diagnosis. Appropriate preventative treatment normally yields gratifying results.

CAUSES OF NEONATAL MORTALITY—A SUMMARY

In view of the above considerations, a revised list of diagnostic categories for use in the investigation of neonatal mortality is proposed below. Its use might make the translation of diagnostic data into preventative action on the farm a little easier:

1. Hypothermia: exposure
2. Hypothermia: starvation
3. Abortion
4. Stillbirth: foetal death
5. Stillbirth: parturient death
6. Infectious disease
7. Congenital defects
8. Accident, predators etc.
9. Unknown.

A flock approach to the prevention of neonatal mortality

As already indicated a reduction of newborn lamb losses in a flock is dependent on the knowledge of the exact causes of loss in the past. In some cases the major cause may be already known e.g. an abortion storm or a period of severe weather during lambing, but in most cases the causes will not be so obvious. To establish the causes two types of records must be kept. The first is a general flock history to include such items as past lambing performances, weather during pregnancy and lambing, and the nutrition and condition of the ewes during pregnancy. Such a record will already be kept by many sheep farmers. The second type of record is a detailed account of every lamb which dies. Such a record need not be laborious to complete. A simple lamb death record sheet is shown in *Table 14.6*, and this could be printed into a pocket notebook using a rubber stamp. The aims of this recording system are:

1. To establish the size of the problem in terms of numbers and cost.
2. To provide diagnostic information which should enable the completion of the revised list of diagnostic categories for the flock.
3. To indicate which types of lambs are most at risk and therefore require more attention in the future.

The record is divided into eight sections as numbered in *Table 14.6*. Sections 1 and 2 require no explanation. Section 3 describes the status of the ewe and section 4 the status of the lamb and the fate of any litter mates. Section 5 refers to abortion. Section 6 describes symptoms observed in

Table 14.6 A LAMB DEATH RECORD SHEET

1. Date:		No:	
2. Weather:		Good/Indifferent/Bad	
3. EWE No:		Age: Condition score:	
Disease:		Milk: None/Some/Plenty	
4. LAMB Type: 1/2/3/4		Sex: M/CM/F	
Age at death: 0, 0–5, 5–12, 12–24, 24–48, 48+ hours			
Licked dry: Yes/No		Size for type: L/M/S	
Litter Mates: Alive – Dead –			
5. Evidence of abortion: Yes/No			
6. SYMPTOMS		7. STILLBIRTH	
Age at examination	:	Decomposition	: Yes/No
Temperature	:	Lung expansion	: Full/Partial/None
Able to stand	: Yes/No	Swelling head/limbs	: Yes/No
Stomach	: Full/empty	Milk in stomach	: Yes/No
Scouring	: Yes/No	Cord shrivelled	: Yes/No
Other		Liver rupture	: Yes/No
		Observed stillbirth	: Yes/No
8. Farm diagnosis:			
Vet diagnosis:			

lambs which have died postnatally and section 7 only applies to lambs which have been identified as stillborn by the shepherd. Section 8 provides for a farm diagnosis to be made by the farmer or shepherd and also provides for a veterinary diagnosis to be made on the basis of the record and a post-mortem examination if this was conducted. Such a recording system is only likely to be useful if it is conscientiously completed for every lamb death and thus the wholehearted cooperation and enthusiasm of both shepherd and farmer are required.

Table 14.7 AN EXAMPLE OF THE TYPE OF INFORMATION WHICH MAY BE GAINED BY THE USE OF THE LAMB DEATH RECORDING SYSTEM

Flock data		
Number of ewes lambing	: 416	
Number of lambs born	: 818	
Mortality in the first week of life	: 146 (18%)	
Causes of death		
Hypothermia—exposure	: 21 lambs	(14%)[a]
Hypothermia—starvation	: 58 lambs	(40%)
Abortion	: None recorded	
Stillbirth—foetal death	: 8 lambs	(6%)
Stillbirth—parturient death	: 18 lambs	(12%)
Stillbirth—type unknown	: 6 lambs	(4%)
Infectious disease	: 14 lambs	(10%)
Congenital defects	: 2 lambs	(1%)
Accidents, predators etc.	: 5 lambs	(3%)
Unknown	: 14 lambs	(10%)
TOTAL	: 146 lambs	(100%)

[a]Figures in parentheses indicate number of lamb losses expressed as a percentage of total lamb losses

Basic data were collected by members of the Pathology Department of the Moredun Research Institute

Table 14.8 FURTHER INFORMATION GAINED FROM THE LAMB DEATH RECORDING SYSTEM FOR THE FLOCK DESCRIBED IN *TABLE 14.7*

Ewes which lost lambs		
Condition score	0:	15 ewes
	1:	63 ewes
	2:	15 ewes
	3:	4 ewes
	Not known:	24 ewes
Lambs which died		
Type	Singles:	8 (10% of all singles born)
	Twins:	79 (15% of all twins born)
	Triplets:	56 (28% of all triplets born)
	Not known:	3
Age at death (excluding stillbirths)		
	0–5 hours:	25 lambs
	5–12 hours:	4 lambs
	12–24 hours:	21 lambs
	24–48 hours:	12 lambs
	48 hours:	36 lambs
	Not known:	16 lambs

Tables 14.7 and *14.8* show some results of the application of this recording system to 146 neonatal deaths in a commercial flock. Examination of the data in *Table 14.7* shows a flock lambing at 197% with losses up to one week of age of 18%. Hypothermia accounted for more than half of these losses and starvation was the major cause. Stillbirth accounted for 22% of the losses and most cases were attributed to parturient death. Infectious disease acquired after birth did not appear to be a serious cause of loss. Examination of the data in *Table 14.8* completes the picture. Most of the dead lambs came from ewes in very poor condition. The mortality rate was higher in lambs from twin and especially triplet litters. Deaths were most common in the first few hours of life and from 12 hours of age onwards with a considerable proportion occurring after 48 hours. It may be of significance that 48 hours was the approximate age at which ewes and their lambs were turned out to remote paddocks where shepherding was limited. With this information a promising approach to the reduction of losses in the future in this flock can be made. The major areas which require attention can be summarized:

1. Ewe nutrition from before tupping onwards.
2. Management of lambing ewes (parturient deaths).
3. Detection and treatment of hypothermia.
4. Management of triplets.
5. Management of lambs 2–7 days after birth.

Role of the veterinary surgeon in the prevention of neonatal mortality

The application of the approach described above requires help from both the agricultural adviser and the veterinary surgeon. This section focuses in particular on the role of the veterinary surgeon. This role can be summarized as follows:

Before lambing

1. Design a recording system and instruct the farm staff on the diagnostic techniques required e.g. the monitoring of rectal temperatures and the examination of lungs from stillborn lambs.
2. Discuss lambing management in general, and regimes for the detection and treatment of sick lambs in particular.

During lambing

1. Check the correct use of the recording system.
2. Perform post-mortem and other examinations as required.
3. Observe lambing from a distance with particular reference to standards of hygiene, the prompt detection of sick lambs, the treatment of sick lambs, the adequacy of labour and supervision, and the condition and feeding of the ewes.
4. Perform normal veterinary duties as required e.g. difficult lambing.

After lambing

1. Collect records, interpret them and summarize the results.
2. Plan for the future with the farmer and the agricultural adviser. It is not unlikely that changes in management may be required from weaning onwards and thus this stage of the scheme should be promptly undertaken.

Veterinary involvement in sheep farming has declined over the last two decades. Much of this decline can be attributed to the comparatively low value of the individual animal and the poor profitability of many sheep enterprises. Recent improvements in the economic returns from sheep farming have been associated with an increased veterinary involvement especially in the field of preventive medicine e.g. sheep health schemes. The further veterinary involvement outlined here is a natural extension. It should be appreciated that the approach described is long-term in nature and though some immediate benefit may be forthcoming e.g. the improved treatment of hypothermic lambs, the full financial benefit of the scheme will only be realized two years after its commencement. It is thus essential that the farmer and his veterinary surgeon should conclude a realistic long-term agreement on fees.

References

BANNATYNE, C.C. (1977). Perinatal losses in traditional hill sheep. In *Perinatal Losses in Lambs*, pp. 17–19. High output lamb production group of the Scottish Agricultural Colleges, Stirling, 1975

DONALD, L.G. (1977). Perinatal loss associated with infective disease. In

Perinatal Losses in Lambs, pp. 38–40. High output lamb production group of the Scottish Agricultural Colleges, Stirling, 1975

EALES, F.A. and SMALL, J. (1980). Summit metabolism in newborn lambs. *Research in Veterinary Science* **29**, 211–218

EALES, F.A. and SMALL, J. (1981). Effects of colostrum on summit metabolic rate in Scottish Blackface lambs at five hours old. *Research in Veterinary Science* **30**, 266–269

EALES, F.A., SMALL, J. and GILMOUR, J.S. (1982). The resuscitation of hypothermic newborn lambs. *Veterinary Record* **110**, 121–123

EALES, F.A., GILMOUR, J.S., BARLOW, R.M. and SMALL, J. (1982). The causes of hypothermia in 89 lambs. *Veterinary Record* **110**, 118–120

HARKER, D.B. (1977). Perinatal diseases in intensively reared lambs. In *Perinatal Losses in Lambs*, pp. 21–24. High output lamb production group of the Scottish Agricultural Colleges, Stirling, 1975

HOUSTON, D.C. and MADDOX, J.G. (1974). Causes of mortality among young Scottish Blackface lambs. *Veterinary Record* **95**, 575

HOWE, K.S. (1976). The cost of mortality in sheep production in the UK, 1971–1974. Report No. 198, Agricultural Economics Unit, University of Exeter

JOHNSTON, W.S. (1977). Caithness sheep loss survey. In *Perinatal Losses in Lambs*, pp. 10–13. High output lamb production group of the Scottish Agricultural Colleges, Stirling, 1975

LINKLATER, K.A. (1979). Abortion in sheep. *In Practice* **1**, 30–33

MEAT AND LIVESTOCK COMMISSION (1976). *Lamb Crops Results. Commercial Sheep Flocks.*

PURSER, A.F. and YOUNG, G.B. (1959). Lamb survival in two hill flocks. *Journal of Animal Science* **1**, 85–91

PURVIS, G.M., OSTLER, D.C., STARR, J., BAXTER, J., BISHOP, J., JAMES, A.D., DUNN, P.G.C., LYNE, A.R., OULD, A. and McCLINTOCK, M. (1979). Lamb mortality in a commercial lowland sheep flock with reference to the influence of climate and economics. *Veterinary Record* **104**, 241–242

SAUNDERS, R.W. (1977). Perinatal lamb mortality associated with lowland grass systems. In *Perinatal Losses in Lambs*, pp. 5–7. High output lamb production group of the Scottish Agricultural Colleges, Stirling, 1975

SPEEDY, A.W., LINKLATER, K.A., MACKENZIE, C.G., MACMILLAN, D.R. and BLANCE, E.W. (1977). A survey of perinatal mortality in upland sheep flocks in South-East Scotland. In *Perinatal Losses in Lambs*, pp. 14–16. High output lamb production group of the Scottish Agricultural Colleges, Stirling, 1975

WATT, A. (1980). Neonatal losses in lambs. *In Practice* **2**, 5–9

WHITELAW, A. (1976). Survey of perinatal losses associated with intensive hill sheep farming. *Veterinary Annual* **16**, 60–65

15

PROBLEMS OF LONG INCUBATION VIRAL DISEASES AND THEIR ERADICATION

R.H. KIMBERLIN
ARC and MRC Neuropathogenesis Unit, West Mains Road, Edinburgh, UK

The early history of virology was focused on acute infections because the rapid tempo and often dramatic nature of the viral–host interactions made them accessible to investigation. Typically, an acute infection leads to multiplication of virus in the host with a consequent stimulation of host defences in the form of interferon production, production of antibodies and development of cell-mediated immunity. Often, but not necessarily, virus multiplication causes cell damage. If this is the case, then the outcome of infection will depend on how effectively the host can clear the virus and replace damaged tissue. Successful vaccination enables the host to respond rapidly so that a potentially debilitating or fatal infection is rendered harmless.

The time between infection and the development of clinically recognizable disease is known as the incubation period. This may occupy a few days or weeks in an acute infection but, at the other end of the spectrum, chronic diseases are known in which incubation can take years. Many are insidious and have only been recognized in circumstances which made their appearance dramatic. The list of chronic virus diseases is growing all the time.

Historically, the most important episode to focus attention on the problems of long incubation viral diseases started in Iceland in 1933 when 20 Karakul rams were imported from Germany (Pálsson, 1976). Two of the rams carried infections which led to the appearance of three viral diseases, of which jaagsiekte (infectious adenomatosis) and maedi affected the lung. Initially, jaagsiekte caused very heavy losses and maedi was largely unnoticed. This was because the incubation period of jaagsiekte, although very long, was shorter than that of maedi which is uncommon in sheep less than four years of age. Six years after the original importation, the incidence of jaagsiekte declined and maedi became prominent. At the peak of the maedi outbreak in the early 1940s, about 60% of the sheep rearing districts in Iceland were affected and the total number of winter-fed sheep had declined by about 40%. The third disease, visna, which affects the CNS, also caused some losses but only in flocks with maedi. It was later shown that both diseases are caused by the same viral infection (see later p.304).

Table 15.1 SOME DEBILITATING OR FATAL, LONG INCUBATION DISEASES CAUSED BY VIRUSES OR VIRUS-LIKE AGENTS

Disease	*Natural host*	*Type of virus*	*Main target organ*
Scrapie	Sheep/goat	Unknown	CNS
Jaagsiekte	Sheep	Retro?	Lung
Maedi	Sheep	Retro	Lung
Visna	Sheep	Retro	CNS
Caprine leucoencephalitis-arthritis	Goat	Retro	CNS, lung, joints
Borna	Sheep/horse	Unknown	CNS
Aleutian disease	Mink	Parvo	Kidneys
Transmissible mink encephalopathy	Mink	Unknown[a]	CNS
Spongiform polioencephalomyelopathy	Mouse	Retro	CNS
Canine distemper demyelinating encephalomyelitis	Dog	Paramyxo	CNS
Creutzfeldt-Jakob disease	Man	Unknown[a]	CNS
Kuru	Man	Unknown[a]	CNS
Progressive multifocal leucoencephalopathy	Man	Papova	CNS
Subacute sclerosing panencephalitis	Man	Paramyxo	CNS
Progressive rubella panencephalitis	Man	Toga	CNS

[a]Unconventional, scrapie-like agents

The source flocks in Germany were clearly infected by these viruses and it is likely that some animals were dying of maedi and jaagsiekte but were not recognized, presumably, because the incidence of disease was low. However the dramatic losses seen in Iceland threatened the entire sheep industry and led to the courageous decision to attempt eradication. This was done by selecting one or more areas each year, slaughtering all sheep contained therein, and restocking from other areas where, fortunately, these diseases had not occurred. The programme was started in 1944 and by 1952, jaagsiekte had been eradicated. Maedi reappeared in some replacement flocks and eradication was not completed until 1965 (Pálsson, 1976).

Field and laboratory studies were made of jaagsiekte, maedi, visna and also of a fourth sheep disease, rida, which was unrelated to the 1933 importation; rida is the Icelandic equivalent of scrapie and had been known for many decades previously. All four diseases were eventually shown to be caused by viruses or virus-like agents but with extremely long incubation periods: usually more than a year. In this respect the diseases were chronic. But whereas many chronic diseases have an unpredictable clinical course, the early clinical signs of maedi, jaagsiekte, visna and rida progress in a predictable way and invariably end in death. The term 'slow infections' was coined to describe this particular group (Sigurdsson, 1954).

The discovery that several chronic debilitating or fatal diseases were caused by viral infection occurring years earlier stimulated the search for others. Ten more can be added to the list, including five affecting man (*Table 15.1*). From a virological point of view, these long incubation period diseases have little else in common, being caused by taxonomically different groups of viruses and with a variety of pathogenetic mechanisms underlying slowness (*Table 15.1*). It must also be emphasized that slowness is not an intrinsic property of either a virus or its host but is a consequence of the interactions between the two: in different circumstances some of the viruses listed in *Table 15.1* can cause acute diseases.

However, from a veterinary standpoint there are common problems associated with long incubation virus diseases and it is convenient to illustrate these by more detailed discussion of two of them, namely scrapie and maedi in sheep. It is emphasized that this is not a critical review for the specialist and references have been limited to selected, recent papers and reviews which cover specific issues in depth. The interested reader who wants information on other long incubation virus diseases should consult the following: Kimberlin (1976), Padgett and Walker (1976), ter Meulen and Katz (1977), ter Meulen and Hall (1978), Prusiner and Hadlow (1979), Martin and Stamp (1980), Cork and Narayan (1980), Verwoerd, Williamson and de Villiers (1980), Sharp (1981) and Swarz, Brooks and Johnson (1981).

Clinical pathology

Until aetiology is established and virus specific criteria become available, recognition of a given disease syndrome usually depends on clinical and pathological criteria.

SCRAPIE

The clinical signs of scrapie start insidiously with non-specific behavioural changes which progress to two kinds of more definite signs (Kimberlin, 1981a). Typically there is marked pruritus associated with compulsive rubbing (hence 'scrapie') and nibbling of the skin and leading to extensive loss of wool and skin lesions. This is accompanied or followed by uncoordinated and exaggerated gait progressing to severe ataxia of the hind limbs and, eventually, recumbency. There is, however, great variation in the clinical signs as reflected in the variety of names used to describe scrapie e.g. la tremblante (trembling), traberkrankheit (trotting disease). It is not uncommon to find affected animals with pruritus and no incoordination and vice versa. Many cases of scrapie are probably unrecognized, especially in hill flocks in which an affected sheep may rapidly become uncoordinated and die from exposure. The diagnosis of isolated cases can be difficult and other causes of intense rubbing (e.g. ectoparasites) and hind-limb ataxia (e.g. visna) have to be considered. Hence histological studies are needed.

The only consistent lesions occur in the CNS. Characteristically, there is an absence of inflammatory changes (in contrast to visna) and of demyelination in scrapie brain. The most striking histological change is neuronal vacuolation accompanied by interstitial spongy degeneration (Fraser, 1976). The presence of these lesions confirms the clinical diagnosis. But in some cases, the histological lesions of scrapie are not well developed. This poses problems because there are no other satisfactory diagnostic criteria for scrapie.

It is interesting that similar lesions are associated with transmissible mink encephalopathy and with kuru and Creutzfeldt-Jakob disease (CJD) of man, indicating the existence of related diseases in other species. Recognition of this (Hadlow, 1959) led directly to the demonstration that CJD and kuru are caused by infectious agents after very long incubation periods (Gajdusek, 1977). The same may eventually be shown for chronic wasting disease of deer which also has scrapie-like lesions in the brain (Williams and Young, 1980). However these relationships should not be overinterpreted. There are only a limited number of ways that brain (or any other target organ) can react to a pathogen and a similar pathology does not necessarily imply a similar aetiological agent. In the absence of other evidence, the occurrence of cerebral amyloid in some experimental models of murine scrapie and in Alzheimer's disease or Down's syndrome should be interpreted with caution (Dickinson, Fraser and Bruce, 1979).

MAEDI

Since both maedi and jaagsiekte affect the lung, it is not surprising to find similarities in clinical signs. Both diseases involve a steadily developing respiratory distress with death due to anoxia, a secondary bacterial pneumonia or other causes. However jaagsiekte is often distinguishable from maedi by moist rales and a frothy mucoid discharge from the nostrils; the latter can be increased by holding the head lower than the chest in the aptly named 'wheelbarrow test'. Pathologically, jaagsiekte is characterized by adenomatous tumours of the lungs and pulmonary lymph nodes. Often the affected parts of the lung are visible to the naked eye and histological examination shows nodular lesions composed of alveoli lined with cuboidal or columnar cells (Martin and Stamp, 1980; Sharp, 1981).

In contrast, the lungs and mediastinal lymph nodes of sheep affected with maedi are greatly increased in weight, often more than double the normal size. The lungs have a spongy consistency and do not collapse fully when the thoracic cavity is opened. Histological examination shows a diffuse thickening of the interalveolar septa caused by infiltration of lymphocytes, monocytes and macrophages. Hypertrophy and hyperplasia of smooth muscles are common and frequently prominent. In many areas of lung tissue these lesions lead to obliteration of the alveoli. The histopathology of maedi shows features in common with other viral pneumonias (Georgsson *et al.*, 1976) but virus-specific, serological tests are available for maedi to clarify diagnosis (see later p.307).

During the epizootics of maedi and jaagsiekte in Iceland, it became clear that both these diseases had been described previously under different names and often with much confusion between the two. However it is now recognized that in terms of clinical-pathological features, maedi (Iceland) is the same as Graaf-Reinet disease (South Africa), chronic progressive pneumonia (North America), zwoegersiekte (Holland) and la bouhite (France) etc. (Pálsson, 1976). The causal viruses of some of these diseases have been isolated. Serological tests and the techniques of nucleic acid hybridization have been used to show that the viruses causing maedi and chronic progressive pneumonia are very similar (Weiss *et al.*, 1976; 1977).

Aetiology

SCRAPIE

Scrapie was a serious problem in 18th century England. This was a time of intensive sheep breeding when many of the present-day breeds were originated. Because of this, and because scrapie often occurred in a familial pattern, it was widely regarded as a disease of genetic origin. This view prevailed until 40–50 years ago when it was shown that scrapie is experimentally transmissible to sheep by injecting brain homogenates from affected animals (Dickinson, 1976). Subsequent studies showed that the disease could be passaged in sheep serially and with diluted or filtered inocula, thus satisfying the main (if not the strictest) criteria for a virus or a virus-like agent (*Table 15.2*). In the last 20 years, scrapie has been experimentally transmitted to several other species notably mice and hamsters in which several experimental models of the disease have been obtained (Kimberlin, 1981b).

Little is known about the molecular nature of the scrapie agent. Although virus-like in many of its biological properties it differs from conventional viruses in several ways, notably in having a high degree of physicochemical stability and no detectable antigenicity. Because of these differences the term 'agent' or 'unconventional virus' is used in relation to

Table 15.2 KOCH'S POSTULATES AS APPLIED TO VIRAL DISEASES

ONE	(a)	Consistent transmission of the disease to experimental animals, or,
	(b)	Consistency in the recovery of virus in cell cultures.
TWO	(a)	Serial transmission of the clinicopathologic process using filtered material and serial dilutions to establish replication of the agent or,
	(b)	Consistent demonstration of recoverable agent (by electron microscopy, immunofluorescence, etc.) in the appropriate cells to explain lesions.
THREE		Studies of normal subjects or subjects with other disease to demonstrate that the suspected causal agent is not ubiquitous or a contaminant.

From Johnson and Gibbs (1974).

scrapie. Recently the group name 'virinos' has been proposed for scrapie and agents which may be related to it (Dickinson and Outram, 1979).

Although scrapie is caused by an infectious agent there is clear evidence that host genes control the development of disease. The nature of this control has been difficult to define in natural scrapie because there is no satisfactory method of assessing the degree of exposure to infection. This problem does not arise with experimental infection. Lines of Herdwick and Cheviot sheep have been selectively bred for increased or decreased incidence of clinical scrapie in response to injection with a standard pool of scrapie known as SSBP/1. The response to SSBP/1 is mainly controlled by a gene called Sip (scrapie incubation period) with two alleles of which the allele conferring susceptibility is dominant for this particular source of agent. The mode of action of the Sip gene is not known for certain but there is evidence that it may have an effect similar to that of the Sinc gene in mice which controls the dynamics of agent replication (Dickinson and Fraser, 1979).

MAEDI

The epidemiological studies of maedi in Icelandic sheep suggested an infectious aetiology, and experimental transmission of maedi by injection of diseased lung extracts supported this (Pálsson, 1976). The virus was later isolated and propagated in cell culture, producing characteristic cytopathic effects. Virus in cell culture provides a convenient source of antigen for a number of specific immunological tests for infection with maedi virus and several different strains have been identified. The relative ease with which maedi virus can be grown and assayed *in vitro* (Thormar, 1976) has led to its purification and identification as a lentivirus in the retrovirus group (Harter *et al.*, 1973; Stowring, Haase and Charman, 1979).

The Icelandic workers noted that visna sometimes occurred in flocks where maedi was common. Transmission experiments showed that both diseases could be produced in sheep by injecting either visna virus or maedi virus. It is now known that these viruses are indistinguishable (Weiss *et al.*, 1976; 1977). Since maedi is the commoner disease, the term maedi-visna virus is used.

The idea of one virus causing more than one clinico-pathologically distinct disease introduces a complexity which it is important to appreciate. Indeed there is recent evidence that infection with progressive pneumonia virus (= maedi-visna) can cause arthritis and synovitis as well as the CNS and lung diseases (Oliver *et al.*, 1981). Serological surveys in the USA have shown that infection with maedi-visna virus is common (Gates *et al.*, 1978; Molitor *et al.*, 1979). Hence maedi-visna virus may cause no disease at all, subclinical disease involving one or more organs, and overt clinical disease depending on which target organ is most affected.

Virus–host interactions

It is clear from the foregoing that pathogenesis of long incubation diseases is subtle and complex. There are, in addition, other constraints which make studies difficult.

SCRAPIE

Experimental infection of sheep with scrapie produces clinical disease in only a proportion of animals and incubation periods can be highly variable. Greater uniformity of response can be achieved by selectively breeding sheep susceptible to a given scrapie inoculum (Dickinson and Fraser, 1979). But the main problem with scrapie is the inapplicability of many of the standard virological methods of study. For example, scrapie agent cannot be identified by electron microscopy, immunological methods or by cytopathic effects in cell culture. Only bioassay methods are available and, for most purposes, these can only be applied satisfactorily to laboratory animals (Kimberlin, 1981b). However, the situation is helped considerably by the existence of many different experimental models of scrapie in mice and also hamsters. Most of these are extremely predictable with standard errors of the mean incubation period varying by only 1 or 2%. Consequently there has been much emphasis on the biological aspects of scrapie, for example on the genetic control of incubation period by the Sinc gene (Dickinson and Fraser, 1979). Also biological methods have been devised to identify different strains of agent and to study mutation and strain selection on serial passage in mice (Bruce and Dickinson, 1979).

Experimental infection in mice (using peripheral routes of injection) leads to agent replication in a variety of lymphoid organs of which spleen and lymph nodes seem to be particularly important. In the very long incubation models of murine scrapie, this stage is preceded by a 'zero-phase' during which the infectious agent cannot be detected by bioassay. Replication of agent proceeds to a plateau concentration in each tissue indicating a restricted process. Little if any agent is detected in blood (Kimberlin, 1979a). A key stage in pathogenesis is entry of agent into the CNS without which disease does not develop. There is suggestive evidence that invasion of the CNS may take place by spread of infection from visceral lymphoid organs along autonomic nerves (Kimberlin and Walker, 1980). Thereafter, agent replicates slowly and continuously in the CNS until clinical disease develops.

Two basic questions about chronic diseases concern the mechanisms of persistence of infection and the nature of the disease process. Persistence of scrapie does not pose any conceptual problems because there is no evidence for a specific, active host response to eliminate it (Outram, 1976). Indeed in one study, unreplicated, infectious agent was present at the injection site a year later (Kimberlin and Walker, 1979). However there is considerable biological evidence for a host restriction on the overall dynamics of agent replication which in mice is controlled by Sinc gene. The restriction seems to involve a limitation in the number of Sinc gene coded sites for agent replication. The best evidence for this comes from experiments in which mice of a given Sinc genotype were injected with two strains of agent of widely differing incubation period in that genotype. The operationally slow agent was injected first followed, after a long interval, by the faster agent which was expected to be the one to kill the mice. In fact, the opposite was found and the slower agent completely blocked the ability of the faster one to produce disease. There was no evidence that immunological processes were involved and, therefore, competition for a limited number of replication sites is the simplest explanation of this phenomenon (Dickinson and Outram, 1979).

In the later stages of incubation, the continued replication of agent in the brain is accompanied by a progressive development of histological lesions (Fraser, 1976). In some models of scrapie, the onset of clinical disease occurs abruptly when agent concentration has reached a certain critical level. These findings suggest that agent replication directly induces progressive tissue damage until the ability of the brain to compensate functionally is exhausted, and overt clinical signs appear leading to death (Outram, 1976). It would be reasonable to assume that the occurrence of histological lesions is the primary cause of clinical disease. But models of scrapie and related diseases are known in which the histological lesions are minimal or even absent. This means that the diagnostically important lesions of scrapie are irrelevant to the primary disease process in the brain, the nature of which remains unknown (Fraser, 1979).

MAEDI

Studies of maedi-visna infection in sheep have been greatly hampered by the difficulties of consistently inducing the disease by injection (Gudnadóttir, 1974). Unfortunately, alternative experimental hosts are not available. However, early transmission experiments with visna showed that histological lesions developed in the brain within a month or two after intracerebral infection. Consequently the pathogenesis of early visna lesions has been studied in detail, the limitations of this model being offset by the availability of several techniques to study the behaviour of virus *in vivo* and the host's responses.

At virtually all times after intracerebral infection, visna virus can be recovered from brain, lymphoid organs and buffy coat cells. However, the virus is nearly always cell-associated and isolation often requires the use of tissue explants and one or more blind passages in culture (Nathanson, Panitch and Pétursson, 1976). It is difficult to find visna virus particles in infected tissues by electron microscopy or immunofluorescence. These findings are in marked contrast to the rapid growth of the virus in cultures of sheep cells and they indicate a restriction of viral multiplication *in vivo*.

Complement fixing antibodies can be detected in serum a few weeks after infection but specific neutralizing antibodies do not develop for a few months (Gudnadóttir, 1974). Titres may remain relatively high for years because of long-term stimulation by viral antigens. Some of these antibodies are produced locally in the brain and can be found in the cerebrospinal fluid in oligoclonal bands of IgG. Although neutralizing antibodies may play a role in limiting the spread of virus it is important to recognize that they do not eliminate infection.

Histological changes can be seen in the brain as early as 2–4 weeks after intracerebral injection of virus but with little progression of lesions after the first month or so. There is a strong correlation between the severity of lesions and the frequency of virus recovery indicating a causal relationship between virus multiplication and lesions. The inflammatory nature of the lesions of visna suggests that they are immunologically mediated. Immunosuppression using antithymocyte serum and cyclophosphamide does not affect the frequency of virus isolation from the CNS but the development of lesions is strikingly suppressed. Similarly, immunopotentiation by injecting high doses of virus in Freund's adjuvant increases the severity of lesions. Hence there is good evidence that the early lesions of visna are caused by immunological attack on cells bearing viral antigens (Pétursson *et al.*, 1979).

Recent studies, using very sensitive methods of nucleic acid hybridization, have revealed an important mechanism by which long-term persistence of infection can occur. In common with other retroviruses, visna replicates via DNA proviral intermediate. It was found that the number of proviral DNA copies/cell *in vivo* was roughly comparable to the number found in highly permissive cells *in vitro*. In contrast, the number of viral RNA copies/cell was about two orders of magnitude lower *in vivo*; in other words the transcription of proviral DNA to viral RNA is considerably blocked and as a consequence very few cells contain viral proteins (Haase *et al.*, 1977; Brahic *et al.*, 1981). This restriction probably slows down the spread of infection and prevents the development of acute clinical disease. It will also allow virus to persist *in vivo* because those cells which do not have viral antigens on their surface will not be eliminated by the immune system.

The rapid development of visna lesions after intracerebral infection does not tally with the long and variable incubation period. However, a plausible explanation for this stems from an observation that some virus isolates, made several years after infecting sheep, were poorly neutralized by antisera collected early after infection (Gudnadóttir, 1974). This and subsequent studies indicate that mutants can arise long after infection, with a sequential development of antibody to parental virus and, later, to each mutant strain (Narayan *et al.*, 1981; Clements *et al.*, 1980). It is possible that each new antigenic variant initially escapes neutralization thus giving new waves of infection and CNS damage. The slow accumulation of irreversible CNS lesions could eventually produce a critical level of damage and clinical disease, but there is no direct evidence for this.

It seems likely that basically similar processes operate in the pathogenesis of maedi lesions. There have been some studies demonstrating that the early immune responses to experimental maedi and visna infection are similar (de Boer, 1975; Sihvonen, 1981). The factors determining the major target organ in the development of disease are unknown but the commoner occurrence of maedi compared to visna could be explained by the evidence that the respiratory tract is an important route of infection (see later p.310).

Possible strategies for control

Vaccination has been successfully used to control many diseases caused by acute viral infection but not with any of the long incubation diseases. Vaccination is inappropriate with scrapie because there appears to be no natural immune response to enhance. With visna, neutralizing antibodies do not eliminate the virus and vaccination can increase the development of CNS lesions.

Since there is a host restriction on the replication of both scrapie agent and maedi virus one can conceive of a chemotherapeutic approach to restrict replication even further. This idea seems plausible in view of the competition experiments carried out with mouse scrapie. However, these experiments depend on the 'blocking' agent being infectious and there is always the risk of it mutating to a more virulent strain (Dickinson and Outram, 1979). A chemical blocking agent would be safer but, even if one was discovered, it might be difficult to achieve both an efficient stable binding to scrapie replication sites and low chronic toxicity.

There is no convincing evidence that maedi is under host genetic control but with scrapie there is (Dickinson and Fraser, 1979). If we can assume that the genetic control of natural scrapie is the same as that seen in sheep experimentally infected with the SSBP/1 source of agent, then nucleus flocks could be bred which might be relatively resistant to disease. However in some models of murine scrapie, incubation can exceed lifespan (Dickinson, Fraser and Outram, 1975). If a parallel situation exists in sheep, then animals which fail to develop clinical disease could still be infected and may possibly be a source of infection to others. Even more important, it is known that Sinc genotypes of mice which are relatively resistant to some strains of scrapie agent (i.e. incubation period is long) are very susceptible to others and this is, in fact, one of the means by which different strains of agent are identified. There is evidence that the same situation applies to sheep infected with different agent strains. Hence genetical methods of control may have a very limited application in scrapie (Kimberlin, 1979b).

We are, therefore, left with methods based on reducing infection by animal management or slaughter. Before discussing these it is necessary to review the processes of natural infection.

Natural infection

SCRAPIE

Several studies have shown that up to 40% of sheep given long-term, natural exposure to scrapie can develop the disease (Kimberlin, 1981b). Contagious spread most likely occurs when sheep are close-penned at lambing or during prolonged winters but the precise routes of infection are not firmly established. Little if any agent is found in body fluids or secretions but many tissues outside the CNS contain agent and could be sources of infection. In particular, the foetal membranes voided at parturition contain quite high titres of agent (Pattison *et al.*, 1974). Sheep can be experimentally infected by oral dosing, scarification and via the conjunctiva and these are likely to be natural routes of infection. The extreme physicochemical stability of scrapie agent is consistent with some circumstantial evidence that sheep may become infected from contaminated buildings or pastures (Kimberlin, 1981b).

Many outbreaks of scrapie have been attributed to the introduction of a new ram to a flock but it is not clear whether the ram himself can be a source of infection; he may in fact introduce genetic factors which cause a pre-existing, subclinical infection in a flock to produce clinical cases.

There is however clear epidemiological evidence that maternal transmission plays a major role in the natural spread of scrapie (Dickinson, 1976). In part this can be due to infection from the dam occurring after parturition and there is evidence that the incidence of scrapie in progeny increases with the time of exposure. However, sheep removed from their dams at birth can also develop scrapie implying infection before or during parturition (Hourrigan *et al.*, 1979). Lambs born to ewes experimentally infected at conception develop scrapie exceptionally early indicating prenatal infection (Dickinson, 1976). This may occur transplacentally since foetal membranes contain appreciable amounts of agent at parturition.

In summary, it appears that a high proportion of sheep with scrapie become infected by maternal transmission before weaning. Since most cases of scrapie develop after 2–3 years of age, an infected ewe can transmit infection to several of her lambs before she herself develops the disease; and in many instances she may never show clinical signs. It is therefore easy to see how scrapie can spread widely and unnoticed. There are several reported outbreaks of scrapie which followed movement of pedigree sheep from one country to another (Hourrigan *et al.*, 1979). As a consequence many countries prohibit the importation of sheep from endemic areas. Quarantine measures are essentially useless because of the time factor involved and because absence of clinical disease is not evidence for absence of infection.

MAEDI

The outbreaks in Iceland are a classic illustration of the spread of infectious disease by the international movement of sheep. More recently the popularity of the Texel breed has contributed to the spread of zwoergersiekte (= maedi) from Holland to other European countries (Krogsrud and

Udnes, 1978); and in the last 3–4 years maedi-virus infection has been found in the UK (Dawson *et al.*, 1979; Spence, Dawson and Markson, 1981).

Observations in Iceland suggested that the most likely method of transmission of maedi was via the respiratory route (Pálsson, 1976). This would certainly have been facilitated by the practice of housing sheep intensively for six months in winter. There is some evidence that maedi-visna virus can be transmitted by faecal contamination of drinking water but indirect transmission from contaminated buildings or pasture seems to be low (Dawson, 1980). This was most clearly shown during the eradication programme in Iceland because, on many occasions, restocking took place only a few weeks after slaughtering the infected flock, and usually premises were not disinfected (Pálsson, 1976).

Infection can be transmitted from the ram as shown by the 1933 importation to Iceland. Maternal transmission is clearly important and lambs born to infected dams will almost certainly become infected. There is some evidence that transplacental infection can occur but this must be set against the fact that clean flocks have been derived by separating lambs and ewes at birth and rearing the lambs in isolation (de Boer *et al.*, 1979; Molitor, Light and Schipper, 1979).

de Boer *et al.* (1979) showed that lambs separated immediately at birth remained free of infection for up to eight years, the maximum time studied. In contrast, 28% of lambs separated after 10 hours developed infection, and in two others groups, separated after six weeks or one year, the incidence of infection was 76% and 81%, respectively. Virus was also isolated from ewe's milk up to five months after lambing. Therefore there is clear evidence for virus transmission from ewe to lamb, and milk is one source of infection.

Serological surveys have shown that the proportion of infected sheep increases with age, from 16–23% in yearlings to 80–83% in animals seven years or older (Gates *et al.*, 1978; Molitor *et al.*, 1979). These data make it likely that virus can be transmitted between adult sheep and that the chance of infection increases with time of exposure.

Eradication and control

SCRAPIE

Eradication of scrapie has only been successful in the special case where the disease was introduced by importation, and the slaughter of all imported animals, their progeny and contacts prevented it becoming endemic. By this means, countries such as Australia and New Zealand are still scrapie-free (Kimberlin, 1981a,b).

The eradication of scrapie from countries where it is endemic seems to be impossible without an efficient diagnostic test to identify infected animals. It was tried in Iceland at the same time as the maedi-visna and jaagsiekte eradication schemes but failed, possibly because of persistence of infection in the environment (Pálsson, 1976; 1979). Vigorous attempts at eradication have been made in the USA by slaughtering sheep in affected and source flocks (Klingsporn, Hourrigan and McDaniel, 1969). This policy has certainly brought the disease under control but has so far failed to eradicate it.

The only practical method of scrapie control in endemic areas is by culling, but economic factors dictate the type of policy used. The slaughter of all sheep in an affected flock and in source flocks depends on adequate compensation for farmers and on there being a good chance of restocking without reintroducing infection. An alternative approach is to slaughter all bloodline relatives of scrapie cases to limit the effects of maternal transmission of infection and to reduce the number of genetically susceptible sheep. But even this could lead to the destruction of an entire flock if one or two breeding rams had been extensively used (Kimberlin, 1981b).

Less disruptive control measures depend on limiting the maternal and lateral spread of infection. This can be done by culling only in the female line and by careful husbandry when sheep are kept in confined spaces. Since foetal membranes are probably a source of infection the risks can be reduced by prompt removal of afterbirth, avoidance of lambing pens, and the use of a different lambing area each year.

Selective culling in the female line will undoubtedly reduce the incidence of scrapie but the method depends on marking lambs at birth and maintaining detailed flock records. This is rarely practised on commercial farms and the long incubation of scrapie means that it will take at least three years between the diagnosis of the first case of scrapie in a flock and the acquisition of sufficient flock records to permit accurate culling. The lack of a diagnostic test for infection is a major limitation to the efficient control of scrapie (Kimberlin, 1981a,b).

MAEDI

Eradication of maedi and jaagsiekte by slaughter was the only means open to the Icelanders faced with such serious losses, but today a course of action like this would be hard to justify in cost-effective terms. Fortunately there are alternative solutions made possible by serological tests for infection.

Two studies have shown that maedi-free lambs can be obtained by removing them from their dams at birth (de Boer *et al.*, 1979; Molitor, Light and Schipper, 1979). The procedure requires strict hygiene at parturition and thereafter, and it is difficult to perform under normal farming conditions. But it might, for example, be a justifiable way of protecting a pedigree flock with a high incidence of infection. A nucleus, clean flock can be obtained fairly quickly and any subsequent appearance of infection can be monitored by serological screening.

A simpler way of reducing maedi-visna infection is to cull all serologically positive animals. A number of tests are available based on the detection of complement fixing or neutralizing antibodies. It is important to recognize that the development of antibodies following infection can take several months and therefore tests on sheep less than one year old will often give erroneous results. Also the tests vary in sensitivity and the use of more than one helps to reduce the problem of false negatives. More sensitive ELISA methods should reduce the problem further (de Boer and Houwers, 1979).

A control programme has been tried out in Holland using a closed flock of five year old sheep and their two year old progeny. The sheep were tested for antibodies every six months and positive reactors eliminated. No seropositive sheep were detected three to seven years after starting the experiment suggesting that maedi infection had been eradicated (de Boer and Houwers, 1979). However it is known that some sheep with maedi lesions are serologically negative and therefore complete eradication may not always be possible (de Boer *et al.*, 1979).

Selective culling is not only efficient but it can be tailored to circumstances. For example the economic losses caused by maedi in a given flock may not be sufficient to justify removal of all infected sheep; in a valuable flock this could be as expensive as doing nothing, at least in the short term. The Norwegians have a control programme based on containment. Animals from an infected flock cannot be sold live, taken to common pastures or to shows. A maedi flock thus restricted is considered unlikely to spread infection and it is up to the owners to decide whether or not to cull infected sheep (Krogsrud and Udnes, 1978). This type of flexibility is important in achieving maximum cost effectiveness.

References

BRAHIC, M., STOWRING, L., VENTURA, P. and HAASE, A.T. (1981). Gene expression in visna virus infection in sheep. *Nature, London* **292**, 240–242

BRUCE, M.E. and DICKINSON, A.G. (1979). Biological stability of different classes of scrapie agent. In *Slow Transmissible Diseases of the Nervous System* (S.B. Prusiner and W.J. Hadlow, Eds.), Vol. 2, pp. 71–86. New York, Academic Press

CLEMENTS, J.E., PEDERSEN, F.S., NARAYAN, O. and HASELTINE, W.A. (1980). Genomic changes associated with antigenic variation of visna virus during persistent infection. *Proceedings of the National Academy of Sciences* **77**, 4454–4458

CORK, L.C. and NARAYAN, O. (1980). The pathogenesis of leucoencephalomyelitis arthritis of goats. 1. Persistent viral infection with progressive pathologic changes. *Laboratory Investigation* **42**, 596–602

DAWSON, M. (1980). Maedi/visna: a review. *Veterinary Record* **106**, 212–216

DAWSON, M., CHASEY, D., KING, A.A., FLOWERS, M.J., DAY, R.H., LUCAS, M.H. and ROBERTS, D.H. (1979). The demonstration of maedi-visna virus in sheep in Great Britain. *Veterinary Record* **105**, 220–223

DE BOER, G.F. (1975). Zwoegerziekte virus, the causative agent for progressive interstitial pneumonia (maedi) and meningo-leucoencephalitis (visna) in sheep. *Research in Veterinary Science* **18**, 15–25

DE BOER, G.F. and HOUWERS, D.J. (1979). Epizootiology of maedi/visna in sheep. In *Aspects of Slow and Persistent Virus Infections* (D.A.J. Tyrrell, Ed.), pp. 198–220. The Hague, Martinus Nijhoff

DE BOER, G.F., TERPSTRA, C., HOUWERS, D.J. and HENDRIKS, J. (1979).

Studies in epidemiology of maedi/visna in sheep. *Research in Veterinary Science* **26**, 202–208

DICKINSON, A.G. (1976). Scrapie in sheep and goats. In *Slow Virus Diseases of Animals and Man* (R.H. Kimberlin, Ed.), pp. 209–241. Amsterdam, North-Holland

DICKINSON, A.G. and FRASER, H. (1979). An assessment of the genetics of scrapie in sheep and mice. In *Slow Transmissible Diseases of the Nervous System*. (S.B. Prusiner and W.J. Hadlow, Eds.), Vol. 1, pp. 367–385. New York, Academic Press

DICKINSON, A.G. and OUTRAM, G.W. (1979). In *Slow Transmissible Diseases of the Nervous System* (S.B. Prusiner and W.J. Hadlow, Eds.), Vol. 2, pp. 13–31. New York, Academic Press

DICKINSON, A.G., FRASER, H. and BRUCE, M.E. (1979). Animal models for the dementias. In *Alzheimer's Disease: Early Recognition of Potentially Reversible Deficits* (A.I.M. Glen and L.J. Whalley, Eds.), pp. 42–45. Edinburgh, Churchill–Livingstone

DICKINSON, A.G., FRASER, H. and OUTRAM, G.W. (1975). Scrapie incubation time can exceed natural lifespan. *Nature, London* **256**, 732–733

FRASER, H. (1976). The pathology of natural and experimental scrapie. In *Slow Virus Diseases of Animals and Man* (R.H. Kimberlin, Ed.), pp. 267–305. Amsterdam, North-Holland

FRASER, H. (1979). Neuropathology of scrapie: the precision of the lesions and their diversity. In *Slow Transmissible Diseases of the Nervous System* (S.B. Prusiner and W.J. Hadlow, Eds.), Vol. 1, pp. 387–406. New York, Academic Press

GAJDUSEK, D.C. (1977). Unconventional viruses and the origin and disappearance of kuru. *Science* **197**, 943–960

GATES, N.L., WINWARD, L.D., GORHAM, J.R. and SHEN, D.T. (1978). Serologic survey of prevalence of ovine progressive pneumonia in Idaho range sheep. *Journal of the American Veterinary Medical Association* **173**, 1575–1579

GEORGSSON, G., NATHANSON, N., PÁLSSON, P.A. and PÉTURSSON, G. (1976). The pathology of visna and maedi in sheep. In *Slow Virus Diseases of Animals and Man* (R.H. Kimberlin, Ed.), pp. 61–96. Amsterdam, North-Holland

GUDNADÓTTIR, M. (1974). Visna-maedi in sheep. *Progress in Medical Virology* **18**, 336–349

HAASE, A.T., STOWRING, L., NARAYAN, O., GRIFFIN, D. and PRICE, D. (1977). Slow persistent infection caused by visna virus: role of host restriction. *Science* **195**, 175–177

HADLOW, W.J. (1959). Scrapie and kuru. *Lancet* **ii**, 289–290

HARTER, D.H., AXEL, R., BURNY, A., GULATI, S., SCHLOM, J. and SPIEGELMAN, S. (1973). The relationship of visna, maedi and RNA tumour viruses as studied by molecular hybridisation. *Virology* **52**, 287–291

HOURRIGAN, J., KLINGSPORN, A., CLARKE, W.W. and DE CAMP, M. (1979). Epidemiology of scrapie in the United States. In *Slow Transmissible Diseases of the Nervous System* (S.B. Prusiner and W.J. Hadlow, Eds.), Vol. 1, pp. 331–356. New York, Academic Press

JOHNSON, R.T. and GIBBS, C.J. Jnr. (1974). Koch's postulates and slow infections of the nervous system. *Archives of Neurology* **30**, 36–38

KIMBERLIN, R.H. (1976). *Slow Virus Diseases of Animals and Man*, (Editor). Amsterdam, North-Holland

KIMBERLIN, R.H. (1979a). Early events in the pathogenesis of scrapie in mice: biological and biochemical studies. In *Slow Transmissible Diseases of the Nervous System* (S.B. Prusiner and W.J. Hadlow, Eds.), Vol. 2, pp. 33–54. New York, Academic Press

KIMBERLIN, R.H. (1979b). An assessment of genetical methods in the control of scrapie. *Livestock Production Science* **6**, 233–242

KIMBERLIN, R.H. (1981a). Scrapie. *British Veterinary Journal* **137**, 106–112

KIMBERLIN, R.H. (1981b). Scrapie as a model slow virus disease: problems, progress and diagnosis. In *Comparative Diagnosis of Viral Diseases* (E. Kurstak and C. Kurstak, Eds.), Vol. 3, pp. 349–390. New York, Academic Press

KIMBERLIN, R.H. and WALKER, C.A. (1979). Pathogenesis of scrapie: agent multiplication in brain at the first and second passage of hamster scrapie in mice. *Journal of General Virology* **42**, 107–117

KIMBERLIN, R.H. and WALKER, C.A. (1980). Pathogenesis of mouse scrapie: evidence for neural spread of infection to the CNS. *Journal of General Virology* **51**, 183–187

KLINGSPORN, A.L., HOURRIGAN, J.L. and McDANIEL, H.A. (1969). Scrapie-eradication and field trial study of the natural disease. *Journal of the American Veterinary Medical Association* **155**, 2172–2177

KROGSRUD, J. and UDNES, H. (1978). Maedi (progressive interstitial pneumonia in sheep). Diagnosis, epizootiology, prevention and control programme in Norway. *Bulletin de l'Office International des Epizooties* **89**, 451–464

MARTIN, W.B. and STAMP, J.T. (1980). Slow virus infections in sheep. *British Veterinary Journal* **136**, 290–295

MOLITOR, T.W., LIGHT, M.R. and SCHIPPER, I.A. (1979). Elevated concentrations in serum immunoglobulins due to infection by ovine progressive pneumonia virus. *American Journal of Veterinary Research* **40**, 69–72

MOLITOR, T.W., SCHIPPER, I.A., BERRYHILL, D.L. and LIGHT, M.R. (1979). Evaluation of the agar-gel immunodiffusion test for the detection of precipitating antibodies against progressive pneumonia virus of sheep. *Canadian Journal of Comparative Medicine* **43**, 280–287

NARAYAN, O., CLEMENTS, J.E., GRIFFIN, D.E. and WOLINSKY, J.S. (1981). Neutralizing antibody spectrum determines the antigenic profiles of emerging mutants of visna virus. *Infection and Immunity* **32**, 1045–1050

NATHANSON, N., PANITCH, H. and PÉTURSSON, G. (1976). Pathogenesis of visna: review and speculation. In *Slow Virus Diseases of Animals and Man* (R.H. Kimberlin, Ed.), pp. 115–131. Amsterdam, North-Holland

OLIVER, R.E., GORHAM, J.R., PARISH, S.F., HADLOW, W.J. and NARAYAN, O. (1981). Ovine progressive pneumonia: pathologic and virologic studies on the naturally occurring disease. *American Journal of Veterinary Research* **42**, 1554–1559

OUTRAM, G.W. (1976). The pathogenesis of scrapie in mice. In *Slow Virus Diseases of Animals and Man* (R.H. Kimberlin, Ed.), pp. 325–357. Amsterdam, North-Holland

PADGETT, B.L. and WALKER, D.L. (1976). New human papovaviruses. *Progress in Medical Virology* **22**, 1–35

PÁLSSON, P.A. (1976). Maedi and visna in sheep. In *Slow Virus Diseases of*

Animals and Man (R.H. Kimberlin, Ed), pp. 17–43. Amsterdam, North-Holland

PÁLSSON, P.A. (1979). Rida (scrapie) in Iceland and its epidemiology. In *Slow Transmissible Diseases of the Nervous System* (S.B. Prusiner and W.J. Hadlow, Eds.), Vol. 1, pp. 357–366. New York, Academic Press

PATTISON, I.H., HOARE, M.N., JEBBETT, J.N. and WATSON, W.A. (1974). Further observations on the production of scrapie in sheep by oral dosing with foetal membranes from scrapie-affected sheep. *British Veterinary Journal* **130**, lxv–lxvii

PÉTURSSON, G., MARTIN, J.R., GEORGSSON, G., NATHANSON, N. and PÁLSSON, P.A. (1979). Visna, the biology of the agent and the disease. In *Aspects of Slow and Persistent Virus Infection* (D.A.J. Tyrrell, Ed.), pp. 165–197. The Hague, Martinus Nijhoff

PRUSINER, S.B. and HADLOW, W.J. (1979). *Slow Transmissible Diseases of the Nervous System* (Editors), Vols. 1 and 2. New York, Academic Press

SHARP, J.M. (1981). Slow virus infections of the respiratory tract of sheep. *Veterinary Record* **108**, 391–393

SIGURDSSON, B. (1954). Observations on three slow infections of sheep. *British Veterinary Journal* **110**, 341–354

SIHVONEN, L. (1981). Early immune responses in experimental maedi. *Research in Veterinary Science* **30**, 217–222

SPENCE, J.B., DAWSON, M. and MARKSON, L.M. (1981). Maedi-visna in Great Britain. *Veterinary Record* **108**, 466

STOWRING, L., HAASE, A.T. and CHARMAN, H.P. (1979). Serological definition of the lentivirus group of retroviruses. *Journal of Virology* **29**, 523–528

SWARZ, J.R., BROOKS, B.R. and JOHNSON, R.T. (1981). Spongiform polioencephalomyelopathy caused by a murine retrovirus. II. Ultrastructural localization of virus replication and spongiform changes in the central nervous system. *Neuropathology and Applied Neurobiology* **7**, 365–380

TER MEULEN, V. and HALL, W.W. (1978). Slow virus infections of the nervous system: virological, immunological and pathogenetic considerations. *Journal of General Virology* **41**, 1–25

TER MEULEN, V. and KATZ, M. (1977). *Slow Virus Infections of the Central Nervous System* (Editors). New York, Spring Verlag

THORMAR, H. (1976). Visna-maedi virus infection in cell cultures and in laboratory animals. In *Slow Virus Diseases of Animals and Man* (R.H. Kimberlin, Ed.), pp. 97–114. Amsterdam, North-Holland

VERWOERD, D.W., WILLIAMSON, A-L. and DE VILLIERS, E-M. (1980). Aetiology of jaagsiekte: transmission by means of subcellular fractions and evidence for the involvement of a retrovirus. *Onderstepoort Journal of Veterinary Research* **45**, 275–280

WEISS, M.J., SWEET, R.W., GULATI, S.C. and HARTER, D.H. (1976). Nucleic acid sequence relationships among slow viruses of sheep. *Virology* **71**, 395–401

WEISS, M.J., ZEELON, E.P., SWEET, R.W., HARTER, D.H. and SPIEGELMAN, S. (1977). Immunological cross-reactions of the major internal protein component from slow viruses of sheep. *Virology* **76**, 851–854

WILLIAMS, E.S. and YOUNG, S. (1980). Chronic wasting disease of captive mule deer: a spongiform encephalopathy. *Journal of Wildlife Diseases* **16**, 89–98

16

EFFECTS OF PARASITISM ON METABOLISM IN THE SHEEP

A.R. SYKES
Department of Animal Science, Lincoln College, Canterbury, New Zealand

A quantitative approach to an understanding of the nutrient requirements of livestock has developed during the last century because, with increasing sophistication of analytical equipment, it has been possible to describe relatively precisely (1) the intake of a particular nutrient, (2) the adaptive changes the animal makes to a range of intakes of that nutrient and, therefore, (3) the critical dietary thresholds for optimum productivity and for survival of the animal.

By comparison, our knowledge of the quantitative effects of parasitism on the metabolism of sheep is rudimentary. It is a complex subject because parasites exist which inhabit virtually all organs of the animal body and either directly or indirectly, through the virus and bacterial organisms they may carry, affect the metabolism of the host. This review is restricted to certain helminths which inhabit the gastrointestinal tract and liver. Even so numerous factors such as site of predilection in the host, feeding habits, resistance of the host to larval development and rate of larval intake determine whether the host is overwhelmed or accommodates infection at considerable metabolic cost.

Assessment of the size and significance of infection

Faecal egg counts allow the detection of gastrointestinal parasitism in an animal but do not describe the effect on performance. Their concentration is susceptible to variation in feed residues passing down the tract as a result of variation in feed intake and quality. In addition, the fecundity of nematodes not only varies between species, but within species may be inversely proportional to population size. Furthermore, depression in wool growth of between 16 and 26% has occurred in animals exposed to infection but in which faecal egg counts were absent because of previously acquired host resistance to larval development (Barger and Southcott, 1975; Steel, Symons and Jones, 1980). In the latter work this situation developed in the later stages of continuous infection. There are, therefore, many deficiencies in this method of assessment.

Frequent anthelmintic therapy, sufficient to eliminate mature worms from the host and suppress faecal egg counts, has been used in field studies to provide 'control' animals against which to quantify effects of infection.

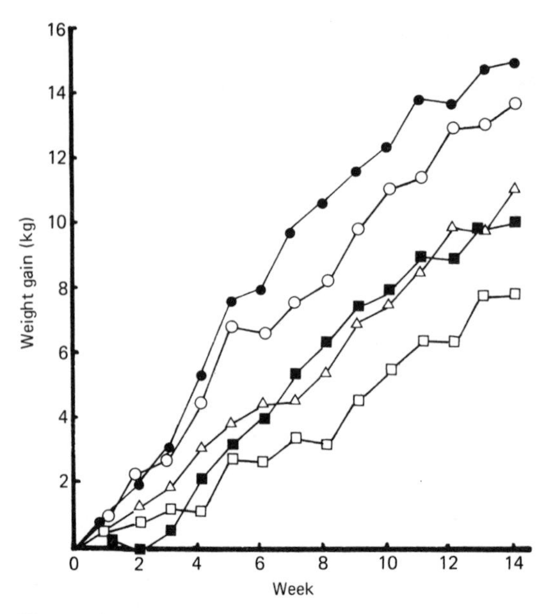

Figure 16.1 The effect of rate of intake of *O. circumcincta* larvae and of regular anthelmintic therapy on growth rate of sheep. Sheep were dosed with 0 (control), ●; 1000, ○; 3000, △; or □, 5000 larvae/day. A further group (■) received 5000 larvae/day and were treated with anthelmintic every 21 days. From Coop, Sykes and Angus (1982)

This is unreliable in continuous infections with *Ostertagia circumcincta* (Coop, Sykes and Angus, 1982; *Figure 16.1*). Sheep infected daily with 5000 larvae grew at half the rate of controls. Anthelmintic dosing at 21-day intervals, which suppressed egg counts in all but the early stages of infection, restored only 20% of the loss in growth rate.

The rate and duration of larval intake probably, therefore, provides the best guide to the likely effect of parasitism on host metabolism.

Effects of parasitism on host productivity

Sudden exposure to large numbers of larvae (>50 000) can cause acute parasitic gastroenteritis and death. Continuous larval intakes of between 2000–5000 larvae/day, which often occur in practice, allow adaptation in the host, and produce relatively stable 'subclinical' conditions. They not only simulate field situations but are necessary for physiological and nutritional studies.

Age and physiological status influence the outcome of infection. Young lambs are highly susceptible to larval intake prior to developing resistance—judged by reduced effect on performance and lower faecal egg counts—with age and larval experience. Reductions in growth rate of between 10–50% as a result of abomasal infections with *O. circumcincta* (*Figure 16.1*) and intestinal infections with *Trichostrongylus colubriformis* and *T. vitrinus* (Sykes and Coop, 1976a; 1977; Sykes, Coop and Angus, 1979; Steel, Symons and Jones, 1980; Steel, Jones and Symons, 1982;

Symons, Steel and Jones, 1981; Jones and Symons, 1982) have been induced by continuous infections with up to 5000 larvae/day. Resistance to infection develops and effects on growth rate tend to diminish after 10–16 weeks of infection (Steel, Symons and Jones, 1980; Steel, Jones and Symons, 1982; Symons, Steel and Jones, 1981), though reduction in wool growth may still occur *vide supra*. Efficiency of food utilization in mature 'resistant' animals has not been rigorously examined since there are deficiencies in measurements based on body weight change.

Parturition is associated with breakdown of 'resistance' to parasitism as judged by development of larvae to maturity and reappearance of eggs in faeces. Few reliable data are available on the effect of larval intake on the performance of the breeding ewe. Very recently R.J. Thomas (personal communication) infected groups of six ewes with either 0 or 2500 larvae of *Haemonchus contortus* each week between nine weeks prepartum and six weeks postpartum. Feed intake was reduced by 11% six weeks after commencement of infection and tended to return to normal during lactation. Milk production of infected sheep during the first six weeks of lactation was reduced by about 25%. On termination of the trial infected sheep were almost 15 kg lighter than control sheep (75 kg).

Our own studies with breeding sheep (Leyva, Henderson and Sykes, 1982) involved four groups of 10 sheep maintained indoors. They were infected with 4000 larvae of *O. circumcincta* each day for six weeks during late pregnancy or early lactation or during both pregnancy and lactation. The four groups were: CC (control during pregnancy and lactation); CI (control:infected); IC (infected:control) and II (infected during both pregnancy and lactation) respectively. The results are summarized in *Table 16.1*. Infection during pregnancy did not affect food intake, lamb birthweight or ewe body weight postpartum. Wool growth was depressed by 11% but not significantly so. During lactation food intake and milk production were reduced. Wool growth and its tensile strength were

Table 16.1 EFFECT OF CHRONIC INFECTION WITH *O. CIRCUMCINCTA* LARVAE (4000/DAY) ON ASPECTS OF PERFORMANCE OF PREGNANT AND LACTATING SHEEP. VALUES ARE GIVEN AS PROPORTIONS OF THOSE IN CONTROL SHEEP

Pregnancy treatment	*Control* (C)		*Infected* (I)	
Feed intake	1.00		0.97	
Wool growth (weeks 1–3)	1.00		1.00	
Wool growth (weeks 4–6)	1.00		0.88	
Body weight postpartum	1.00		0.98	
Lamb birthweight	1.00		0.95	
Lactation treatment	CC	CI	IC	II
Feed intake	1.00	0.77	0.92	0.87
Milk production (weeks 1–3)	1.00	0.97	1.00	0.95
Milk production (weeks 4–6)	1.00	0.77	0.93	0.79
Wool growth (weeks 1–3)	1.00	0.86	0.95	0.87
Wool growth (weeks 4–6)	1.00	0.70	0.87	0.77
Tensile strength of wool grown during pregnancy and lactation	1.00	0.55	0.86	0.77

From Leyva, Henderson and Sykes (1982)

markedly reduced and body weight loss exaggerated by infection. The effects of infection during lactation appeared to be greatest in sheep not infected during pregnancy (CI). The performance of lactating ewes, therefore, appears to be very susceptible to intake of nematode larvae.

Parasite development and pathogenesis of infection

The major nematode and trematode parasites of the gastrointestinal tract, their location, feeding habits and the pathological damage caused have been described by Sykes and Coop (1979) and are set out in *Table 16.2*. The nematode parasites, with the exception of *Strongyloides papillosus* which invades the host via the skin, have similar life cycles. Infective third

Table 16.2 SOME IMPORTANT HELMINTH PARASITES OF SHEEP, THEIR LOCATION IN THE HOST AND FEEDING ACTIVITIES

Site	*Species*	*Direct blood feeders*	*Causing blood loss*
Abomasum	*Haemonchus contortus*	*	
	Ostertagia circumcincta		
	Trichostrongylus axei		
Duodenum and proximal small intestine	*Nematodirus battus*		
	Trichostrongylus colubriformis		
	Trichostrongylus vitrinus		
	Cooperia curticei		
	Strongyloides papillosus		
	Capillaria longipes		
	Bunostomum trigonocephalum	*	
Caecum, colon and distal ileum	*Oesophagostomum venulosum*		*
	Oesophagostomum columbianum		*
	Chabertia ovina		*
	Trichuris ovis		*
Liver	*Fasciola hepatica*	*	

stage larvae, ingested with herbage, ex-sheath and burrow in the glands or epithelium of the tract and subsequently emerge to live as adults in close association with the mucosa or burrow beneath the surface epithelium.

In infections in the abomasum, particularly with *O. circumcincta* and *H. contortus*, parietal and chief cells are damaged by fourth stage larvae as they emerge from the gastric glands during development to adults. Acid secretion is impaired and pH can increase from normal values of 2–3 to 6–7. The major lesions in intestinal parasitism are extensive villus atrophy, mucosal thickening and stunting of micro-villi (*Figure 16.2*).

The pathology of infections in the large intestine have not been well described.

Fasciola hepatica ex-cysts in the rumen, penetrates the intestinal mucosa and swims on serosal surfaces until it reaches the liver capsule. It subsequently undergoes a period of growth, tunnelling and feeding in the liver parenchyma for about 8–10 weeks before entering the bile duct as an

(*a*)

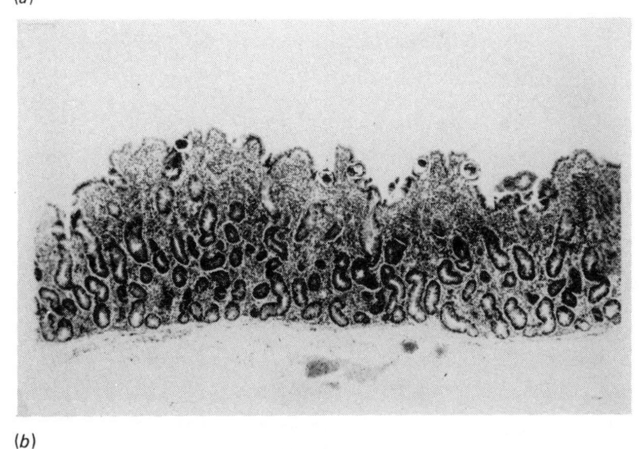

(*b*)

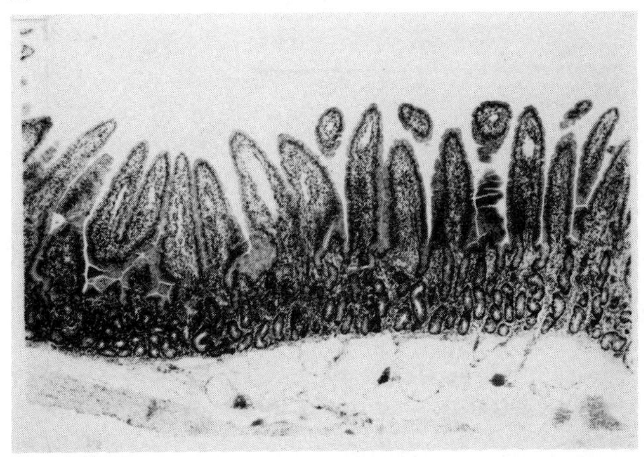

Figure 16.2 The effect of infection with *T. colubriformis* on the morphology of the proximal small intestine: (a) control; (b) infected. By courtesy of Sykes and Coop (1976b)

adult to commence egg laying. In sheep mature fluke accumulate in the bile ducts and can cause biliary congestion.

The majority of parasites appear to feed on general tissue exudate. Others cause special problems by feeding on red blood cells (e.g. *H. contortus*) or by causing such extensive damage that leakage of red blood cells occurs (e.g. *B. trigonocephalum*). In fascioliasis blood loss occurs initially through haemorrhage of the liver parenchyma during migration of the fluke and subsequently as a result of the haematophagic activities of the adult fluke in the bile duct. The aetiology of the anaemias associated with these infections has been described by Dargie (1975) and Jennings (1976).

Increased losses of plasma into the gastrointestinal tract have been observed in infections of the abomasum with *O. circumcincta* (Holmes and McLean, 1971; Symons, Steel and Jones, 1981; Steel, Jones and Symons, 1982) and *H. contortus* (Dargie, 1975) and in infections of the small intestine with *T. colubriformis* (Barker, 1973; Steel, Symons and Jones, 1980; Symons, Steel and Jones, 1981). Leakage of plasma proteins appears

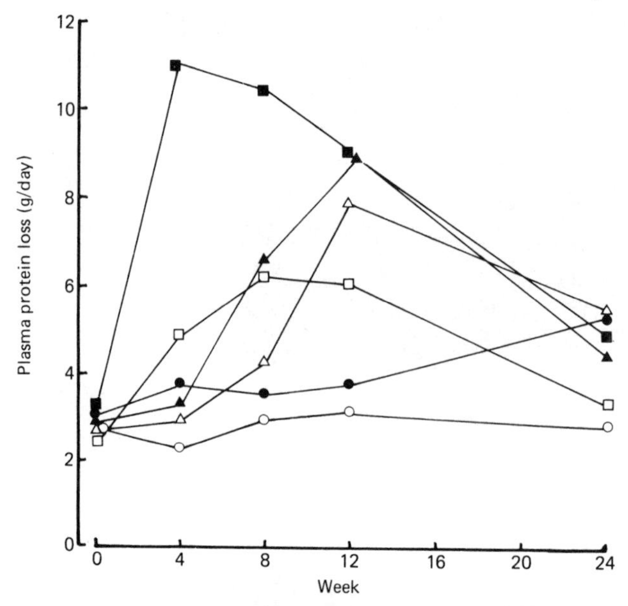

Figure 16.3 The effect of infection with *T. colubriformis* on loss of plasma protein into the alimentary tract. Sheep were infected with 0, ○; 300, ●; 900, △; 3000, ▲; 9000, □; and 30 000, ■ larvae each week. By courtesy of Steel (1978)

to be related to larval intake (*Figure 16.3*). Irreversible loss of albumin may double; albumin pool size decreases but turnover rate of albumin increases and rate of albumin synthesis may increase by 6–8 g/day (Steel, Symons and Jones, 1980; Symons, Steel and Jones, 1981).

The effect of infection on the endocrine balance of the host has been little studied. Gastrin concentration in plasma increases as a result of abomasal infection (Titchen, 1982) and may have general effects on secretions and motility of the gastrointestinal tract.

Little is known of the effect of parasitism on secretion of hormones with more generalized metabolic activity. In an acute infection with *T. colubriformis* plasma cortisol concentration increased and T_4 concentration decreased (Pritchard, Hennessey and Griffiths, 1974) and in both chronic intestinal and abomasal parasitism plasma T_4 concentrations decreased (A.R. Sykes and R.L. Coop, unpublished).

Site of infection and nutrient absorption

Site of infection may have consequences for absorption of particular nutrients. The extent will, however, depend on the capacity for compensatory absorption at other sites.

ABOMASAL PARASITISM

Infection with *O. circumcincta* reduces apparent digestibility of dietary nitrogen (Parkins, Holmes and Bremner, 1973; Sykes and Coop, 1977;

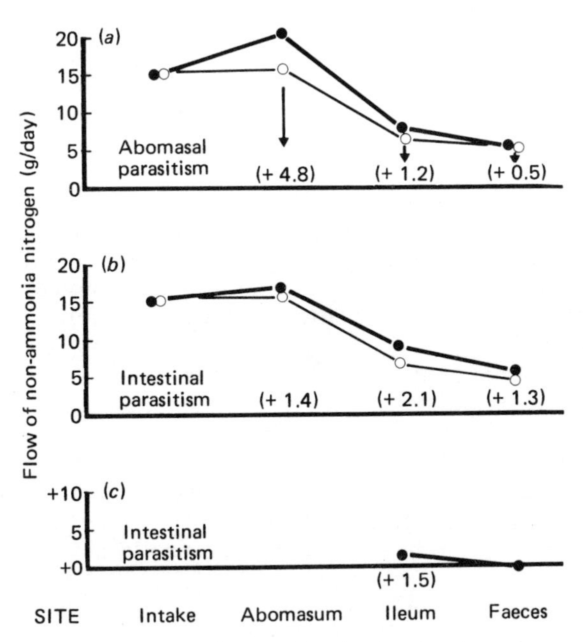

Figure 16.4 The effect of infection in the abomasum and small intestine on nitrogen transactions in the digestive tract. ● and ○ denote infected and control sheep respectively. From (a) Steel (1978); (b) Steel and Hogan (1972); (c) Poppi *et al.* (1981)

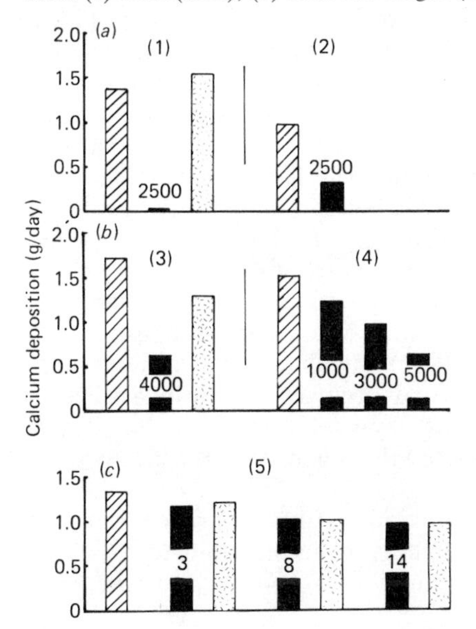

Figure 16.5 Effect of parasitism in (a) the small intestine (*T. colubriformis* and *T. vitrinus*), (b) the abomasum (*O. circumcincta*) and (c) the liver (*F. hepatica*) on rate of calcium deposition in the body compared with that in controls offered food *ad libitum* or pair-fed. All values are related to data from sheep killed at the commencement of infection. (1) Sykes and Coop, 1976; (2) Sykes, Coop and Angus, 1979; (3) Sykes and Coop, 1977; (4) Coop, Sykes and Angus, 1982; (5) Sykes, Coop and Rushton, 1980. Daily larval intakes are indicated on histograms, ■, infected; ▨, control fed *ad libitum* and ▧, control pair-fed

Steel, Symons and Jones, 1980; Coop, Sykes and Angus, 1982). The effect is greatest in the first few weeks of infection and tends to diminish after 12–16 weeks (Sykes and Coop, 1977; Coop, Sykes and Angus, 1982; Steel, Jones and Symons, 1982). Steel (1978) demonstrated increased quantities of non-ammonia nitrogen leaving the abomasum, presumably of endogenous origin (*Figure 16.4*). These were largely, though not completely reabsorbed before the caecum, providing evidence for normal digestion and absorption. Return of apparent digestibility towards normality in the later stages of infection could indicate reductions in secretions with increasing host resistance, increased intestinal digestion and absorption or increased caecal degradation of protein. The latter is certainly a feature of abomasal parasitism (*Figure 16.4*).

Increase in abomasal pH could have implications for pH sensitive processes in distal regions of the tract, such as calcium absorption. Duodenum pH may increase to 6.5 in abomasal infections (Mapes and Coop, 1970). Direct evidence for failure of absorption of calcium or phosphorus is not available, though skeletal calcium and phosphorus deposition are impaired (*Figure 16.5*). Unpublished data (W. Wilson and A.C. Field, personal communication) have suggested phosphorus absorption, at least, not to be impaired.

INTESTINAL PARASITISM

Evidence for reduction in apparent digestibility of nitrogen is conflicting. Reductions were observed by Franklin, Gordon and MacGregor (1946), Steel and Hogan (1972), Barger (1973), Steel, Symons and Jones (1980), Sykes, Coop and Angus (1979), but under similar circumstances Sykes and Coop (1976a) and Poppi *et al.* (1981) failed to demonstrate similar effects. Steel and Hogan (1972) and Poppi *et al.* (1981) (*Figure 16.4*) demonstrated increased flow rates of nitrogen at the ileum. The latter workers demonstrated that digestion and absorption of ^{35}S-labelled rumen bacteria was not impaired, an observation which confirmed the findings of Symons and Jones (1970), in several species, for normal absorption of casein and of ^{14}C-labelled *Chlorella* protein in intestinal infections. On the assumption that other protein entering the small intestine had the same digestion and absorption coefficients, Poppi *et al.* (1981) calculated that an additional secretion of 5 g endogenous nitrogen had occurred. Additional plasma protein loss was estimated as only 1 g/day and they argued that cell sloughing and mucus secretion contributed the balance. This aspect will be discussed further later (p.329).

There is direct evidence for a reduction in disappearance rates of phosphorus and calcium across the small intestine in these infections (D.P. Poppi and J.C. MacRae, personal communication; W. Wilson and A.C. Field, personal communication).

LARGE INTESTINE

The effect of parasitism on function of the large intestine has not been much investigated.

LIVER

Fascioliasis has been little studied in relation to liver function, and there is no evidence that it results in impaired feed digestion and absorption (Peters and Weingartner, 1971; Sinclair, 1975; Sykes, Coop and Rushton, 1980).

Energy digestibility

There is no consensus on the extent to which energy digestibility is modified by parasitism. Reductions have been observed as a result of abomasal parasitism (Sykes and Coop, 1977; Coop, Sykes and Angus, 1982) and intestinal parasitism (Sykes and Coop, 1979; MacRae *et al.*, 1979), but contrary evidence is available (Sykes and Coop, 1976a). Reductions have often been transitory and small (up to 5% digestibility units) and detected only by continuous balances using markers in feed (Sykes, Coop and Angus, 1979; Coop, Sykes and Angus, 1982). Their quantitative significance is difficult to assess.

Effects on growth, food intake and nutrient utilization

BONE GROWTH

Infections of the abomasum and small intestine reduce skeletal growth in young lambs. This can be demonstrated in terms of skeletal mineral deposition (*Figure 16.5*) or in detailed bone chemical pathology (*Figure 16.6*) (Sykes, Coop and Angus, 1975; 1977; 1979; Coop *et al.*, 1981; Coop, Sykes and Angus, 1982; Sykes and Coop, 1976a; 1977). Common lesions in abomasal and intestinal parasitism include reduction in growth of the external dimensions of bones, measured as tibial length or as volume of bone, reduced activity of bone-forming cells (osteoblasts) and reductions in the degree of mineralization of bone matrix, the latter being particularly the case in intestinal parasitism.

There are several possible explanations for these lesions (Sykes, 1982). Reduced bone growth could result from reduced availability of bone matrix-forming substances (matrix osteoporosis) as occurs in protein and energy deficiencies, induced calcium or phosphorus deficiency (mineral osteoporosis) or changes in the endocrine environment, viz. increased corticosteroid secretion which may reduce osteoblastic activity. It is not possible at this stage to separate these possibilities. There is, however, a stronger argument for an induced mineral deficiency in intestinal parasitism based on reduced phosphorus uptake (D.P. Poppi and J.C. MacRae, personal communication) and in consistent reductions in plasma phosphorus concentrations (Coop, Sykes and Angus, 1976; Coop, Angus and Sykes, 1979). These latter changes in plasma mineral concentrations are not as pronounced in abomasal parasitism (Coop, Sykes and Angus, 1977; Coop *et al.*, 1981). Reductions in bone matrix deposition in both intestinal and abomasal parasitism may be related to changes in the partitioning of amino

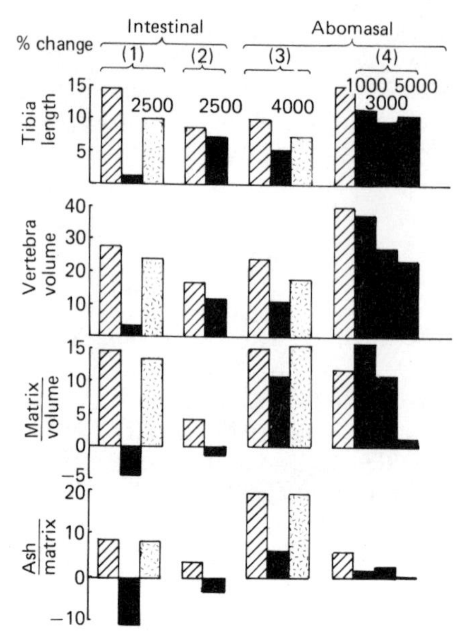

Figure 16.6 Effect of parasitism in the abomasum (*O. circumcincta*) and small intestine (*T. colubriformis* and *T. vitrinus*) on components of bone growth in sheep compared with controls offered food *ad libitum* or pair-fed. All values are related to data from animals killed at the commencement of infection. (1) Sykes, Coop and Angus, 1975; (2) Sykes, Coop and Angus, 1979; (3) Sykes, Coop and Angus, 1977; (4) Coop *et al.*, 1981. Daily larval intakes are indicated on histograms. ■, infected; ▨, control fed *ad libitum* and ▩, control pair-fed

acid and energy away from body growth and into the viscera and homeostasis. The evidence for reduced, rather than increased activity of Haversian systems in bone in abomasal parasitism (Coop *et al.*, 1981) supports an hypothesis of matrix rather than mineral osteoporosis. This aspect of bone growth has not yet been examined in intestinal parasitism in which a combined matrix and mineral osteoporosis may occur.

Body calcium deposition in fascioliasis appears to be reduced in direct proportion to the reduction in feed intake (*Figure 16.5*) rather than as a specific consequence of infection.

FEED INTAKE

Reduction in feed intake is a consequence of acute and chronic parasitism of all sites. This has ranged from virtual complete anorexia in the former type of infection to small reductions of the order of 10–20% in the latter in young growing lambs and in mature breeding ewes (Sykes and Coop, 1976a; 1977; Steel, Symons and Jones, 1980; Steel, Jones and Symons, 1982; Symons, Steel and Jones, 1981; Leyva, Henderson and Sykes, 1982). Symons and Hennessey (1981) demonstrated a relationship between reduction in feed intake and increase in plasma concentrations of cholecystekinin on infection with *T. colubriformis*. Infusion of cholecystekinin reduced food intake and may be considered to be one mediator of appetite

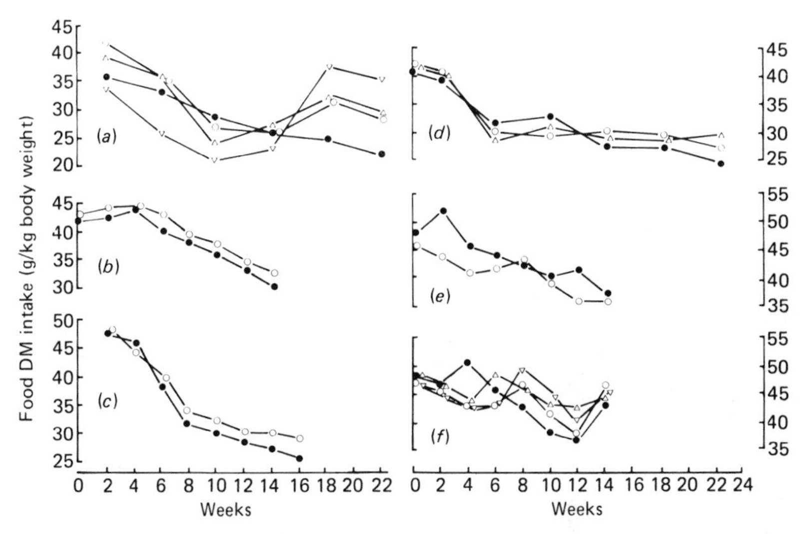

Figure 16.7 Feed DM intake (g/kg body weight) of sheep chronically infected in the small intestine with (a) *T. colubriformis* at the rates of ●, 0; ○, 450; △, 1500 and ▽, 4500 larvae/day, (b) ●, 0 or ○, 2500 *T. colubriformis* larvae/day (Sykes and Coop, 1976) (c) ●, 0 or ○, 2500 *T. vitrinus* larvae/day (Sykes, Coop and Angus, 1979) or in the abomasum with *O. circumcincta* at the rate of (d) ●, 0; ○, 5500; △, 17 000 larvae/day (Symons, Steel and Jones, 1981) (e) ●, 0 or ○, 4000 larvae/day (Sykes and Coop, 1977), (f) ●, 0; ○, 1000; △, 3000; or ▽, 5000 larvae/day (Coop, Sykes and Angus, 1982)

in parasitic infection. There are, however, many other possibilities. For example if, as a consequence of induced mineral deficiency, skeletal growth is impaired, nutrient demand for growth of bone and muscle tissue would presumably be reduced. Reduction in feed intake is an early consequence of phosphorus and calcium deficiency in growing lambs (Field, Suttle and Nisbet, 1975).

On the other hand, when data for feed intake of chronically infected and control sheep from the literature are expressed in relation to body weight (g/kg W), reductions in feed intake may be relatively small (*Figure 16.7*). Indeed, in some cases feed intake may tend to be higher in the earlier stages of light infestations and is almost invariably higher in the later stages of infection than in control sheep. This suggests that positive compensatory mechanisms may exist in certain infectious states. In their studies with mature breeding sheep Leyva, Henderson and Sykes (1982) observed reductions in feed intake during lactation but not during pregnancy (*Table 16.1*). They suggested on the basis of evidence (in elevated plasma pepsinogen concentrations) for abomasal damage during both pregnancy and lactation, that intake may be particularly sensitive to infection when the animal is operating close to physiological limits, as is usually the case in rapidly growing lambs and lactating ewes.

NUTRIENT UTILIZATION

Comparative slaughter trials (*Figure 16.8*) have demonstrated clearly that gross efficiency of use of metabolizable energy (ME) for growth in young lambs can be impaired by up to 50%. There are conceptual problems in the use of ME data derived from normal animals in diseased animals, particularly if parasitism changes the proportion of dietary protein absorbed as amino acid, or reduces the ability of the animal to incorporate absorbed amino acid into body protein. The errors in its use are probably small, however, in relation to those involved in the measurement of digestible energy (DE).

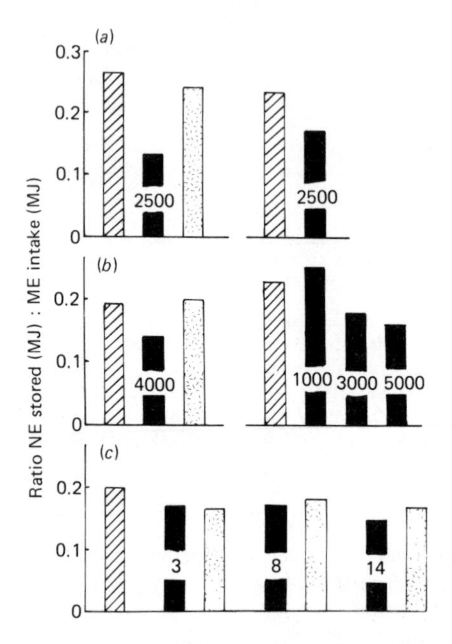

Figure 16.8 Efficiency of use of metabolizable energy (net energy stored:metabolizable energy intake—NE (MJ)/ME (MJ)) in sheep chronically infected in (a) the small intestine (*T. colubriformis* and *T. vitrinus*), (b) the abomasum (*O. circumcincta*) or (c) the liver (*F. hepatica*). Data from Sykes and Coop, 1976; Sykes, Coop and Angus, 1979; Sykes and Coop, 1977; Coop, Sykes and Angus, 1982; Sykes, Coop and Rushton, 1980 . Daily larval intakes are indicated on histograms. ■, infected; ▨, control fed *ad libitum* and ▩, control pair-fed

A full understanding of the mechanism of a reduction in efficiency of utilization of ME requires more detailed information on the changes in protein synthetic rates in tissues as a result of parasitism and particularly in the viscera. These studies (Symons and Jones, 1975; Jones and Symons, 1982) have not yet been sufficiently comprehensive. Jones and Symons (1982) demonstrated that tyrosine flux, expressed per unit of body weight or per g protein intake, was depressed by infection. However, compared with that in pair-fed controls, tyrosine flux was increased which, on the assumption that all tyrosine leaving the plasma pool was incorporated into protein, represented synthesis of an additional 50 g protein/day. Synthesis of fixed liver proteins increased by 50%, fractional synthetic rate of albumin by a factor of 3, while muscle protein synthesis was reduced by 25%. The relative size of the liver and muscle components of the body, however, make it unlikely that these shifts themselves could explain the change in whole body protein synthesis. Very recent studies in guinea pigs infected with *T. colubriformis* (L.E.A. Symons and W.O. Jones, personal communication; Steel and Symons, 1982) have demonstrated marked increases in fractional synthetic rates of protein in gastrointestinal tissue, not only at the site of infection viz. proximal small intestine (+21%) but also in the unaffected distal small intestine (+70%) and large intestine (+30%). These findings are supported by the observation of increased mitotic rates of crypt cells in the distal ileum and colon of rats with infection in the proximal small intestine (Symons, 1978). The phenomenon is probably very significant in the explanation of the large reduction in efficiency of use of ME because the viscera is a major contributor to body protein synthesis (Edmunds, Buttery and Fisher, 1980). In calculating the quantitative significance of these changes for efficiency of use of ME it is difficult to judge whether a value close to the calculated true cost of protein synthesis (3–13 kJ ME/g protein) (Buttery and Boorman, 1978; Webster *et al.*, 1980) or to the net cost of protein deposition in the whole body (45–60 kJ ME/g protein) (Kielanowski, 1976) should be used. However, if a value of 30 kJ ME/g protein is assumed, it can be calculated that the 50% reduction in efficiency of utilization of ME recorded in intestinal parasitism by Sykes and Coop (1976a) would be accounted for by additional synthesis of 30 g protein/day. This is in good agreement with the estimate of Jones and Symons (1982) for increased synthesis of 50 g protein/day and of an additional endogenous loss of 5 g nitrogen/day calculated by Poppi *et al.* (1981), both studies using similar infections.

Table 16.3 EFFECT OF CHRONIC FASCIOLIASIS ON LIVER WEIGHT (g/kg BODY WEIGHT) IN SHEEP. AL3, AL8 AND AL14 WERE INFECTED WITH 3, 8 AND 14 METACERCARIAE OF *F. HEPATICA* ON FIVE DAYS EACH WEEK FOR 22 WEEKS AND FED *AD LIBITUM*. PF3, PF8 AND PF14 WERE CONTROL ANIMALS PAIR-FED TO INFECTED SHEEP; CONTROL SHEEP (ALC) WERE OFFERED FOOD *AD LIBITUM* AND CI WERE SLAUGHTERED AS CONTROLS AT THE OUTSET

	Liver weight (g/kg body weight)		
Controls	CI	ALC	
	15.1	15.3	
Infected and *ad libitum* fed	AL3	AL8	AL14
	19.2	21.2	27.1
Pair-fed controls	PF3	PF8	PF14
	13.7	15.5	13.7

From Sykes, Coop and Rushton (1980)

Effects of fascioliasis on efficiency of food utilization have not been demonstrated (*Figure 16.8*) despite major damage to the liver. The liver exhibits compensatory hypertrophy (*Table 16.3*), apparently functional (Sykes, Coop and Rushton, 1980), which presumably is achieved with a relatively small increase in total protein synthesis.

MODERATING FACTORS

The outcome of infection will be determined by factors which reduce worm establishment on the one hand and which minimize the effect of a resident worm population on the other. Limited evidence suggests that breed (Preston and Allonby, 1978; Le Jambre, 1978), plane of nutrition (Gibson, 1963; Brunsdon, 1962; 1964) and possibly dietary protein (Preston and Allonby, 1978; Berry and Dargie, 1976), cobalt (Downey, 1965) and phosphorus (A.C. Field, personal communication) contents may be important in moderating infection, but much further study is required on these aspects.

Conclusions

Gastrointestinal parasitism probably has a much more marked effect on the metabolism and performance of sheep than has hitherto been recognized. The effects are comprehensive and include influences on appetite, skeletal growth, haematopoiesis, mineral and protein metabolism, and indeed can be confused with and may exaggerate the effects of several nutritional deficiencies. The precise mechanism of effect is not entirely clear and will probably vary markedly with the physiological status of the animal and its need for and ability to buffer supply of a particular nutrient. It is clear that a consequence is a dramatic reduction in efficiency of utilization of feed. Parasitism probably contributes to some of the large differences often observed between the feed requirements of animals kept indoors and those of their free grazing contemporaries. The potential benefit of *effective* control measures is therefore very large.

References

BARGER, I.A. (1973). *Trichostrongylosus* and wool growth. 1. Feed digestibility and mineral absorption in infected sheep. *Australian Journal of Experimental Agriculture and Animal Husbandry* **13**, 42–47

BARGER, I.A. and SOUTHCOTT, W.H. (1975). Trichostrongylosus and wool growth. 3. The wool growth response of resistant grazing sheep to larval challenge. *Australian Journal of Experimental Agriculture* **15**, 167–172

BARKER, I.K. (1973). A study of the pathogenesis of *Trichostrongylus colubriformis* infection in lambs with observations on the contribution of gastrointestinal plasma loss. *International Journal of Parasitology* **3**, 743–757

BERRY, C.I. and DARGIE, J.D. (1976). The role of host nutrition in the pathogenesis of ovine fascioliasis. *Veterinary Parasitology* **2**, 317–332

BRUNSDON, R.V. (1962). The effect of nutrition on age resistance of sheep to infestation with *Nematodirus* spp. *New Zealand Veterinary Journal* **10**, 123–127

BRUNSDON, R.V. (1964). The effect of nutrition on the establishment and persistence of trichostrongyle infestation. *New Zealand Veterinary Journal* **12**, 108–111

BUTTERY, P.J. and BOORMAN, K.N. (1978). The energetic efficiency of amino acid metabolism. In *Protein Metabolism and Nutrition*, pp. 197–200. Publication 16, European Association for Animal Production. London and Boston, Butterworths

COOP, R.L., ANGUS, K.W. and SYKES, A.R. (1979). Chronic infection with

Trichostrongylus vitrinus in sheep. Pathological changes in the small intestine. *Research in Veterinary Science* **26**, 363–371

COOP, R.L., SYKES, A.R. and ANGUS, K.W. (1976). Subclinical trichostrongylosis in growing lambs produced by continuous larval dosing. The effect on performance and plasma constituents. *Research in Veterinary Science* **21**, 253–258

COOP, R.L., SYKES, A.R. and ANGUS, K.W. (1977). The effect of a daily intake of *Ostertagia circumcincta* larvae on bodyweight, food intake and concentration of serum constituents in sheep. *Research in Veterinary Science* **23**, 76–83

COOP, R.L., SYKES, A.R. and AITCHISON, G.U. (1982). The effect of three levels of intake of *Ostertagia circumcincta* larvae on growth rate, food intake and body composition of growing lambs. *Journal of Agricultural Science, Cambridge* (in press).

COOP, R.L., SYKES, A.R., SPENCE, J.A. and ANGUS, K.W. (1981). *Ostertagia circumcincta* infection of lambs, the effect of different intakes of larvae on skeletal development. *Journal of Comparative Pathology* **91**, 521–530

DARGIE, J.D. (1975). Applications of radioisotopic techniques to the study of red cell and plasma protein metabolism in helminth diseases of sheep. *Symposium of the British Society for Parasitology* **13**, 1–26

DOWNEY, N.E. (1965). Some relationships between Trichostrongylid infestation and cobalt status in lambs: 1. *Haemonchus contortus* infestation. *British Veterinary Journal* **121**, 362–370

EDMUNDS, B.K., BUTTERY, P.J. and FISHER, C. (1980). Protein and energy metabolism in the growing pig. In *Energy Metabolism* (L.E. Mount, Ed.), pp. 129–133. Publication 26, European Association for Animal Production. London and Boston, Butterworths

FIELD, A.C., SUTTLE, N.F. and NISBET, D.I. (1975). Effects of diets low in calcium and phosphorus on the development of growing lambs. *Journal of Agricultural Science, Cambridge* **84**, 435–442

FRANKLIN, M.C., GORDON, H.McL. and MacGREGOR, C.H. (1946). A study of nutritional and biochemical effects in sheep of infestation with *Trichostrongylus colubriformis. Journal of the Council for Scientific and Industrial Research* **19**, 46–60

GIBSON, T.E. (1963). The influence of nutrition on the relationship between gastrointestinal parasites and their hosts. *Proceedings of the Nutrition Society* **22**, 15–19

HOLMES, P.H. and McLEAN, J.M. (1971). The pathophysiology of ovine ostertagiasis: a study of the changes in plasma protein metabolism following single infections. *Research in Veterinary Science* **12**, 265–271

JENNINGS, F.W. (1976). The anaemias of parasitic infections. In *Pathophysiology of Parasitic Infection* (E.J.L. Soulsby, Ed.), pp. 41–67. New York, San Francisco, London, Academic Press

JONES, W.O. and SYMONS, L.E.A. (1982). Protein synthesis in the whole body, liver, skeletal muscle and kidney cortex of lambs infected by the nematode *Trichostrongylus colubriformis. International Journal for Parasitology* **12**, 295–301

KIELANOWSKI, J. (1976). Energy cost of protein deposition. In *Protein Metabolism and Nutrition*, pp. 207–215. Publication 16, European Association for Animal Production. London and Boston, Butterworths

LE JAMBRE, L.F. (1978). Host factors in helminth control. In *The Epidemiology and Control of Gastrointestinal Parasites of Sheep in Australia* (A.D. Donald, W.H. Southcott and J.H. Dineen, Eds.), pp. 207–215. CSIRO Division of Animal Health

LEYVA, V., HENDERSON, A.E. and SYKES, A.R. (1982). The effect of daily infection with *O. circumcincta* larvae on food intake, milk production and wool growth in sheep. *Journal of Agricultural Science, Cambridge* **99**, 249–259

MACRAE, J.C., SHARMAN, G.A.M., SMITH, J.S., EASTON, J.F. and COOP, R.L. (1979). Preliminary observations on the effects of *Trichostrongylus colubriformis* infestation on the energy and nitrogen metabolism of lambs. *Animal Production* **28**, 456

MAPES, C.J. and COOP, R.L. (1970). The interaction of infections of *Haemonchus contortus* and *Nematodirus battus* in lambs. *Journal of Comparative Pathology* **80**, 123–136

PARKINS, J.J., HOLMES, P.H. and BREMNER, K.C. (1973). The pathophysiology of ovine ostertagiasis: some nitrogen balance and digestibility studies. *Research in Veterinary Science* **14**, 21–28

PETERS, E. and WEINGARTNER, E. (1971). Einfluss einer experimentellen Fasciolose and Magnesium-stoffwechsel, Futterverdauflichkeit und Gewichtsentwicklung beim Schaf. *Deutsche tierarztliche Wochenschrift* **78**, 535–537

POPPI, D.P., MACRAE, J.C., CORRIGAL, W. and COOP, R.L. (1981). Nitrogen digestion in sheep infected with intestinal parasites. *Proceedings of the Nutrition Society* **40**, 116A

PRESTON, J.M. and ALLONBY, E.W. (1978). The influence of breed on the susceptibility of sheep and goats to a single experimental infection with *Haemonchus contortus. Veterinary Record* **103**, 509–512

PRITCHARD, R.K., HENNESSEY, D.R. and GRIFFITHS, D.A. (1974). Endocrine responses of sheep to infection with *Trichostrongylus colubriformis. Research in Veterinary Science* **17**, 182–188

SINCLAIR, K.B. (1975). The resistance of sheep to *Fasciola hepatica*: studies on the pathophysiology of challenge infections. *Research in Veterinary Science* **19**, 296–303

STEEL, J.W. (1978). Inter-relationships between gastrointestinal helminth infection, nutrition and impaired productivity in the ruminant. In *Recent Advances in Animal Nutrition* (D.J. Farrell, Ed.), p. 98. Armidale, NSW, University of New England Press

STEEL, J.W. and HOGAN, J.P. (1972). Pathophysiology of helminth infections. *Digestive Physiology and Metabolism in Ruminants*, p. 75. Annual Report, CSIRO Division of Animal Health

STEEL, J.W. and SYMONS, L.E.A. (1982). Nitrogen metabolism in nematodosis of sheep in relation to productivity. In *Biology and Control of Endoparasites: Proceedings of the McMaster Animal Health Laboratory 50th Jubilee Symposium* (L.E.A. Symons, A.D. Donald, and J.K. Dineen, Eds.), pp. 235–256. Sydney, Academic Press

STEEL, J.W., JONES, W.O. and SYMONS, L.E.A. (1982). Effects of a concurrent infection of *Trichostrongylus colubriformis* on the productivity and physiological and metabolic responses of lambs infected with *Ostertagia circumcincta. Australian Journal of Agricultural Research* **33**, 131–140

STEEL, J.W., SYMONS, L.E.A. and JONES, W.O. (1980). Effects of level of larval intake on the productivity and physiology and metabolic responses of lambs infected with *Trichostrongylus colubriformis*. *Australian Journal of Agricultural Research* **31**, 821–838

SYKES, A.R. (1982). Nutritional and physiological aspects of helminthiasis in sheep. In *Biology and Control of Endoparasites: Proceedings McMaster Laboratory 50th Jubilee Symposium* (L.E.A. Symons, A.D. Donald and D.K. Dineen, Eds.), pp. 345–347. Sydney, Academic Press

SYKES, A.R. and COOP, R.L. (1976a). Intake and utilization of food by growing lambs with parasitic damage to the small intestine caused by daily dosing with *Trichostrongylus colubriformis* larvae. *Journal of Agricultural Science, Cambridge* **86**, 507–515

SYKES, A.R. and COOP, R.L. (1976b). Chronic parasitism and animal efficiency. *Agricultural Research Council Research Review* **3**, 41–46

SYKES, A.R. and COOP, R.L. (1977). Intake and utilization of food by growing sheep with abomasal damage caused by daily dosing with *Ostertagia circumcincta* larvae. *Journal of Agricultural Science, Cambridge* **88**, 671–677

SYKES, A.R. and COOP, R.L. (1979). Effects of parasitism on host metabolism. In *The Management and Diseases of Sheep* (J.M.M. Cunningham, J.T. Stamp and W.B. Martin, Eds.), pp. 345–347. Slough, Commonwealth Agricultural Bureau

SYKES, A.R., COOP, R.L. and ANGUS, K.W. (1975). Experimental production of osteoporosis in growing lambs by continuous dosing with *Trichostrongylus colubriformis* larvae. *Journal of Comparative Pathology* **85**, 549–559

SYKES, A.R., COOP, R.L. and ANGUS, K.W. (1977). The influence of chronic *Ostertagia circumcincta* infection on the skeleton of growing sheep. *Journal of Comparative Pathology* **87**, 521–529

SYKES, A.R., COOP, R.L. and ANGUS, K.W. (1979). Chronic infection with *Trichostrongylus vitrinus* in sheep. Some effects on food utilisation, skeletal growth and certain serum constituents. *Research in Veterinary Science* **26**, 372–377

SYKES, A.R., COOP, R.L. and RUSHTON, B. (1980). Chronic subclinical fascioliasis in sheep: effect on food intake, food utilization and blood constituents. *Research in Veterinary Science* **28**, 63–70

SYMONS, L.E.A. (1978). Protein Metabolism. 5: *Trichostrongylus colubriformis*. Changes of host body mass and protein synthesis in guinea pigs with light to heavy infections. *Experimental Parasitology* **44**, 7–13

SYMONS, L.E.A. and HENNESSEY, D.R. (1981). Cholecystokinin and anorexia in sheep infected by the intestinal nematode *Trichostrongylus colubriformis*. *International Journal of Parasitology* **11**, 55–58

SYMONS, L.E.A. and JONES, W.O. (1970). *Nematospiroides dubius, Nippostrongylus brasiliensis* and *Trichostrongylus colubriformis*: Protein digestion in infected animals. *Experimental Parasitology* **27**, 496–506

SYMONS, L.E.A. and JONES, W.O. (1975). Skeletal muscle, liver and wool protein synthesis by sheep infected by the nematode *Trichostrongylus colubriformis*. *Australian Journal of Agricultural Research* **26**, 1063–1072

SYMONS, L.E.A., STEEL, J.W. and JONES, W.O. (1981). Effects of level of larval intake on the productivity and physiological and metabolic responses of lambs infected with *Ostertagia circumcincta*. *Australian Journal of Agricultural Research* **32**, 139–148

TITCHEN, D.A. (1982). Summary of hormonal and physiological changes in helminth infection. In *Biology and Control of Endoparasites: Proceedings McMaster Laboratory 50th Jubilee Symposium* (L.E.A. Symons, A.D. Donald and D.K. Dineen, Eds.), pp. 257–275. Sydney, Academic Press

WEBSTER, A.J.F., LOBLEY, G.E., REEDS, P.J. and PULLAR, J.D. (1980). Protein mass, protein synthesis and heat loss in the Zucker rat. In *Energy Metabolism* (L.E. Mount, Ed.), pp. 125–128. Publication 26, European Association for Animal Production. London and Boston, Butterworths

17

INTEGRATION OF SHEEP INTO MIXED FARMING SYSTEMS

A.W. SPEEDY
Department of Agricultural and Forest Sciences, University of Oxford, Agricultural Science Building, Parks Road, Oxford, UK
and
A. GIBSON
Animal Production Advisory and Development Department, East of Scotland College of Agriculture, Bush Estate, Penicuik, Midlothian, UK

A sheep flock usually forms part of a larger farming system with several different enterprises. On arable farms the sheep are mainly intended to utilize short-term leys in the arable rotation. On lowland farms they may also be run in conjunction with finishing beef cattle or dairy cows; on upland farms they are normally associated with suckler cows. Only on hill farms are sheep found as the sole main enterprise, with few cattle, but the farming system may be integrated with forestry. Maxwell, Sibbald and Eadie (1979) have described an objective method of planning the integration of forestry and sheep farming in the hills. The main objective of this chapter, however, is to discuss the planning of integrated farming systems on lowland and marginal mixed farms.

Advantages of integration

The major benefit of mixed farming systems comes from the ability to alternate or rotate the pattern of land use so that different crops or stock utilize given areas each year. A grass break in the arable rotation is advantageous in the maintenance of soil fertility and structure and, in the short term, reduces the dependence on chemical fertilizers to maximize yield of the following crop. Current recommendations (East of Scotland College of Agriculture, 1977) are that nitrogen applications for wheat can be reduced from 145 kg/hectare to 75 kg/hectare (in spring only), following a 1–2 year grazed ley, and alleviate the need for autumn dressings of phosphate (P_2O_5) and potash (K_2O) of 40 kg/hectare each. The rates of application for spring barley can be reduced from 110 kg nitrogen, 50 kg P_2O_5 and 50 kg K_2O/hectare to 50 kg nitrogen only. Levels of nitrogen and P_2O_5 for potatoes can be reduced by 30 kg/hectare and K_2O by 40 kg/hectare after a ley. With longer grass leys and more than 50% of grass in the rotation, some reduction in fertilizer requirements is possible for two years or more.

Alternation of the different livestock systems, sheep and cattle, on longer term leys and permanent pasture has advantages to both animal species. Grassland management can be improved because the different patterns of demand of cattle and sheep can be combined to give more uniform utilization. Finishing cattle, dairy cows and autumn calving suckler cows are housed for up to six months of the year and the grazing of cattle areas with sheep in autumn and early winter ensures removal of excess herbage and avoidance of winter kill of grasses. The sheep system benefits from the scope to extend the grazing area at this time when grass growth is declining to zero and flock requirements are increased during the mating season (Speedy and Clark, 1981).

The greatest benefit of alternate grazing by cattle and sheep comes from the control of parasite infection. Because the intestinal worm parasites of sheep and cattle are almost always host-specific, grazing by cattle in the previous year provides 'clean' areas for ewes and lambs in spring (Rutter, 1975). Similarly, young cattle can be grazed on land that carried sheep in the previous year—a system that is particularly beneficial to young calves born in the previous winter or early spring. Adult ruminants develop immunity to parasites but this may take up to 12 months to be fully effective (Armour, 1979) and both lambs and young calves are likely to suffer increasing parasite problems if they are regularly grazed at high stocking rates on the same fields.

The deleterious effects of parasites on animal performance has already been discussed by Sykes in Chapter 16. It is clear that effective parasite control has considerable financial benefit.

Methods of parasite control

The epidemiological pattern of worm infection by the major species of nematodes (*Nematodirus*, *Ostertagia*, *Trichostrongylus*, *Haemonchus* and *Cooperia* spp.) is one of two seasonal peaks of different larval activity, both of which are of practical significance (Thomas and Boag, 1972). The first is usually observed as increased pasture larval contamination in April and increased faecal egg counts from lambs in May. This is due to the overwintering of not only *Nematodirus* spp. but also *Ostertagia* and *Trichostrongylus* spp. (Rose, 1965). The second appears as a further peak of infection in mid-summer, resulting from the development of worm eggs deposited earlier by the ewes after lambing, referred to as the post-parturient rise. It is due to the breakdown of host resistance and development of arrested (hypobiotic) larvae in the gut (Armour, 1979).

Control of parasitic disease is normally attempted by the use of anthelmintics; ewes are dosed in spring and lambs are dosed in May and regularly thereafter. Gibson and Everett (1967) have recommended that the treatment of ewes should be given one to three weeks after lambing to prevent pasture contamination (particularly with *Ostertagia* spp.). This is probably more critical when levamisole and thiabendazole are used but, even with bendimidazoles which are effective against arrested larvae, the value of drenching will be limited if the ewes are returned to infected pasture. Treatment of ewes in spring is highly effective against *Trichostrongylus* and *Cooperia* spp. but ineffective in controlling levels of *Nematodirus* spp. in lambs. Indeed, there may be significantly greater levels of *Nematodirus* infection of lambs from treated ewes (Sewell, 1973).

The only method of alleviating the problem of overwintered infection is to graze lambs on pasture that was not contaminated in the previous year (Black, 1960). Anthelmintic treatment of lambs on contaminated pasture (while probably essential in therapeutic cases) is likely to be of limited value in preventing chronic parasitism. Rutter *et al.* (1976) found that lamb performance was reduced, despite either routine or strategic use of anthelmintic, when lambs were grazed on contaminated pasture compared with lambs grazed on 'clean' pasture (grazed by cattle in the previous year). Furthermore, recent evidence from Australia (Pritchard *et al.*,1980) and the Netherlands (Boersema and Lewing-van der Wiel, 1982) has suggested that dependence on benzimidazole compounds may be unsatisfactory because of the possible development of parasite resistance. Hence the use of anthelmintics should be directed towards the avoidance of pasture contamination rather than prophylaxis; a similar recommendation has been made for young cattle by Michel *et al.* (1981). The basis of intensive grazing systems should be the grazing of susceptible animals on uncontaminated pasture each year.

Systems of clean grazing

Acceptance of these principles must be accompanied by the development of flexible systems of farm management which enable the conditions for annual provision of 'clean' pasture to be achieved. A complete system was first demonstrated by Rutter *et al.* (1975) at the East of Scotland College of Agriculture, involving a three-year cycle of sheep, hay and suckler cows. The pasture was maintained for seven years, to be followed by a crop of potatoes and then reseeded under barley. Sheep always grazed young grass or followed cows in the rotation and 'clean' aftermaths were available for lambs and calves in July.

Alternative systems have been described by Speedy (1980a). The simplest system, which pertains to arable farms, is to graze ewes and lambs on short-term leys so that they are provided with young, reseeded pasture each spring. In this system, it is important to use undersown stubbles for ewes in the preceding autumn and not to contaminate them with finishing lambs. This can be effected by timing the date of lambing so that the majority of lambs are finished by weaning and only a few remain to graze the small area of aftermath from the hay or silage fields needed for the sheep flock. On arable farms without cattle there is a case for mainly one-year leys but the majority of farmers prefer to operate a two-year system. It is even more important in that case to avoid pasture contamination in the autumn but it is still likely that lambs from ewes grazing second year grass will require prophylactic treatment with anthelmintic. The additional annual costs of grass establishment with short-term leys can be set against the improved performance and possible higher stocking rates on first year grass.

On permanent pasture which cannot be cut for hay or silage, sheep should be alternated with cattle on a biennial basis, with conservation on a separate area (within a lowground arable rotation). This system has also been operated on the East of Scotland College farm on a mixture of steep and level fields. Ewes and lambs were alternated with either cows or weaned calves and conservation was made where possible on the 'cattle' side. A biennial system of alternate sheep and cattle is also appropriate for lambs and '18-month-beef' cattle, when both are susceptible to parasites. Conservation may be included in each of the alternated areas with a 'two-field' or '1-2-3' system for cattle (Hood and Bailie, 1973) and hay and aftermath grazing on a smaller proportion of the sheep side.

Broadbent *et al.* (1980) offer various plans for different systems of beef and sheep management. They suggest a system for sheep and semi-intensive beef with a rotation: ewes and lambs, calves, conservation with

Three year rotation

Alternate sheep/cattle with separate conversion

Figure 17.1 Alternative systems for providing clean grazing. From Speedy (1980a)

Three-year ley

75% sheep

Four-year ley

33% sheep

Figure 17.2 Systems to accommodate different proportions of cattle and sheep on clean grazing

lambs and calves on aftermath and a second year of conservation with calves only on the aftermath. For a 'sheep only' system, they propose a three-year ley with sheep the first year, conservation in the second (with aftermath grazed by ewes) and sheep again in the third year. They also describe various 'safe' methods where 'clean' grazing is not possible but some evasive action can be taken to avoid severe parasite problems.

Further development of these systems requires that they are adaptable to a variety of individual farm situations. *Figures 17.1* and *17.2* summarize possible variations to accommodate different proportions of sheep and cattle (Speedy, 1980a).

The grazing and crop rotation plan for a farm is likely to be based on one of the above systems or combine two or more alternatives. The precise balance will depend on the types of land and their proportion, field sizes and other physical constraints, stock numbers, systems of production, forage requirements and fodder crop yields.

Stocking rates of sheep and cattle

Kilkenny *et al.* (1978) have used a system of land classification to indicate the potential stocking rates of cattle under different soil and climatic conditions, with different levels of nitrogen fertilizer. Similar stocking rates of sheep, on a livestock unit basis (*Table 17.1*) should be achievable in

Table 17.1 TARGET STOCKING RATES

Site quality	*Fertilizer nitrogen* (kg/ha)	*Animals/hectare*				
		Dairy cows	*Suckler cows with calves*	*Cattle from 200 kg*	*Cattle from 350 kg*	*Ewes with lambs*
Poor	150	2.7	2.2	5.0	3.4	10
Average	150	3.6	3.0	6.7	4.5	14
	300	4.2	3.5	7.7	5.3	16
Excellent	150	4.5	3.7	8.3	5.7	17
	300	5.2	4.4	9.7	6.6	20

From Holmes, Craven and Kilkenny (1980)

practice, with clean grazing. It has been suggested that stocking rates should vary with different sizes of ewe (MLC, 1978) but size is usually confounded with lambing percentage and some allowance must be made for the greater number of lambs from larger crossbred ewes. Very high stocking rates, with levels of nitrogen fertilizer above 250 kg/hectare have not been demonstrated in commercial flocks and it is not yet known whether, with clean grazing, higher summer stocking rates, comparable with intensive dairy cows, can be achieved.

Maximum stocking rates used are around 18 ewes/hectare (Fell, 1979) at 187 kg nitrogen/hectare on ryegrass (*Lolium perenne* and *L. multiflorum*) swards. Rutter *et al.* (1975) applied 210 kg nitrogen/hectare to support a stocking rate of 17.5 Scottish halfbred ewes and 29 lambs/hectare. These figures suggest that 10–12 kg nitrogen/ewe/hectare are required, where this is divided into 3–4 applications throughout the season. Consideration of fertilizer levels on a headage basis, rather than by land area, is a sensible approach to fertilizer requirements at different stocking rates and results in similar forage costs/ewe, even when higher levels of nitrogen/hectare are employed. Lowman, Swift and Graham (1978) have recommended 1 kg nitrogen/10 kg liveweight at turnout for beef cattle.

In planning grazing requirements, adjustments to stocking rate must be made for different classes of land within a farm, and fertilizer levels adjusted accordingly.

Balance of conservation and grazing

The different proportions of land required for various livestock systems have been calculated from the Scottish Agricultural Colleges (1981) in *Table 17.2*. Autumn calving suckler cows, for example, require 45% of the

Table 17.2 FORAGE REQUIREMENTS OF LIVESTOCK (HECTARES PER HEAD)

	Lowland suckler cows		*Beef production*		*Lowland ewes*
	Spring calving	*Autumn calving*	*18–20 mo: Sep–Oct born*	*22–24 mo: Apr–May born*	*Fat and store lamb production off grass*
Silage and aftermath	0.15	0.27	0.18	0.24	—
Hay and aftermath	—	—	0.01	0.02	0.008
Swedes	—	—	—	—	0.008
Grazing	0.27	0.33	0.14	0.15	0.085
Total forage (ha)	0.42	0.60	0.33	0.41	0.101

From Scottish Agricultural Colleges (1981)

grass area for silage and an 18-month beef system (autumn born calves) requires 58% for silage and hay. Sheep, however, when wintered on hay and swedes outside, only require 8% of their total grass area for hay with an *additional* area of 8% for swedes.

Silage and hay yields will vary, however, depending on the number of cuts and level of fertilizer applied. With 80 kg nitrogen/hectare, silage will yield 10–15 tonnes/hectare with one cut and hay will yield 6–7 tonnes/hectare (Frame, 1980); with two cuts of silage (120 + 100 kg nitrogen/hectare) a yield of 35–40 tonnes/hectare is obtained.

The figures for stocking rate, winter feed requirements and forage crop yield for an individual farm can be used to produce a grass budget for the total livestock enterprise.

Stock numbers and the balance of enterprises

The size of individual farm enterprises may be determined by physical limitations, availability of fixed resources, economic performance, capital or individual preference. To assess a limited number of proposals, gross margins may be summed, fixed costs deducted and projected net farm

Table 17.3 RELATIVE GROSS MARGINS AND CAPITAL REQUIREMENTS FOR DIFFERENT LIVESTOCK ENTERPRISES STOCKED AT 2 LIVESTOCK UNITS (LU)/HECTARE

	Dairy cows	*Suckler cows*	*18-month beef*	*Sheep*	*Barley*
Gross margin/hectare (2 LU/ha)	1000	300	450	400	250–450
Capital/hectare (excl. buildings)	1050	1100	850	900	150
Capital/hectare (incl. buildings)	2450	1800	1750	1150	—

profits obtained. *Table17.3* shows the relative gross margins for different livestock enterprises at equivalent stocking rates of two livestock units/ hectare. It also shows the capital involvement, which will depend on whether additional housing is required to permit an increase in stock. More sophisticated techniques, such as linear programming, may be applied to determine the optimum balance of enterprises (Barnard and Nix, 1979); but it is important to translate the theoretical solution into a workable farm plan on a field basis. This should take account of field restrictions, rotational considerations and parasite control in livestock.

It may be desirable, with the introduction of clean grazing and higher stocking rates, to *gradually* increase livestock numbers; it is important to avoid unbalancing of the age distribution in a flock or herd. The complication of changing livestock numbers, with different requirements for grazing and conservation, involves a detailed planning operation which projects several years ahead.

Computer application

To facilitate the planning operation described above, a computer program ('CLEANGRAZE') has been written in IMP (Interactive Multi-access Programming language) on the ICL 2972 computer at Edinburgh. It is designed to provide a field by field rotation plan to accommodate sheep on clean grass with cattle, cereals and the necessary hay and silage production. The first step is to obtain a farm plan, an example of which is illustrated in *Figure 17.3*, with current land use records. The user is then asked to specify field names, areas and current use; stock numbers (which may vary over the five year plan) and stocking rates of different classes of stock; forage crop yields and forage requirements; maximum years that crops and grass may be grown on each land class and projected enterprise gross margins.

The program then allocates the appropriate use to each field for each year, allowing for the numbers of animals, stock-carrying capacity of the different land classes and the total requirements for silage, hay and root crops. The first step is to give preference to dry stock on rough grazing and unmowable grass, preferably divided between ewe-hoggs and heifers (or dry cows); ewes and single lambs or suckler cows may be placed on permanent pasture if there is no dry stock. Remaining ewe-hoggs and heifers will be allocated to the next land class, mowable permanent pasture, and then ewes and lambs, hay/silage or cows depending on the present use (i.e. sheep follow cattle). The class 2 arable fields are used for

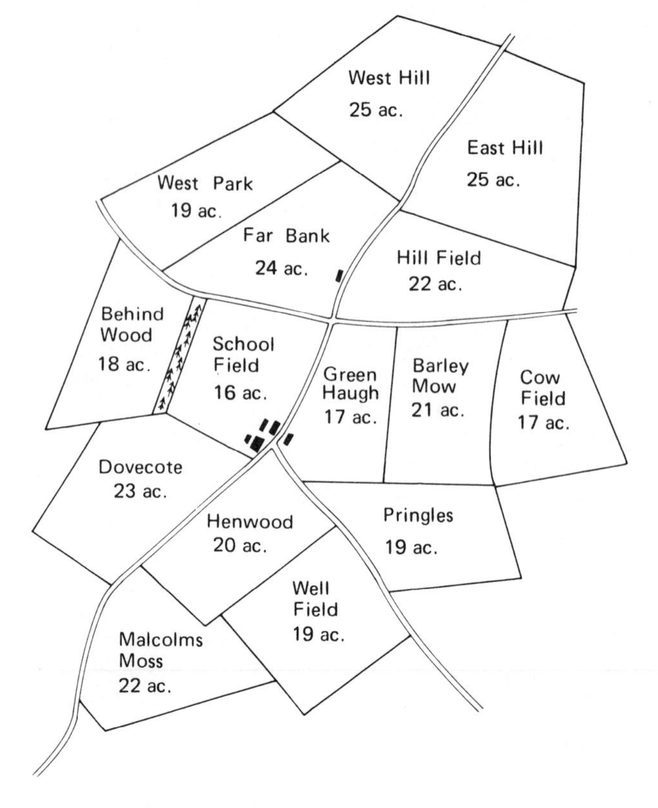

Figure 17.3 Example of a farm plan

sheep, conservation, fattening cattle or cows, barley and *Brassica* root crops. The maximum number of years in crop or grass and the total remaining stock of each class will determine the particular use of a field. Fields currently grazed by sheep will be used first for silage or hay, then cattle, and eventually be sown with barley, and the roots are inserted at a predetermined point in the crop rotation. A similar allocation is then performed for the class 1 arable fields.

In this way, all the stock are allocated with the number in each field determined by the area and appropriate stocking rate. Dry stock, then suckling and finishing animals are allocated first to permanent grass and then rotation grass, in that order, with the remainder devoted to crops. Several runs may be performed with different specifications to obtain the desired solution.

Examples of a printout, relating to the example farm, are shown in *Tables 17.4, 17.5* and *17.6*. This is a simple example but the program is able to deal with more complex problems, with varying field sizes, and allows changes in stock numbers over the five year plan. The method demonstrates the principles applied to the planning of an integrated system for a mixed farm. The use of the computer program allows rapid solutions which can, if necessary, be modified in discussion on advisory cases.

Table 17.4 EXAMPLE OF DATA SUMMARY FROM THE COMPUTER PROGRAM CLEANGRAZE

DATA SUMMARY

FIELD	SIZE	TYPE	USE	CROP YEAR
WEST HILL	25	UNMOWABLE	1	
EAST HILL	25	PERMANENT GRASS	2	
HILL FIELD	22	MOWABLE	1	
FAR BANK	24	PERMANENT GRASS	6	
WEST PARK	19		2	
WELL FIELD	19	CLASS 2 ARABLE	1	GRASS 1
GREEN HAUGH	17		6	GRASS 2
BARLEY MOW	21		2	GRASS 3
SCHOOL FIELD	16		1	GRASS 4
DOVECOTE	23		6	GRASS 5
BEHIND WOOD	18		8	BARLEY 1
COW FIELD	17		8	BARLEY 2
PRINGLES	19		9	SWEDES/TURNIPS 1
MALCOLMS MOSS	22		8	BARLEY 4
HENWOOD	20		8	BARLEY 5

STOCK NUMBERS

EWES	475
SUCKLER COWS	70

GRASS STOCKING RATES
UNITS: ANIMALS/ACRES

	PERMANENT GRASS	ARABLE CLASS 2
EWES	5.0	7.0
SUCKLER COWS	1.0	1.3

FORAGE CROP YIELDS
UNITS: TONNES/ACRES

	PERMANENT GRASS	ARABLE CLASS 2
HAY	2.0	2.0
SILAGE	10.0	10.0
SWEDES		25.0

MAXIMUM NUMBER OF YEARS IN GRASS:
ARABLE CLASS 2– 5 YEARS

MINIMUM NUMBER OF YEARS IN BARLEY:
ARABLE CLASS 2– 4 YEARS

MAXIMUM NUMBER OF YEARS IN BARLEY:
ARABLE CLASS 2– 5 YEARS

SWEDES/TURNIPS GROWN IN CROP YEAR 3

FORAGE REQUIREMENTS
UNITS: TONNES

	HAY	SWEDES
EWES	0.10	0.50
SUCKLER COWS	1.00	3.00

ENTERPRISE GROSS MARGINS
UNITS: £/ANIMAL

EWES	25.00
SUCKLER COWS	100.00

UNITS: £/ACRE

BARLEY-CLASS 2	120.00

Table 17.5 A FIELD USE PLAN BY THE COMPUTER PROGRAM CLEANGRAZE

FIELD USE PLAN

FIELD	SIZE (acre)	Stock numbers or forage crop yields (t/acre) 1981	1982	1983	1984	1985
UNMOWABLE PERMANENT GRASS						
WEST HILL	25	SUCKLERS	EWES	SUCKLERS	EWES	SUCKLERS
		25 25[a]	125 125	25 25	125 125	25 25
FAST HILL	25	EWES	SUCKLERS	EWES	SUCKLERS	EWES
		125 125	25 25	125 125	25 25	125 125
MOWABLE PERMANENT GRASS						
HILL FIELD	22	HAY	SUCKLERS	EWES	HAY	SUCKLERS
		44 44	22 22	110 110	44 44	22 22
FAR BANK	24	SUCKLERS	EWES	HAY	SUCKLERS	EWES
		24 24	120 120	48 48	24 24	120 120
WEST PARK	19	EWES	HAY	SUCKLERS	EWES	HAY
		95 95	38 38	19 19	95 95	38 38
WELL FIELD	19	HAY	SUCKLERS	EWES	HAY	BARLEY
		38 38	23 25	133 133	38 38	
		GRASS 1	GRASS 3	GRASS 4	GRASS 5	YEAR 1
GREEN HAUGH	17	SUCKLERS	EWES	HAY	BARLEY	BARLEY
		21 22	119 119	34 34		
		GRASS 3	GRASS 4	GRASS 5	YEAR 1	YEAR 2
BARLEY MOW	21	EWES	HAY	BARLEY	BARLEY	ROOTS
		147 147	42 42			448 525
		GRASS 4	GRASS 5	YEAR 1	YEAR 1	YEAR 1
SCHOOL FIELD	16	HAY	BARLEY	BARLEY	ROOTS	BARLEY
		32 32			400 400	
		GRASS 5	YEAR 1	YEAR 2	YEAR 1	YEAR 4
DOVECOTE	23	BARLEY	BARLEY	ROOTS	BARLEY	BAR U/S
				448 575		
		YEAR 1	YEAR 2	YEAR 1	YEAR 4	YEAR 5

BEHIND WOOD	18	BARLEY	ROOTS	BARLEY	BAR U/S	EWES
			448 450			126 126
		YEAR 2	YEAR 1	YEAR 4	YEAR 5	GRASS 1
COW FIELD	17	ROOTS	BARLEY	BAR U/S	EWES	HAY
		425 425			119 119	34 34
		YEAR 1	YEAR 4	YEAR 5	GRASS 1	GRASS 2
PRINGLES	19	BARLEY	BAR U/S	EWES	HAY	SUCKLERS
				107 133	36 38	23 25
		YEAR 4	YEAR 5	GRASS 1	GRASS 2	GRASS 3
MALCOLMS MOSS	22	BAR U/S	EWES	HAY	SUCKLERS	EWES
			111 154	36 44	21 29	104 154
		YEAR 5	GRASS 1	GRASS 2	GRASS 3	GRASS 4
HENWOOD	20	EWES	HAY	SUCKLERS	EWES	HAY
		108 140	38 40	26 26	136 140	40 40
		GRASS 1	GRASS 2	GRASS 3	GRASS 4	GRASS 5

[a] The first figure indicates the number of stock accommodation or the tonnage of forage crop provided; the second gives the potential for that field.

Table 17.6 A STOCK AND CROP SUMMARY PRODUCED BY THE COMPUTER PROGRAM CLEANGRAZE

STOCK AND CROP SUMMARY	1981	1982	1983	1984	1985
EWES	475	475	475	475	475
SUCKLER COWS	70	70	70	70	70
UNITS: TONNES					
HAY	114	120	126	120	112
(SURPLUS/DEFICIT)	–4	3	9	3	–6
SWEDES/TURNIPS	425	450	575	400	525
(SURPLUS/DEFICIT)	–23	3	128	–48	78
UNITS: ACRES					
BARLEY CLASS 2	82	75	72	79	75
BARLEY UNDERSOWN	22	19	17	18	23
GROSS MARGIN SUMMARY					
EWES & LAMBS	11875.00	11875.00	11875.00	11875.00	11875.00
SUCKLER COWS	7000.00	7000.00	7000.00	7000.00	7000.00
BARLEY–CLASS 2	9840.00	9000.00	8640.00	9480.00	9000.00
TOTAL GROSS MARGIN £	28715.00	27875.00	27515.00	28355.00	27875.00

The development of integrated systems on commercial farms

The successful development of integrated farming systems which combine high productivity from the sheep flock and other livestock enterprises with high overall farm output is illustrated by the following examples.

A system of management for an arable farm in Lincolnshire is described in detail by Fell (1979). The area is suitable for the production of sugar beet and peas, as well as cereal crops, winter wheat and barley. The heavy clay land benefits from grass leys to improve and maintain soil structure but it has been necessary to improve drainage to allow satisfactory conditions for both grazing and crops.

A flock of 1700 ewes is run on 135 hectares of grass and forage crops, utilizing two-year leys in the arable rotation. 570 ewe-hoggs and 176 rams are also accommodated and the stocking rate in 1979, mentioned above, was 18 ewes/hectare. An important feature of the system is the provision of winter housing for the sheep flock, which avoids pasture damage and allows high stocking rates in spring and summer; as a result, sheep numbers have been gradually built up since 1961 to the present number in 1975. The author (Fell, 1979) emphasizes the importance, in an arable context, of restricting the life of the leys to two years and notes that it is particularly in a third year that (parasite) problems begin to develop.

The description demonstrates the potential for management of a large, intensive flock on a productive lowland farm, with high output/hectare (£672 gross output/hectare in 1979) and beneficial effects on associated crop production.

The successful development of integrated livestock and crop production on a large upland farm in south-east Scotland was reported by Gray (1980). It refers to a 450 hectare mixed arable and stock rearing farm with 208 hectares arable ground and the balance, permanent and hill grazing. The intention, in 1973, was to increase the output of the farm business by intensification, and clean grazing was adopted. Because the aim was to utilize the better land for barley and wheat with limited rotational grass for conservation as silage, the permanent grass was mainly used for grazing.

It was necessary to improve some of the permanent grazing and this was done by pioneer cropping and reseeding. The increase in stock numbers demanded an increase in winter fodder production and this was accompanied by a change from hay to silage. The greater numbers were also associated with improvements in stocking rates and, despite the larger flock and herd sizes, the crop area has also been increased. The changes in land use following the implementation of the plan are shown in *Table 17.7* and the livestock changes are given in *Table 17.8*. The increase in livestock

Table 17.7 CHANGES IN LAND USE ON AN UPLAND FARM

	1973 (hectares)	*1980* (hectares)
Cereals	109	123
Swedes/Rape	18	21
Silage	9	35
Hay	28	16
Rotational grazing	34	11
Permanent pasture	92	86
Rough/Hill	120	120
Other	39	46

From Gray (1980)

Table 17.8 CHANGES IN STOCK NUMBERS ON AN UPLAND FARM

	1973	*1980*
Suckler cows	125	170
Ewes	550	700
Ewe hoggs	—	170
Bulling heifers	12	17
Bulls	4	6
Tups	12	18

From Gray (1980)

Table 17.9 CHANGES IN STOCKING RATE ON AN UPLAND FARM

	1973	*1980*
Total adjusted grazing hectares	200	188
Total adjusted livestock units	277	383
Livestock units/hectare	1.39	2.04

From Gray (1980)

has been achieved within the capabilities of existing labour and mostly using existing buildings; 30 of the 170 cows are spring calved to allow outwintering while the remaining 140 are housed following autumn calving; a plastic tunnel building is used for wintering 100 older ewes. Attention was given to fencing, which is essential to control stock for clean grazing. Fertilizer usage for grazing in 1980 was 190–210 kg nitrogen, 30 kg

P_2O_5 and 24 kg K_2O per hectare but this relatively high level is in proportion to the stocking rate which reached two livestock units/adjusted hectare (*Table 17.9*). Good animal performance was maintained at the higher stocking rate.

It is clear that considerable increases in output are possible on this type of upland, mixed farm, through intensification of grazing. Emphasis has been placed on maximizing the use of permanent grass with livestock, relying on clean grazing to ensure high individual production, and the crop area has consequently increased.

Several other examples of the application of clean grazing and intensification on upland farms have been cited previously by Speedy (1980b). On a steep upland farm in the Borders, the sheep flock was increased from 459 to 606 and suckler cow numbers from 113 to 163. Clean grazing was achieved simply by dividing the farm in two: ewes and lambs, ewe-hoggs, hay and forage crops for sheep on one side and cows and calves, heifer replacements and further conservation on the other. With reseeding, greater use of nitrogen fertilizer, and increased stocking rates, improved lamb performance was obtained from the sheep flock.

Methods of application of these improved farm systems varied but the average increases in livestock numbers and crop production were 21% more ewes, 14.5% more cattle and a 23.5% increase in the area of arable crops. On steep upland farms, the emphasis was placed on stock numbers, while on mixed arable and livestock farms, the opportunity was taken to reduce the grass area and increase arable cropping. In most cases, the tendency was to move sheep and cattle onto permanent pasture and release the better land for crops. Increases were made within the constraints of capital, labour and the availability of buildings so that improvements in total farm gross margins were made with only minor changes in fixed costs.

Finally, mention should be made of a small hill farm in Perthshire, with limited inbye ground. The farm carried 250 Blackface ewes and 30 suckler cows and, up to 1978, sheep performance had been poor, culminating in a lambing percentage of 74 in that year with weaning weights averaging 22 kg. Fortunately, the decision had been taken to improve two areas (16 hectares each) of the lower hill ground, to be used strategically in conjunction with the unimproved hill, according to the Hill Farming Research Organisation's 'two-pasture system' (Armstrong, Eadie and Maxwell, 1977). General improvements were also made in ewe nutrition, the health programme and the wintering arrangements for ewe-hoggs. In 1979, the lambing percentage was still only 87% but, with the sheep run on clean, improved pasture, the weaning weights averaged 33 kg. In the following year, the lambing percentage was raised to 125% with lambs averaging 30 kg at weaning. The sheep flock derived a double benefit from the introduction of improved pasture and the avoidance of parasite problems. There may therefore be some scope for the benefits of integrated cattle and sheep systems to be obtained on improved hill farms.

Conclusions

The integration of sheep into mixed farming systems requires planning over several years to achieve the objective, to increase overall farm output.

Provision of clean grazing, stocking rates, winter feed production and rotation of crops and grass must allow for varied production systems on individual farms. The use of a simple, interactive computer program can facilitate development plans in advisory cases.

Emphasis in practice will tend towards greater concentration of livestock on permanent grass areas with consequent increases in cropping on better land. Increases in stocking rate, based on clean grazing, and the planned rotation of both crops and grass use result in the maintenance of individual animal performance and an overall increase in farm output.

The development of improved systems, combining more effective use of grass, higher stocking rates, parasite control, good animal performance and increased crop production, has considerable scope for application on commercial farms.

References

ARMOUR, J. (1979). Recent advances in the epidemiology of sheep endoparasites. In *The Management and Diseases of Sheep*, pp. 339–344. Papers presented at a British Council Special Course, Edinburgh, 1978. Slough, Commonwealth Agricultural Bureaux

ARMSTRONG, R.H., EADIE, J. and MAXWELL, T.J. (1977). The development and assessment of a modified hill sheep system at Sourhope, in the Cheviot Hills (1968–1976). *Hill Farming Research Organisation, Seventh Report*, pp. 69–97

BARNARD, G.S. and NIX, J.S. (1979). *Farm Planning and Control*, 2nd edition. Cambridge, Cambridge University Press

BLACK, W.J.M. (1960). Control of *Nematodirus* diseases by grassland management. *Proceedings of the 8th International Grassland Congress*, pp. 723–726

BOERSEMA, J.H. and LEWING-VAN DER WIEL, P.J. (1982). Benzimidazole resistance in a field strain of *Haemonchus contortus* in the Netherlands. *Veterinary Record* **110**, 203–204

BROADBENT, J.S., LATHAM, J.O., MICHEL, J.F. and NOBLE, J. (1980). Grazing plans for the control of stomach and intestinal worms in sheep and in cattle. Ministry of Agriculture, Fisheries and Food, Booklet 2154

EAST OF SCOTLAND COLLEGE OF AGRICULTURE (1977). *Fertilizer Recommendations*, Revised Edition, Bulletin No. 18. Edinburgh, The East of Scotland College of Agriculture

FELL, H.R. (1979). *Intensive Sheep Management*. Ipswich, Farming Press

FRAME, J. (1980). Grassland farming—scope for improvement? In *Farming Under Pressure*, Papers given at the Twentieth Annual Conference. Edinburgh, The East of Scotland College of Agriculture

GIBSON, T.E. and EVERETT, G. (1967). The absence of the post-parturient rise in faecal egg count and its effect on the control of gastro-intestinal nematodes in lambs. *British Veterinary Journal* **123**, 247–251

GRAY, P.H.B. (1980). Using grass on my farm. In *Farming under Pressure*, Papers given at the Twentieth Annual Conference. Edinburgh, The East of Scotland College of Agriculture

HOLMES, W., CRAVEN, J. and KILKENNY, J.B. (1980). Application on the farm. In *Grass—its Production and Utilization* (W. Holmes, Ed.), pp. 239–268. Oxford, British Grassland Society and Blackwell Scientific Publications

HOOD, A.E.M. and BAILIE, J.H. (1973). A new grazing system for beef cattle—the two-field system. *Journal of the British Grassland Society* **28**, 101–108

LOWMAN, B.G., SWIFT, G. and GRAHAM, H.R. (1978). *Grazing Management for Beef Cattle*, Technical Note Number 198 A/C. Edinburgh, The East of Scotland College of Agriculture

KILKENNY, J.B., HOLMES, W., BAKER, R.D., WALSH, A. and SHAW, P.G. (1978). *Grazing Management*, Beef Production Handbook No. 4. Milton Keynes, The Meat and Livestock Commission

MAXWELL, T.J., SIBBALD, A.R. and EADIE, J. (1979). Integration of forestry and agriculture—a model. *Agricultural Systems* **4**, 161–188

MEAT AND LIVESTOCK COMMISSION (1978). *Sheep Facts. A Manual of Economic Standards*. Milton Keynes, The Meat and Livestock Commission

MICHEL, J.F., LATHAM, J.O., CHURCH, B.M. and LEECH, P.K. (1981). Use of anthelmintics for cattle in England and Wales during 1978. *Veterinary Record* **108**, 252–258

PRITCHARD, R.K., HALL, C.A., KELLY, J.D., MARTIN, I.C.A. and DONALD, A.D. (1980). The problem of anthelmintic resistance in nematodes. *Australian Veterinary Journal* **56**, 239–251

ROSE, J.H. (1965). The rested pasture as a source of lung-worm and gastro-intestinal infection for lambs. *Veterinary Record* **77**, 749–752

RUTTER, W. (1975). *Sheep from Grass*, Bulletin No. 13. Edinburgh, The East of Scotland College of Agriculture

RUTTER, W., BLACK, W.J.M., FITZSIMONS, J., CARSON, I.S. and SWIFT, G. (1975). Grassland based sheep systems. Edinburgh School of Agriculture, Annual Report, pp. 48–49

RUTTER, W., HUNTER, A.G., MATHIESON, A.O., BLACK, W.J.M. and WATT, J.A.A. (1976). Intestinal parasites. Edinburgh School of Agriculture, Annual Report, pp. 48–49

SCOTTISH AGRICULTURAL COLLEGES (1981). *Farm Management Handbook, 1981/82*. Edinburgh, Aberdeen and Auchincruive, The Scottish Agricultural Colleges

SEWELL, M.M.H. (1973). The influence of anthelmintic treatment of ewes on the relative proportions of gastrointestinal parasites in their lambs. *Veterinary Record* **92**, 371–372

SPEEDY, A.W. (1980a). *Sheep Production—Science into Practice*. London and New York, Longman

SPEEDY, A.W. (1980b). Upland sheep production. In *The Effective Use of Forage and Animal Resources in the Hills and Uplands* (J. Frame, Ed.), pp. 69–81. British Grassland Society, Occasional Symposium No. 12

SPEEDY, A.W. and CLARK, C.F.S. (1981). Lowland sheep: the nutrition and management cycle. *Veterinary Record* **108**, 493–496

THOMAS, R.J. and BOAG, B. (1972). Epidemiological studies on gastro-intestinal nematodes of sheep. *Research in Veterinary Science* **13**, 61–69

V

Reproduction

18

ENDOCRINE CONTROL OF REPRODUCTION IN THE EWE

W. HARESIGN, B.J. McLEOD and G.M. WEBSTER
University of Nottingham School of Agriculture, Sutton Bonington, Loughborough, Leicestershire, UK

The seasonal nature of breeding activity in the ewe dictates that, with few exceptions, it is possible to achieve only a single pregnancy each year, and an inherently low ovulation rate imposes a further limit on the biological efficiency of sheep by restricting litter size. Removal of one or both of these limitations would, therefore, markedly increase the levels of production. Teleologically, seasonal breeding may be explained as an evolutionary adaptation which ensures the most beneficial climatic and nutritional environment for survival of the newborn lamb in the wild. However, the wide range of environmental conditions in which domestic sheep have been kept over the years has resulted in a wide range in reproductive performance between breeds, in terms of both litter size and length of the breeding season.

In temperate climates the breeding season is limited to the autumn and winter months, but may be considerably modified by genetic selection, geographical location and level of nutrition (Robinson, 1951; Eckstein and Zuckerman, 1956; Kelly, Allison and Shackell, 1976; Robertson, 1977). Moreover, the depth of seasonal anoestrus varies between breeds and is positively correlated with the length of the non-breeding interval (Robinson, 1951).

In spite of the fact that it has been known for many years that the breeding season is controlled by the light:dark ratio (Yeates, 1949), little progress has been made in successfully manipulating breeding activity in the ewe. This has been due to inadequate knowledge of the hormonal mechanisms controlling the oestrous cycle, and the endocrine causes of seasonal anoestrus. Considerable advances in both of these areas have been made in recent years, due largely to the development of appropriate hormone assay techniques. It is not intended here to give an exhaustive review of the hormonal control of the oestrous cycle, but rather to highlight those recent developments which may lend themselves to the practical manipulation of breeding activity in the ewe.

The hormonal control of the oestrous cycle

HYPOTHALAMIC CONTROL OF PITUITARY GONADOTROPHIN SECRETION

The normal sequence of hormonal changes responsible for the control of the oestrous cycle is governed principally by the hypothalamic–pituitary

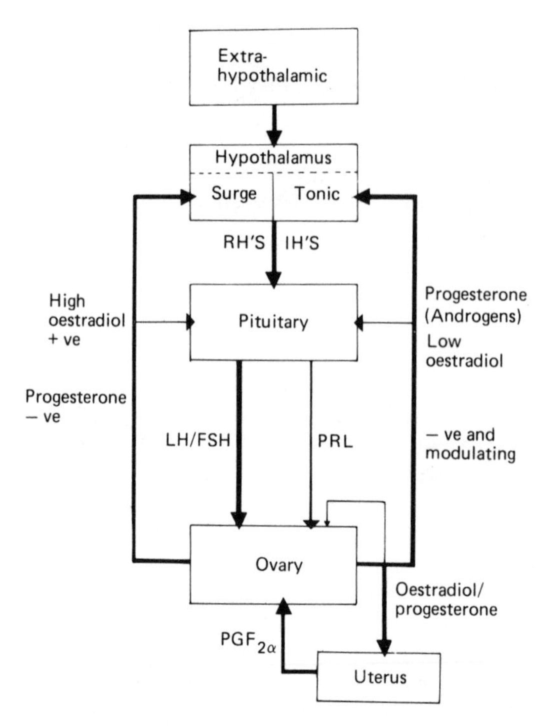

Figure 18.1 Schematic diagram indicating the endocrine mechanisms involved in the control of the oestrous cycle of the ewe

axis, although this, in turn, is modified by hormonal feedback mechanisms involving ovarian steroid hormones, which themselves are produced as a result of gonadotrophic stimulation. The overall control of reproduction, therefore, involves a delicate interplay between the hypothalamic–pituitary axis, the ovaries and the uterus, which is involved in regulating the lifespan of the corpus luteum (*Figure 18.1*). In addition, extra-hypothalamic centres impinge on the hypothalamus to further modulate its activity (see Ellendorff, 1978 for review).

The control that the hypothalamus exerts on pituitary gonadotrophin secretion is mediated by releasing hormones—or an inhibitory hormone in the case of prolactin (PRL)—produced and secreted by hypothalamic neurones and passed to the anterior pituitary via a portal blood system (Daniel and Prichard, 1957). Evidence for the existence of two separate releasing hormones, one for luteinizing hormone (LH) and one for follicle stimulating hormone (FSH), is equivocal. Highly purified or synthetic gonadotrophin releasing hormone (GnRH) induces release of both LH and FSH from the anterior pituitary (Debeljuk, Arimura and Schally, 1972; Crighton and Foster, 1972; Redding *et al.*, 1972; Pelletier, 1976) and the suppression of endogenous GnRH concentrations, by active immunization of ewes against synthetic GnRH, causes a substantial reduction in pituitary and plasma concentrations of both LH and FSH (Clarke, Fraser and McNeilly, 1978; Jeffcoate, Foster and Crighton, 1978; Narayana and

Dobson, 1979). Furthermore, attempts to separate LH and FSH releasing activity from hypothalamic extracts have proved largely unsuccessful.

In the rat the gonadotrophin releasing activity of the hypothalamus has been localized and shows two distinct regions which appear to be involved in the control of LH release—the arcuate nucleus/median eminence region of the medial basal hypothalamus and the suprachiasmatic nucleus/preoptic area (Crighton, Schneider and McCann, 1970; McCann, 1974). FSH releasing activity has, however, been detected in only one area, the medial basal region (Watanabe and McCann, 1968; McCann, 1974). A two-level concept of hypothalamic control of pituitary LH secretion has therefore been postulated, in which the median eminence/arcuate nucleus region is concerned with basal and/or episodic LH secretion while the preoptic/suprachiasmatic area promotes the preovulatory LH surge at oestrus. Limited evidence in the ewe supports this concept, since Jackson *et al.* (1978) have demonstrated that basal and episodic LH secretion does not require the preoptic/suprachiasmatic connections with the basomedial hypothalamus, whereas the preovulatory LH surge is eliminated if these connections are transected.

The response of the anterior pituitary gland to GnRH varies considerably with the reproductive status of the ewe. The response is greatest when the natural preovulatory gonadotrophin release is imminent (Reeves, Arimura and Schally, 1974; Foster and Crighton, 1976a,b) or following oestradiol pretreatment (Reeves, Arimura and Schally, 1971; Haresign and Lamming, 1978; *Table 18.1*), whereas it is lower following a prolonged

Table 18.1 LH RELEASE FOLLOWING INJECTION OF 150 μg GnRH INTO ANOESTROUS EWES WITH AND WITHOUT PRETREATMENT WITH 50 μg OESTRADIOL BENZOATE (ODB) GIVEN 7 HOURS BEFORE GnRH, AND COMPARISON WITH THE NATURAL LH RELEASE AT OESTRUS.

Treatment	*Height of LH peak* (ng/ml plasma)	*Duration of LH peak* (hours)	*Area under LH peak* (mm^2)	*Area as % of natural peak at oestrus*
150 μg GnRH alone	98.2±11.8^a	4.40±0.12^a	1318±139^a	25.5
50 μg ODB + 150 μg GnRH	178.0±48.1^b	6.63±0.16^b	3523±871^b	68.1
Natural oestrus	151.0±22.8^b	8.30±0.54^c	5172±373^c	100

^{a,b,c}Different superscripts within column indicate that the means differ significantly ($P<0.05$).
From Haresign and Lamming (1978)

period of exposure to progesterone (Jenkin, Heap and Symons, 1977). In addition, the response of the anterior pituitary to repeated exposure to GnRH differs, depending on the time interval between injections and the dose level of GnRH employed. Relatively high doses of GnRH result in a sensitization of the pituitary to further stimulation by GnRH if the interval between two successive injections is less than 3 hours, but thereafter the pituitary becomes refractory to further stimulation for a period of 72–96 hours (Crighton and Foster, 1976, 1977; Symons, Cunningham and Saba, 1974). However, repeated injections of much lower doses of GnRH do not result in refractoriness of pituitary response, but an episodic release of LH follows each injection (Lincoln, 1979; McLeod, Haresign and Lamming, 1982a,b; see *Figure 18.8*).

PITUITARY GONADOTROPHIN SECRETION

The gonadotrophins, LH and FSH, are synthesized and stored as secretory granules within the basophil cells of the anterior pituitary (Hutchinson, 1979), and are later released by exocytosis in response to GnRH stimulation (Fawcett, Long and Jones, 1969). Early estimates of plasma LH concentrations reported low levels throughout the oestrous cycle, with a large surge on the day of oestrus (Geschwind and Dewey, 1968; Niswender *et al.*, 1968). Plasma concentrations were reported to rise abruptly from basal values of <10 ng/ml within 16 hours after the onset of oestrus, to reach peak values of 60–300 ng/ml within 5–6 hours, and then return to basal values again 10–12 hours later (Niswender *et al.*, 1968; Goding *et al.*, 1969; Wheatley and Radford, 1969).

Recently, the development of much more sensitive radioimmunoassay techniques, together with more frequent sampling regimes (blood sampling interval 10–15 minutes) has shown that in the ewe the natural pattern of LH secretion between successive preovulatory LH surges is episodic in nature. Basal levels of 0.1–2.0 ng/ml are interspersed with small, short-lived episodes, each of 5–15 ng/ml and of about 30–45 minutes duration. During the luteal phase of the oestrous cycle these LH episodes occur irregularly at 3–12 hour intervals (Baird, Swanston and Scaramuzzi, 1976; Hauger, Karsch and Foster, 1977; Yuthasastrakosol, Palmer and Howland, 1977; B.R. Friman and W. Haresign, unpublished observations). However, following regression of the corpus luteum, starting some two to three days prior to oestrus, the frequency of these LH episodes gradually increases until they occur at a rate of approximately one episode/1–2 hours immediately prior to the preovulatory LH surge (Yuthasastrakosol, Palmer and Howland, 1977; Baird, 1978). This change in LH episode frequency results in the increase in mean circulating LH concentrations apparent at this time (*Figure 18.2*). That this increase in mean LH concentrations is related to falling plasma progesterone concentrations can be seen from experiments involving the abrupt withdrawal of progesterone implants used either to extend the luteal phase of the oestrous cycle (Karsch *et al.*, 1978) or to synchronize ovulation (B.R. Friman and W. Haresign, unpublished data; *Figure 18.2*). Indeed, the very much more rapid reduction in plasma progesterone concentrations following implant withdrawal compared with natural luteolysis results in a very much more rapid increase in mean LH concentrations.

The patterns of plasma FSH concentrations in the ewe are much less well defined, due largely to difficulties in developing specific and sensitive radioimmunoassays for this hormone, as well as to the considerable variation in estimates of FSH potencies between laboratories. Notwithstanding these difficulties, two peaks in plasma FSH concentrations have been reported to occur, the first coincident with the preovulatory LH peak at oestrus and the second one day later (L'Hermite *et al.*, 1972; Salamonsen *et al.*, 1973; Pant, Fitzpatrick and Hopkinson, 1973; McNeilly *et al.*, 1976; Pant, Hopkinson and Fitzpatrick, 1977). However, a recent report has suggested the presence of only a single FSH peak, coincident with the preovulatory LH surge at oestrus (Goodman, Pickover and Karsch, 1981).

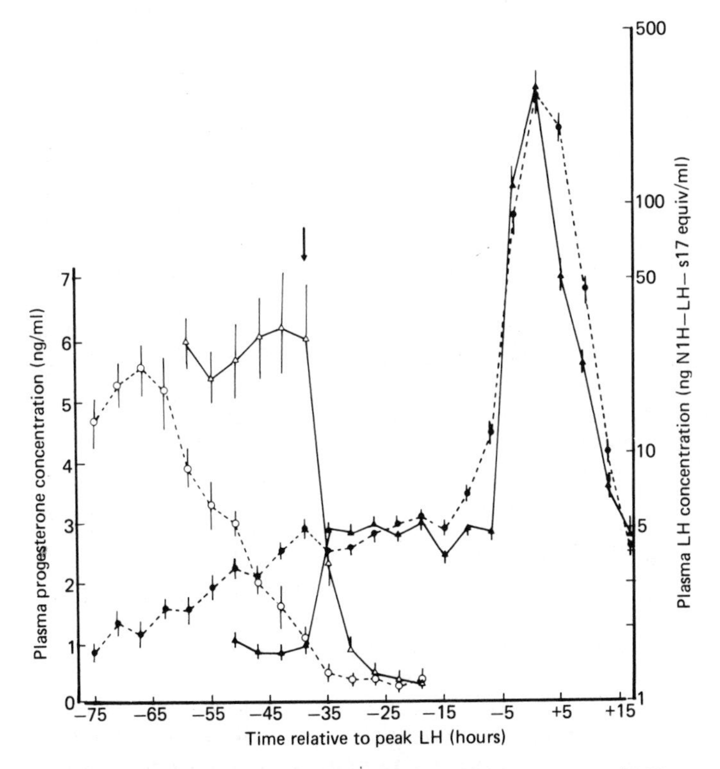

Figure 18.2 Changes in mean ±S.E.M. plasma progesterone and LH concentrations in ewes at a natural oestrus (○-- ○, progesterone; ●-- ●, LH) or following progesterone implant withdrawal (△——△, progesterone, ▲——▲, LH). Note the much more rapid decrease in progesterone and increase in LH concentrations after progesterone implant withdrawal (↓). From B.R. Friman and W. Haresign (unpublished observations)

Throughout the remainder of the cycle most groups report that FSH concentrations are characterized by short troughs of low concentrations followed by extended periods of variable duration, when FSH levels are elevated (Salamonsen *et al.*, 1973; Pant, Hopkinson and Fitzpatrick, 1977). Goodman, Pickover and Karsch (1981), by contrast, have reported that FSH concentrations remain slightly elevated for 1–2 days after the preovulatory peak at oestrus, but then decline to a mid-luteal phase nadir, remaining there until the preovulatory FSH surge at the next oestrus.

THE ROLE OF GONADOTROPHIC HORMONES IN OVULATION AND FOLLICLE GROWTH

Ovulation has been shown to occur at a constant time interval of 21–26 hours after the preovulatory gonadotrophin peaks at oestrus (Cumming *et al.*, 1971; 1973) thereby implicating these surges in the process of ovulation. In addition, the administration of a dose of GnRH which is sufficient to induce an immediate preovulatory-type surge of both LH and FSH will result in ovulation in seasonally anoestrous ewes (Foster and Crighton, 1974; Haresign *et al.*, 1976), lactating ewes (Restall and Radford, 1974) and during the luteal phase of the oestrous cycle (Foster, 1974; Haresign, 1976). Although these data establish a clear relationship between the preovulatory release patterns of the two gonadotrophic hormones and ovulation, they do not determine which of them is implicated. However, when the natural gonadotrophin surges in cyclic ewes at oestrus were blocked with chlorpromazine, infusion of purified LH alone was sufficient to induce ovulation. Moreover, infusion of purified FSH alone was ineffective, and there was no apparent benefit from infusing FSH along with LH, suggesting that LH alone is responsible for the induction of ovulation (Robertson, Rakha and Buttle, 1966). In spite of its clear association with the process of ovulation, there does not appear to be any relationship between the magnitude of the preovulatory LH surge and ovulation rate in the ewe (Thimonier and Pelletier, 1971; Land *et al.*, 1973; Quirke, Hanrahan and Gosling, 1979, 1981; Haresign, 1981a).

The classical view of the control of follicle growth assumed that FSH was the hormone principally involved. Early studies on the growth of ovarian follicles in the ewe indicated a steady growth throughout the oestrous cycle, culminating in the final preovulatory enlargement just prior to ovulation (Kammlade *et al.*, 1952; Santolucito, Clegg and Cole, 1960). However, later studies suggested that marked growth of one or two follicles occurred between the onset of oestrus and day 5 of the cycle (oestrus = day 0), with little or no further increase in volume until the final few hours prior to ovulation (Robertson and Hutchinson, 1962; Hutchinson and Robertson, 1966). This view of follicle growth was, however, inconsistent with observations that unilateral ovariectomy as late as day 14 of the cycle allowed full compensation of ovulation rate from the remaining ovary (Findlay and Cumming, 1977) and that pregnant mare's serum gonadotrophin (PMSG) induced the greatest superovulatory response when given between days 12–14 of the oestrous cycle (Cumming and McDonald, 1967). Clarification of this dilemma has been provided by recent data which demonstrate that only those large follicles present on the ovaries some three days prior to ovulation actually proceed through to ovulation (Smeaton and Robertson, 1971; Dufour *et al.*, 1972; Bherer, Matton and Dufour, 1977). This has led to the suggestion that there are at least three phases of follicle growth in the ewe during the oestrous cycle, with only those large follicles resulting from the final growth phase undergoing full development and ovulation, whereas large follicles resulting from the earlier growth phases become atretic before they can undergo the full maturational changes needed for ovulation.

However, attempts to correlate these phases of follicle growth with changes in FSH secretion during the oestrous cycle have proved unsuccessful. Indeed, FSH concentrations during the 2–3 days prior to ovulation do if anything decrease, rather than increase. However, the involvement of FSH in the process of follicle growth is supported by the fact that growing follicles show a proliferation of FSH receptors, whereas atretic follicles are characterized by a loss of these receptors (Carson, Findlay and Burger, 1979). The answer to this problem is perhaps suggested by data from the rat. In this species it has been shown that the early stages of follicle growth are FSH dependent while the final growth and maturation of the follicle depends to a progressively greater extent on LH secretion (Richards, Rao and Ireland, 1978). Support for such a system operating in the ewe can be gleaned from a number of experiments. Firstly, administration of exogenous LH to the culture medium, rather than FSH, is responsible for the production of oestrogen from sheep follicles maintained *in vitro* (Hay and Moor, 1973; McIntosh and Moor, 1973; Weiss *et al.*, 1976). Secondly, there is a parallel increase in both episodic LH secretion and oestradiol levels during the follicular phase of the oestrous cycle when the final phase of follicle growth and maturation is known to occur (Baird, 1978; Karsch *et al.*, 1978). Furthermore, episodic LH secretion during the luteal phase of the oestrous cycle and during seasonal anoestrus is immediately followed by an increase in oestradiol secretion rate (Baird, Swanston and Scaramuzzi, 1976; Scaramuzzi and Baird, 1979). Circumstantial evidence, therefore, points to the possibility that, as in the rat, only the early phases of follicle growth are FSH dependent, with episodic LH secretion becoming important in the later maturational changes, and this is supported by data given on p.368 of this chapter (McLeod, Haresign and Lamming, 1982). The precise role of the FSH peaks at oestrus, therefore, remain unclear although they may be involved in inducing the early stages of growth and development of those follicles destined to ovulate at the next oestrus.

STEROID HORMONE LEVELS THROUGHOUT THE OESTROUS CYCLE

The two major steroids secreted by the sheep ovary during the oestrous cycle are progesterone and oestradiol-17β. In addition, there is also secretion of the androgens, principally testosterone and androsteredione, although it is unclear whether these two steroids have any direct physiological role other than as intermediates in the biosynthesis of oestrogens (Peters and McNatty, 1980).

Plasma progesterone concentrations are basal (<0.5 ng/ml) at oestrus and for the first 3–4 days of thecycle, but thereafter increase to reach peak concentrations of 1–7 ng/ml by day 9 or 10. These high concentrations are then maintained until about day 14 or 15 of the cycle when they decline rapidly, to reach basal values 24–48 hours prior to the onset of the next oestrus (Thorburn, Basset and Smith, 1969; McNatty, Revfeim and Young, 1973; Pant, Hopkinson and Fitzpatrick, 1977). Synthesis of progesterone is largely confined to the luteinized granulosa cells after ovulation, although thecal tissue may produce some progesterone, particularly during the early stages of the oestrous cycle (Hay and Moor, 1975b, 1978). In the pregnant ewe, plasma progesterone concentrations do not fall on day 14–15 of the cycle, but are maintained at luteal phase levels until 60–80 days of pregnancy, when they increase steadily as a result of placental synthesis until just prior to parturition (Basset *et al.*, 1969;

Challis, Harrison and Heap, 1971). In spite of the fact that the corpus luteum is the major source of progesterone during the cycle and early pregnancy, there is little relationship between ovulation rate or foetal numbers and progesterone concentrations, due primarily to the fact that increments in ovulation rate are associated with disproportionately small increases in plasma progesterone concentrations (Quirke and Gosling, 1975; Quirke, Hanrahan and Gosling, 1979).

The function of the corpus luteum in the ewe is under the dual control of both trophic and lytic influences which regulate its growth and regression. The maintenance of the corpus luteum of the cycle is achieved by a luteotrophic complex involving PRL and small amounts of LH (Kaltenbach *et al.*, 1967; Denamur, Martinet and Short, 1973), although there is evidence that the corpus luteum of pregnancy may be sustained by luteotrophins produced by the placenta (Perry *et al.*, 1976).

The regression of the corpus luteum results from the secretion of prostaglandin $F_{2\alpha}$ ($PGF_{2\alpha}$) from the uterine endometrium. $PGF_{2\alpha}$ levels increase on day 14 of the oestrous cycle in the non-pregnant ewe and effect a marked decline in progesterone output (Thorburn *et al.*, 1972). However, the presence of an embryo at this time prevents luteal regression either by secreting antiluteolytic substances or by preventing the release of $PGF_{2\alpha}$ (Moor, 1968; Thorburn *et al.*, 1972). Moreover, it has been clearly demonstrated that the luteolytic stimulus is dominant to the luteotrophic one. The synthesis of $PGF_{2\alpha}$ appears to be a result of a period of progesterone dominance, while its release may be the result of an increase in oestradiol secretion (Caldwell *et al.*, 1972b).

The secretion of oestrogens is limited to the largest one or two non-atretic, developing follicles (Moor, 1973; Hay and Moor, 1975a, 1978), and its synthesis appears to require an intimate association between the granulosa and theca interna layers (Baird, 1977; Hay and Moor, 1978). Oestradiol-17β concentrations rise from basal levels of 5–10 pg/ml as progesterone concentrations decline towards the end of the oestrous cycle, reaching peak concentrations of approximately 20 pg/ml during pro-oestrus or very early oestrus, and then return to basal values within 24 hours (Pant, Hopkinson and Fitzpatrick, 1977). Other increases in plasma oestradiol concentrations have been reported to occur between days 3–4 (Scaramuzzi, Caldwell and Moor, 1970; Mattner and Braden, 1972; Baird and Scaramuzzi, 1976), days 6–9 and days 11–12 of the cycle (Scaramuzzi, Caldwell and Moor, 1970; Cox *et al.*, 1971; Mattner and Braden, 1972), but these additional peaks are of smaller magnitude than the one at pro-oestrus.

More recently it has been shown that there are small increases in the oestradiol secretion rate in response to each LH episode, both during the luteal phase of the oestrous cycle (Baird, Swanston and Scaramuzzi, 1976) and during seasonal anoestrus (Scaramuzzi and Baird, 1979; *Figure 18.3*). This pattern of secretion is very similar to that which occurs in the ram for LH and testosterone (Haynes and Schanbacher, Chapter 22).

The successive decline in progesterone concentrations followed by the rising titre of oestradiol at the end of the oestrous cycle are an essential prerequisite for the expression of oestrous behaviour in the ewe (Robinson, 1951; 1954) as well as facilitating changes in the histology of the

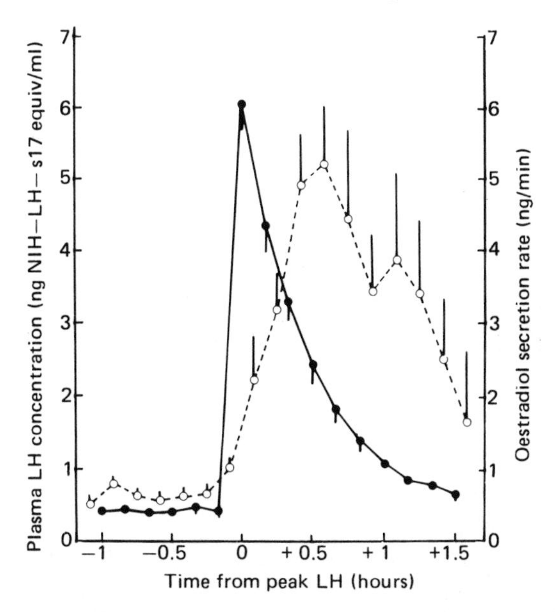

Figure 18.3 The relationship between episodic LH secretion and oestradiol secretion rate in six seasonally anoestrous ewes. The data have been grouped around the peak LH concentration of an LH episode. From Scaramuzzi and Baird (1979)

endometrium and cervix to promote sperm and ovum transport and to facilitate implantation. In addition, the continuance of luteal phase progesterone concentrations are necessary for the maintenance of pregnancy.

HORMONAL FEEDBACK MECHANISMS REGULATING THE OESTROUS CYCLE

Evidence for a negative feedback mechanism controlling pituitary gonadotrophin secretion, which involves ovarian steroids, emanates from studies involving ovariectomy. The withdrawal of ovarian steroids in this manner results in the typical post-castration rise in plasma LH and FSH concentrations (Diekman and Malvern, 1973; Karsch and Foster, 1975; Karsch *et al.*, 1978). Moreover, the inverse relationship observed during the normal oestrous cycle between progesterone and tonic LH concentrations (*Figure 18.4*) suggests that progesterone may be an important component of this feedback regulation (Karsch *et al.*, 1978). This inverse relationship is even more evident over the period immediately following withdrawal of progesterone implants (*Figure 18.2*), and has led Karsch *et al.* (1978) to postulate that progesterone is the principal negative feedback hormone, although they do acknowledge the need for other ovarian steroids. However, this hypothesis fails to take account of a number of important findings. Firstly, as shown in *Figure 18.3* there is a very close association between short-term changes in LH and oestradiol secretion, typical of a self-regulating feedback mechanism. Indeed, the sequence of hormonal changes suggests that episodic LH is responsible for driving oestrogen secretion from developing follicles. Secondly, progesterone

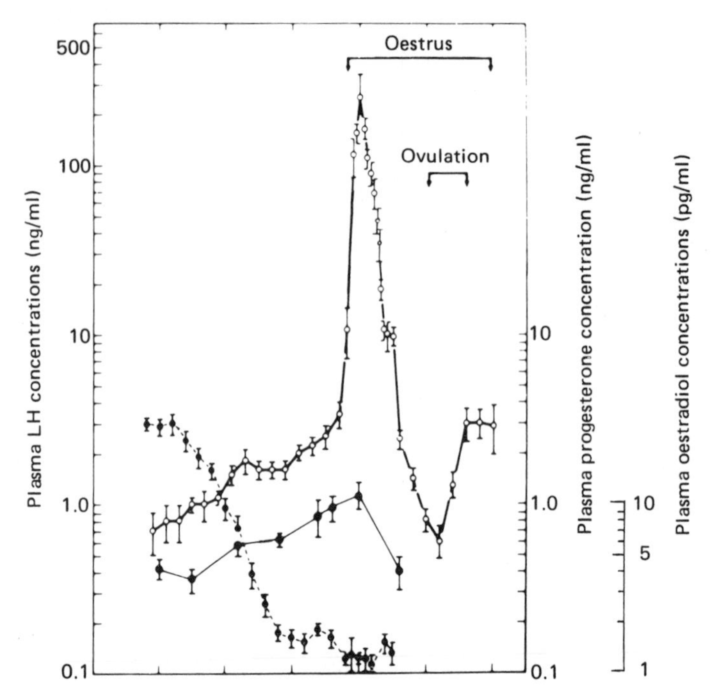

Figure 18.4 Temporal changes in LH (○——,○), oestradiol-17β (●——●) and progesterone (●---●) concentrations in peripheral plasma of ewes at a natural oestrus. The data have been normalized around the timing of the preovulatory LH peak. From Karsch *et al.* (1978)

alone will only temporarily suppress the typical post-castration rise in LH concentrations, being effective for a period of only 3–4 days (Karsch *et al.*, 1978). Finally, oestradiol alone is an effective negative feedback controller of tonic LH secretion during seasonal anoestrus (see *Figure 18.5*) and it seems unlikely that negative feedback control is mediated by one steroid during the breeding season and another during seasonal anoestrus. Rather, a more likely explanation is that oestradiol is the major negative feedback hormone, but during the breeding season it requires the presence of progesterone to potentiate its feedback activity. Since injections of very small doses of GnRH will induce LH episodes, it seems probable that the site of action of this negative feedback control is the arcuate nucleus/ median eminence area of the hypothalamus responsible for the control of episodic LH secretion (see earlier section, p.355).

The regulation of tonic FSH secretion is less well understood, and indeed cannot be fully accounted for by the negative feedback effects of the steroid hormones, oestrogen and progesterone (Goodman, Pickover and Karsch, 1981). It has been suggested that the ovary might, therefore, secrete a non-steroidal hormone which contributes to the control of FSH concentrations, in much the same way that inhibin does in the male (Main, Davies and Setchell, 1979).

Evidence suggests that the preovulatory gonadotrophin surges are controlled by the positive feedback action of rising oestrogen titres during the follicular phase of the oestrous cycle (Goding *et al.*, 1969; Radford, Wallace and Wheatley, 1970; Bclt, Kelley and Hawk, 1971; Beck and Reeves, 1973; Jonas *et al.*, 1973), although the presence of high progesterone concentrations will completely block this response (Piper and Foote, 1968; Scaramuzzi *et al.*, 1971; Hooley *et al.*, 1974). This blockade can, however, be overcome by the administration of GnRH (Foster, 1974; Haresign, 1976) indicating that the major site of the positive feedback action of oestradiol is the hypothalamus, and in particular the preoptic/suprachiasmatic area. A period of pretreatment with oestradiol will increase pituitary responsiveness to GnRH stimulation (Haresign and Lamming, 1978) and, in addition, the response to GnRH is also much higher at oestrus, when oestradiol levels are elevated (Reeves, Arimura and Schally, 1974; Foster and Crighton, 1976a,b), ensuring maximal LH release as a result of positive feedback.

The foregoing discussions allow a working hypothesis to be developed to explain the hormonal regulation of the oestrous cycle in the ewe. During the luteal phase of the cycle when progesterone levels are elevated, they potentiate the negative feedback action of oestrogen on episodic LH secretion, limiting LH episode frequency to only one/3–12 hours. This frequency is not sufficient to promote the final stages of follicle growth and maturation. However, the fall in progesterone concentrations attendant on luteal regression liberates episodic LH secretion from full negative feedback control, with the result that episode frequency increases, and this in turn promotes the final phases of follicle growth and development. The consequent increase in oestrogen secretion from these developing follicles promotes oestrus, the preovulatory LH surge and ovulation. The cellular reorganization within the ovulated follicle to produce the corpus luteum, in conjunction with the luteotrophic stimulus, results in an increase in plasma progesterone concentrations, and this effects the full negative feedback on tonic LH secretion, preventing the final stages of follicle growth and development until luteal regression occurs prior to the next oestrus.

THE POTENTIAL FOR MANIPULATION OF THE OESTROUS CYCLE

Since progesterone can be considered as the 'organizer' of the oestrous cycle, manipulation of the progesterone status of the animal provides a convenient means of controlling the cycle. Indeed, those techniques for synchronization of the oestrous cycle which have been developed revolve around the artificial shortening or lengthening of the period of progesterone dominance. The synchronous removal of progesterone by these means results in a high degree of synchrony in the timing of oestrus and ovulation, and this will be discussed in much more detail in Chapter 19 by Cognie and Mauleon.

The important finding that episodic LH secretion appears to be responsible for the final phases of follicle growth and maturation in the ewe (Baird, 1978; McLeod, Haresign and Lamming, 1982a,b) raises the interesting question of whether this component of gonadotrophin secretion may be responsible for determining ovulation rate and thereby litter size; particularly since Hay and Moor (1978) have shown that injection of PMSG late in the oestrous cycle induces its superovulation effect by preventing the late atresia of follicles. As was discussed earlier, there appears to be no correlation between ovulation rate and the size of the preovulatory LH surge. Indeed the amount of LH released at oestrus is far in excess of that required to induce ovulation (Foster, 1974; Haresign, 1976). Circumstantial evidence from the data of Haresign (1981a,b) would support an involvement of tonic LH secretion in the prevention of the late atresia of follicles and possibly, therefore, ovulation rate. However, more direct evidence is shown by the data in *Table 18.2*. The minor steroids involved

Table 18.2 THE EFFECT OF TREATING WELSH MOUNTAIN EWES WITH AN ANTISERUM TO EITHER OESTRONE (E_1), OESTRADIOL (E_2), ANDROSTENEDIONE (A) OR TESTOSTERONE (T) BETWEEN DAYS 10–12 OF THE OESTROUS CYCLE ON LH EPISODE FREQUENCY AND OVULATION RATE.

Treatment	*Mean number of LH episodes/18 hours*	*Ovulation rate*
Anti-E_1	3.61	1.50
Anti-E_2	4.51	2.29
Anti-A	3.48	1.56
Anti-T	3.40	1.56
Controls	2.00	1.00

From Pathiraja *et al.* (1980)

are all intermediates in the biosynthesis of oestradiol, and it is possible that immunization against them reduces the level of negative feedback, thereby allowing the observed increase in episodic LH secretion. The potential, therefore, exists for manipulating litter size in the ewe, either by removing part of the steroid negative feedback components to promote higher tonic LH secretion, or by using GnRH to directly stimulate increased LH production from the anterior pituitary, although further work is required before either of these techniques can be applied in the field.

The endocrinology of seasonal anoestrus

During seasonal anoestrus plasma progesterone concentrations remain basal due to the absence of preovulatory gonadotrophin surges and, therefore, ovulation (Roche *et al.*, 1970; Yuthasastrakosol, Palmer and Howland, 1973). It is pertinent, therefore, to examine the detailed hormone changes of anoestrus and their interrelationships to establish why oestrous cycles are absent.

It has been suggested that seasonal anoestrus may be due to a reduction in the positive feedback sensitivity of the hypothalamic–pituitary axis since it has been shown by some workers that there is seasonal variation in the response to exogenous oestrogen (Land, Wheeler and Carr, 1976; Land, Carr and Thompson, 1979), although others have found no such changes (Friman, Haresign and Lamming, 1979). However, the changes in sensitivity, if present, appear to be so small that they are unlikely to account for the complete absence of preovulatory gonadotrophin surges.

The ovaries of the anoestrous ewe are not totally inactive; indeed periods of early follicle growth and regression occur throughout this period of reproductive quiescence, so much so that follicles as large as those found during the luteal phase of the oestrous cycle may be present (Hutchinson and Robertson, 1966; Matton, Bherer and Dufour, 1977). Moreover, hypophysectomy during anoestrus results in the immediate regression of the ovary, indicating that some gonadotrophic support is present (Denamur and Mauleon, 1963). Although early studies suggested that pituitary and plasma LH and FSH concentrations were similar to those found during the luteal phase of the oestrous cycle (Goding *et al.*, 1969; Roche *et al.*, 1970), more recent work has shown that the frequency of episodic LH secretion is lower than that found during the luteal phase of the oestrous

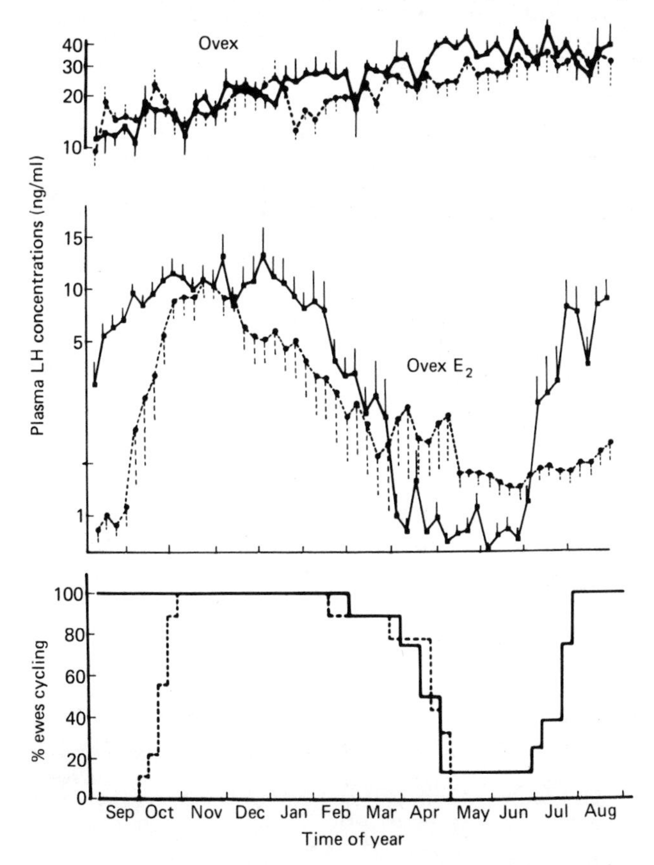

Figure 18.5 Mean ±S.E.M. LH concentrations throughout the year in ovariectomized Welsh Mountain (----) and Dorset Horn (——) ewes with either empty subcutaneous implants (OVEX, N = 2 per breed) or subcutaneous implants containing oestradiol-17β (OVEX E_2, N = 10 per breed) designed to maintain plasma oestradiol concentrations at levels similar to those found during the luteal phase of the oestrous cycle. The low part of the figure shows cyclical activity in 10 entire ewes of each breed maintained under the same conditions of natural daylength and temperature. Note the seasonal changes in LH concentrations in OVEX E_2 ewes but not OVEX ewes, and that the timing of these changes coincided with the changes in cyclical activity of entire ewes of each breed but at different times of the year in the two breeds. From Webster and Haresign (1983)

cycle, particularly in those breeds with a deep seasonal anoestrus (Scaramuzzi and Martenz, 1975; Scaramuzzi and Baird, 1979; see also *Figure 18.7*). Similar changes in episodic LH secretion with season have also been shown to occur in the ram (Lincoln, 1979; Haynes and Schanbacher, Chapter 22).

The data presented in *Figures 18.5* and *18.7* indicate that the seasonal change in LH episode frequency is attributable to a change in the negative feedback responses of the hypothalamic–pituitary axis to the same circulating levels of oestradiol as those present during the luteal phase of the oestrous cycle. Of particular interest is the change in mean circulating LH concentrations in ovariectomized, oestrogen-treated ewes at the same time as the changes in cyclical activity of entire ewes of the same breed, even though the length of the breeding season was very different for the two breeds. Similar results have also been reported by Karsch *et al.* (1978), and collectively they support the hypothesis that seasonal anoestrus represents a period of active inhibition of cyclical activity, rather than a period of passive inactivity. Goodman and Karsch (1981) have also shown a seasonal change in LH episode frequency in untreated ovariectomized ewes, indicating that there may also be a reduction in 'positive drive' of the hypothalamic–pituitary axis during seasonal anoestrus. However, these

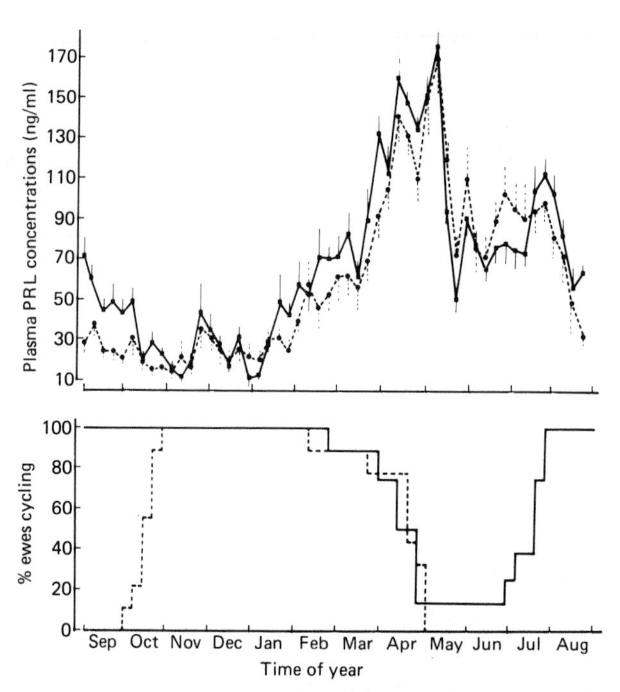

Figure 18.6 Mean ±S.E.M. plasma PRL concentrations throughout the year in the same ovariectomized Welsh Mountain (----; $N = 10$) or Dorset Horn (——; $N = 10$) ewes with subcutaneous oestradiol implants as those represented in *Figure 18.5*. Also shown is the cyclical activity of 10 entire ewes of each breed. Note that though there is a seasonal change in PRL concentrations, this occurred at the same time of year in both breeds: unlike the changes in LH concentrations shown in *Figure 18.5*

changes are small compared with the oestrogen-modulated seasonal change in LH episode frequency.

Since progesterone appears to potentiate the negative feedback action of oestradiol during the breeding season it is of interest to consider whether some other hormone which shows a seasonal pattern of change might facilitate the negative feedback of oestradiol during seasonal anoestrus. PRL concentrations had previously been shown to be positively correlated with daylength (Lamming, Moseley and McNeilly, 1974; Walton *et al.*, 1977), and to have an antigonadotrophic effect in the human female (Thorner, 1978). Although the data in *Figures 18.5* and *18.6* indicate an inverse relationship between mean circulating levels of LH and PRL, it is noticeable that PRL concentrations of both breeds showed an identical temporal pattern of change whereas changes in LH occurred at very different times of the year. However, the inverse relationship between LH and PRL was much more evident in profiles from individual ewes (*Figure 18.7*).

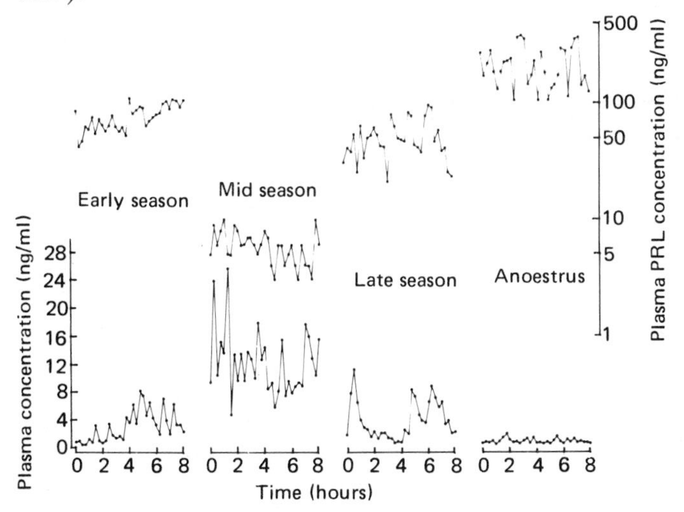

Figure 18.7 Plasma LH and PRL concentrations in an individual ovariectomized ewe with an oestradiol implant designed to maintain circulating oestradiol concentrations at levels similar to those found during the luteal phase of the oestrous cycle. Blood samples were collected at 15 minute intervals at times corresponding to stages of breeding activity listed for entire ewes of the same breed. From G.M. Webster and W. Haresign (unpublished data)

It has been suggested that melatonin secretion from the pineal gland may be involved in the transfer of information concerning photoperiod to the hypothalamic–pituitary–gonadal axis in both the ewe (Kennaway, Gilmore and Seamark, 1981; 1982) and the ram (Lincoln, 1979), although its precise mode of action remains unclear. Of particular interest in these data is the observation that melatonin feeding of seasonally anoestrous ewes resulted in a marked reduction in PRL concentrations, and this was then followed by an earlier onset of the breeding season. Whether these, and the data previously described, represent an antigonadotrophic effect of PRL in the ewe or are merely coincidental changes requires further clarification.

Since an increase in episodic LH secretion appears to be important in stimulating the final phases of follicle growth and maturation and the rise in oestradiol concentrations during the follicular phase of the oestrous cycle, it seems likely that the lack of ovulation during seasonal anoestrus results from an inadequate pattern of episodic LH secretion (see *Figure 18.7*). Because follicle development is not stimulated, there is no rise in oestradiol secretion to trigger the preovulatory gonadotrophin surge and induce ovulation.

THE POTENTIAL FOR INDUCING OUT-OF-SEASON BREEDING IN THE EWE

Attempts to induce out-of-season breeding in the ewe by directly stimulating a preovulatory-type gonadotrophin surge using either oestradiol (Haresign, 1976) or synthetic GnRH (Haresign *et al.*, 1976) have been unsuccessful. Although such treatments resulted in ovulation, this was neither accompanied by oestrous behaviour nor followed by normal luteal function, and invariably produced only a single ovulation. The failure of these treatments appears to be in the lack of adequate follicle maturation prior to the induction of ovulation (Haresign and Lamming, 1978).

The injection of PMSG following a period of progesterone priming will induce a fertile oestrus in seasonally anoestrous ewes, although this is associated with highly variable ovulation rates. Moreover, the confounding inhibitory influences of seasonal anoestrus, the postpartum period and lactation combine to result in very poor conception rates in all but a few select breeds (see Haresign, 1978). Such factors are of paramount importance when attempts are being made to increase lambing frequency.

Since seasonal anoestrus appears to be associated with an inadequate (reduced) pattern of episodic LH secretion, the possibility exists that manipulation of this component might induce follicle development and ovulation. Using a dose level of GnRH designed to induce LH episodes (250 ng GnRH/injection), rather than a preovulatory-type LH surge (150 μg GnRH/injection), it has been shown that two-hourly injections of GnRH increased LH episode frequency from a pretreatment value of one/8–12 hours up to a figure of one/2 hours (McLeod, Haresign and Lamming, 1982a,b; *Figure 18.8*) and this was eventually followed by a

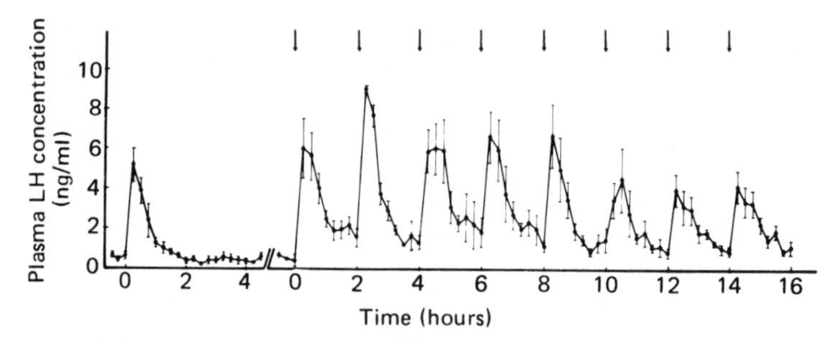

Figure 18.8 Mean ±S.E.M. plasma LH concentrations in five progesterone-primed seasonally anoestrous ewes before and during the first 16 hours of treatment with two-hourly intravenous injections (↓) of 250 ng GnRH. The LH data from the pretreatment period have been normalized about a natural LH episode. Progesterone implants were removed at the time of the second GnRH injection. From McLeod, Haresign and Lamming (1982b)

Table 18.3 RESPONSE OF SEASONALLY ANOESTROUS CLUN FOREST EWES TO INJECTION OF 250 ng GnRH EVERY TWO HOURS FOR 48 HOURS AFTER A 12-DAY PRETREATMENT PERIOD WITH PROGESTERONE IMPLANTS.

Number of ewes treated	15
Number in oestrus	15
Time of oestrus after first injection	36.3±2.7 hours
Number ovulating	15
Ovulation rate	1.67±0.12
Number with normal luteal function	15
Number conceiving and lambing	11
Litter size	1.45±0.16

B.J. McLeod and W. Haresign (unpublished data)

preovulatory LH surge. Since the ewes primed with progesterone came into oestrus, it is likely that the LH surge was due to the positive feedback action of oestradiol from developing follicles, rather than as a direct result of GnRH treatment. The ovulation rate and fertility of ewes treated in this way paralleled that of this breed mated naturally during the breeding season and are given in *Table 18.3*. Although GnRH will release FSH as well as LH, it is likely that the responses observed were due to LH alone, since Lincoln and Short (1980) have shown that doses of similar magnitude do not effect any change in FSH secretion in the ram until some four days after the start of treatment, whereas the preovulatory surge in these ewes occurred within 48 hours of the start of treatment. In support of this suggestion it has been shown that the induction of ovulation following the introduction of the ram to seasonally anoestrous ewes is also followed by an immediate increase in episodic LH secretion prior to the preovulatory LH surge (Oldham and Lindsay, 1980; Martin, Oldham and Lindsay, 1980).

Although these data clearly support the hypothesis that seasonal anoestrus is due to an inadequate pattern of episodic LH secretion, they do not determine whether it is the episodic pattern *per se* or the effective increase in mean circulating LH levels which is important. Indeed, it was suggested in an earlier section (p.361) that an episodic pattern may reflect nothing other than a component of a self-regulating endocrine mechanism. In an attempt to clarify this problem, and also to find a more practical mode of GnRH administration, progesterone-primed, seasonally anoestrous ewes were infused intravenously with either 125 or 250 ng GnRH/hour, continuously for 48 hours. This produced a sustained elevation in plasma LH concentrations (*Figure 18.9*), and eventually a preovulatory LH surge and ovulation, the LH response to the higher dose level (250 ng GnRH/hour) being significantly greater ($P<0.05$) than that to the lower level (125 ng GnRH/hour). All ewes came into oestrus at a mean time of 37.0±1.2 hours after the start of infusion, and produced viable corpora lutea. There was also evidence of a marked effect of dose level on ovulation rate, ewes infused at the 125 ng/hour dose level having a mean ovulation rate of 1.27±0.12, while those infused at a rate of 250 ng/hour had a mean ovulation rate of 1.75±0.21. This supports the suggestion made in an earlier section that tonic LH secretion may be an important determinant of ovulation rate.

Further work is required to investigate whether similar results can be achieved with slow-release, subcutaneous GnRH implants, a more convenient route of administration for practical application.

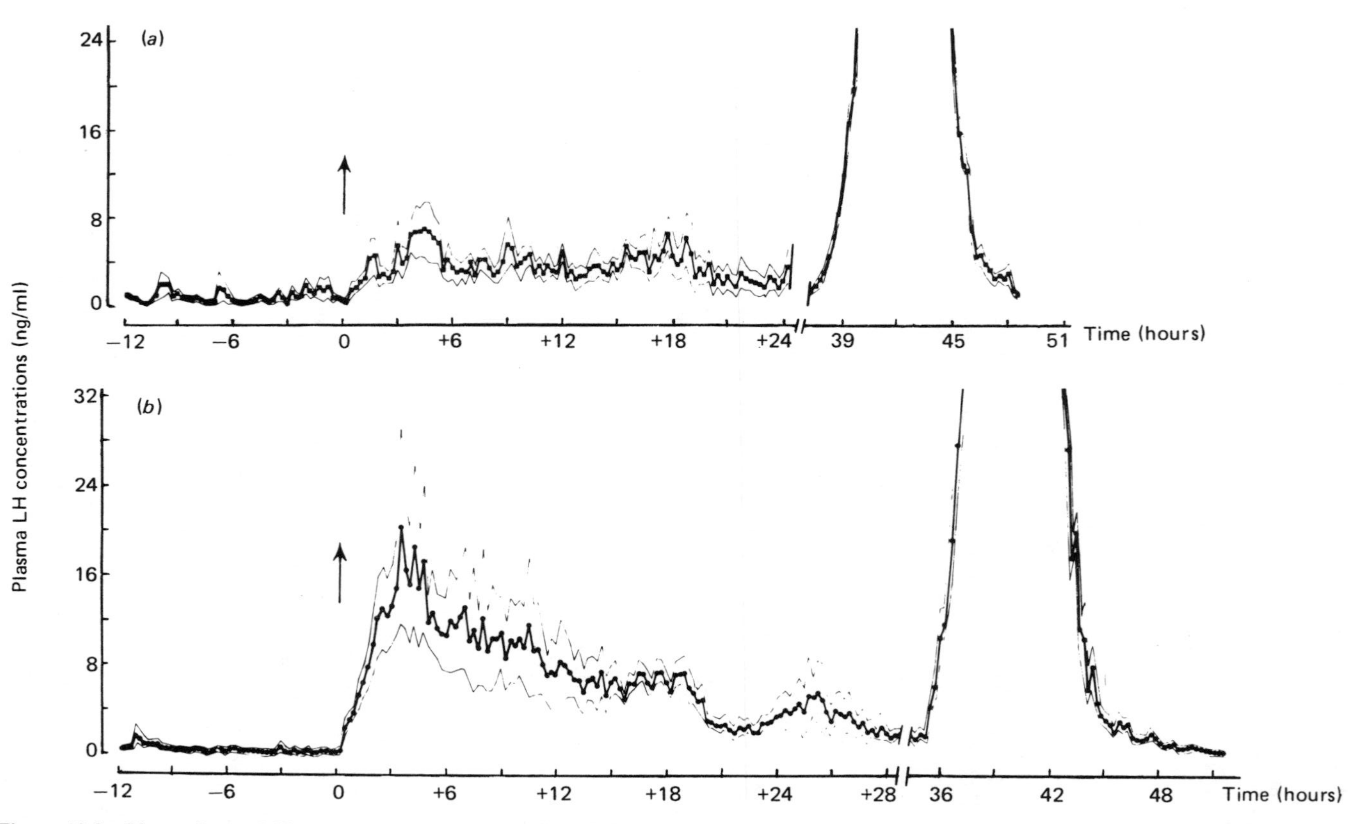

Figure 18.9 Mean plasma LH concentrations of two groups of five progesterone-primed seasonally anoestrous ewes infused intravenously with either (a) 125 ng/hour or (b) 250 ng/hour GnRH for 48 hours following removal of progesterone implants (arrows). The solid line indicates the group mean and the S.E.M. is indicated by the area bounded by the fine lines. From McLeod and Haresign (unpublished data)

Conclusions

It is clear from the foregoing sections that considerable advances have, and still are, being made in our understanding of the hormonal control of the oestrous cycle and the underlying causes of seasonal anoestrus. It is hoped that this will facilitate the development of more viable treatment regimes to manipulate reproductive performance in ewes than those methods available in the past, which were based largely on empirical results.

References

BAIRD, D.T. (1977). Evidence *in vivo* for the two cell hypothesis of oestrogen synthesis by the sheep Graafian follicle. *Journal of Reproduction and Fertility* **50**, 183–185

BAIRD, D.T. (1978). Pulsatile secretion of LH and ovarian estradiol during the follicular phase of the sheep estrous cycle. *Biology of Reproduction* **18**, 359–364

BAIRD, D.T. and SCARAMUZZI, R.J. (1976). Changes in the secretion of ovarian steroids and pituitary luteinizing hormone in the peri-ovulatory period in the ewe: The effect of progesterone. *Journal of Endocrinology* **70**, 237–245

BAIRD, D.T., SWANSTON, I. and SCARAMUZZI, R.J. (1976). Pulsatile release of LH and secretion of ovarian steroids in sheep during the luteal phase of the estrous cycle. *Endocrinology* **98**, 1490–1496

BASSET, J.M., OXBORROW, T.J., SMITH, I.D. and THORBURN, G.D. (1969). The concentration of progesterone in the peripheral plasma of the pregnant ewe. *Journal of Endocrinology* **45**, 449–457

BECK, T.W. and REEVES, J.J. (1973). Serum luteinizing hormone (LH) levels in ewes treated with various dosages of 17β-estradiol at three stages of the anestrous season. *Journal of Animal Science* **36**, 566–570

BHERER, J., MATTON, P. and DUFOUR, J. (1977). Fate of the two largest follicles in the ewe after injection of gonadotrophins at two stages of estrous cycle. *Proceedings of the Society of Experimental Biology and Medicine* **154**, 412–414

BOLT, D.J., KELLY, H.E. and HAWK, H.W. (1971). Release of LH by estradiol in cycling ewes. *Biology of Reproduction* **4**, 35–40

CALDWELL, B.V., AULETTE, F.J., GORDON, J.W. and SPEROFF, L. (1972a). Further studies on the role of prostaglandins in reproductive physiology. *Prostaglandins in Fertility Control* **2**, 217–234

CALDWELL, B.V., TILLSON, S.A., BROCK, W.A. and SPEROFF, L. (1972b). The effects of exogenous progesterone and estradiol on prostaglandin F levels in ovariectomised ewes. *Prostaglandins* **1**, 217–228

CARSON, R.S., FINDLAY, J.K. and BURGER, H. (1979). Receptors for gonadotrophins in the ovine follicle during growth and atresia. In *Ovarian Follicular and Corpus Luteum Function* (C.P. Channing, J.M. Marsh and W.A. Sadler, Eds.), pp. 89–94. New York, Plenum

CHALLIS, J.R.G., HARRISON, F.A. and HEAP, R.B. (1971). Uterine production of oestrogens and progesterone at parturition in the sheep. *Journal of Reproduction and Fertility* **25**, 306–307

CLARKE, I.J., FRASER, H.M. and McNEILLY, A.S. (1978). Active immunization of ewes against LH-RH, and its effect on ovulation and gonadotrophin, prolactin and ovarian steroid secretion. *Journal of Endocrinology* **78**, 39–47

COX, R.I., MATTNER, P.E., SHUTT, D.A. and THORBURN, G.D. (1971). Ovarian secretion of oestradiol during the oestrous cycle in the ewe. *Journal of Reproduction and Fertility* **24**, 133–134

CRIGHTON, D.B. and FOSTER, J.P.(1972). The effects of a synthetic preparation of gonadotrophin releasing factor on gonadotrophin release from the ovine pituitary *in vivo* and *in vitro*. *Journal of Endocrinology* **55**, xxiii–xxiv

CRIGHTON, D.B. and FOSTER, J.P. (1976). Effects of duplicate injections of synthetic luteinizing hormone releasing hormone at various intervals on luteinizing hormone release in the anoestrous ewe. *Journal of Endocrinology* **29**, 36P–37P

CRIGHTON, D.B. and FOSTER, J.P. (1977). Luteinizing hormone release after two injections of luteinizing hormone releasing hormone in the ewe. *Journal of Endocrinology* **72**, 59–67

CRIGHTON, D.B., SCHNEIDER, H.P.G. and McCANN, S.M. (1970). Localization of LH-releasing factor in the hypothalamus and neurohypophysis as determined by an *in vitro* method. *Endocrinology* **87**, 323–329

CUMMING, I.A. and McDONALD, M.F. (1967). The production of ova by New Zealand Romney ewes following hormonal stimulation. *New Zealand Journal of Agricultural Research* **10**, 226–236

CUMMING, I.A., BROWN, J.M., DE-BLOCKEY, M.A., WINFIELD, C.G., BAXTER, R.W. and GODING, J.R. (1971). Constancy of interval between luteinizing hormone release and ovulation in the ewe. *Journal of Reproduction and Fertility* **24**, 134–135

CUMMING, I.A., BUCKMASTER, J.M., DE-BLOCKEY, M.A., GODING, J.R., WINFIELD, C.G. and BAXTER, R.W. (1973). Constancy of interval between luteinizing hormone release and ovulation in the ewe. *Biology of Reproduction* **9**, 24–29

DANIEL, P.M. and PRICHARD, M.M.L. (1957). The vascular arrangements of the pituitary gland of the sheep. *Quarterly Journal of Experimental Physiology* **42**, 237–248

DEBELJUK, L., ARIMURA, A. and SCHALLY, A.V. (1972). Studies on the pituitary responsiveness to luteinizing hormone-releasing hormone (LH-RH) in intact male rats of different ages. *Endocrinology* **90**, 585–588

DENAMUR, R. and MAULEON, P. (1963). Effects de l'hypophysectomie sur la morphologie et l'histologie du corps jaune des ovins. *Comptes Rendus de l'Academie des Sciences, Paris, Series D* **257**, 264–267

DENAMUR, R., MARTINET, J. and SHORT, R.V. (1973). Pituitary control of the ovine corpus luteum. *Journal of Reproduction and Fertility* **32**, 207–220

DIEKMAN, M.A. and MALVERN, P.V. (1973). Effect of ovariectomy and estradiol on LH patterns in ewes. *Journal of Animal Science* **37**, 562–567

DUFOUR, J., WHITMORE, H.L., GINTHER, O.J. and CASIDA, L.E. (1972). Identification of the ovulating follicle by its size on different days of the estrous cycle in heifers. *Journal of Animal Science* **34**, 85–87

ECKSTEIN, P. and ZUCKERMAN, S. (1956). The oestrous cycle in the mammalia. In *Marshall's Physiology of Reproduction* (A.S. Parkes, Ed.), Vol. 1, pp. 245–249. London, Longmans, Green and Co.

ELLENDORFF, F. (1978). Extra-hypothalamic centres involved in the control of ovulation. In *Control of Ovulation* (D.B. Crighton, N.B. Haynes, G.R. Foxcroft and G.E. Lamming, Eds.), pp. 7–19. London, Butterworths

FAWCETT, D.W., LONG, J.A. and JONES, A.L. (1969). The ultrastructure of endocrine glands. *Recent Progress in Hormone Research* **25**, 315–368

FINDLAY, J.K. and CUMMING, I.A. (1977). The effect of unilateral ovariectomy on plasma gonadotrophin levels, estrus and ovulation rate in sheep. *Biology of Reproduction* **17**, 178–183

FOSTER, J.P. (1974). Luteinizing hormone release and ovulation in the ewe. PhD Thesis. University of Nottingham

FOSTER, J.P. and CRIGHTON, D.B. (1974). Luteinizing hormone (LH) release after single injections of a synthetic LH-releasing hormone (LH-RH) in the ewe at three different reproductive stages and comparison with natural LH release at oestrus. *Theriogenology* **2**, 87–100

FOSTER, J.P. and CRIGHTON, D.B. (1976a). Luteinizing hormone release after injection of synthetic luteinizing hormone releasing hormone at various stages of the oestrous cycle in the sheep. *Journal of Endocrinology* **71**, 41P–42P

FOSTER, J.P. and CRIGHTON, D.B. (1976b). Pituitary responsiveness to a single injection of synthetic luteinizing hormone-releasing hormone before and after the natural preovulatory plasma luteinizing hormone peak in sheep. *Journal of Endocrinology* **71**, 269–270

FRIMAN, B.R., HARESIGN, W. and LAMMING, G.E. (1979). The effect of oestradiol on LH release in the ewe. *Journal of Reproduction and Fertility* **57**, 537

GESHWIND, I.I. and DEWEY, R. (1968). Dynamics of luteinizing hormone secretion in the cycling ewe: a radioimmunoassay study. *Proceedings of the Society for Experimental Biology and Medicine* **129**, 451–455

GODING, J.R., CATT, K.J., BROWN, J.M., KALTENBACH, C.C., CUMMING, I.A. and MOLE, B.J. (1969). Radioimmunoassay for ovine luteinizing hormone. Secretion of luteinizing hormone during estrus and following estrogen administration in the sheep. *Endocrinology* **85**, 133–142

GOODMAN, R.L. and KARSCH, F.J. (1981). A critique of the evidence on the importance of steroid feedback to seasonal changes in gonadotrophin secretion. *Journal of Reproduction and Fertility, Supplement* **30**, 1–13

GOODMAN, R.L., PICKOVER, S.M. and KARSCH, F.J. (1981). Ovarian feedback control of follicle-stimulating hormone in the ewe: Evidence for selective suppression. *Endocrinology* **108**, 772–777

HARESIGN, W. (1976). Control of ovarian function in the ewe. PhD Thesis. University of Nottingham

HARESIGN, W. (1978). Ovulation control in the sheep. In *Control of Ovulation* (D.B. Crighton, N.B. Haynes, G.R. Foxcroft and G.E. Lamming, Eds.), pp. 435–451. London, Butterworths

HARESIGN, W. (1981a). The influence of nutrition on reproduction in the ewe. 1. Effects on ovulation rate, follicle development and luteinizing hormone release. *Animal Production* **32**, 197–202

HARESIGN, W. (1981b). The influence of nutrition on reproduction in the ewe. 2. Effects of undernutrition on pituitary responsiveness to luteinizing hormone-releasing hormone stimulation. *Animal Production* **32**, 257–260

HARESIGN, W. and LAMMING, G.E. (1978). Comparison of LH release and luteal function in cyclic and LH-RH treated anoestrous ewes pretreated with PMSG or oestrogen. *Journal of Reproduction and Fertility* **52**, 349–353

HARESIGN, W., FOSTER, J.P., HAYNES, N.B., CRIGHTON, D.B. and LAMMING, G.E. (1976). Progesterone levels following treatment of seasonally anoestrous ewes with synthetic LH-releasing hormone. *Journal of Reproduction and Fertility* **43**, 269–279

HAUGER, R.L., KARSCH, F.J. and FOSTER, D.L. (1977). A new concept for control of the estrous cycle in the ewe based on the temporal relationships between luteinizing hormone, estradiol and progesterone in peripheral serum and evidence that progesterone inhibits tonic LH secretion. *Endocrinology* **101**, 807–817

HAY, M.F. and MOOR, R.M. (1973). The Graafian follicle of the sheep: Relationships between gonadotrophins, steroid production, morphology and oocyte maturation. *Annales de Biologie Animale Biochemie Biophysique, Suppl.* **13**, 241–247

HAY, M.F. and MOOR, R.M. (1975a). Functional and structural relationships in the Graafian follicle population of the sheep ovary. *Journal of Reproduction and Fertility* **45**, 583–593

HAY, M.F. and MOOR, R.M. (1975b). Distribution of Δ^5-3β-hydroxysteroid dehydrogenase activity in the Graafian follicle of the sheep. *Journal of Reproduction and Fertility* **43**, 313–322

HAY, M.F. and MOOR, R.M. (1978). Changes in the Graafian follicle population during the follicular phase of the oestrous cycle. In *Control of Ovulation* (D.B. Crighton, N.B. Haynes, G.R. Foxcroft and G.E. Lamming, Eds.), pp. 177–196. London, Butterworths

HOOLEY, R.D., BAXTER, R.W., CHAMLEY, W.A., CUMMING, I.A., JONAS, H.A. and FINDLAY, J.K. (1974). FSH and LH response to gonadotrophin releasing hormone during the ovine estrous cycle and following progesterone administration. *Endocrinology* **95**, 937–942

HUTCHINSON, J.S.M. (1979). Gonadotrophic hormones. In *The Hypothalamo–Pituitary Control of the Ovary* (D.F. Horrobin, Ed.), pp. 4–5. Quebec, Eden Press

HUTCHINSON, J.S.M. and ROBERTSON, H.A. (1966). The growth of the follicle and corpus luteum in the ovary. *Research in Veterinary Science* **7**, 17–24

JACKSON, G.L., KUEHL, D., McDOWELL, K. and ZALESKI, A. (1978). Effects of hypothalamic deafferentation on the secretion of luteinizing hormone in the ewe. *Biology of Reproduction* **18**, 808–819

JEFFCOATE, I.A., FOSTER, J.P. and CRIGHTON, D.B. (1978). Effect of active immunization of ewes against synthetic luteinizing hormone releasing hormone. *Theriogenology* **10**, 323–335

JENKIN, G., HEAP, R.B. and SYMONS, D.B.A. (1977). Pituitary responsiveness to synthetic LH-RH and pituitary LH content at various reproductive stages in the sheep. *Journal of Reproduction and Fertility* **49**, 207–214

JONAS, H.A., SALAMONSEN, L.A., BURGER, H.G., CHAMLEY, W.A., CUMMING, I.A., FINDLAY, J.K. and GODING, J.R. (1973). Release of FSH after administration of gonadotrophin releasing hormone or estradiol to the anestrous ewe. *Endocrinology* **92**, 862–865

KALTENBACH, C.C., COOK, B., NISWENDER, G.D. and NALBANDOV, A.V. (1967). Effects of pituitary hormones on progesterone synthesis in ovine luteal tissue *in vitro*. *Endocrinology* **81**, 1407–1409

KAMMLADE, W.G., WELCH, J.A., NALBANDOV, A.V. and NORTON, H.W. (1952). Pituitary activity of sheep in relation to the breeding season. *Journal of Animal Science* **11**, 646–655

KARSCH, F.J. and FOSTER, D.L. (1975). Sexual differentiation of the mechanisation controlling the preovulatory discharge of luteinizing hormone in sheep. *Endocrinology* **97**, 373–379

KARSCH, F.J., LEGAN, S.J., RYAN, K.D. and FOSTER, D.L. (1978). The feedback effects of ovarian steroids on gonadotrophin secretion. In *Control of Ovulation* (D.B. Crighton, N.B. Haynes, G.R. Foxcroft and G.E. Lamming, Eds.), pp. 29–48. London, Butterworths

KELLY, R.W., ALLISON, A.J. and SHACKELL, G.H. (1976). Seasonal variation in oestrus and ovarian activity in five breeds of ewe in Otago. *New Zealand Journal of Experimental Agriculture*. **4**, 209–214

KENNAWAY, D.J., GILMORE, T.A. and SEAMARK, R.F. (1981). The effect of melatonin feeding on serum prolactin, gonadotrophins and the onset of estrous activity in seasonally anestrous ewes. *Biology of Reproduction* **24**, *Supplement 1*, 109A

KENNAWAY, D.J., GILMORE, T.A. and SEAMARK, R.F. (1982). The effect of melatonin feeding on serum prolactin and gonadotrophin levels and the onset of seasonal estrous cyclicity in sheep. *Endocrinology* **110**(5), 1766–1772

LAMMING, G.E., MOSELEY, S.R. and McNEILLY, J.R. (1974). Prolactin release in the sheep. *Journal of Reproduction and Fertility* **40**, 151–168

LAND, R.B., CARR, W.R. and THOMPSON, R. (1979). Genetic and environmental variation in the LH response of ovariectomized sheep to LH-RH. *Journal of Reproduction and Fertility* **56**, 243–248

LAND, R.B., WHEELER, A.G. and CARR, W.R. (1976). Seasonal variation in the oestrogen-induced LH discharge of ovariectomized Finnish Landrace and Scottish Blackface ewes. *Annales de Biologie Animale Biochemie Biophysique* **16**, 521–528

LAND, R.B., PELLETTER, J., THIMONIER, J. and MAULEON, P. (1973). A quantitative study of genetic differences in the incidence of oestrus, ovulation and plasma LH concentration in the sheep. *Journal of Endocrinology* **58**, 305–317

L'HERMITE, M., NISWENDER, G.D., REICHERT, L.E. and MIDGLEY, A.R. (1972). Serum follicle stimulating hormone in sheep as measured by radioimmunoassay. *Biology of Reproduction* **6**, 325–332

LINCOLN, G.A. (1979). Differential control of luteinizing hormone and follicle stimulating hormone by luteinizing hormone releasing hormone in the ram. *Journal of Endocrinology* **80**, 133–140

LINCOLN, G.A. and SHORT, R.V. (1980). Seasonal breeding: nature's contraceptive. *Recent Progress in Hormone Research* **36**, 1–52

McCANN, S.M. (1974). Regulation of the secretion of follicle stimulating hormone and luteinizing hormone. In *Handbook of Physiology, Section 7* (E. Knobil and W.H. Sawyer, Eds.), pp. 489–517. Washington, American Physiological Society

McINTOSH, J.E.A. and MOOR, R.M. (1973). Regulation of steroid secretion in sheep ovarian follicles. *Journal of Reproduction and Fertility* **35**, 605–606

McLEOD, B.J., HARESIGN, W. and LAMMING, G.E. (1982a). The induction of ovulation and luteal function in seasonally anoestrous ewes treated with small dose multiple injections of GnRH. *Journal of Reproduction and Fertility* **65**, 215–221

McLEOD, B.J., HARESIGN, W. and LAMMING, G.E. (1982b). Response of seasonally anoestrous ewes to small dose multiple injections of GnRH with and without progesterone pretreatment. *Journal of Reproduction and Fertility* **65**, 223–230

McLEOD, B.J., HARESIGN, W. and LAMMING, G.E. (1983). Induction of ovulation in seasonally anoestrus ewes by continuous infusion of low doses of GnRH. *Journal of Reproduction and Fertility* (in press)

McNATTY, K.P., REVFEIM, K.J.A. and YOUNG, A. (1973). Peripheral plasma progesterone concentrations in sheep during the oestrous cycle. *Journal of Endocrinology* **58**, 219–225

McNEILLY, J.R., McNEILLY, A.S., WALTON, J.S. and CUNNINGHAM, F.J. (1976). Development and application of a heterologous radioimmunoassay for ovine follicle stimulating hormone. *Journal of Endocrinology* **70**, 69–79

MAIN, S.J., DAVIES, R.V. and SETCHELL, B.P. (1979). The evidence that Inhibin must exist. *Journal of Reproduction and Fertility, Supplement* **26**, 3–14

MARTIN, G.B., OLDHAM, C.M. and LINDSAY, D.R. (1980). Increased plasma LH levels in seasonally anovular Merino ewes following the introduction of rams. *Animal Reproduction Science* **3**, 125–132

MATTNER, P.E. and BRADEN, A.W.H. (1972). Secretion of oestradiol-17β by the ovine ovary during the luteal phase of the oestrous cycle in relation to ovulation. *Journal of Reproduction and Fertility* **28**, 136P–137P

MATTON, P., BHERER, J. and DUFOUR, J.J. (1977). Morphology and responsiveness of the two largest ovarian follicles in anestrous ewes. *Canadian Journal of Animal Science* **57**, 459–464

MOOR, R.M. (1968). Effect of the embryo on corpus luteum function. *Journal of Animal Science, Supplement* **27**, 97–118

MOOR, R.M. (1973). Oestrogen production by individual follicles explanted from ovaries of the sheep. *Journal of Reproduction and Fertility* **32**, 545–548

MOOR, R.M., HAY, M.F., DOTT, H.M. and CRAN, D.G. (1978). Macroscopic identification and steroidogenic function of atretic follicles in sheep. *Journal of Endocrinology* **77**, 309–318

NARAYANA, K. and DOBSON, H. (1979). Effect of administration of antibody against GnRH on preovulatory LH and FSH surges in the ewe. *Journal of Reproduction and Fertility* **57**, 65–72

NISWENDER, G.D., ROCHE, J.F., FOSTER, D.L. and MIDGLEY, A.R. (1968). Radioimmunoassay of serum levels of luteinizing hormone during the cycle and early pregnancy in ewes. *Proceedings of the Society for Experimental Biology and Medicine* **129**, 901–904

OLDHAM, C.M. and LINDSAY, D.R. (1980). Laparoscopy in the ewe: a photographic record of the ovarian activity of ewes experiencing normal or abnormal oestrous cycles. *Animal Reproduction Science* **3**, 119–124

PANT, H.C., FITZPATRICK, R.J. and HOPKINSON, C.R.N. (1973). Interrelationships between plasma follicle stimulating hormone (FSH) and 17βoestradiol in sheep during the oestrous cycle. *Acta Endocrinologica Copenhagen. Supplement* **177**, 12

PANT, H.C., HOPKINSON, C.R.N. and FITZPATRICK, R.J. (1977). Concentration of oestradiol, progesterone, luteinizing hormone and follicle stimulating hormone in the jugular venous plasma of ewes during the oestrous cycle. *Journal of Endocrinology* **73**, 247–255

PATHIRAJA, N., CARR, W.R., LAND, R.B. and MORRIS, B. (1980). Ovulation rate and tonic LH of sheep treated with antisera to androstenedione, testosterone, oestradiol-17β and oestrone. *Journal of Reproduction and Fertility* **61**, 251

PELLETIER, J. (1976). Influence of LH-RF on LH and FSH release in domestic animals. *Annales de Biologie Animale Biochemie Biophysique* **16**, 213–234

PERRY, J.S., HEAP, R.B., GADSBY, J.E. and BURTON, R.D. (1976). Endocrinology of the blastocyst and its role in the establishment of pregnancy. *Journal of Reproduction and Fertility, Supplement* **25**, 85–104

PETERS, H. and McNATTY, K.P. (1980). Control of follicular growth. In *The Ovary* (C.A. Finn, Ed.), pp. 60–74. London, Granada Publishing

PIPER, E.L. and FOOTE, W.C. (1968). Ovulation and corpus luteum maintenance in ewes treated with 17β-oestradiol. *Journal of Reproduction and Fertility* **16**, 253–259

QUIRKE, J.F. and GOSLING, J. (1975). Progesterone concentration in the peripheral plasma of Galway and Finnish Landrace sheep during the oestrous cycle. *Irish Journal of Agricultural Research* **14**, 49–53

QUIRKE, J.F., HANRAHAN, J.P. and GOSLING, J.P. (1979). Plasma progesterone levels throughout the oestrous cycle and release of LH at oestrus in sheep with different ovulation rates. *Journal of Reproduction and Fertility* **55**, 37–44

QUIRKE, J.F., HANRAHAN, J.P. and GOSLING, J.P. (1981). Duration of oestrus, ovulation rate, time of ovulation and plasma LH, total oestrogen and progesterone in Galway adult ewes and ewe lambs. *Journal of Reproduction and Fertility* **61**, 265–272

RADFORD, H.M., WALLACE, A.L. and WHEATLEY, I.S. (1970). LH release, ovulation and oestrus following the treatment of anoestrous ewes with ovarian steroids. *Journal of Reproduction and Fertility* **21**, 371–373

REDDING, T.W., SCHALLY, A.V., ARIMURA, A. and MATSUO, H. (1972). Stimulation of release of luteinizing hormone (LH) and follicle stimulating hormone (FSH) in tissue cultures of rat pituitaries in response to natural and synthetic LH and FSH releasing hormone. *Endocrinology* **90**, 764–770

REEVES, J.J., ARIMURA, A. and SCHALLY, A.V. (1971). Changes in pituitary responsiveness to luteinizing hormone-releasing hormone (LH-RH) in anestrous ewes pretreated with estradiol benzoate. *Biology of Reproduction* **4**, 88–92

REEVES, J.J., ARIMURA, A. and SCHALLY, A.V. (1974). Pituitary responsiveness to purified luteinizing hormone-releasing hormone (LH-RH) at various stages of the estrous cycle in sheep. *Journal of Animal Science* **32**, 123–126

RESTALL, B.J. and RADFORD, H.M. (1974). The induction of reproductive activity in lactating ewes with gonadotrophin-releasing hormone (Gn-RH). *Journal of Reproduction and Fertility* **36**, 475–476

RICHARDS, J.S., RAO, M.C. and IRELAND, J.J. (1978). Actions of pituitary gonadotrophins on the ovary. In *Control of Ovulation* (D.B. Crighton, N.B. Haynes, G.R. Foxcroft and G.E. Lamming, Eds.), pp. 197–216. London, Butterworths

ROBERTSON, H.A. (1977). Reproduction in the ewe and goat. In *Reproduction in Domestic Animals* (H.H. Cole and P.T. Cupps, Eds.), pp. 475–498. New York, Academic Press

ROBERTSON, H.A. and HUTCHINSON, J.S.M. (1962). The levels of FSH and LH in the pituitary of the ewe in relation to follicular growth and ovulation. *Journal of Endocrinology* **24**, 143–151

ROBERTSON, H.A., RAKHA, A.M. and BUTTLE, H.L. (1966). *loc cit* H.A. ROBERTSON. The endogenous control of estrus and ovulation in sheep, cattle and swine. *Vitamins and Hormones* **27**, 91–130

ROBINSON, T.J. (1951). Reproduction in the ewe. *Biological Review* **26**, 121–157

ROBINSON, T.J. (1954). The necessity for progesterone with estrogen for the induction of recurrent estrus in the ovariectomised ewe. *Endocrinology* **55**, 403–408

ROCHE, J.F., FOSTER, D.L., KARSCH, F.J., COOK, B. and DJUIK, P.J. (1970). Levels of luteinizing hormone in the sera and pituitary of ewes during the estrous cycle and anestrus. *Endocrinology* **86**, 568–572

SALAMONSEN, L.A., JONAS, H.A., BURGER, H.G., BUCKMASTER, J.M., CHAMLEY, W.A., CUMMING, I.A., FINDLAY, J.K. and GODING, J.R. (1973). A heterologous radioimmunoassay for follicle stimulating hormone: application to measurement of FSH in the ovine estrous cycle and in several other species including man. *Endocrinology* **93**, 610–618

SANTOLUCITO, J.A., CLEGG, M.T. and COLE, H.H. (1960). Pituitary gonadotrophins in the ewe at different stages of the estrous cycle. *Endocrinology* **66**, 273–279

SCARAMUZZI, R.J. and BAIRD, D.T. (1979). Ovarian steroid secretion in sheep during anoestrus. In *Sheep Breeding* (2nd Edition) (G.J. Tomes, D.E. Robertson, R.J. Lightfoot and W. Haresign, Eds.), pp. 463–470. London, Butterworths

SCARAMUZZI, R.J. and MARTENZ, N.D. (1975). Effects of active immunization against androstenedione on luteinizing hormone levels in the ewe. In *Immunization with Hormones in Reproductive Research* (E. Nieschlag, Ed.), p. 141. Amsterdam, North Holland Publishing Co.

SCARAMUZZI, R.J., CALDWELL, B.V. and MOOR, R.M. (1970). Radioimmunoassay of LH and estrogen during the estrous cycle of the ewe. *Biology of Reproduction* **3**, 110–119

SCARAMUZZI, R.J., TILLSON, S.A., THORNEYCROFT, I.H. and CALDWELL, B.V. (1971). Action of exogenous progesterone and estrogen on behavioural estrus, and luteinizing hormone levels in the ovariectomised ewe. *Endocrinology* **88**, 1184–1189

SMEATON, T.C. and ROBERTSON, H.A. (1971). Studies on the growth and atresia of Graafian follicles in the ovary of the sheep. *Journal of Reproduction and Fertility* **25**, 243–252

SYMONS, A.M., CUNNINGHAM, N.F. and SABA, N. (1974). The gonadotrophic hormone response of anoestrous and cyclic ewes to synthetic luteinizing hormone-releasing hormone. *Journal of Reproduction and Fertility* **39**, 11–12

THIMONIER, J. and PELLETIER, J. (1971). Difference genetique dans la discharge ovulante (LH) chez des brebis de race Ile-de-France; Relations avec le nombre d'ovulations. *Annales de Biologie Animale Biochemie Biophysique* **11**, 559–567

THORBURN, G.D., BASSET, J.M. and SMITH, I.D. (1969). Progesterone concentrations in the peripheral plasma of sheep during the oestrous cycle. *Journal of Endocrinology* **45**, 459–469

THORBURN, G.D., COX, R.I., CURRIE, W.B., RESTALL, B.J. and SCHNEIDER, W. (1972). Prostaglandin F and progesterone concentrations in utero-ovarian venous plasma of the ewe during the oestrous cycle and early pregnancy. *Journal of Reproduction and Fertility, Supplement* **18**, 151–158

THORNER, M.O. (1978). Hyperprolactinaemia and ovulation. In *Control of Ovulation* (D.B. Crighton, N.B. Haynes, G.R. Foxcroft and G.E. Lamming, Eds.), pp. 397–409. London, Butterworths

WALTON, J.S., McNEILLY, J.R., McNEILLY, A.S. and CUNNINGHAM, F.J. (1977). Changes in concentrations of follicle-stimulating hormone, luteinizing hormone, prolactin and progesterone in the plasma of ewes during the transition from anoestrus to breeding activity. *Journal of Endocrinology* **75**, 127–136

WATANABE, S. and McCANN, S.M. (1968). Localization of FSH-releasing factor in the hypothalamus and neurohypophysis as determined by *in vitro* assay. *Endocrinology* **82**, 664–673

WEBSTER, G.M. and HARESIGN, W. (1983). Seasonal changes in LH and prolactin concentrations in ewes of two breeds. *Journal of Reproduction and Fertility* **67** (in press)

WEISS, T.J., SEAMARK, R.F., McINTOSH, J.E.A. and MOOR, R.M. (1976). Cyclic AMP in sheep ovarian follicles: Site of production and response to gonadotrophins. *Journal of Reproduction and Fertility* **46**, 347–353

WHEATLEY, I.S. and RADFORD, H.M. (1969). Luteinizing hormone secretion during the oestrous cycle of the ewe as determined by radioimmunoassay. *Journal of Reproduction and Fertility* **19**, 211–214

YEATES, N.T.M. (1949). The breeding season of the sheep with particular reference to its modification by artificial means using light. *Journal of Agricultural Science, Cambridge* **39**, 1–42

YUTHASASTRAKOSOL, P., PALMER, W.M. and HOWLAND, D.E. (1973). Hormone levels in the anoestrous and cycling ewe. *Journal of Animal Science* **37**, 334

YUTHASASTRAKOSOL, P., PALMER, W.M. and HOWLAND, D.E. (1977). Release of LH in anoestrous and cyclic ewes. *Journal of Reproduction and Fertility* **50**, 319–321

19

CONTROL OF REPRODUCTION IN THE EWE

Y. COGNIE and P. MAULEON
INRA, Station de Physiologie de la Reproduction, 37380 Nouzilly, France

Techniques using progesterone and pregnant mare's serum gonadotrophin (PMSG) to induce a fertile oestrus in the ewe no matter what its physiological state (postpartum, dry or milking ewes) and at any time of the year are no longer an innovation for French breeders. These methods are used to devise production systems adapted to environmental conditions and to take into account the genetic potential of the animals. If intensification of production systems is required to improve the efficiency of breeding systems, it is necessary to define the optimum way of using the techniques of oestrus control.

Historical development of oestrus control techniques

Three main steps have marked the development of the control of the reproductive cycle in the sheep:

1. The demonstration, in the fifties, of the role of progesterone to inhibit ovulation during the period of its administration and to induce a behavioural, synchronized oestrus after the end of treatment during the sexual season (Dutt and Casida, 1948; Dauzier *et al.*, 1953).
2. The finding that, during the seasonal anoestrus, the injection of PMSG after a period of progesterone priming is necessary to induce a fertile oestrus (Dauzier *et al.*, 1954).
3. The availability, in the sixties, of synthetic progestagens, more active than progesterone itself, which could be administered by routes other than the intramuscular route.

Fluorogestone acetate (FGA) was chosen among many synthetic progestagens as being the most useful because its physiological characteristics were similar to those of progesterone and Robinson (1964) proposed administration by the vaginal route.

In the last ten years, even though this technique has not been widely adopted elsewhere, the INRA, with the help of research and advisory staffs, has defined the conditions necessary for using the artificial control of the oestrous cycle in different systems of production.

Three main reasons led French breeders to use progestagens:

1. To induce early mating of female lambs when 9–11 months of age after reaching an adequate body development, the management of the flock being made easier and more efficient by the synchronized lambing of groups of young and adult ewes;
2. To increase the annual rhythm of lambing by mating ewes again in the spring (three lambings in two years);
3. To facilitate the practice of artificial insemination (AI), either to take advantage of genetic improvement (e.g. for milk production in Aveyron), or to facilitate mating of ewes in which oestrus had been synchronized without recourse to the need for keeping large numbers of rams.

For these reasons, since the technique was introduced in 1966, the use of controlled breeding in sheep flocks has continued to increase: 300 000 ewes treated in 1974 and 1 200 000 in 1981, i.e. more than 15% of the national flock. This technique is also presently developing around the Mediterranean area in Spain, Sardinia and Greece.

The synthetic analogues of prostaglandin $PGF_{2\alpha}$, by their luteolytic activity, also control the time of ovulation in the ewe, but only when an active corpus luteum exists at the time of the treatment (Haresign, 1977; Thimonier, 1981). Since in most instances oestrus control is needed in anoestrous females, prostaglandins are of limited use.

Improvements of techniques for controlling the oestrous cycle of the ewe

STANDARDIZATION OF PREPARATIONS

After overcoming problems of standardization of the vaginal sponge production and of assessing the activity of each batch of PMSG, the treatment has been adapted to the breed and the physiological state of the animals (*Table 19.1*). The dose of PMSG may also be modified according to the response obtained in that flock during the previous year.

Table 19.1 PROGESTAGEN/PMSG TREATMENTS USED TO INDUCE OESTRUS IN EWES

Physiological state	*Breeding season*			*Seasonal anoestrus*		
	FGA treatment		*PMSG*[a] Dose (iu)	*FGA treatment*		*PMSG*[a] Dose (iu)
	Dose (mg)	*Duration* (days)		*Dose* (mg)	*Duration* (days)	
Dry ewes	40	12–14	400	30	12	500–600
Nursing ewes[b]	40	12–14	500	30	12	600–700
Ewe lambs (8–12 months)	40	12–14	400	40	12–14	500

[a]The PMSG dose is modified according to breed (for example, decrease the dose by 100 iu for F_1 crosses of the Romanov breed)
[b]Lambing to mating interval: 45 days in autumn, 60 days in spring
From Thimonier and Cognie (1971); Colas *et al.* (1973)

In the initial trials fertility at the induced oestrus was quite variable, and this variation was partly due to an underestimation of the importance of factors associated with the male (Lauferon, 1975). Impairment by the synthetic progestagens of both transport and survival of spermatozoa in the female genital tract (Quinlivan and Robinson, 1969) made it necessary to modify some of the components of the treatments; mating had to be delayed to the second half of heat or AI limited to 55±1 hours after the end of the treatment (Colas, 1975), and when AI was used it required the deposition of at least 500×10^6 motile spermatozoa (Colas, 1975).

The recognition of the importance of the male component of fertility, particularly after the use of controlled breeding, has led to changes in the management of rams at the induced oestrus when these are used in preference to AI. Firstly, the rams are properly prepared by adequate nutrition prior to mating and secondly the number of ewes/ram is restricted to one ram/5–7 ewes. By spreading the timing of withdrawal of vaginal sponges it is possible for one male to serve 48–64 females without any decrease in fertility (Galindez, Prud'hon and Reboul, 1977).

In addition, to limit the reduction in fertility associated with the use of synthetic progestagens due to its effects on sperm transport, it was necessary to modify the dose levels used and the duration of treatment for ewes in different physiological conditions.

TREATMENT OF ANOESTROUS EWES

The more powerful the steroid, the lower the dose level necessary to block gonadotrophin release and, consequently, the quicker its elimination from the system following sponge withdrawal. This is probably the reason for a better degree of synchronization of oestrus after the end of treatment and a higher level of fertility at the induced oestrus obtained after administration of fluorogestone acetate (FGA, Intervet) compared with medroxy-progesterone acetate (MAP, Upjohn) (*Table 19.2*).

The administration of progesterone to cyclic ewes for 10 days was shown by Bray, Hecker and Wodzicka-Tomaszewska (1976) to allow regression of

Table 19.2 SYNCHRONIZATION OF OESTRUS AND FERTILITY IN MILKING EWES (MANCHEGA OR MANCHEGA-× BREED) AFTER VAGINAL SPONGE TREATMENT (EITHER FGA (30 mg), INTERVET, OR MAP (60 mg), UPJOHN) AND INJECTION OF 550 iu OF PMSG AT THE END OF PROGESTAGEN TREATMENT IN FEBRUARY

Group[(a)]	*Treatment*	*Number of ewes*	*% ewes in oestrus between 24 and 60 hours*	*Ewes lambing as % of those in oestrus*
I	FGA (30 mg)	79	98.7	50.6
	MAP (60 mg)	84	91.7	33.7
II	FGA (30 mg)	75	96.0	64.9
	MAP (60 mg)	73	83.6	46.4
I + II	FGA (30 mg)	154	97.4 $P<0.01$	57.5 $P<0.01$
	MAP (60 mg)	157	87.9	39.5

[(a)]Group I: milked ewes; Group II: dry ewes
From Cognie *et al.* (1978)

Table 19.3 INFLUENCE OF THE DOSE AND DURATION OF FGA TREATMENT ON CONCEPTION RATE AND PROLIFICACY OF PREALPES EWES DURING ANOESTRUS (APRIL)

Treatment			*% females in oestrus*	*Conception rate* (%)	*Prolificacy* (%)
FGA doses	*Duration* (days)	*No. treated*			
30 mg +	12	55	98.2	70.9	149
500 iu PMSG	10	54	100.0	77.8	169
20 mg +	12	55	94.5	60.0	167
500 iu PMSG	10	54	96.3	57.4	154

Effect of dose 20 mg vs. 30 mg: $P<0.01$
Effect of duration: N.S.

corpus lutea and produce synchronization of heat. The influence of a shortening in the duration of vaginal sponge treatment from 12 to 10 days and its interaction with the dose of FGA was studied in Prealpes ewes during anoestrus (*Table 19.3*). The lower progestagen dose was sufficient to induce a synchronized oestrus, but fertility was lowered. However, the decrease in duration of treatment from 12 to 10 days did not have any significant effect on fertility. Increasing the duration of sponge treatment from 12 up to 16 days does reduce fertility at the induced oestrus (Gonzalez-Lopez, personal communication).

TREATMENT OF LACTATING EWES

Experiments in Ile-de-France on ewes nursing their lambs after a spring lambing, showed that reducing either the dose of progestagen (*Table 19.4*) or the duration of treatment (*Table 19.5*) does not improve fertility at the oestrus induced by the progestagen/PMSG treatment. Both treatments were unable to break the ovarian inactivity of lactating ewes.

Table 19.4 INFLUENCE OF DOSE OF FGA ON CONCEPTION RATE AND PROLIFICACY OF LACTATING ILE-DE-FRANCE EWES INSEMINATED 45 DAYS AFTER LAMBING IN THE SPRING

FGA dose (mg)	*No. treated*	*% females in oestrus*	*Conception rate* (%)	*Prolificacy* (%)
20	96	85.4	17.7	205.9
40	99	83.8	22.2	145.5

Table 19.5 INFLUENCE OF THE DURATION OF VAGINAL SPONGE TREATMENT ON CONCEPTION RATE AND PROLIFICACY OF LACTATING ILE-DE-FRANCE EWES INSEMINATED 50–80 DAYS AFTER LAMBING IN THE SPRING

Duration of treatment (days)	*No. treated*	*Conception rate* (%)		*Prolificacy* (%)
		Day 18[a]	*Term*[a]	
12	73	70	48	171
6	70	70	48	153

[a]Pregnancy diagnosed by both plasma progesterone concentrations of >1 ng/ml 18 days after mating and the production of lambs at term.

INDUCTION OF PUBERTY IN THE EWE LAMB

A reduction in the duration of progestagen administration from 14 days to 10 days does not modify the response to the treatment or the fertility of ewe lambs (*Table 19.6*). An alternative to the vaginal route of progestagen

Table 19.6 OESTRUS AND FERTILITY FOLLOWING VAGINAL (40 mg/DAY) OR ORAL (8 mg/DAY) ADMINISTRATION OF FGA TO FEMALE LAMBS OF LACAUNE × ROMANOV BREED (TWO AIs AT 48 AND 60 HOURS)

Treatment	*Duration* (days)	*No. treated*	*Oestrus* (%)	*Conception rate* (%)	*Prolificacy* (%)
Vaginal	14	27	100	70.4	1.84
	10	32	100	81.3	2.04
Oral	14	30	87	70.0	1.48
	10	28	100	60.7	1.71

administration was studied in an attempt to avoid either difficulties of sponge insertion into ewe lambs or the irritation and bacterial proliferation which occur while the sponges are *in situ*. Bacterial proliferation can be minimized by dusting the sponges with antibiotic powder prior to insertion (Brice and Jardon, 1981).

ORAL ADMINISTRATION OF PROGESTAGENS

Since it has been shown that sperm removal from the female tract is more rapid following vaginal compared with oral administration of progestagens (Hawk and Cowley, 1971), it was decided to investigate whether oral administration of FGA would result in higher fertility in treated ewes. FGA was mixed with the concentrate and fed at doses of 6–8 mg/ewe/day. The time of the onset of heat was compared with that found after use of vaginal sponges, with and without injection of PMSG (*Table 19.7*). The

Table 19.7 TIMING OF OESTRUS IN EWES TREATED WITH PROGESTAGENS BY DIFFERENT ROUTES DURING THE BREEDING SEASON

Progestagen treatment[(a)]	*PMSG*[(b)] (iu)	*No. treated animals*	*Interval end of treatment–onset of oestrus* (% treated animals)					
			24	*36*	*48*	*60*	*72*	*84*
(1) Vaginal								
40 mg/day FGA	0	42	5	79	17			
for 14 days	400	42	21	62	17			
(2) Oral								
8 mg/day FGA	0	24			8	50	38	4
for 12 days	400	24			29	54	4	4
(3) Subcutaneous								
3 mg/day SC 21009								
for 10 or 12 days	500	37	70	27				

[(a)] Groups (1) and (2) = Ile-de-France ewes; Group (3) = Aragonaise ewes
[(b)] PMSG is injected the last day (Groups (1) and (3)) or one day after the end of progestagen treatment (Group (2))

Table 19.8 INFLUENCE OF THE ROUTE OF ADMINISTRATION AND DURATION OF FGA TREATMENT IN LACTATING EWES TREATED 54 DAYS (EXPERIMENT 1, PREALPES BREED) OR 43 DAYS (EXPERIMENT 2, ILE-DE-FRANCE) AFTER LAMBING IN THE SPRING

Experiment	*Treatment*[a]			*No. treated animals*	*Oestrus* (%)	*Conception rate* (%)	*Prolificacy* (%)
	Route	*Dose* (mg)	*Duration* (days)				
I (Feb.)	Oral	8/day	6	30	97	66.7	2.25
	Vaginal	30	6	27	93	55.6	2.13
II (April)	Oral	8/day	6	90	—	47.8	2.07
	Vaginal	30	12	147	—	48.3	1.72

[a]600 iu (Experiment 1) or 650 iu (Experiment 2) of PMSG were injected, either 24 hours after the last administration of FGA orally, or at withdrawal of vaginal sponges

results were much better than those reported by Lindsay *et al.* (1967) following oral administration of MAP. Even if the fertility of ewes receiving oral administration of FGA is as good as that with vaginal sponges (*Table 19.8*), the quantity of progestagen required/ewe (80 mg vs. 40 mg respectively) is higher and the treatment costs therefore twice as high. However, in lactating ewes, oral administration can be shortened to six days when given during the non-breeding season without any loss in fertility (*Table 19.8*).

SUBCUTANEOUS ADMINISTRATION OF PROGESTAGENS

The strong progestagen activity of Norgestomet (SC 21009, Searle) allows the use of very low doses to block ovulation in the ewe. A subcutaneous implant of 3 mg results in a very rapid onset of oestrus after the end of treatment (*Table 19.7*). The mean interval from end of treatment to ovulation is about 55 hours with the implant compared with 62 hours after use of vaginal sponges impregnated with FGA (Cognie, Mariana and Thimonier, 1970; Cognie, Folch and Alonso, 1976). To get similar levels of fertility with these two types of treatment it is necessary to inseminate 49±1 hours after Norgestomet implant removal compared with 55±1 hours after FGA treatment. With such modifications the treatment appears to be as effective in adult ewes but not in young ewes (*Table 19.9*). However, with each age group the prolificacy was lower after administration of SC 21009.

Table 19.9 INFLUENCE OF THE PROGESTAGEN TYPE AND THE AGE OF EWES ON THE FERTILITY OF LACAUNE (FROM OVITEST, UNPUBLISHED DATA)

Treatment[a]	*Age of ewes*	*No. animals treated*	*Conception rate* (%)		*Prolificacy* (%)	*Lambing*[b] (%)
Vaginal	<2 years	614	62.1	N.S.	1.53	0.95
(30 mg FGA)	>2 years	1072	61.4		1.65	1.01
Subcutaneous implant	<2 years	1253	53.5	$P<0.01$	1.43	0.77
(3 mg SC 21009)	>2 years	2230	66.8		1.53	1.02

[a]Both treatments were given with 400–450 iu PMSG
[b]Number of lambs expressed as % of ewes mated

Development and limits of techniques for controlling reproduction in the ewe

Over the last ten years in France, since these techniques have been available, there have been several reasons for their uptake at the farm level. In addition to its use in AI programmes for genetic improvement, progestagen/PMSG treatment has been employed to simplify flock management and to increase lambing frequency. Appropriate management decisions designed to obtain increased levels of profit and to simplify flock management involve greater attention to two important factors at two different periods during the reproductive cycle (Aguer, 1981; Lehen, 1981): firstly preparation of males and females immediately prior to mating and, secondly, adequate supervision of ewes at lambing to reduce perinatal mortality. This latter point is even more important as a result of the 30% increase in prolificacy after progestagen/PMSG treatment. It is also important to adapt the nutritional status of the ewe during late pregnancy and lactation, particularly in ewes producing more than one lamb (Robinson, 1981). In addition, it has been estimated that 29% and 18% of lambs from primiparous Ile-de-France and Prealpes ewes, respectively, die at or soon after birth without adequate shepherding (Poindron, 1981), but this can be much reduced by penning such animals for a short period after lambing to allow development of maternal instincts. A further benefit is the reduction in weekend working by inducing parturition. All of these factors are aided by the synchronization of oestrus.

An increase in flock productivity as a result of increasing the lambing frequency is more possible with progestagen/PMSG treatment started one month after parturition in the breeding season or one month later when both postpartum and lactational effects are confounded with seasonal anoestrus (Cognie, Cornu and Mauleon, 1974). Since an earlier return to cyclical ovarian activity after parturition is dependent on a reduction in the suckling stimulus and a lower prolactin release (Kann, Martinet and Schirar, 1977), the conception rate can be increased in treated ewes by reducing the number of lambs suckled (Cognie, Cornu and Mauleon, 1974; Cognie, Hernandez Barreto and Saumande, 1975).

In order to increase the percentage of ewes ovulating after progestagen/PMSG treatment, it is necessary to increase the dose of PMSG in lactating ewes (*Table 19.1*). However, higher doses of PMSG increase not only the variation in ovulation rate between ewes but also the variation in time intervals from end of treatment to ovulation (Cognie and Pelletier, 1976). Low plasma oestradiol levels, reduced egg recovery and fertilization rates (Cognie, Hernandez Barreto and Saumande, 1975; Quirke *et al.*, 1981) illustrate the inadequate ovarian response of lactating ewes treated during seasonal anoestrus.

POSSIBILITIES OF IMPROVING FERTILITY OF LACTATING EWES DURING THE SPRING

Accelerating the rate of uterine involution by inducing ovulation soon after parturition does not improve and, sometimes, depresses the response of

Table 19.10 FERTILITY OF SUCKLING EWES SYNCHRONIZED WITH PROGESTAGEN/PMSG TREATMENT IN APRIL WITH OR WITHOUT OESTRADIOL (EXPERIMENT 1) OR PMSG (EXPERIMENT 2) PRETREATMENT

Treatments[a]	*No. animals treated*	*% of ewes showing oestrus*		*% of ewes lambing*	
Experiment 1					
Control	56	67	N.S.	29	N.S.
Pre-treated with 50 μg oestradiol	52	67		37	
Experiment 2					
Control	136	78	$P<0.01$	23	$P<0.01$
Pretreated with 600 iu PMSG	76	47		5	

[a]Treatments: Day 20 after lambing, 50 μg oestradiol benzoate (Experiment 1) or 600 iu PMSG (Experiment 2) pretreatment (treated ewes only)
Day 24 after lambing, FGA pessaries (all ewes)
Day 36 after lambing, pessaries removal + PMSG (700 or 800 iu) (all ewes)
Day 38 after lambing, AI (all ewes)

lactating ewes to progestagen/PMSG treatment (*Table 19.10*). Moreover, there is evidence that such practices may increase the incidence of persistent corpora lutea and the refractoriness to PMSG. Although it is possible to improve the ovulatory response by injecting oestradiol benzoate or GnRH 24 hours after sponge removal, this does not improve fertility. The synchronization of ovulation is perfect but the quality of ovulated eggs is poor (Pelletier and Thimonier, 1975).

More recent work has focused on attempts to obtain a better degree of follicular growth before the induction of ovulation. However, sequential application of PMSG six and three days before and on the day of sponge removal is followed by a reduction in the preovulatory levels of oestradiol (F. Gayerie, personal communication) and a decrease in the percentage of ewes ovulating compared with controls in both lactating and seasonally anoestrous ewes (*Table 19.11*). Furthermore, injection of PMSG two days before sponge removal in lactating ewes advances the onset of oestrus without any improvement in reproductive performance compared with ewes treated at the end of the progestagen treatment (*Table 19.12*).

Table 19.11 OVULATION AFTER SINGLE OR SEQUENTIAL INJECTION OF PMSG IN LACTATING OR SEASONAL ANOESTROUS EWES IN MARCH

Physiological state of ewes	*Treatment*[a]	*No. animals treated*	*% ewes with corpora lutea*	*No. of ewes with different numbers of corpora lutea*				
				1	*2*	*3*	*4*	*5+*
Lactating	Control	12	66.7	2	1	2	2	1
	Sequential	14	21.4	1	2			
Dry	Control	19	89.5	1	11	3	1	1
	Sequential	17	52.9	4	5			

[a]Control treatment: 800 and 600 iu PMSG injected respectively to lactating and dry ewes at sponge removal.
Sequential treatment: 200, 200, 400 or 150, 150, 300 iu PMSG respectively injected into lactating and dry ewes six and three days before and on the day of sponge removal.

Table 19.12 INTERVAL BETWEEN SPONGE REMOVAL AND ONSET OF OESTRUS AND REPRODUCTIVE PERFORMANCE OF ILE-DE-FRANCE LACTATING EWES TREATED 27±3 DAYS AFTER LAMBING (MARCH)

PMSG treatment		*No. animals treated*	*Time from removal of sponges* (hours)				*% ewes lambing*	*Lambs born/ ewe lambing*
Dose (iu)	*Day*[a]		*24*	*36*	*48*	*60*		
			(% ewes in oestrus)					
800	D_{-2}	62	32	84	100		44.4	1.61
800	D_0	63	2	36	83	100	43.5	1.52

[a] D_0 = day of sponge removal

Preliminary results obtained with injection of follicle stimulating hormone (FSH) every two hours for 24 hours followed by luteinizing hormone (LH) 36 hours later in anovulatory Ile-de-France ewes (Oussaid and Cognie, unpublished data) or with twice daily injections of FSH during the final three days of the progestagen treatment in ewe lambs (Wright *et al.*, 1981) suggest that the poor ovulatory response is due to inadequate FSH levels in lactating ewes. However, further work is required to determine if this can be translated into a viable treatment regime. Since hyperprolactinaemia due to suckling does not block the LH response of lactationally anovular ewes at teasing (Poindron *et al.*, 1980), it may be possible to combine the 'ram effect' with treatments to improve fertility. The possibilities offered by the 'male effect' in progestagen-primed ewes to induce a synchronized oestrus and ovulations have been studied (Cognie *et al.*, 1980) and appear to be a sufficient stimulus to break shallow anoestrous conditions and minimize the cost of treatment, although inappropriate for ewes in deep anoestrus. While the majority of lactating ewes treated in the spring rapidly return to anoestrus before conceiving when kept under conditions of natural daylength, fertility is reported to be higher in ewes kept artificially under short days (Robinson, 1981). Furthermore, lactating Ile-de-France ewes exposed to declining daylength from February maintained a higher level of ovarian activity compared with control ewes.

Conclusions

In spite of the many recent attempts to improve our understanding of endocrine control of reproduction (see Haresign, Chapter 18), comparatively little work aimed at studying folliculogenesis and ovarian inactivity during seasonal and lactational anoestrus in the ewe have been conducted. While it has been suggested that alterations to the ordered sequence of developmental changes in the follicle result in atresia (Richards, Rao and Ireland, 1978), relatively small antral follicles do appear to be capable of being mobilized for oestrogen production within just a few hours if given sequential LH injections (McNatty *et al.*, 1981). However, it is not yet apparent whether this is applicable to all anoestrous conditions and breeds of ewe, nor whether sequential applications of purified gonadotrophins are necessary. More work is required to understand ovarian inactivity in lactating ewes so that more effective treatments can be developed. In spite of the fact that the interference of seasonal anoestrus with the unfavourable influences of suckling limits the efficiency of frequent lambing programmes, reasonable levels of success can be achieved by the choice of genotypes with high reproductive performance.

References

AGUER, D. (1981). La synchronisation des chaleurs. Pourquoi? Comment? *Pâtre* **287**, 25–28

BRAY, A.R., HECKER, J.F. and WODZICKA-TOMASZEWSKA, M. (1976). Progestin induction of short oestrous cycles in ewes. *VIIIth International Congress on Animal Reproduction and Artificial Insemination, Krakow*, pp. 443–446

BRICE, Y. and JARDON, C. (1981). Préparation des mâles et des femelles à la lutte. *Pâtre* **287**, 11–19

COGNIE, Y. (1972). Utilisation du F.G.A. administré par voie orale pour la synchronisation de l'oestrus et l'insémination artificielle des brebis. *VIIth International Congress on Animal Reproduction and Artificial Insemination, Munich*, pp. 971–976

COGNIE, Y. and PELLETIER, J. (1976). Preovulatory LH release and ovulation in dry and lactating ewes after progestagen and PMSG treatment during the seasonal anoestrus. *Annales de Biologie Animale Biochimie Biophysique* **16**, 529–536

COGNIE, Y., CORNU, C. and MAULEON, P. (1974). The influence of lactation on fertility of ewes treated during post-partum anoestrus with vaginal sponges impregnated with FGA (Chronogest). *International Symposium on Physiopathology of Reproduction and Artificial Insemination in Small Ruminants, Thessaloniki*, pp. 33–36

COGNIE, Y., FOLCH, J. and ALONSO DE MIGUEL, M. (1976). Utilisation des implants sous-cutanés de SC 21009 pour la synchronisation des chaleurs chez la brebis. *2ièmes Journées de la Recherche Ovine et Caprine*, pp. 288–294

COGNIE, Y., HERNANDEZ BARRETO, M. and SAUMANDE, J. (1975). Low fertility in nursing ewes during the non-breeding season. *Annales de Biologie Animale Biochimie Biophysique* **15**, 329–343

COGNIE, Y., MARIANA, J.C. and THIMONIER, J. (1970). Etude du moment d'ovulation chez la brebis normale ou traitée par un progestagène associé ou non à une injection de PMSG. *Annales de Biologie Animale Biochimie Biophysique* **10**, 15–24

COGNIE, Y., SALLERAS, J.M. and SIERRA ALFRANCA, I. (1978). Intensificacion reproductiva mediante metodos hormonales en ovejas de Raza Manchega y cruce Cadzon & Manchega. 1. Incidencia de la lactacion sobre la fertilidad. *Anales de la Facultad de Veterinaria de Zaragoza* **11–12**, 625–640

COGNIE, Y., GAYERIE, F., OLDHAM, C.M. and POINDRON, P. (1980). Increased ovulation rate at the ram-induced ovulation and its commercial application. *Proceedings of the Australian Society of Animal Production* **13**, 80–81

COLAS, G. (1975). The use of progestagen SC 9880 as an aid for artificial insemination in ewes. *Annales de Biologie Animale Biochimie Biophysique* **15**, 317–327

COLAS, G., THIMONIER, J., COUROT, M. and ORTAVANT, R. (1973). Fertilité, prolificité et fecondité pendant la saison sexuelle des brebis inseminées artificiellement après traitement à l'acétate de fluorogestone. *Annales de Zootechnie* **22**, 441–451

DAUZIER, L., ORTAVANT, R., THIBAULT, C. and WINTENBERGER, S. (1953). Recherches expérimentales sur le rôle de la progestérone dans le cycle sexuel de la brebis et de la chèvre. *Annales d'Endocrinologie* **4**, 553–559

DAUZIER, L., ORTAVANT, R., THIBAULT, C. and WINTENBERGER, S. (1954). Résultats nouveaux sur la gestation à contre-saison chez la brebis et chez la chèvre, possibilité d'utilisation pratique. *Annales de Zootechnie* **2**, 89–94

DUTT, R.H. and CASIDA, L.E. (1948). Alteration of the estrual cycle in sheep by use of progesterone and its effect upon subsequent ovulation and fertility. *Endocrinology* **43**, 208–217

GALINDEZ, F.J., PRUD'HON, M. and REBOUL, G. (1977). Reproductive performance of group synchronized Merinos d'Arles and Romanov crossbred ewes. I. A note on the effect of lactation on fecundity. *Animal Production* **24**, 113–116

HARESIGN, W. (1977). Ovulation control in the sheep. In *Control of Ovulation* (D.B. Crighton, N.B. Haynes, G.R. Foxcroft and G.E. Lamming, Eds.), pp. 435–451. London, Butterworths

HAWK, H.W. and COWLEY, H.H. (1971). Loss of spermatozoa from the reproductive tract of the ewe and intensification of sperm 'breakage' by progestagen. *Journal of Reproduction and Fertility* **27**, 339–347

KANN, G., MARTINET, J. and SCHIRAR, A. (1977). Hypothalamic–pituitary control during lactation in sheep. In *Control of Ovulation* (D.B. Crighton, N.B. Haynes, G.R. Foxcroft and G.E. Lamming, Eds.), pp. 319–333. London, Butterworths

LAUFERON, M. (1975). Synchronisation des chaleurs chez les ovins. *Searle Colloquium, Montpelier*, pp. 9–15

LINDSAY, D.R., MOORE, N.W., ROBINSON, T.J., SALOMON, S. and SHELTON, J.N. (1967). The evaluation of an oral progestagen (Provera, MAP) for the synchronization of oestrus in the entire cyclic Merino ewe. In *The Control of the Ovarian Cycle in the Sheep* (T.J. Robinson, Ed.), pp. 3–13

LEHEN, A. (1981). Raisonner ses luttes: un délicat équilibre entre une recherche d'amélioration de son revenu et des conditions de travail. *Pâtre* **287**, 31–35

McNATTY, R.P., GIBB, M., DOBSON, C. and THURLEY, D.C. (1981). Evidence that changes in luteinizing hormone secretion regulate the growth of the preovulatory follicle in the ewe. *Journal of Endocrinology* **90**, 375–389

PELLETIER, J. and THIMONIER, J. (1975). Interactions between ovarian steroids or progestagens and LH release. *Annales de Biologie Animale Biochimie Biophysique* **15**, 131–146

POINDRON, P. (1981). Contribution à l'étude des mécanismes de régulation du comportement maternel chez la brebis (*Ovis aries*). Thèse Doct. es Sci. Nat., University of Provence

POINDRON, P., COGNIE, Y., GAYERIE, F., ORGEUR, P., OLDHAM, C.M. and RAVAULT, J.P. (1980). Changes in gonadotrophins and prolactin levels in isolated (seasonally or lactationally) anovular ewes associated with ovulation caused by the introduction of rams. *Physiology and Behaviour* **25**, 227–236

QUINLIVAN, T.D. and ROBINSON, R.J. (1969). Numbers of spermatozoa in the genital tract after artificial insemination of progestagen treated ewes. *Journal of Reproduction and Fertility* **19**, 73–86

QUIRKE, J.F., HANRAHAN, J.P., SHEEHAN, W. and GOSLING, J.P. (1981). Effect of lactation on some aspects of reproduction in progestagen –PMSG treated ewes during the non-breeding season. *Irish Journal of Agricultural Research* **20**, 1–8

RICHARDS, J.S., RAO, M.C. and IRELAND, J.J. (1978). Actions of pituitary gonadotrophins on the ovary. In *Control of Ovulation* (D.B. Crighton, N.B. Haynes, G.R. Foxcroft and G.E. Lamming, Eds.), pp. 197–216. London, Butterworths

ROBINSON, T.J. (1964). Synchronization of oestrus in sheep by intravaginal and subcutaneous application of progestin impregnated sponges. *Proceedings of the Australian Society of Animal Production* **5**, 47–52

ROBINSON, J.J. (1981). Photoperiodic and nutritional influences on the reproductive performance of ewes in accelerated lambing systems. *Proceedings of 32nd Annual Meeting, EAAP, Zagreb*

THIMONIER, J. (1981). Practical uses of prostaglandines in sheep and goats. *Acta Veterinaria Scandinavica, Supplement* **77**, 193–208

THIMONIER, J. and COGNIE, Y. (1971). Accélération des mise-bas et conduite d'élevage chez les ovins. *Bulletin Technique d'Information du Ministère de l'Agriculture* **257**, 187–196

WRIGHT, R.W., BONDIOLI, K., GRAMMER, J., KUZAN, F. and MENINO, A. (1981). FSH or FSH plus LH superovulation in ewes following estrus synchronisation with medroxyprogesterone acetate pessaries. *Journal of Animal Science* **52**, 115–117

20

THE INFLUENCE OF ENVIRONMENTAL FACTORS ON THE ATTAINMENT OF PUBERTY IN EWE LAMBS

Ó.R. DÝRMUNDSSON
Agricultural Society of Iceland, Reykjavík, Iceland

Early breeding of female lambs is a means of enhancing the lifetime productivity of the ewe. Thus, increasingly, attention is being paid to factors affecting sexual development and reproductive performance of ewe lambs, and their potential contribution to systems of intensified sheep production has become widely accepted.

Most commonly, *puberty* is defined as the time at which reproduction first becomes possible, characterized by ovulation in the female, whereas *sexual maturity* is not reached until the animal expresses its full reproductive power (Asdell, 1946). In ewe lambs such a distinction between puberty and sexual maturity is pertinent, since ewes do not acquire their complete reproductive capacity until the adult stage is reached. Moreover, the mere attainment of physiological puberty in ewe lambs need not be concomitant with the ability to conceive and carry a foetus to term.

Puberty in ewe lambs is determined by most observers in terms of age at first behavioural oestrus (Joubert, 1963; Dýrmundsson, 1973). Other methods include studies on the formation of ovarian follicles and the development of the reproductive tract (Allen and Lamming, 1961); ovulation has been determined by laparotomy (Sutham, Hulet and Botkin, 1971) and blood hormone levels have been monitored (Downing, 1980). Although it is normally assumed that ovulation accompanies first oestrus, ovulation without oestrus (Foote, Sefidbakht and Madsen, 1970), or oestrus without ovulation, can occur in ewe lambs (Chu and Edey, 1978; Downing, 1980). Edey, Kilgour and Bremner (1978) have, in fact, argued that puberty would be more adequately defined by including both the requirement to ovulate and to allow insemination by the ram.

Genetic background

Sexual development is affected by both genetic and environmental factors and the interaction between these (Land, 1978). The scientific literature provides a large number of references to breed and strain differences in the incidence, and in the age and body weight at first oestrus (Hafez, 1952; Dýrmundsson, 1973). There may also be considerable genetic variability among individuals of the same breed (Hafez, 1953; Dýrmundsson, 1972; Quirke, 1977). Crossbred ewe lambs tend to have better reproductive performance than purebreds (Hight, Lang and Jury, 1973; Hohenboken and Cochran, 1976) and heterosis may contribute to earlier sexual development (Dickerson and Laster, 1975; Jakubec, 1977).

Regarding the incidence of oestrus in female sheep in their first year of life, breed and strain variation is evident, virtually ranging from 0–100% in the respective groups reported on (Watson and Gamble, 1961; Dýrmundsson and Lees, 1972a; Bichard *et al.*, 1974; Land, Russell and Donald, 1974; Thimonier, 1975; Tierney, 1976; Sierra, 1979). Genetic effects on puberty are, however, obscured to a varying extent by environmental factors, such as the plane of nutrition and the season of birth. Furthermore, published results on early breeding may not fully indicate the actual breed potential in each case.

Pubertal age and body weight

There are clearly great variations worldwide, both between and within breeds, with respect to both age and body weight at puberty. Since mean values quoted in the literature for a certain breed or crossbreed, in a certain locality, may be partly determined by environmental conditions, care should be taken in making direct and unqualified comparisons between breeds on the basis of this information. In a review presented by Dýrmundsson (1973) most of the mean values cited were within a range of 6–18 months. Exceptional cases of late summer or autumn born ewe lambs attaining puberty and even conceiving as early as 3–4 months of age have been reported, for example, by Robinson and Ørskov (1975) and Dýrmundsson (1979). A wide variation is also found to exist in body weight at first oestrus, mainly within the range of 30–50 kg. The above findings are supported by several reports published in recent years, for example by Murray (1972), Bichard *et al.* (1974), Dickerson and Laster (1975), Hawker (1976), Cedillo, Hohenboken and Drummond (1977), Younis *et al.* (1978) and Quirke (1978a). On the whole, breeds of British origin normally attain puberty at a lower age than breeds of the Merino type. Much information has recently become available on highly precocious and prolific breeds, namely the Finnish Landrace (Maijala and Österberg, 1977), the Romanov (Ricordeau *et al.*, 1978), and their crosses (Cedillo, Hohenboken and Drummond, 1977; Jakubec, 1977; Ricordeau *et al.*, 1978; Sierra, 1979).

Care must be taken in generalizing about the minimum age and body weight required for the attainment of puberty. Similarly, the average liveweight at puberty, expressed as a percentage of the adult weight, is not constant. It would seem, however, that in many cases first oestrus in ewe lambs is attained at weights ranging from 50–70% of adult body weight (Hafez, 1952; 1953; Dýrmundsson, 1973).

Natural factors affecting sexual development

NUTRITION

The close association between general body growth and the growth and development of reproductive organs is well established (Pálsson and Vergés, 1952). The effects of the plane of nutrition on sexual development are also well documented in the literature (Lamming, 1969). Thus underfeeding of immature animals may seriously retard pubertal development whereas high-plane feeding may advance puberty. Several workers have demonstrated that ewe lambs growing at faster rates will exhibit their first oestrus and are more likely to conceive at a lower age and heavier body weight than ewe lambs growing at slower rates (Allen and Lamming, 1961; Burfening *et al.*, 1971; Keane, 1975a; Quirke, 1979). Twin lambs tend to experience their first oestrus at a higher mean age and a lower mean body weight than single lambs (Mounib, Ahmed and Hamada, 1956; Dýrmundsson and Lees, 1972a,b). Results obtained by Downing (1980) show that slower growth rates lower the percentage of ewe lambs attaining puberty, but in those doing so, nutritional and liveweight differences have only small effects on date and age at first oestrus. She suggested that body proportions, such as total body protein, might be a more precise parameter of sexual development than total body weight.

While earlier results indicated that nutritional flushing prior to mating did not have a clear effect on ovulation rate in ewe lambs (Williams, 1954; Allen and Lamming, 1961), such an effect has, however, been reported by Keane (1974) and Downing (1980). In practice, however, flushing is not advocated since twinning in ewe lambs is generally not considered desirable, mainly because it contributes to heavy lamb losses. It is noteworthy that, according to Stoerger *et al.* (1976), the feeding of very high energy diets prior to breeding may be associated with an increased incidence of barrenness in ewe lambs, possibly due to overfatness. As a rule, the provision of adequate nutrition to promote normal growth is emphasized, particularly when ewe lambs are bred (Nedkvitne, 1975; Dýrmundsson, 1976). Gunn (1977) has found that the early growth pattern of ewes can affect their reproductive potential. This stresses the need to pay special attention to the level of feeding since in many locations nutrition would seem to be the environmental factor most limiting to successfully exploiting the breeding potential of ewe lambs.

Results obtained by Robinson *et al.* (1971), and Quirke, Sheehan and Lawlor (1978) point to differences in the pattern of nutrient utilization of pregnant ewe lambs compared with that of adult ewes. Whilst maternal body weight gain was found to increase, lamb birthweight tended to decrease with increasing protein intake. Here it would seem that, in addition to the overall level of feeding, special consideration should be given to the protein:energy ratio in the diet. Adalsteinsson (1972) found that the shearing of housed, well-fed ewe lambs during gestation was associated with an increase in the birthweight of their progeny. According to Nedkvitne (1979) pre-mating shearing of ewe lambs is associated with greater feed intake and better reproductive performance. Lees (1978a) pointed out that generous feeding of immature ewe lambs during the very latest stages of pregnancy may result in the birth of offspring which are too large, thereby causing lambing difficulties. The proper level of late pregnancy nutrition may, in fact, be even more critical in the ewe lamb than in the adult ewe.

SEASON OF BIRTH AND DAYLIGHT

There is ample evidence to indicate that seasonality is an important factor in the attainment of puberty in ewe lambs (Dýrmundsson, 1973). Hammond (1944) first demonstrated a relationship between date of birth and date of first oestrus. Thus, there is neither a fixed age, body weight nor time of year at which ewe lambs experience their first oestrus, owing to the complex interaction between these factors and time of birth. Although lambs born early tend to attain puberty earlier in the season than lambs born later, they normally do so at a higher age and heavier body weight (Dýrmundsson and Lees, 1972b; Downing, 1980).

It appears that ewe lambs of most breeds studied so far in the northern hemisphere will experience their first oestrus in the autumn and winter months irrespective of time of birth (Hafez, 1952; Burfening, Hoversland and Horn, 1974; Dýrmundsson, 1978). This is normally also the case in the southern hemisphere (Ch´ang and Raeside, 1957; Watson and Gamble, 1961; Joubert, 1962). In breeds of tropical origin, however, it would seem that ewe lambs may experience a less well-defined seasonal onset of first oestrus than ewe lambs at higher latitudes (Mounib, Ahmed and Hamada, 1956; Younis *et al.*, 1978). While it has been suggested that the daylight environment experienced by the lamb during rearing may be a critical factor in regulating breeding activity (Dýrmundsson and Lees, 1972a,b; Foster and Ryan, 1981), nevertheless, Ducker, Bowman and Temple (1973) found that various artificial light treatments only modified the occurrence of oestrus in ewe lambs. The breeding pattern still remained basically seasonal in nature, irrespective of light treatment, providing evidence of an inherent rhythm controlling oestrous activity in the lambs. In summary, the daylight environment at higher latitudes would seem to be the most important factor determining the attainment of puberty in those ewe lambs sufficiently heavy to show oestrus (Dufour, 1975; Downing, 1980).

TEMPERATURE

There appears to be a total lack of information in the scientific literature on the direct effect of temperature on sexual development in ewe lambs. As pointed out by Sadleir (1969), the effect of temperature changes on puberty reported in other mammals is known to be very closely related to the effect of temperature on body growth. This may be of special importance in subtropical and tropical regions. The removal of the fleece towards the end of anoestrus was found to advance the time of first oestrus in adult ewes (Lees, 1967) whereas autumn shearing treatments did not have any clear effect on the attainment of puberty and cyclic activity in ewe lambs in Wales (Dýrmundsson and Lees, 1972c).

RAM EFFECT

The onset of breeding activity in ewes may be influenced by the introduction of the ram (Schinckel, 1954). In ewe lambs, however, information is very limited regarding the so-called 'ram effect' on the attainment of puberty. Dýrmundsson and Lees (1972d) reported that the sudden introduction of rams to Clun ewe lambs in the transition from non-breeding to breeding activity resulted in a high degree of synchronization of first matings. Moore and McMillan (1981), however, failed to detect any effect of ram introduction on ovulation in pubertal Romney ewe lambs.

SEXUAL BEHAVIOUR AND CYCLIC ACTIVITY

Behavioural signs of oestrus in ewe lambs are usually weak and the intensity of oestrus is less evident than in yearlings and adult ewes (Hafez, 1951). Thus ewe lambs in heat tend to make little or no attempt to approach the ram, but accept service when the latter makes sexual advances. Limited evidence suggests that ewe lambs exhibit a degree of preference for rams of their own breed (Lees, 1971). The duration of overt oestrus is also normally shorter in ewe lambs than in older females (Hafez, 1952; Dýrmundsson, 1978; Edey, Kilgour and Bremner, 1978). In a few cases, however, oestrus has been reported of similar length in ewes and ewe lambs (Land, 1970; Boshoff, Burger and Cronje, 1975). Rams may vary both in their efficiency in detecting ewe lamb oestrus and in their urge and ability to mate with them, both under pen and paddock conditions (Lees, 1978b). Here the ratio of rams to ewe lambs may be of importance. Moreover, in view of the fact that rams tend to show preference for the more mature females (Downing, 1980), mating ewe lambs separately from older ewes, and using active rams only, is sound general advice.

The first oestrus is experienced later and the breeding season is of a shorter duration in ewe lambs than in yearlings and adult ewes (Hafez, 1952). That the breeding season is evenly spaced about the shortest day (Hammond, 1944; Dýrmundsson and Lees, 1972a) does not appear to be a general rule; in fact some reports show that the mid-point has occurred either before (Ch´ang and Raeside, 1957; Keane, 1975b) or after (Hafez, 1952; Dýrmundsson, 1978) the shortest day. Ewe lambs which attain puberty early normally experience their last oestrus late in the season. Thus earlier birth dates and faster body growth rates favour extended breeding activity in ewe lambs (Dýrmundsson and Lees, 1972a; Quirke, 1978a; Moore and Smeaton, 1980). The number of oestrous cycles may vary from 1–11 during the breeding season, depending on the breed, environment and sexual development of the lambs (Dýrmundsson, 1973; Burfening, Hoversland and Horn, 1974; Keane, 1975b; Dýrmundsson, 1978; Ricordeau *et al.*, 1978; Moore and Smeaton, 1980).

Ewe lambs exhibit less regular oestrous cycles and tend to have slightly shorter ones than adults. In ewe lambs there may be a relatively high incidence of silent heats, particularly in individuals with low growth rates (Hafez, 1952; Mounib, Ahmed and Hamada, 1956; Badawy, El-Bashary and Mohsen, 1973; Chu and Edey, 1978). It would seem that the greater the number of heats exhibited the greater is the regularity of cycling in ewe lambs (Foote, Sefidbakht and Madsen, 1970).

Reproductive efficiency

In his review, Dýrmundsson (1973) tabulated references to lambing performance in ewe lambs of several breeds of sheep indicating a wide variation, both between and within breeds. While the proportion of mated ewe lambs actually lambing is commonly in the range of 60–80% a number of references quoted in the review show much lower rates. These findings are supported by more recent reports, for example, by Bichard *et al.* (1974), Allison *et al.* (1975), Hulet and Price (1975), Tyrrell (1976), Quirke (1978b), Dýrmundsson and Hallgrímsson (1978) and Veress *et al.* (1979). Thus ewe lambs have, as a rule, lower conception rates than older ewes. Their lambing rates are lower too (Forrest and Bichard, 1974; Dickerson and Glimp, 1975; Baker *et al.*, 1978), twinning is not common and triplet births are extremely rare in ewe lambs, with the exception of highly prolific breeds and their crosses such as the Finnish Landrace (Maijala and Österberg, 1977; Price and Ercanbrack, 1975; Jakubec, 1977) and the Romanov (Ricordeau *et al.*, 1978; Sierra, 1979).

In ewe lambs greater body weight and higher age at mating are generally associated with better lambing performance, both in terms of ewes lambing and number born (Bowman, 1966; Christenson, Laster and Glimp, 1976; Dýrmundsson, 1973, 1981; Downing, 1980). However, such a relationship may not exist once the ewe lambs are above a certain threshold of body weight (Keane, 1974; Dýrmundsson, 1976) and sometimes apparently well grown individuals show disappointing performance. Crossbreeding may enhance reproductive efficiency in ewe lambs (Jakubec, 1977). In general, the incidence of reproductive failure is higher in ewe lambs than in adults (Forrest and Bichard, 1974), yet there is a dearth of information on the several causes of infertility in mated ewe lambs. Recent studies have, however, attempted to diagnose certain areas of reproductive loss in ewe lambs (Quirke, 1977; Edey *et al.*, 1977; Edey, Kilgour and Bremner, 1978; Chu and Edey, 1978; A.J. Allison, personal communication, 1979; Hamra and Bryant, 1979, 1982; Downing, 1980), including deficient sexual behaviour, anovulatory oestrus, fertilization failure, embryonic mortality and foetal abortion. Quirke (1981) and Quirke, Adams and Hanrahan in Chapter 21 have reviewed the available evidence on the sources of prenatal mortality in ewe lambs.

The length of gestation in ewe lambs seems to be similar to, or, in some cases, somewhat shorter than that observed in yearlings and adult ewes (Gordon, 1967; Dýrmundsson, 1973).

Several authors have reported higher lamb mortality in ewes lambing at one year of age than in older ewes, particularly in the neonatal period (Dýrmundsson, 1973; 1976; Bichard *et al.*, 1974; Tyrrell, 1976). The births of large single lambs may cause dystocia (Laster, Glimp and Dickerson, 1972) and twin births may increase losses due to low birthweights and poor viability of lambs (Lees, 1971; Quirke, 1979). Several reports indicate that singles from ewe lambs have a birthweight similar to that of twin lambs born to adult ewes (Dýrmundsson, 1973). While some ewe lambs may exhibit poor mothering behaviour following parturition (Eltan, 1974), it seems that yearlings previously bred as ewe lambs tend to be more reliable breeders, better mothers and to have fewer lambing troubles (Lewis, 1959; Lees, 1978a). Downing (1980) recorded the same mean shepherding time at lambing for ewe lambs and older ewes but pointed out, however, that poorly grown mothers needed extra care at lambing, especially if udder development and milk production was deficient. Although information is scant on the milk production of ewes lambing at one year of age, their yield is generally found to be lower than that of older ewes (Treacher, 1977). Body growth rates of lambs reared by ewe lambs normally compare favourably with those of twins reared by yearlings and older ewes but they tend to grow at a slower rate than singles reared by mature ewes (Dýrmundsson, 1976; Baker *et al.*, 1978). Reproductive performance of ewe lambs reared by one year old ewes appears to be similar to that of ewe lambs reared by older ewes (Downing, 1980).

Early breeding and lifetime performance

While most ewes are bred for the first time at the yearling stage, the practice of breeding from ewe lambs is common in certain breeds and localities with favourable conditions (Williams, 1954; Dýrmundsson, 1973). Certainly, breeding from ewe lambs appears to be gaining wider acceptance among sheep farmers.

Several authorities have established that breeding from ewe lambs may considerably check their growth and development at this early stage of life, even if only temporarily. Pálsson (1953) demonstrated that pregnancy in Icelandic ewe lambs substantially retarded growth and development of fat, the last maturing tissue of the body, and that lactation from 12–16 months of age resulted in a lighter carcass, narrower in both the hind quarters and the chest, than if the ewe were barren. He concluded that ewes bred as ewe lambs and those bred first as yearlings reach the same body weight at about 2–2.5 years of age, and similar results have been reported for several other breeds under a wide range of conditions (Spencer *et al.*, 1942; Williams, 1954; Keane, 1974; Steine, 1974; Tyrrell, 1976). The fleece weight of bred ewe lambs is usually lower than that of unmated ones but this difference rarely persists in later years (Southam, Hulet and Botkin, 1971; Tyrrell, 1976; Levine *et al.*, 1978; Ponzoni, Azzarini and Walker, 1979).

Although some early work suggested that breeding from ewe lambs adversely affected subsequent fertility and longevity (Briggs, 1936; Smirnov, 1935), it is now universally accepted that early breeding of adequately nourished and properly managed ewe lambs does not have detrimental effects on lifetime performance (Yalcin and Bichard, 1964; Cannon and Bath, 1969; Hohenboken *et al.*, 1977; Steine, 1979). Moreover, there is a mounting evidence that early breeding may result in enhanced subsequent fertility and a worthwhile increase in the overall productivity of the ewe (Dýrmundsson, 1973; Evans *et al.*, 1975; Levine *et al.*, 1978; Baker *et al.*, 1978; Collyer, 1981).

Selection based on early fertility

It is well documented that estimates of heritability of fertility traits in the ewe vary considerably in different breeds of sheep, but they tend to be rather low (Turner, 1969), and genetic improvement in fertility by direct selection can generally be expected to be a relatively slow process. Against this background some attention has been paid to ewe lamb characters, such as body weight, oestrous performance, ovulatory activity and lambing rate, in an attempt to relate these traits to subsequent fertility. Spencer *et al.* (1942) suggested that earlier sexual development might indicate a generally higher level of fertility in the ewe and Ch'ang and Raeside (1957) proposed that some genetic progress could be made by indirect selection for fertility at the ewe lamb stage.

Wiggins (1955), Hulet, Wiggins and Ercanbrack (1969) and Burfening *et al.* (1972) found that range ewe lambs in the USA, which cycled as ewe lambs, showed better lifetime performance, resulting in greater cumulative production, than those not exhibiting oestrus in their first year. In New Zealand, Hight and Jury (1976) and Moore, Knight and Whyman (1978) have reported on such positive associations. Moreover, Ch'ang and Rae (1972) found that both body weight at 14 months of age and the number of oestrous cycles exhibited during the first autumn of life were genetically correlated with subsequent reproductive efficiency in Romney ewes, indicating that these traits may be useful indirect criteria. However, further estimates obtained by Baker *et al.* (unpublished results cited by Dalton and Baker, 1979) do not support this conclusion and Hight and Jury (1976) found that culling of ewe hoggets on the basis of their body weight resulted in small gains in lifetime lambing performance. Although cyclic activity of ewe lambs may be positively related to ovulation rate at 1.5–2 years of age (Hight and Jury, 1976; Mayer and French, 1979) and possibly at higher ages, it is important to realize the confounding effects of level of nutrition and body condition on such relationships (Barlow and Hodges, 1976; Moore and Smeaton, 1980).

Since faster body growth and more intense oestrous activity are associated with earlier attainment of puberty it would seem reasonable to assume, with reference to the above findings, that sexual precocity in ewe lambs may be related to reproductive performance in subsequent years. Certainly ewes of highly prolific breeds such as the Finnish Landrace (Maijala and Österberg, 1977) and the Romanov (Ricordeau *et al.*, 1978) are also known to undergo early sexual maturity. Results obtained recently from an analysis of data from recorded commercial sheep flocks in the UK (Collyer, 1981) support the evidence presented above. The data relate to Clun Forest, Poll Dorset and crossbred ewe lambs with the mean number of lambs born/100 ewe lambs mated of 82%, 119% and 68%, respectively. While breed differences were apparent, indicating genetic variation, there was ample evidence of a positive relationship between ewe lamb performance and subsequent reproductive efficiency. Thus, within the Clun breed, where this relationship was strongest, the ewes giving birth to twins as ewe lambs produced in their first five years of life an average of 37% more lambs than ewes producing singles as ewe lambs, which in turn had a 19% advantage over those not lambing in their first year of life. Collyer concluded that the most prolific ewes proved to be those producing twins as ewe lambs. Furthermore, according to unpublished data from extensive flock recording in Iceland (J.V. Jónmundsson, 1982) twin births in ewe lambs appear to be associated with a higher mean lambing rate and a relatively high incidence of triplet births in subsequent years. There is also evidence of a strong genetic correlation between lambing rate at one and two years of age (Eikje, 1975).

Apart from the potential of including ewe lamb reproductive performance in the selection of replacements in recorded commercial flocks, it is clear that mating ewe lambs in breeding programmes will shorten the generation interval thus speeding up genetic improvement (Donald, Read and Russell, 1968). In this context Gjedrem (1969) pointed out the advantages of early breeding when rams are progeny tested for the prolificacy and mothering ability of their daughters.

References

ADALSTEINSSON, S. (1972). Experiments on winter shearing of sheep in Iceland. *Acta Agriculturae Scandinavica* **22**, 93–96

ALLEN, D.M. and LAMMING, G.E. (1961). Some effects of nutrition on the growth and sexual development of ewe lambs. *Journal of Agricultural Science, Cambridge* **57**, 87–95

ALLISON, A.J., KELLY, R.W., LEWIS, J.S. and BINNIE, D.B. (1975). Preliminary studies on the efficiency of mating ewe hoggets. *Proceedings of the New Zealand Society for Animal Production* **35**, 83–90

ASDELL, S.A. (1946). *Patterns of Mammalian Reproduction*. Ithaca, New York, Comstock

BADAWY, A.A., EL-BASHARY, A.S. and MOHSEN, M.K.M. (1973). A study of the sexual behaviour of the female Barky sheep. *Alexandria Journal of Agricultural Research* **21**, 1–9

BAKER, R.L., STEINE, T.A., VÅBENO, A.W., BEKKEN, A. and GJEDREM, T. (1978). Effect of mating ewe lambs on lifetime productive performance. *Acta Agriculturae Scandinavica* **28**, 203–217

BARLOW, R. and HODGES, C.J. (1976). Reproductive performance of ewe lambs; genetic correlation with weaning weight and subsequent reproductive performance. *Australian Journal of Experimental Agriculture and Animal Husbandry* **16**, 321–324

BICHARD, M., YOUNIS, A.A., FORREST, P.A. and CUMBERLAND, P.H. (1974). Analysis of production records from a lowland sheep flock. 4. Factors influencing the incidence of successful pregnancy in young females. *Animal Production* **19**, 177–191

BOSHOFF, D.A., BURGER, F.J.L. and CRONJE, J.A. (1975). Sexual activity of Romanov–Karakul crosses under semi arid conditions. *South African Journal of Animal Science* **5**, 91–94

BOWMAN, J.C. (1966). Meat from sheep. *Animal Breeding Abstracts* **34**, 293–319

BRIGGS, H.M. (1936). Some effects of breeding ewe lambs. *Bulletin of the North Dakota Agricultural Experiment Station* **285**, 2 pp.

BURFENING, P.J., HOVERSLAND, A.S. and HORN, J.L. van (1974). Occurrence and frequency of estrus in range ewe lambs. *Proceedings of the American Society of Animal Science (West. Sect.)* **25**, 211–213

BURFENING, P.J., HOVERSLAND, A.S., DRUMMOND, J. and HORN, J.L. van (1971). Supplementation for wintering range ewe lambs; effect on growth and estrus as ewe lambs. *Journal of Animal Science* **33**, 711–714

BURFENING, P.J., HOVERSLAND, A.S., DRUMMOND, J. and HORN, J.L. van

(1972). Effect of estrus as a ewe lamb on lifetime production. *Journal of Animal Science* **34**, 889–890

CANNON, D.J. and BATH, J.G. (1969). Effect of age at first joining on lifetime production by Border Leicester × Merino ewes. *Australian Journal of Experimental Agriculture and Animal Husbandry* **9**, 467, 477–481

CEDILLO, R.M., HOHENBOKEN, W. and DRUMMOND, J. (1977). Genetic and environmental effects on age at first estrus and on wool and lamb production of crossbred ewe lambs. *Journal of Animal Science* **44**, 948–957

CH'ANG, T.S. and RAE, A.L. (1972). The genetic basis of growth, reproduction and maternal environment in Romney ewes. II. Genetic covariation between hogget characters, fertility and maternal environment in the ewe. *Australian Journal of Agricultural Research* **23**, 149–165

CH'ANG, T.S. and RAESIDE, J.I. (1957). A study on the breeding season of Romney ewe lambs. *Proceedings of the New Zealand Society for Animal Production* **17**, 80–87

CHRISTENSON, R.R., LASTER, D.B. and GLIMP, H.A. (1976). Influence of dietary energy and protein on reproductive performance of Finn-cross ewe lambs. *Journal of Animal Science* **42**, 448–452

CHU, T.T. and EDEY, T.N. (1978). Reproductive performance of ewe lambs at puberty. *Proceedings of the Australian Society for Animal Production* **12**, 251

COLLYER, A.B. (1981). Early breeding and subsequent reproductive performance in the ewe. BSc (Hons) Dissertation, University of Nottingham

DALTON, D.C. and BAKER, R.L. (1979). Selection experiments with beef cattle and sheep. In *Selection Experiments in Laboratory and Domestic Animals*, (A. Robertson, Ed.), pp. 131–143. Slough, Commonwealth Agricultural Bureaux

DICKERSON, G.E. and GLIMP, H.A. (1975). Breed and age effects on lamb production of ewes. *Journal of Animal Science* **40**, 397–408

DICKERSON, G.E. and LASTER, D.B. (1975). Breed, heterosis and environmental influences on growth and puberty in ewe lambs. *Journal of Animal Science* **41**, 1–9

DONALD, H.P., READ, J.L. and RUSSELL, W.S. (1968). A comparative trial of crossbred ewes by Finnish Landrace and other sires. *Animal Production* **10**, 413–421

DOWNING, J.M. (1980). Studies on the effects of date of birth and plane of nutrition on attainment of puberty and reproductive performance in Clun Forest ewe lambs. PhD Thesis, University of Wales

DUCKER, M.J., BOWMAN, J.C. and TEMPLE, A. (1973). The effect of constant photoperiod on the expression of oestrus in the ewe. *Journal of Reproduction and Fertility, Supplement* **19**, 143–150

DUFOUR, J.J. (1975). Effects of seasons on postpartum characteristics of sheep being selected for year-round breeding and on puberty of their female progeny. *Canadian Journal of Animal Science* **55**, 487–492

DÝRMUNDSSON, Ó.R. (1972). Studies on the attainment of puberty and reproductive performance in Clun Forest ewe and ram lambs. PhD Thesis, University of Wales

DÝRMUNDSSON, Ó.R. (1973). Puberty and early reproductive performance in sheep. I. Ewe lambs. *Animal Breeding Abstracts* **41**, 273–289

DÝRMUNDSSON, Ó.R. (1976). Breeding from ewe lambs—a common practice in Iceland. *Bulletin of Research Institute Nedri-Ás, Hveragerdi, Iceland* **24**, 12 pp

DÝRMUNDSSON, Ó.R. (1978). Studies on the breeding season of Icelandic ewes and ewe lambs. *Journal of Agricultural Science, Cambridge* **90**, 275–281

DÝRMUNDSSON, Ó.R. (1979). Kynbroski og fengitími íslenska saudfjárins (English abstract). *Náttúrufrædingurinn* **49**, 278–288

DÝRMUNDSSON, Ó.R. (1981). Natural factors affecting puberty and reproductive performance in ewe lambs: a review. *Livestock Production Science* **8**, 55–65

DÝRMUNDSSON, Ó.R. and HALLGRÍMSSON, S. (1978). Reproductive efficiency of Icelandic sheep. *Livestock Production Science* **5**, 231–234

DÝRMUNDSSON, Ó.R. and LEES, J.L. (1972a). Attainment of puberty and reproductive performance in Clun Forest ewe lambs. *Journal of Agricultural Science, Cambridge* **78**, 39–45

DÝRMUNDSSON, Ó.R. and LEES, J.L. (1972b). A note on factors affecting puberty in Clun Forest female lambs. *Animal Production* **15**, 311–314

DÝRMUNDSSON, Ó.R. and LEES, J.L. (1972c). Effect of autumn shearing on breeding activity in Clun Forest ewe lambs. *Journal of Agricultural Science, Cambridge* **79**, 431–433

DÝMUNDSSON, Ó.R. and LEES, J.L. (1972d). Effect of rams on the onset of breeding activity in Clun Forest ewe lambs. *Journal of Agricultural Science, Cambridge* **79**, 269–271

EDEY, T.N., KILGOUR, R. and BREMNER, K. (1978). Sexual behaviour and reproductive performance of ewe lambs at and after puberty. *Journal of Agricultural Science, Cambridge* **90**, 83–91

EDEY, T.N., CHU, T.T., KILGOUR, R., SMITH, J.F. and TERVIT, H.R. (1977). Estrus without ovulation in pubertal ewes. *Theriogenology* **7**, 11–15

EIKJE, E.D. (1975). Studies on sheep production records. VII. Genetic, phenotypic and environmental parameters for productivity traits of ewes. *Acta Agriculturae Scandinavica* **25**, 242–252

ELTAN, Ö. (1974). Peri-parturient behaviour in ewes and early post-natal behaviour and performance in lambs. PhD Thesis, University of Wales

EVANS, A.D., ANDRUS, K., NIELSEN, J.R., GARDNER, R.W., PARK, R.L. and WALLENTINE, M.V. (1975). Early development and breeding of ewe lambs. *Journal of Animal Science* **41**, 266

FOOTE, W.C., SEFIDBAKHT, N. and MADSEN, M.A. (1970). Pubertal estrus and ovulation and subsequent estrus cycle patterns in the ewe. *Journal of Animal Science* **30**, 86–90

FORREST, P.A. and BICHARD, M. (1974). Analysis of production records from a lowland sheep flock. 2. Flock statistics and reproductive performance. *Animal Production* **19**, 25–32

FOSTER, D.L. and RYAN, K.D. (1981). Endocrine mechanisms governing transition into adulthood in female sheep. *Journal of Reproduction and Fertility, Supplement* **30**, 75–80

GJEDREM, T. (1969). Some attempts to increase the efficiency of sheep selection. *Acta Agriculturae Scandinavica* **19**, 116–126

GORDON, I. (1967). Aspects of reproduction and neonatal mortality in ewe lambs and adult sheep. *Journal of the Department of Agriculture, Republic of Ireland* **64**, 76–127

GUNN, R.B. (1977). The effects of two nutritional environments from 6 weeks prepartum to 12 months of age on lifetime performance and reproductive potential of Scottish Blackface ewes in two adult environments. *Animal Production* **25**, 155–164

HAFEZ, E.S.E. (1951). Mating behaviour in sheep. *Nature, London* **167**, 777–778

HAFEZ, E.S.E. (1952). Studies on the breeding season and reproduction of the ewe. *Journal of Agricultural Science, Cambridge* **42**, 189–265

HAFEZ, E.S.E. (1953). Puberty in female farm animals. *Empire Journal of Experimental Agriculture* **21**, 217–225

HAMMOND, J., Jr. (1944). On the breeding season in the sheep. *Journal of Agricultural Science, Cambridge* **34**, 97–105

HAMRA, A.M. and BRYANT, M.J. (1979). Reproductive performance during mating and early pregnancy in young female sheep. *Animal Production* **28**, 235–243

HAMRA, A.M. and BRYANT, M.J. (1982). The effects of level of feeding during rearing and early pregnancy upon reproduction in young female sheep. *Animal Production* **34**, 41–48

HAWKER, H. (1976). The effect of age on sheep production in an arid environment. PhD Thesis, University of New South Wales (*Dissertation Abstracts* (1977), **38**, 2)

HIGHT, G.K. and JURY, K.E. (1976). Hill country sheep production. VII. Relationship of hogget and two-year-old oestrus and ovulation rate to subsequent fertility in Romney and Border Leicester × Romney ewes. *New Zealand Journal of Agricultural Research* **19**, 281–288

HIGHT, G.K., LANG, D.R. and JURY, K.E. (1973). Hill country sheep production. V. Occurrence of oestrus and ovulation rate of Romney and Border Leicester × Romney ewe hoggets. *New Zealand Journal of Agricultural Research* **16**, 509–517

HOHENBOKEN, W. and COCHRAN, P.E. (1976). Heterosis for ewe lamb productivity. *Journal of Animal Science* **42**, 819–823

HOHENBOKEN, W., VAVRA, M., PHILLIPS, R. and McARTHUR, J.A.B. (1977). The effect of age at first lambing on production and longevity of Columbia and Targhee ewes. *Technical Bulletin of the Agricultural Experiment Station* No. 138

HULET, C.V. and PRICE, D.A. (1975). Effects of feed, breed and year of pregnancy in ewe lambs. *Theriogenology* **3**, 15–20

HULET, C.V., WIGGINS, E.L. and ERCANBRACK, S.K. (1969). Estrus in range lambs and its relationship to lifetime reproductive performance. *Journal of Animal Science* **28**, 246–252

JAKUBEC, V. (1977). Productivity of crosses based on prolific breeds of sheep. *Livestock Production Science* **4**, 379–392

JÓNMUNDSSON, J.V. (1982). Segir frjósemi gemlingsárid eitthvad um frjósemi ánna sídar? *Freyr* **78**, 578–579 [Paper in Icelandic]

JOUBERT, D.M. (1962). Sex behaviour of purebred and crossbred Merino and Blackhead Persian ewes. *Journal of Reproduction and Fertility* **3**, 41–49

JOUBERT, D.M. (1963). Puberty in female farm animals. *Animal Breeding Abstracts* **31**, 295–306

KEANE, M.G. (1974). Effect of bodyweight on attainment of puberty and reproductive performance in Suffolk × Galway ewe lambs. *Irish Journal of Agricultural Research* **13**, 263–274

KEANE, M.G. (1975a). Effect of age and plane of nutrition during breeding on the reproductive performance of Suffolk × Galway ewe lambs. *Irish Journal of Agricultural Research* **14**, 91–98

KEANE, M.G. (1975b). The duration of the breeding season in Suffolk × Galway ewe lambs. *Journal of Agricultural Science, Cambridge* **85**, 569–570

LAMMING, G.E. (1969). Nutrition and reproduction. In *The Science of Nutrition of Farm Livestock, Part I* (D. Cuthbertson, Ed.), pp. 411–453. London, Pergamon Press

LAND, R.B. (1970). A relationship between the duration of oestrus, ovulation rate and litter size of sheep. *Journal of Reproduction and Fertility* **23**, 49–53

LAND, R.B. (1978). Reproduction in young sheep: some genetic and environmental sources of variation. *Journal of Reproduction and Fertility* **52**, 427–436

LAND, R.B., RUSSELL, W.S. and DONALD, H.P. (1974). The litter size and fertility of Finnish Landrace and Tasmanian Merino sheep and their reciprocal crosses. *Animal Production* **18**, 265–271

LASTER, D.B., GLIMP, H.A. and DICKERSON, G.E. (1972). Factors affecting reproduction in ewe lambs. *Journal of Animal Science* **35**, 79–83

LEES, J.L. (1967). Effect of time of shearing on the onset of breeding activity in the ewe. *Nature, London* **214**, 743–744

LEES, J.L. (1971). Some aspects of reproductive efficiency in sheep. *Veterinary Record* **88**, 86–95

LEES, J.L. (1978a). Factors affecting puberty and mating behaviour in sheep. In *The Management and Diseases of Sheep* (British Council, Ed.), pp. 124–151. British Council and Commonwealth Agricultural Bureaux

LEES, J.L. (1978b). Functional infertility in sheep. *Veterinary Record* **102**, 232–236

LEVINE, J.M., VAVRA, M., PHILLIPS, R. and HOHENBOKEN, W. (1978). Ewe lamb conception as an indicator of future production in farm flock Columbia and Targhee ewes. *Journal of Animal Science* **46**, 19–25

LEWIS, K.H.C. (1959). Invermay investigation of effects of hoggets mating on lifetime performance of ewes. *New Zealand Journal of Agriculture* **99**, 537–542

MAIJALA, K. and ÖSTERBERG, S. (1977). Productivity of pure Finnsheep in Finland and abroad. *Livestock Production Science* **4**, 355–377

MAYER, H.H. and FRENCH, R.L. (1979). Hogget liveweight–oestrus relationship among sheep breeds. *Proceedings of the New Zealand Society of Animal Production* **39**, 56–62

MOORE, R.W. and MCMILLAN, W.H. (1981). The effect of nutrition and the time of ram introduction on the onset of puberty in Romney ewe lambs. Annual Report 1980/81, Whatawhata Hill Country Research Station

MOORE, R.W. and SMEATON, D.C. (1980). Effects of different growth paths from 4 to 11 months of age on Romney hogget oestrus and subsequent reproduction. *Proceedings of the New Zealand Society of Animal Production* **40**, 27–33

MOORE, R.W., KNIGHT, T.W. and WHYMAN, D. (1978). Influence of hogget oestrus on subsequent ewe fertility. *Proceedings of the New Zealand Society of Animal Production* **38**, 90–96

MOUNIB, M.S., AHMED, I.A. and HAMADA, M.K.O. (1956). A study of sexual behaviour of the female Rahmany sheep. *Alexandria Journal of Agricultural Research* **4**, 85–108

MURRAY, R.M. (1972). Age of onset of puberty in Merino ewes in semi-arid tropical Queensland. *Proceedings of the Australian Society of Animal Production* **9**, 181–185

NEDKVITNE, J.J. (1975). Fóring av lam som skal lamma årsgamle. Mimeo No. **50**, 17 pp. Agricultural University of Norway (NLH)

NEDKVITNE, J.J. (1979). Effect of nutrition and of shearing before the mating season on the breeding activity of female lambs. *Proceedings EAAP, Sheep and Goat Commission, Harrogate*

PÁLSSON, H. (1953). Áhrif fangs á fyrsta vetri á vöxt og broska ánna (English abstract). *Atvinnudeild Háskólans, Rit Landbúnadardeildar*, B-Flokkur No. **5**, 84 pp

PÁLSSON, H. and VERGÉS, J.B. (1952). Effects of plane of nutrition on growth and development of carcase quality in lambs. *Journal of Agricultural Science, Cambridge* **42**, 1–149

PONZONI, R.W., AZZARINI, M. and WALKER, S.K. (1979). Production in mature Corriedale ewes first mated at 7 to 11 or 18 months of age. *Animal Production* **29**, 385–391

PRICE, D.A. and ERCANBRACK, S.K. (1975). Lamb production of Finnsheep crossbred ewe lambs. *Journal of Animal Science* **41**, 255–256

QUIRKE, J.F. (1977). Studies related to the reproductive performance of adult and immature sheep. PhD Thesis, National University of Ireland, Dublin

QUIRKE, J.F. (1978a). Onset of puberty and oestrous activity in Galway, Finnish Landrace and Finn-cross ewe lambs during their first breeding season. *Irish Journal of Agricultural Research* **17**, 15–23

QUIRKE, J.F. (1978b). Reproductive performance of Galway, Finnish Landrace and Finn-cross ewe lambs. *Irish Journal of Agricultural Research* **17**, 25–32

QUIRKE, J.F. (1979). Effect of body weight on the attainment of puberty and reproductive performance of Galway and Fingalway female lambs. *Animal Production* **28**, 297–307

QUIRKE, J.F. (1981). Regulation of puberty and reproduction in female lambs: a review. *Livestock Production Science* **8**, 37–53

QUIRKE, J.F., SHEEHAN, W. and LAWLOR, M.J. (1978). The growth of pregnant female lambs and their progeny in relation to dietary protein and energy during pregnancy. *Irish Journal of Agricultural Research* **17**, 33–42

RICORDEAU, G., TCHAMITCHIAN, L., THIMONIER, J., FLAMANT, J.C. and THERIEZ, M. (1978). First survey of results obtained in France on reproductive and maternal performance in sheep, with particular reference to the Romanov breed and crosses with it. *Livestock Production Science* **5**, 181–201

ROBINSON, J.J. and ØRSKOV, E.R. (1975). An integrated approach to improving the biological efficiency of sheep meat production. *World Review of Animal Production* **11**, 63–76

ROBINSON, J.J., FRASER, C., CORSE, E.L. and GILL, J.C. (1971). Reproductive performance and protein utilization in pregnancy of sheep conceiving at eight months of age. *Animal Production* **13**, 653–660

SADLEIR, R.M.F.S. (1969). *The Ecology of Reproduction in Wild and Domestic Mammals*. London, Methuen

SCHINCKEL, P.G. (1954). The effect of the ram on the incidence and occurrence of oestrus in ewes. *Australian Veterinary Journal* **30**, 189–195

SIERRA ALFRANCA, I. (1979). Mejora de los caracteres reproductivos de la raza Rasa Aragonesa por cruzamiento con la Romanov (English abstract). *Zootechnia* **28**, 9–34

SMIRNOV, L. (1935). Prolificacy of the Romanov sheep (in Russian). *Problemў Zhivotnovodstva* **8**, 7–19 (*Animal Breeding Abstracts* **4**, 195–196)

SOUTHAM, E.R., HULET, C.V. and BOTKIN, M.P. (1971). Factors influencing reproduction in ewe lambs. *Journal of Animal Science* **33**, 1282–1287

SPENCER, D.A., SCHOTT, R.G., PHILLIPS, R.W. and AUNE, B. (1942). Performance of ewes first bred as lambs compared with ewes bred first as yearlings. *Journal of Animal Science* **1**, 27–33

STEINE, T.A. (1974). Verknaden av lamming ved 1 års alder på produksjonseigenskaper hjå sau (English abstract). Report No. **347**, 11 pp. Agricultural University of Norway (NLH)

STEINE, T.A. (1979). Effect of early breeding on production performance of ewes. *Proceedings EAAP, Sheep and Goat Commission, Harrogate*

STOERGER, M.F., HINDS, F.G., LEWIS, J.M., WALLACE, M. and DZUIK, P.J. (1976). Influence of dietary roughage level on reproductive rate in ewe lambs. *Journal of Animal Science* **43**, 952–958

THIMONIER, J. (1975). Etude de la puberté et de la saison sexuelle chez les races prolifiques et leurs croisements avec des races Francaises. *lères Journées de la Recherche Ovine et Caprine* **11**, 18–37

TIERNEY, M.L. (1976). Genetic aspects of puberty in Merino ewes. In *Sheep Breeding*, pp. 322–329. Proceedings of International Congress, Muresk and Perth, W. Australia

TREACHER, T.T. (1977). The effects on milk production of the number of lambs suckled and age, parity and size of ewe. *Proceedings of EAAP, Sheep and Goat Commission, Brussels*

TURNER, H.N. (1969). Genetic improvement of reproduction rate in sheep. *Animal Breeding Abstracts* **37**, 545–563

TYRRELL, R. (1976). Some effects of pregnancy in eight-month-old Merino ewes. *Australian Journal of Experimental Agriculture and Animal Husbandry* **16**, 458–461

VERESS, L., LOVAS, L., RADNAI, L., VÉGH, J. and TURAI, I. (1979). Influence of the beginning of puberty on ewe's performance. *Proceedings of EAAP, Sheep and Goat Commission, Harrogate*

WATSON, R.H. and GAMBLE, L.C. (1961). Puberty in the Merino ewe with special reference to the influence of season of birth upon its occurrence. *Australian Journal of Agricultural Research* **12**, 124–138

WIGGINS, E.L. (1955). Estrus in range ewe lambs and its relation to subsequent reproduction. *Journal of Animal Science* **14**, 1260

WILLIAMS, S.M. (1954). Fertility in Clun Forest sheep. *Journal of Agricultural Science, Cambridge* **45**, 202–228

YALCIN, B.C. and BICHARD, M. (1964). Crossbred sheep production. I. Factors affecting production from the crossbred ewe flock. *Animal Production* **6**, 73–84

YOUNIS, A.A., EL-GABOORY, I.A., EL-TAWIL, E.A. and EL-SHOBOKSHY, A.S. (1978). Age at puberty and possibility of early breeding in Awassi ewes. *Journal of Agricultural Science, Cambridge* **90**, 255–260

21

ARTIFICIAL INDUCTION OF PUBERTY IN EWE LAMBS

J.F. QUIRKE[†,‡], T.E. ADAMS[†] and J.P. HANRAHAN[‡]
Department of Animal Science, University of California, Davis, California, USA[†] and Agricultural Institute, Belclare, Co. Galway, Eire[‡]

Sheep producers in Britain, and in many other countries as well, traditionally manage their ewes to lamb first at two years of age. Lifetime production, however, can be enhanced by breeding ewes to produce a lamb crop in their first year. First year lambing has assumed increased importance in the sheep industry today due to increased economic pressures to intensify production and improve productivity.

Interest in controlled breeding in ewe lambs arises for a number of reasons. The breeding season commences later for ewe lambs than for mature ewes (Hafez, 1952) which can result in an undesirable extension of the lambing period when ewe lambs are included in the breeding flock. Also, the attainment of puberty depends on a variety of factors, including breed, liveweight, date of birth and age (Dýrmundsson, 1981). Therefore, there can be great variation both in the timing of first oestrus and in the proportion of ewe lambs which are mated in the first season. Spring-born female lambs that fail to attain puberty before 7–8 months of age are, of course, automatically excluded from the breeding ewe population in their first year and, because of seasonal factors, are unlikely to lamb before they are two years old. In view of these considerations, the development of a simple and effective humoral method for the induction and synchronization of oestrus and ovulation in ewe lambs could be of considerable practical advantage in the management of breeding flocks.

Some consideration of the physiology of puberty and the endocrine regulation of the oestrous cycle is essential in the considered manipulation of reproductive processes in very young ewes through the use of exogenous hormones. This chapter will focus on the endocrinology of the pubertal process, the methods used to induce and synchronize cycles in ewe lambs, and the factors which influence the occurrence of oestrus, ovulation and pregnancy. Reproductive failure following mating is more common in ewe lambs than in mature ewes (Dýrmundsson, 1973; Quirke, 1981) and the possible reasons for this phenomenon will also be examined. The endocrinology of the ovine oestrous cycle, and the physiological basis of the various treatments which have been applied to control ovulation in the sheep, have been dealt with extensively elsewhere (Cumming, 1979; Haresign, 1978; Thimonier, 1979). It is not proposed, therefore, to consider either of these aspects in any detail here.

Endocrinology of puberty

THE REPRODUCTIVE SYSTEM DURING FOETAL DEVELOPMENT

Attainment of reproductive competence in sheep, as in other mammalian species, is not an abrupt transition from quiescence to full function but is, rather, the consequence of the gradual functional maturation of the gonads, adenohypophysis, and hypothalamus (Grumbach, 1980; Ojeda *et al.*, 1980; Foster, 1980). The evolution of these elements of the reproductive process to the functional synchrony evident in the adult begins early in prenatal life. The call to reproductive development is heard first in the embryo of the genetic male. Within four weeks of intrauterine life the condensation of mesoderm along the genital ridge comes to assume the morphologic character of the embryonic testis (Mauleon, 1978). Secretion of testosterone and other androgens from this primitive organ is evident by 30–35 days of embryonic life (Attal, 1969; Sklar *et al.*, 1978; Levasseur, 1979). Attending the secretion of testosterone in the genetic male is suppression of the development of the Müllerian duct system, the primordia of the female genitalia (Mauleon, 1978). In the absence of testicular differentiation and testosterone secretion the mesoderm of the genital ridge differentiates to form the ovary, the Wolffian ducts regress and the Müllerian ducts begin the process of differentiation which will eventually lead to the primary reproductive organs of the female. Development of female reproductive structures is evident by the 35th day of intrauterine life in the foetal lamb (Mauleon, 1978). By most accounts the foetal ovary does not assume an active secretory function, although this tissue does contain a full complement of steroidogenic enzymes (Mauleon, Bezard and Terqui, 1977; Levasseur, 1979). Oestrogens (principally 17α-oestradiol, oestrone, and 17β-oestradiol) are evident in the circulation of the foetal lamb but these are thought to be of maternal origin (Findlay and Seamark, 1973).

By morphological criteria the adenohypophysis of the foetal lamb is functionally differentiated by the seventh week of intrauterine life (Alexander *et al.*, 1973). Indeed, gonadotrophic hormones are evident in both the foetal pituitary and the embryonic circulation by day 55 of gestation (Foster *et al.*, 1972a). The temporal changes in circulating gonadotrophins through the course of gestation is similar in male and female foetuses, with the concentration of luteinizing hormone (LH) increasing to a maximum at about mid gestation and declining thereafter while follicle stimulating hormone (FSH) continues to increase through the third quarter of gestation before eventual decline (Foster *et al.*, 1972a; Sklar *et al.*, 1978; 1981). The pituitary gonadotrophin content continues to increase throughout gestation (Foster *et al.*, 1972a). Secretion of gonadotrophins from the foetal pituitary is augmented during the same temporal frame in which oogenesis and the initial stages of folliculogenesis occur (Levasseur, 1979). This temporal relationship has led to the presumption that gonadotrophic hormones of foetal origin assume a critical role in the functional development of the ovary (Levasseur, 1979; Grumbach, 1980). Lending credence to this postulate is the observation that ovarian development is attenuated after foetal decapitation *in utero* (Mauleon, 1978).

That the hypothalamus also assumes a secretory function early in foetal development is indicated by the observation that the hypothalamic hormone which regulates gonadotrophin secretion in the adult i.e. gonadotrophin releasing hormone (GnRH) has been identified in the hypothalamus of the foetal lamb by the 58th day of gestation (Mueller *et al.*, 1978). Furthermore, hypothalamic extracts or synthetic GnRH will induce gonadotrophin secretion from the pituitary glands of foetal lambs (Foster *et al.*, 1972b; Mueller *et al.*, 1981). Indeed, the response of the foetal pituitary to exogenous GnRH increases progressively through gestation to an apex by the third quarter of gestation and is reduced thereafter, in a pattern markedly similar to the pattern of gonadotrophin secretion during normal foetal development (Mueller *et al.*, 1981). These observations lead to the presumption that secretion of GnRH from the hypothalamus of the ovine foetus is relatively unfettered through the first three quarters of gestation but GnRH secretion is reduced through the remainder of gestation coincident with the rise of gonadal steroids in the foetal circulation and the maturation of the negative feedback control of the hypothalamic–hypophyseal unit (Grumbach, 1980; Foster, 1980; Gluckman *et al.*, 1979).

POSTNATAL DEVELOPMENT OF THE REPRODUCTIVE SYSTEM

Reproductive development in the neonatal and postnatal lamb is keyed to the relative impact of the gonadal steroids on the hypothalamic–hypophyseal locus. In the ewe lamb the attainment of reproductive competence represents the climax of protracted interplay between the negative and positive feedback effects of oestradiol on gonadotrophin secretion (Ryan and Foster, 1980; Foster, 1980). Although the negative feedback effects of oestradiol on the hypothalamic–hypophyseal axis have been noted *in utero* (Gluckman *et al.*, 1979), the suppressive effects of oestradiol come to full flower only after the fifth week of neonatal life (Foster, Cook and Nalbandov, 1972; Foster, Jaffe and Niswender, 1975). After this period in development removal of the steroid generator (ovariectomy) is recognized by the hypothalamic–hypophyseal unit which compensates accordingly with augmented gonadotrophin secretion. That the hypothalamic–hypophyseal locus in the prepubertal animal is extraordinarily sensitive to the suppressive influences of oestradiol is indicated by the observation that the post-castration increase in gonadotrophins can be negated by a level of exogenous oestradiol which is ineffective in the adult ovariectomized ewe in this regard (Foster and Ryan, 1979). Furthermore, the potency of oestradiol as a negative regulator of the hypothalamic–hypophyseal locus is markedly reduced in the ovariectomized lamb at 28–32 weeks of postnatal development (Ryan and Foster, 1980; Foster and Ryan, 1979). Perhaps not coincidentally this is about the time of first ovulation in intact ewe lambs. These phenomena have been taken to indicate that the hypothalamic–hypophyseal axis is inordinately attuned to the suppressive effects of oestradiol during much of the prepubertal period, but becomes much less sensitive to the negative feedback effects of oestradiol during the peripubertal period (Ryan and Foster, 1980). For full expression the period of transition in steroid sensitivity must coincide with the normal breeding season. In the absence of such coincidence at 28–32 weeks of age the transition in steroid sensitivity and, thus, puberty are delayed until onset of the subsequent breeding season (Foster, 1981). Similar hypotheses have been presented to account for the induction of puberty in the primate and rodent (Grumbach, 1980; Ojeda *et al.*, 1980).

The positive feedback effects of oestradiol are also evident early in postnatal life. By three to five weeks of age the positive feedback mechanisms are in place and exogenous oestradiol will induce a surge of gonadotrophin secretion (Land, Thimonier and Pelletier, 1970; Foster and Karsch, 1975). Apparently the positive feedback mechanism gradually matures during the prepubertal period because the magnitude of oestradiol-induced gonadotrophin secretion increases progressively through this period (Foster and Karsch, 1975). The hypothalamic–hypophyseal unit is relatively less sensitive to the positive feedback effects of oestradiol than to the suppressive effects of this steroid and, therefore, a much higher concentration of oestradiol is required to permit expression of the gonadotrophin surge than is required to suppress secretion of the gonadotrophins (Foster and Karsch, 1975; Foster and Ryan, 1979).

Secretion of gonadotrophins in the adult ewe is characterized by two phases: tonic gonadotrophin secretion during the luteal and follicular stages of the oestrous cycle and the surge secretion required for ovulation (Hauger, Karsch and Foster, 1977). In the developing ewe lamb the tonic mode of gonadotrophin secretion is functional through the bulk of the prepubertal period. The maturation of the surge mechanism of gonadotrophin secretion heralds the onset of puberty in the young sheep (Ryan and Foster, 1980). The tonic level of FSH secretion increases about two-fold during the first ten weeks of life and then remains relatively constant through the remainder of the pubertal process (Foster, Jaffe and Niswender, 1975; Foster *et al.*, 1975). In contrast, the secretion of LH in the prepubertal ewe lamb is highly dynamic, both temporarily and quantitatively, reflecting the great impact of the steroidal regulatory mechanisms on the secretion of this gonadotrophin and the rapid rate of decay of this glycoprotein. The characteristic pattern of LH secretion in the prepubertal ewe lamb is pulsatile, with the frequency and amplitude of the LH quanta particularly increased during the peripubertal period, coincident with the lessened effects of the suppressive influence of oestradiol on the hypothalamic–hypophyseal locus (Foster *et al.*, 1975; Ryan and Foster, 1979). An hourly pulse frequency (*circhoral rhythm*; Knobil, 1980) seems to be the threshold which permits continuation of the maturational development of the follicles leading to the first ovulation. Indeed, ovulation in the prepubertal lamb can be induced by the hourly administration of exogenous LH; however, LH administration at three-hourly intervals is ineffective in this regard (Ryan and Foster, 1980). The functional changes triggered by the circhoral LH stimulation in the prepubertal lamb have not been precisely defined but it is not unreasonable to postulate that such a frequency of stimulation is required to induce the synthesis of receptors for LH in the developing follicles and thereby enhance the sensitivity of the developing Graafian follicles to the LH stimuli. Increases in LH receptor concentration have been noted in the developing follicles of the rodent during the peripubertal period (Ojeda *et al.*, 1980). Furthermore, such increases in LH receptor concentration are associated with augmented oestradiol secretion in response to a quantum of exogenous LH or HCG (human chorionic gonadotrophin). Moreover, although FSH is an active modulator of LH receptor concentration, LH and FSH in tandem are much more potent in this regard (Richards, Rao and Ireland, 1978).

It is perhaps pertinent to note that ewe lambs ovariectomized prior to the induction of puberty exhibit LH pulses with a circhoral period (Foster, Jaffe and Niswender, 1975). This observation has led to the presumption that the high sensitivity of the hypothalamic–hypophyseal unit to the negative feedback effects of oestradiol results in suppression of the frequency and magnitude of GnRH release from the hypothalamus of the intact prepubertal lamb. When freed of the constraints imposed by oestradiol, as in the ovariectomized lamb or in the intact lamb during the peripubertal period, the hypothalamus assumes an intrinsic circhoral rhythm of GnRH release. Thus the development of the circhoral pattern of LH release heralds release of the hypothalamus from the suppressive effects of oestradiol and signals the imminence of the first gonadotrophin surge and ovulation. Indeed, the circhoral pattern of LH secretion in the intact ewe lamb is followed within 48–72 hours by the first gonadotrophin surge, ovulation, and the attainment of reproductive competence (Ryan and Foster, 1980).

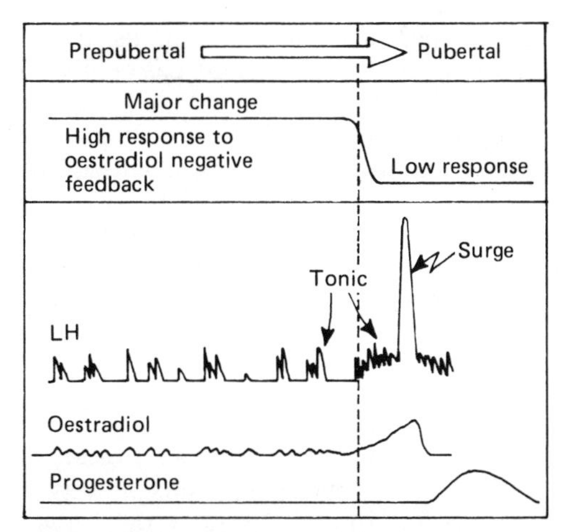

Figure 21.1 Hypothetical model for onset of the pubertal process in the female lamb. From Foster and Ryan (1980)

A hypothetical model for puberty in the ewe lamb is shown in *Figure 21.1.* The onset of puberty in the lamb is the climax of a gradual process of maturation that begins early in foetal life and which represents a complex interplay between factors from the hypothalamus, anterior pituitary and the gonads. Impinging upon this schema may be environmental factors, such as photoperiod, temperature, nutrition, or internal factors, such as humoral inputs from the adrenal gland or the pineal. Puberty could well be regarded as the final goal in the contest between two opposing forces, the negative and positive feedback effects of oestradiol. The suppressive effects of oestradiol dominate through most of the contest but in the end the suppressive effects are markedly weakened and the positive feedback effect of oestradiol eventually becomes indomitable, thus resulting in the first gonadotrophin surge and ovulation.

Induction of puberty

PHYSIOLOGICAL BASIS FOR INDUCTION OF PUBERTY

The induction and control of ovulation and oestrus in sheep is achieved either by duplicating or terminating the secretory activity of the corpus luteum. Progesterone or synthetic progestagens administered by oral, subcutaneous or intravaginal routes simulate the action of the corpus luteum. They block the positive feedback effect of oestradiol and prevent the LH discharge which leads to ovulation and the formation of a new corpus luteum. The gonadotrophins which are released spontaneously following the withdrawal of the progestagen treatment stimulate the maturation of preovulatory follicles which produce sufficient oestradiol to induce behavioural oestrus and the LH release required to trigger ovulation. In prepubertal ewe lambs and seasonally anoestrous ewes the progestagen treatment also increases hypothalamic sensitivity to oestradiol thereby facilitating the expression of behavioural oestrus. It is necessary in these classes of females, however, to supplement the endogenous release of gonadotrophins following cessation of the progestagen treatment with pregnant mare's serum gonadotrophin (PMSG).

Another approach to the control of oestrus is to use either prostaglandin $F_{2\alpha}$ or one of its potent analogues to terminate the secretory activity of the corpus luteum. The prostaglandins, however, are only luteolytic after day 4–5 of the oestrous cycle in sheep and are unable to induce oestrus and ovulation in non-cyclic females (Thimonier, 1981) and are thus unable to induce ovulation in prepubertal ewe lambs.

PRACTICAL ASPECTS OF TREATMENT

Progestagens, administered by intramuscular injection or as feed additives have been used in conjunction with PMSG to induce precocious puberty in ewe lambs (Burfening and van Horn, 1970; Gordon, 1967a; Wright *et al.*, 1976). However, these approaches have clear limitations from the labour and management points of view and are not extensively used under practical conditions. Administration of progestagens via the intravaginal route, which is now so widely used in the treatment of mature ewes (Gordon, 1975; Thimonier and Cognie, 1977), has not been widely applied in ewe lambs because of apprehension concerning difficulties in inserting and removing the sponges. Although these operations can sometimes be difficult in ewe lambs for anatomical reasons, problems can be avoided if care is exercised. It is important to avoid damage to the vaginal tissues, particularly in lambs which present a ring-like constriction in the vagina. This condition, which has been described by Inkster (1958), can be quite common in young maiden ewes. In such animals, the sponge should be pushed past the constriction using an index finger rather than the applicator. The sponge loss rate in ewe lambs is usually within the range considered normal for mature ewes and the incidence of failure in sponge removal is also low (Keane, 1974; Quirke, 1979a; Quirke and Gosling, 1981).

Subcutaneous implants are an alternative to the intravaginal sponge and have given acceptable results in trials with ewe lambs (Keane, 1974; Southam, Hulet and Botkin, 1971; Tsakalof, Vlachos and Latousakis, 1977). Although the risk of damage to the vagina, inherent in the use of sponges in young ewes, can be avoided by the use of implants, they are more likely to be lost if incorrectly inserted and can become surrounded by localized infections. Therefore, they are not widely used in sheep at any age.

TYPES OF STEROIDS AND DOSES USED

The most commonly used progestagens in sheep sponges are cronolone (G.D. Searle Co.) and medroxyprogesterone acetate (MAP) (Upjohn Co., Ltd.). Comparative trials have shown that sponges containing these compounds are equally effective in inducing and synchronizing oestrus, in both adult ewes and ewe lambs at doses of 30 and 60 mg, respectively (Gordon, 1975; Quirke, 1979a). The optimum dose of progestagen to employ in sponges for ewe lambs has received scant attention. In mature ewes treated during the breeding season improvements have been obtained in both conception rate and litter size by increasing the dose of cronolone from 30 to 40 mg (Colas *et al.*, 1973). French researchers favour the higher dose of cronolone (40 mg) for ewe lambs and recommend a 14-day treatment duration (Thimonier and Cognie, 1977). Treatment protocols for inducing puberty in ewe lambs which include HCG and oestrogens have been examined in several experimental situations, but it would appear unlikely that any additional benefit can be expected from the use of such hormones (Keane, 1975a; Quirke, 1981; Thimonier *et al.*, 1968).

USE OF PMSG

In both mature ewes and ewe lambs which have attained puberty, a treatment with progestagen sponges is sufficient to synchronize cycles during the breeding season (Keane, 1974; Quirke, 1979b). However, it is necessary to use PMSG in conjunction with the sponges when induced puberty is desired in advance of the normal breeding season. It is also known that a low dose of PMSG given at the time of sponge removal will effect a more reliable and precise synchronization of oestrus, ovulation, and improved fertility in mature ewes (Haresign, 1978). For this reason and because it may not always be known which lambs have attained puberty at any given time during the breeding season, PMSG should be routinely used as part of the treatment protocol.

The dose of PMSG to use can be influenced by several factors.

Table 21.1 EFFECT OF INCREASING THE DOSE OF PMSG AT SPONGE REMOVAL ON OVULATION AND PREGNANCY RATES AND NUMBER OF EMBRYOS IMPLANTED IN 8-MONTH OLD EWE LAMBS

	Dose of PMSG (iu)		
	500	*750*	*1000*
Mean ovulation rate[a]	1.78	2.01	2.38
% ewe lambs pregnant[b]	47.2	32.7	37.2
Mean number of corpora lutea in pregnant ewe lambs[b]	1.90	2.72	3.27
Mean number of embryos[b]	1.27	1.39	1.68

[a] At 5–8 days post sponge removal.
[b] At 28–33 days post sponge removal.
From Quirke (1979b)

Incremental levels of PMSG, given to 7–8 month old ewe lambs at the time of sponge removal, can lead to an increase in ovulation rate and number of embryos implanted (Quirke, 1979a; *Table 21.1*). However, treatment with high levels of PMSG should be avoided since this may result in an increase in the incidence of multiple births and can bring excessively high lamb mortality in its train (Dýrmundsson, 1973; Quirke, Sheehan and Lawlor, 1978; Quirke and Gosling, 1981; Quirke, 1979c). Under practical farming conditions, the ewe lamb with twins can be more of a liability than an asset to the producer when the high mortality and low growth rate of the twin progeny, together with the additional maternal stress involved, are considered. Any aspect of the hormone treatment which can influence the twinning rate should therefore be carefully assessed. There is substantial genetic variation in the ovulation rate of ewe lambs in response to PMSG (Quirke, 1978). The optimum level of PMSG to use may vary according to breed and should be the minimum necessary for a good oestrous response and a normal ovulation rate. Thimonier and Cognie (1977) recommend PMSG in the range 250–500 iu for 7–8 month old lambs during the breeding season, and a somewhat higher level (400–600 iu) during the seasonal anoestrous period.

NUTRITION AND LIVEWEIGHT

Inadequate nutrition appears to adversely affect both the incidence and timing of oestrus following progestagen–PMSG treatment (Keane, 1975b; *Table 21.2*). Although the nutritional restriction imposed in this experiment was severe, the results, nevertheless, highlight the impact of inadequate nutrition on reproductive performance in the progestagen–PMSG treated ewe lamb. This is a matter which can be easily overlooked during the autumn period when the quality and availability of pasture diminish rapidly.

The influence of liveweight *per se* on the reproductive performance of ewe lambs following progestagen–PMSG treatment has received little attention. Studies of the effects of body weight on the progression to normal puberty in ewe lambs have yielded equivocal results (Dýrmundsson, 1981; Quirke, 1981). The limited number of trials conducted with progestagen–PMSG treated lambs have also yielded inconclusive results

Table 21.2 EFFECT OF LEVEL OF NUTRITION ON THE INCIDENCE AND TIME OF OCCURRENCE OF OESTRUS IN PROGESTAGEN–PMSG TREATED EWE LAMBS

	Nutritional level	
	Ad libitum	*Restricted*
No. of animals	40	40
No. which showed heat (%)[(a)]	39 (97.5)	27 (67.5)
No. on heat at 40 hours[(b)] (%)[(c)]	28 (71.8)	6 (22.2)
No. on heat at 48 hours (%)	10 (25.6)	11 (40.7)
No. on heat at 72 hours (%)	1 (2.6)	3 (11.1)
No. on heat at 96 hours (%)	—	7 (25.9)

[(a)]Of total group.
[(b)]Following pessary withdrawal.
[(c)]Of those that showed heat.
From Keane (1975a)

Table 21.3 INFLUENCE OF LIVEWEIGHT ON THE FERTILITY OF EWE LAMBS OF THE LACAUNE BREED FOLLOWING TREATMENT WITH PROGESTAGEN SPONGES AND 400iu PMSG

Body weight class (kg)	*Percentage fertility*	
	Induced oestrus	*Two consecutive oestrous periods*
30.5–40	47.4	55.3
40.5–50	65.2	78.3

From Thimonier *et al.* (1968)

(Al-Wahab and Bryant, 1978a; Keane, 1974; Thimonier *et al.*, 1968). The most definitive data are those of Thimonier *et al.* (1968) which show a substantial beneficial effect of increased liveweight on reproductive performance (*Table 21.3*). It has been emphasized by a number of researchers that hormone therapy should not be applied to poorly grown animals (Dýrmundsson, 1973). Under practical conditions, only lambs which are heavier than 60% of their mature body weight and more than seven months of age should be considered for treatment.

AGE AND SEASONAL EFFECTS

Although it is possible to activate many of the components of the reproductive process in female lambs at a very early age with exogenous hormones (Quirke, 1981), it is unrealistic, from a practical animal management viewpoint, to consider inducing mating earlier than 20 weeks. After this time it would appear that age is not a factor of critical significance in determining the occurrence of oestrus and ovulation following treatment with progestagen sponges and PMSG (Al-Wahab and Bryant, 1978a; 1978b). However, fertility improves significantly with advancing age and approach of the onset of the breeding season (Keane, 1974; Thimonier *et al.*, 1968; *Table 21.4*).

It is difficult to dissociate the effects of age, season and liveweight on lambing performance, and the influence of these factors is confounded in most studies. The continuation of ovarian cyclicity following hormonal stimulation can be markedly influenced by these variables. This is a matter of practical significance since the return to service of lambs which fail to become pregnant at the induced oestrus affords such animals a second opportunity to conceive, thus enhancing the overall effectiveness of the treatment. Under field conditions in Britain the probability of continued ovarian cyclicity subsequent to progestagen–PMSG treatment appears to be very low (11%) in 20-week old lambs during August, moderately high (65%) in 28-week old lambs during September and is maximal (87–100%) during the breeding season in October (Al-Wahab and Bryant, 1978b). Photostimulation at a young age can, apparently, promote cyclicity but does not augment fertility at the induced oestrus (Al-Wahab and Bryant, 1978a).

Table 21.4 INFLUENCE OF TIME OF YEAR AND APPROACH OF THE BREEDING SEASON ON THE FERTILITY OF LACAUNE EWE LAMBS FOLLOWING PROGESTAGEN–PMSG TREATMENT

Time of year	*% lambs in oestrus*	*Percentage fertility*	
		Induced oestrus	*Two consecutive oestrous periods*
15–31 July	100	9.1	9.1
1–15 August	95.1	39.0	51.2
16–31 August	98.5	66.6	74.5

From Thimonier *et al.* (1968)

The weight of experimental evidence indicates that pregnancy is most likely to be established and maintained when the progestagen–PMSG treatment is applied either close to the time of onset of, or during, the breeding season. Delaying the time of treatment from the breeding season in December to the early stages of the first anoestrum in January or to the period of complete anoestrum in March reduces its effectiveness in terms of both oestrous response and lambing outcome (Thimonier *et al.*, 1968).

MATING MANAGEMENT

The use of natural service as the method of insemination following the application of oestrous control measures in mature ewes requires a ewe to ram ratio of not more than 10:1 to ensure an adequate conception rate. Some form of supervision of mating, either hand mating or introduction of the rams 48 hours after sponge removal, has also been shown to improve fertility (Boland and Gordon, 1979; Gordon, 1975; Jennings and Crowley, 1972). One might assume that similar measures are necessary to ensure optimum fertility at the controlled oestrus in ewe lambs (Keane, 1975a).

Artificial insemination has been successfully employed in ewe lambs in France (Colas, 1979; *Table 21.5*). Unlike natural service, this procedure permits one to monitor semen quality carefully when inseminations are

Table 21.5 LAMBING RATE OF PROGESTAGEN–PMSG TREATED EWE LAMBS (EIGHT MONTHS OLD) FOLLOWING ARTIFICIAL INSEMINATION

Breed	*Season*	
	Spring	*Autumn*
Ile-de-France	53.6 (56)	57.2 (98)
Prealpes	77.7 (36)	88.8 (18)

Figures in parentheses indicate numbers of ewe lambs inseminated.
From Colas (1979)

required outside the normal breeding season; it can also overcome the consequences of any deficiencies in the behavioural responses of ewe lambs to rams at oestrus (Edey, Kilgour and Bremner, 1978) which might result in failure of insemination under natural mating conditions. A single insemination with 500×10^6 spermatozoa, in a volume of 0.25 ml, at 52 hours after sponge withdrawal is adequate for optimum fertility in ewe lambs according to Colas (1979).

GENETIC EFFECTS

A number of studies in Britain and Ireland have revealed substantial differences among breeds in the reproductive potential of ewe lambs following mating under natural conditions (Donald, Read and Russel, 1968; Gordon, 1967b; Quirke, 1979c). There is also evidence for genetic variation in reproductive performance following hormone treatment (Quirke, 1979b; Quirke and Gosling, 1981; Southam, Hulet and Botkin, 1971). Furthermore, although progestagen–PMSG can effectively induce puberty and synchronize oestrus in pubertal lambs of many breeds, it is unlikely to improve the overall level of fertility for any breed beyond that which can be obtained under natural conditions during the breeding season (Quirke, 1978; Thimonier *et al.*, 1968). The results obtained by Quirke and Gosling (1981; *Table 21.6*) serve as an example of both the level of

Table 21.6 COMPARISON OF THE REPRODUCTIVE PERFORMANCE OF EIGHT-MONTH-OLD GALWAY AND SUFFOLK × GALWAY EWE LAMBS AFTER PROGESTAGEN–PMSG TREATMENT DURING THE BREEDING SEASON

	Galway	*Suffolk × Galway*
No. of ewe lambs	66	66
Body weight at mating (kg)	48.5	48.4
No. mated (%)	66 (100)	66 (100)
No. returned to service (%)	16 (24.2)	16 (24.2)
Conceptions:		
1st service (%)	29 (43.9)	41 (62.1)
2nd service (%)	8 (50)	6 (37.5)
1st and 2nd services (%)	37 (56.1)	47 (71.2)
Births:		
Single-bearing ewes (%)	26 (70.3)	37 (78.7)
Twin-bearing ewes (%)	11 (29.7)	10 (21.3)
Litter size	1.19	1.12

From Quirke and Gosling (1981)

performance, and the variation thereof, which can be obtained with lowland breed types in Ireland; the animals in this study were treated with 40 mg cronolone sponges and 400 iu PMSG during October and subsequently exposed to rams for a 25-day mating period.

The problem of subfertility in ewe lambs

It is a common finding in experiments on breeding ewe lambs, both under natural conditions and following hormone therapy, that the conception and lambing rates obtained are seldom comparable with those which might be expected from mature ewes (Dýrmundsson, 1973; 1981; Quirke, 1979b; 1981). Although there is great variation both within and between breeds, the reproductive failure rate observed is usually within the range 20–40%. That this problem exists under field conditions in Britain is apparent from the flock records collected by the Meat and Livestock Commission in 1977. These records show that 37% of young female sheep mated in their first year failed to produce a lamb. Some research effort has been directed towards this problem in recent years, particularly in progestagen–PMSG treated animals.

OCCURRENCE OF OESTRUS AND OVULATION

If conception is to occur when natural mating is employed following hormone treatment, it is essential that the occurrence of oestrus and ovulation be closely coordinated. However, anomalies can occur which may reduce the effectiveness of the treatment (Quirke, 1979a; *Table 21.7*). Thus, although the incidence of oestrus following progestagen–PMSG treatment is normally 90% or more (Quirke, 1981), conception can

Table 21.7 FREQUENCY OF COINCIDENT OESTRUS AND OVULATION IN PROGESTAGEN–PMSG TREATED GALWAY EWE LAMBS

Number of ewe lambs treated	362
Oestrous and ovulatory responses:	
No response (%)	25(6.9)
Silent ovulation (%)	11(3.0)
Anovulatory oestrus (%)	24(6.6)
Coincident oestrus and ovulation (%)	302(83.4)

From Quirke (1979a)

evidently be precluded in a significant proportion of the mated animals because of ovulation failure (*Table 21.7*). The expression of oestrus unaccompanied by ovulation is not restricted to hormone-treated animals, however, and Edey *et al.* (1977) have reported the phenomenon in between 7% and 33% of untreated pubertal lambs of three breeds.

It is well known that in mature ewes ovulation occurs around the end of heat (Robinson, 1959; Holst and Braden, 1972) and that the duration of oestrus in ewe lambs is shorter than for mature ewes (Edey, Kilgour and Bremner, 1978; Hafez, 1952; Hanrahan and Quirke, 1975). In adult ewes the preovulatory LH release commences shortly after the onset of oestrus, and the interval between the beginning of the LH discharge and ovulation is known to be fairly constant at 21–26 hours (Cumming *et al.*, 1973). It is possible, therefore, that ovulation does not occur in many ewe lambs until well after the end of behavioural oestrus. Thus the temporal relationship between insemination by the ram and release of the egg from the follicle may be sub-optimal in many animals. The only investigation into this possibility has been with progestagen–PMSG treated ewe lambs and the results indicate that the majority of such animals ovulated shortly before the end of heat (Quirke, Hanrahan and Gosling, 1981; *Table 21.8*).

Table 21.8 TIME OF OVULATION IN GALWAY EWE LAMBS FOLLOWING PROGESTAGEN–PMSG TREATMENT

Time after onset of oestrus[a] (hours)	*No. of lambs examined*	*No. with corpora lutea*	*Percentage ovulating*
17	18	0	0
22	18	3	17
27(i)	16	8	50
(ii)[b]	17	12	71
32(i)	17	16	94
(ii)[c]	18	18	100
37	15	15	100

[a]Duration of oestrus = 29.9 hours
[b]Previously examined at 17 hours
[c]Previously examined at 22 hours
From Quirke, Hanrahan and Gosling (1979)

However, in this study the duration of oestrus was not different in ewe lambs and similarly treated adult ewes; the question, therefore, deserves further investigation in other breeds under both natural conditions and following hormonal stimulation.

INSEMINATION AND FERTILIZATION FAILURE

Many of the behavioural responses of mature ewes to rams at oestrus may be either absent or infrequently displayed in ewe lambs. The responsiveness to rams of females from these two age groups has been rated by Edey, Kilgour and Bremner (1978) who observed that only 27% of ewe lambs displayed the full range of behavioural oestrous activities typical of mature ewes (*Table 21.9*). Thus, in comparison with mature ewes, lambs have a

Table 21.9 DISTRIBUTION OF OESTROUS BEHAVIOUR RESPONSE RATINGS (%) FOR MATURE EWES AND EWE LAMBS

	Rating of ewe responsiveness				
	1 Full adult	*2 Near adult*	*3 Firm standing*	*4 Partial standing*	*5 Weak response*
Ewe lambs	27	14	29	14	16
Mature ewes	95	5	—	—	—

From Edey, Kilgour and Bremner (1978)

reduced ability to maintain the attention of the ram during oestrus and while this may not influence the rate of detection of oestrus, the frequency of service can be reduced (Edey, Kilgour and Bremner, 1978) and insemination failure can occur in many animals (Allison *et al.*, 1975; Killeen and Quirke, 1979). The benefit to fertility of an increased number of services in ewe lambs has been demonstrated by Allison *et al.* (1975). It would appear that most of the mating activity in ewe lambs takes place very early in oestrus and that few services occur during the final stages of the oestrous period (Edey, Kilgour and Bremner, 1978). The possible effect of this on conception has been referred to earlier in the context of the timing of ovulation in relation to oestrus.

Simple flock management practices, such as keeping the age groups separated at mating time, can also influence the mating outcome. When ewe lambs and mature ewes are in a mixed group and enter oestrus simultaneously, rams prefer older ewes and the likelihood of mating failure in the younger ewes is sharply increased (Keane, 1976; *Table 21.10*).

Table 21.10 SEXUAL BEHAVIOUR OF ADULT EWES AND EWE LAMBS

	Separately		*Mixed*	
	Lambs	*Adults*	*Lambs*	*Adults*
No. observed	18	17	9	9
Total mounts	245	166	38	166
Total matings	22	60	2	17
No. mated (%)	11 (61.1)	17 (100)	2 (22.2)	8 (88.9)

Observation period: Separately—for the duration of oestrus; Mixed—12 hours
From Keane (1976)

The estimates of the rates of insemination and fertilization which are available vary widely and depend on the conditions under which the various experiments were performed (Allison *et al.*, 1975; Al-Wahab and Bryant, 1978a; 1978b; Keane, 1975c; Killeen and Quirke, 1979; Quirke, 1981). The most extensive data for progestagen–PMSG treated ewe lambs are those of Killeen and Quirke (1979) and these show that the fertilization rate is high (92%) provided there is evidence that sperm had been deposited in the vagina. This would indicate that sperm transport within the female tract is unlikely to be a major factor limiting the fertility of very young ewes. The management of mating is, however, clearly a factor of paramount importance. The incidence of mating failure can be reduced by avoiding mixing ewe lambs with older ewes, keeping both the size of the mating group and mating area small and using a low ewe to ram ratio in order to minimize any selectivity on the part of the rams.

EMBRYONIC MORTALITY

Embryonic mortality in sheep is a subject which has received considerable attention during the last twenty years. However, most studies in this connection have focused on mature ewes and there is comparatively little information available regarding embryonic mortality in ewe lambs. The basal wastage rate of fertilized eggs in adult ewes is normally in the range of 20–30% (Edey, 1969; 1979). The few estimates available for ewe lambs (Al-Wahab and Bryant, 1978a; 1978b; Hamra and Bryant, 1979; Quirke, 1979a) are considerably in excess of this range which suggests that high embryonic mortality may be a significant factor contributing to the lower lambing rate of ewe lambs compared with adult ewes. The greatest losses evidently occur between days 5 and 21 after mating (Killeen and Quirke, 1979), and it has been estimated that by 26–28 days post-mating 63% of fertilized eggs are not represented by viable embryos in progestagen–PMSG treated ewe lambs of the Galway breed (Quirke, 1979a).

A number of recent studies have attempted to identify the factors which contribute to the very high level of embryonic mortality in young ewes. In both progestagen–PMSG treated and untreated lambs, plasma progesterone concentrations during the oestrous cycle and in early pregnancy are within the normal limits for mature ewes (Quirke and Gosling, 1979; 1981; Quirke, Hanrahan and Gosling, 1981; Smith *et al.*, 1977). Inadequate function of the corpus luteum is, therefore, unlikely to be a major factor limiting the establishment and maintenance of pregnancy. This is apparent also from the results of two egg transfer experiments in which the survival to term of one or more fertilized eggs from mature ewes, transferred around day 4, was shown to be similar in the uteri of ewe lamb and adult ewe recipients (Quirke, 1979b; Quirke, Hanrahan and Gosling, 1978; *Table 21.11*).

However, there is evidence of a difference in the quality of embryos from ewe lambs and mature ewes (Quirke and Hanrahan, 1977). In this experiment (*Table 21.12*) genetically marked 8–16 cell eggs from progestagen–PMSG treated lambs and adults were transferred into the uteri of

Table 21.11 SURVIVAL OF ONE AND TWO FERTILIZED EGGS FROM ADULT EWES IN THE UTERI OF EWE LAMB AND ADULT EWE RECIPIENTS

	Type of recipient	
	Adult ewe	*Ewe lamb*
Single egg transfers:		
No. of recipients	54	43
No. recipients lambed (%)	31(57.4)	26(60.5)
Two egg transfers:		
No. of recipients	54	34
No. recipients lambed (%)	40(74.1)	23(67.6)
No. single births (%)	11(27.5)	9(39.1)
No. twin births (%)	29(72.5)	14(60.9)

From Quirke, Hanrahan and Gosling (1978)

Table 21.12 COMPARISON OF THE SURVIVAL *IN UTERI* OF ADULT EWES OF CLEAVED EGGS FROM ADULT EWES AND EWE LAMBS

	Source of fertilized eggs	
	Adult ewes	*Ewe lambs*
Number of eggs transferred	48	48
Number of lambs born	35	16
Percentage eggs survived	72.9	33.3

From Quirke and Hanrahan (1977)

mature ewes; each recipient received two eggs, one from an adult ewe and the other from a ewe lamb. More than 70% of the adult ewe eggs developed to term whereas only 33% of the ewe lamb eggs did so. This survival rate of lamb eggs compares with a value of 37% estimated in a slaughter study using progestagen–PMSG treated ewe lambs of the same breed (Quirke, 1979a). It is interesting to note in this regard that eggs from prepubertal lambs have a reduced potential for continued development *in vitro* when compared with those from mature ewes (Wright *et al.*, 1976). The underlying reasons for the apparent lower viability of eggs from ewe lambs are not understood. The problem, however, clearly arises as a result of events which occur in the female reproductive tract before the 8–16 cell developmental stage of the embryo; perhaps even in the preovulatory follicle. The latter possibility has been suggested because of some evidence concerning differences between progestagen–PMSG treated lambs and adults in oestrogen levels *in vivo*, around the time of oestrus, and in the pattern of secretion of oestrogens *in vitro* by explanted follicles (Quirke, Hanrahan and Gosling, 1981; Trounson, Willadsen and Moor, 1977).

References

ALEXANDER, D.P., BRITTON, H.G., CAMERON, E., FOSTER, C.L. and NIXON, D.A. (1973). Adenohypophysis of foetal sheep: correlation of ultrastructure with functional activity. *Journal of Physiology* **230**, 10P–12P

ALLISON, A.J., KELLY, R.W., LEWIS, J.S. and BINNIE, D.B. (1975). Preliminary studies on the efficiency of mating of ewe hoggets. *Proceedings of the New Zealand Society of Animal Production* **35**, 83–90

AL-WAHAB, R.H.M. and BRYANT, M.J. (1978a). The effect of reduction in daylength, level of feeding and age on the reproduction of young female sheep mated at an induced ovulation. *Animal Production* **26**, 317–324

AL-WAHAB, R.H.M. and BRYANT, M.J. (1978b). Reproduction in young female sheep induced to breed at various ages. *Animal Production* **26**, 309–316

ATTAL, J. (1969). Levels of testosterone, androstenedione, estrone and estradiol-17β in the testes of fetal sheep. *Endocrinology* **85**, 280–289

BOLAND, M.P. and GORDON, I. (1979). Effect of timing of ram introduction in progestagen–PMSG treated anoestrous ewes. *Journal of Agricultural Science, Cambridge* **92**, 247–249

BURFENING, P.J. and VAN HORN, P.J. (1970). Induction of fertile oestrus in prepubertal ewes during the anoestrous season. *Journal of Reproduction and Fertility* **23**, 147–150

COLAS, G. (1979). Fertility in the ewe after artificial insemination with fresh and frozen semen at the induced oestrus, and influence of the photoperiod on the semen quality in the ram. *Livestock Production Science* **6**, 153–166

COLAS, G., THIMONIER, J., COUROT, M. and ORTAVANT, R. (1973). Fertilité, prolificité et fecondité pendant la saison sexualle des brebis inseminées artificiellement après traitement à l'acetate de fluorogestone. *Annales de Zootechnie* **22**, 441–451

CUMMING, I.A. (1979). Synchronization of ovulation. In *Sheep Breeding*, Second Edition (G.J. Tomes, D.E. Robertson, R.J. Lightfoot and W. Haresign, Eds.), pp. 403–421. London, Butterworths

CUMMING, I.A., BUCKMASTER, J.M., BLOCKEY, M.A. DE B., GODING, J.R., WINFIELD, C.G. and BAXTER, R.W. (1973). Constancy of interval between luteinizing hormone release and ovulation in the ewe. *Biology of Reproduction* **9**, 24–29

DONALD, H.P., READ, J.L. and RUSSEL, W.S. (1968). A comparative trial of crossbred ewes by Finnish Landrace and other sires. *Animal Production* **10**, 413–421

DYRMUNDSSON, O.R. (1973). Puberty and early reproductive performance in sheep. 1. Ewe lambs. *Animal Breeding Abstracts* **41**, 273–289

DYRMUNDSSON, O.R. (1981). Natural factors affecting puberty and reproductive performance; a review. *Livestock Production Science* **8**, 55–65

EDEY, T.N. (1969). Prenatal mortality in sheep. *Animal Breeding Abstracts* **37**, 173–190

EDEY, T.N. (1979). Embryo mortality. In *Sheep Breeding*, Second Edition (G.J. Tomes, D.E. Robertson, R.J. Lightfoot and W. Haresign, Eds.), pp. 315–325. London, Butterworths

EDEY, T.N., KILGOUR, R. and BREMNER, K. (1978). Sexual behaviour and reproductive performance of ewe lambs at and after puberty. *Journal of Agricultural Science, Cambridge* **90**, 83–91

EDEY, T.N., CHU, T.T., KILGOUR, R., SMITH, J.F. and TERVIT, H.R. (1977). Estrus without ovulation in pubertal ewes. *Theriogenology* **7**, 11–15

FINDLAY, J.K. and SEAMARK, R.F. (1973). The occurrence and metabolism of oestrogens in the sheep foetus and placenta. In *The Endocrinology of Pregnancy and Parturition—Experimental Studies in the Sheep* (C.G. Perrepoint, Ed.), pp. 54–70. Cardiff, Alpha Omega Alpha Publishing Co.

FOSTER, D.L. (1980). Comparative development of mammalian females: proposed analogues among patterns of LH secretion in various species. In *Problems in Pediatric Endocrinology* (C. La Cauza and A.W. Root, Eds.), pp. 193–210. London, Academic Press

FOSTER, D.L. (1981). Mechanism for delay of first ovulation in lambs born in the wrong season (Fall). *Biology of Reproduction* **25**, 85–92

FOSTER, D.L. and KARSCH, F.J. (1975). Development of the mechanism regulating the preovulatory surge of luteinizing hormone in sheep. *Endocrinology* **97**, 1205–1209

FOSTER, D.L. and RYAN, K.D. (1979). Endocrine mechanisms governing transition into adulthood: A marked decrease in inhibitory feedback action of estradiol on tonic secretion of luteinizing hormone in the lamb during puberty. *Endocrinology* **105**, 896–904

FOSTER, D.L. and RYAN, K.D. (1980). Mechanisms governing onset of ovarian cyclicity at puberty in the lamb. *Annales de Biologie Animale Biochemie Biophysique* **19**, 1369–1380

FOSTER, D.L., COOK, B. and NALBANDOV, A.V. (1972). Regulation of luteinizing hormone (LH) in the fetal and neonatal lamb: effects of castration during the early postnatal period on levels of LH in sera and pituitaries of neonatal lambs. *Biology of Reproduction* **6**, 253–257

FOSTER, D.L., JAFFE, R.B. and NISWENDER, G.D. (1975). Sequential patterns of circulating LH and FSH in female sheep during the early postnatal period. Effect of gonadectomy. *Endocrinology* **96**, 15–22

FOSTER, D.L., LEMONS, J.A., JAFFE, R.A. and NISWENDER, G.D. (1975). Sequential patterns of circulating luteinizing hormone and follicle-stimulating hormone in female sheep from early postnatal life through the first estrous cyces. *Endocrinology* **97**, 985–994

FOSTER, D.L., ROCHE, J.F., KARSCH, F.J., NORTON, H.W., COOK, B. and NALBANDOV, A.V. (1972a). Regulation of luteinizing hormone in the fetal and neonatal lamb. 1. LH concentrations in blood and pituitary. *Endocrinology* **90**, 102–112

FOSTER, D.L., CRUZ, I.A., JACKSON, G.L., COOK, B. and NALBANDOV, A.V. (1972b). Regulation of luteinizing hormone in the fetal and neonatal lamb. III. Release of LH by the pituitary *in vivo* in response to crude ovine hypothalamic extract or purified porcine gonadotrophic releasing factor. *Endocrinology* **90**, 673–683

GLUCKMAN, P.D., MARTI-HENNEBERG, C., KAPLAN, S.L., RUDOLPH, A.M. and GRUMBACH, M.M. (1979). The ontogeny of negative feedback by estrogen on gonadotropin secretion in the ovine fetus. *Endocrinology* **104**, 120 A (Abstract)

GORDON, I. (1967a). Progesterone–PMS therapy in the induction of pregnancy in anoestrous ewes and ewe lambs. *Journal of the Department of Agriculture, Republic of Ireland* **64**, 38–50

GORDON, I. (1967b). Aspects of reproduction and neonatal mortality in ewe lambs and adult sheep. *Journal of the Department of Agriculture, Republic of Ireland* **64**, 76–127

GORDON, I. (1975). Hormonal control of reproduction in sheep. *Proceedings of the British Society of Animal Production* **4** (New series), 79–93

GRUMBACH, M.M. (1980). The neuroendocrinology of puberty. *Hospital Practice* **15**, 51–60

HAFEZ, E.S.E. (1952). Studies on the breeding season and reproduction of the ewe. *Journal of Agricultural Science, Cambridge* **42**, 189–265

HAMRA, A.M. and BRYANT, M.J. (1979). Reproductive performance during mating and early pregnancy in young female sheep. *Animal Production* **28**, 235–244

HANRAHAN, J.P. and QUIRKE, J.F. (1975). Repeatability of the duration of oestrus and breed differences in the relationship between duration of oestrus and ovulation rate of sheep. *Journal of Reproduction and Fertility* **45**, 29–36

HARESIGN, W. (1978). Ovulation control in the sheep. In *Control of Ovulation* (D.B. Crighton, N.B. Haynes, G.R. Foxcroft and G.E. Lamming, Eds.), pp. 433–451. London, Butterworths

HAUGER, R.L., KARSCH, F.J. and FOSTER, D.L. (1977). A new concept for control of the estrous cycle of the ewe based on the temporal relationships between luteinizing hormone, estradiol, and progesterone in peripheral serum and evidence that progesterone inhibits tonic LH secretion. *Endocrinology* **101**, 807–817

HOLST, P.J. and BRADEN, A.W.H. (1972). Ovum transport in the ewe. *Australian Journal of Biological Science* **25**, 167–173

INKSTER, I. (1958). Current ideas on raising the lambing percentage of sheep. *Proceedings of the Ruakura Farmers Conference Week*, pp. 74–79

JENNINGS, J.J. and CROWLEY, J.P. (1972). The influence of mating management on fertility in ewes following progesterone–PMS treatment. *Veterinary Record* **90**, 495–498

KEANE, M.G. (1974). Effect of progestagen–PMS treatment on reproduction in ewe lambs. *Irish Journal of Agricultural Research* **13**, 39–48

KEANE, M.G. (1975a). Effect of 17β-oestradiol pretreatment and system of mating on the reproductive performance of progestagen–PMS treated non-cyclic ewe lambs. *Irish Journal of Agricultural Research* **14**, 7–13

KEANE, M.G. (1975b). Effect of nutrition and dose level of PMS on oestrous response and ovulation rate in progestagen treated non-cyclic Suffolk × Galway ewe lambs. *Journal of Agricultural Science, Cambridge* **84**, 507–511

KEANE, M.G. (1975c). Induction of oestrus and pregnancy in Autumn-born ewe lambs. *Irish Journal of Agricultural Research* **14**, 81–84

KEANE, M.G. (1976). Breeding from ewe lambs. *Farm and Food Research* **7**, 10–12

KILLEEN, I.D. and QUIRKE, J.F. (1979). Reproductive wastage in Galway ewe lambs. *Animal Production Research Report,* pp. 87–88. Dublin, An Foras Taluntais

KNOBIL, E. (1980). The neuroendocrine control of the menstrual cycle. *Recent Progress in Hormone Research* **36**, 53–88

LAND, R.B., THIMONIER, J. and PELLETIER, J. (1970). Possibilité d'induction d'une decharge de LH par une injection d'oestrogène chez l'agneau femelle en fonction de l'age. *Comptes Rendus de l'Academie des Sciences, Paris, Series D* **271**, 1549–1551

LEVASSEUR, MARIE-CLAIRE (1979). Thoughts on puberty. The gonads. *Annales de Biologie Animale Biochemie Biophysique* **19**, 321–335

MAULEON, P. (1978). Ovarian development in young mammals. In *Control of Ovulation* (D.B. Crighton, N.B. Haynes, G.R. Foxcroft and G.E. Lamming, Eds.), pp. 141–158. London, Butterworths

MAULEON, P., BEZARD, J. and TERQUI, M. (1977). Very early and transient 17β-estradiol secretion by fetal sheep ovary. *In vitro* study. *Annales de Biologie Animale Biochemie Biophysique* **17**, 399–401

MEAT AND LIVESTOCK COMMISSION (1977). *Data Sheets on Upland and Lowland Sheep Production*. Bletchley, Milton Keynes, Meat and Livestock Commission

MUELLER, P.L., SKLAR, C.A., GLUCKMAN, P.D., KAPLAN, S.L. and GRUMBACH, M.M. (1981). Hormone ontogeny in the ovine fetus. IX. Luteinizing hormone and follicle-stimulating hormone response to luteinizing hormone-releasing hormone factor in mid and late gestation and in the neonate. *Endocrinology* **108**, 881–886

MUELLER, P.L., GLUCKMAN, P.D., SKLAR, C.A., KAPLAN, S.L., RUDOLPH, A.M. and GRUMBACH, M.M. (1978). The ontogeny of the pituitary response to LRF in the ovine fetus. *Endocrinology* **102**, 172A (Abstract)

OJEDA, S.R., ANDREWS, W.W., ADVIS, J.P. and SMITH WHITE, S. (1980). Recent advances in the endocrinology of puberty. *Endocrine Reviews* **1**, 228–257

QUIRKE, J.F. (1978). Reproductive performance of Galway, Finnish Landrace and Finn-cross ewe lambs. *Irish Journal of Agricultural Research* **17**, 25–32

QUIRKE, J.F. (1979a). Oestrus, ovulation, fertilization and early embryo mortality in progestagen–PMSG treated Galway ewe lambs. *Irish Journal of Agricultural Research* **18**, 1–11

QUIRKE, J.F. (1979b). Control of reproduction in adult ewes and ewe lambs, and estimation of reproductive wastage in ewe lambs following treatment with progestagen impregnated sponges and PMSG. *Livestock Production Science* **6**, 295–305

QUIRKE, J.F. (1979c). Effect of bodyweight on the attainment of puberty and reproductive performance of Galway and Fingalway female lambs. *Animal Production* **28**, 297–307

QUIRKE, J.F. (1981). Regulation of puberty and reproduction in female lambs; a review. *Livestock Production Science* **8**, 37–53

QUIRKE, J.F. and GOSLING, J.P. (1979). Prepubertal plasma luteinizing hormone concentrations and progesterone concentrations during the oestrous cycle and early pregnancy in Galway and Fingalway ewe lambs. *Animal Production* **28**, 1–12

QUIRKE, J.F. and GOSLING, J.P. (1981). Reproductive performance and plasma progesterone concentrations in ewe lambs and mature ewes following treatment for synchronization of oestrus during the breeding season. *Irish Journal of Agricultural Research* **20**, 9–20

QUIRKE, J.F. and HANRAHAN, J.P. (1977). Comparison of the survival in the uteri of adult ewes of cleaved ova from adult ewes and ewe lambs. *Journal of Reproduction and Fertility* **51**, 487–489

QUIRKE, J.F., HANRAHAN, J.P. and GOSLING, J.P. (1978). Reproduction in ewe lambs. *Animal Production Research Report*, pp. 155–159. Dublin, An Foras Taluntais

QUIRKE, J.F., HANRAHAN, J.P. and GOSLING, J.P. (1981). Duration of oestrus, ovulation rate, time of ovulation and plasma LH, total oestrogen and progesterone in Galway adult ewes and ewe lambs. *Journal of Reproduction and Fertility* **61**, 265–272

QUIRKE, J.F., SHEEHAN, W. and LAWLOR, M.J. (1978). The growth of pregnant female lambs and their progeny in relation to dietary protein and energy during pregnancy. *Irish Journal of Agricultural Research* **17**, 33–42

RICHARDS, J.S., RAO, M.C. and IRELAND, J.J. (1978). Actions of pituitary gonadotrophins on the ovary. In *Control of Ovulation* (D.B. Crighton, N.B. Haynes, G.R. Foxcroft and G.E. Lamming, Eds.), pp. 197–215. London, Butterworths

ROBINSON, T.J. (1959). The oestrous cycle of the ewe and doe. In *Reproduction in Domestic Animals*, First Edition (H.H. Cole and P.T. Cupps, Eds.), pp. 291–333. New York, Academic Press

RYAN, K.D. and FOSTER, D.L. (1980). Neuroendocrine mechanisms involved in onset of puberty in the female: concepts derived from the lamb. *Federation Proceedings* **39**, 2372–2377

SKLAR, C.A., MUELLER, P.L., GLUCKMAN, P.D., KAPLAN, S.L., RUDOLPH, A.M. and GRUMBACH, M.M. (1978). The ontogeny of gonadotropins and sex steroids in the sheep fetus. *Pediatric Research* **12**, 420 (Abstract)

SKLAR, C.A., MUELLER, P.L., GLUCKMAN, P.D., KAPLAN, S.L., RUDOLPH, A.M. and GRUMBACH, M.M. (1981). Hormone ontogeny in the ovine fetus. VII. Circulating luteinizing hormone and follicle-stimulating hormone in mid and late gestation. *Endocrinology* **108**, 874–880

SMITH, J.F., DROST, H., FAIRCLOUGH, R.J., PETERSON, A.J. and TERVIT, H.R. (1977). Effect of age on peripheral levels of progesterone and oestradiol-17β, and duration of oestrus in Romney Marsh ewes. *New Zealand Journal of Agricultural Research* **19**, 277–280

SOUTHAM, E.R., HULET, C.V. and BOTKIN, M.P. (1971). Factors affecting reproduction in ewe lambs. *Journal of Animal Science* **33**, 1282–1287

THIMONIER, J. (1979). Hormonal control of oestrous cycle in the ewe (a review). *Livestock Production Science* **6**, 39–50

THIMONIER, J. (1981). Practical uses of prostaglandins in sheep and goats. In *Prostaglandins in Animal Reproduction* (L. Edqvist and H. Kindahl, Eds.), pp. 193–208. *Acta Veterinaria Scandinavica, Supplementum* **77**

THIMONIER, J. and COGNIE, Y. (1977). Application of control of reproduction of sheep. In *Management of Reproduction in Sheep and Goats Symposium*, pp. 109–118. Madison, USA, University of Wisconsin

THIMONIER, J., MAULEON, P., COGNIE, Y. and ORTAVANT, R. (1968). Declenchement de l'oestrus et obtention precoce de gestations chez des agnelles à l'aide d'eponges vaginales impregnées d'acétate de fluorogestone. *Annales de Zootechnie* **17**, 275–288

TROUNSON, A.O., WILLADSEN, S.M. and MOOR, R.M. (1977). Reproductive function in prepubertal lambs: ovulation, embryo development and ovarian steroidogenesis. *Journal of Reproduction and Fertility* **49**, 69–75

TSAKALOF, P., VLACHOS, N. and LATOUSAKIS, D. (1977). Observations on the reproductive performance of ewe lambs synchronised for oestrus. *Veterinary Record* **100**, 380–382

WRIGHT, R.W., ANDERSON, G.B., CUPPS, P.T., DROST, M. and BRADFORD, G.E. (1976). *In vitro* culture of embryos from adult and prepubertal ewes. *Journal of Animal Science* **42**, 912–917

22

THE CONTROL OF REPRODUCTIVE ACTIVITY IN THE RAM

N.B. HAYNES
University of Nottingham School of Agriculture, Sutton Bonington, Loughborough, Leicestershire, UK
and
B.D. SCHANBACHER
US Department of Agriculture, Clay Centre, Nebraska, USA

Until recently, studies on the factors which control reproductive activity in sheep, with emphasis on development of methods to overcome their normal seasonality of breeding, have concentrated to a large extent upon the ewe. It has become apparent however, that the problems of subfertility which can occur in controlled reproduction programmes, particularly those involving out-of-season breeding, may not be confined to the female but could be contributed to by the ram which also has seasonal fluctuations in breeding activity (Schanbacher, 1979; Lincoln and Short, 1980). Allison and Robinson (1971) have demonstrated that some 120 $\times$ 10^6 sperm are required to fertilize a ewe efficiently. With estimates of 50–300 $\times$ 10^6 sperm/ejaculate (in rams which are not mated excessively; Allison, 1972) it is obvious that unless the production rate of sperm is kept maximal, the ram can certainly be a limiting factor and this has been demonstrated in field trials (Lindsay and Signoret, 1980). For straightforward reasons and more subtle ones discussed later (p.443), efficient sexual behaviour is also necessary to obtain good fertility with natural mating. Gordon (1977) stated that 'the capability of the ram (libido and fertility) must be given very careful consideration in controlled oestrus applications. First and foremost, the ram should be sexually experienced and have a good record of achievement in producing pregnancies'. In this respect Jennings and Crowley (1970; 1972) concluded that, since near normal conception rates in progestagen–PMSG treated sheep could be achieved by hand-mating, much of the subfertility reported for such sheep was a reflection of inadequacies among the rams, either in libido or fertility, in coping with a situation in which a group of ewes were in oestrus simultaneously. This situation can be aggravated by impairment of sperm transport and survival after progestagen treatment. On bases such as these, the aim of maximization of semen production and male sexual behaviour particularly at times of year when they are sexually depressed, becomes important. Both of these factors are modulated by the brain–pituitary–testicular axis. Luteinizing hormone (LH), released from the pituitary under the influence of the hypothalamic hormone, gonadotrophin releasing hormone (GnRH), acts upon the Leydig cells in the interstitial compartment of the testes to facilitate testosterone production, and testosterone has a major part to play in the maintenance of the secondary sex organs, spermatogenesis and sexual behaviour. Follicle stimulating hormone (FSH) has a somewhat less well defined action, but it appears to interact with the Sertoli cell component of the seminiferous tubules to initiate and aid in the maintenance of spermatogenesis. A third pituitary hormone, prolactin, is involved in testicular function in some species and synergizes with LH in stimulating steroidogenesis. Thus, the endocrine and exocrine functions of the testis are regulated by two or three pituitary gonadotrophins which in turn are regulated by the brain. In addition this regulation is modified by feedback from secretory products of the testis. Hence, the control of the components of reproductive activity in the ram, namely sexual behaviour and semen production depends, within genetic constraints, upon the ontogeny of the brain–pituitary–testicular axis with the production of appropriate concentrations of pituitary and testicular hormones. This chapter summarizes our current understanding in this area and points to features which merit particular study in regard to maximizing reproductive efficiency in the ram.

Development of the brain–pituitary–testicular axis

Age of puberty in rams, like other species, is difficult to quantify and occurs between 12 and 45 weeks of age depending upon which criteria are used to define it. Behavioural parameters are not much help since forms of sexual behaviour are present soon after birth and full expression of such behaviour develops gradually (Howles, Webster and Haynes, 1980). Age of puberty is usually assessed in relation to such things as changes in testicular morphology and appearance of spermatozoa in the seminiferous tubules, epidydimides or ejaculates (cf. Carmon and Green, 1952; Watson, Sapsford and McCance, 1956; Courot, 1967; Crim and Geschwind, 1972; Schanbacher, Gomes and Van Demark, 1974; Lee *et al.*, 1976a). Taking the appearance of spermatozoa as a major criterion, puberty is still difficult to define in more than a general sense. For instance, sperm samples could be obtained from Clun rams by electroejaculation at about five months of age and sperm were found in the epidydimis somewhat earlier than this, but with an age range extending from 99–176 days (Dýrmundsson and Lees, 1972). Louda *et al.* (1981) reported that Romanov and Finnish ram lambs were capable of impregnating females by five months of age but it was not possible to obtain samples by artificial vagina until around 11 months. On the other hand, spermatozoa were not found in the seminiferous tubules of Merino/Borridale rams until around 10 months (Lee *et al.*, 1976b). These ranges are not surprising since there are no doubt differences due to breed, and within breeds; things such as nutrition and growth rate are modifying factors in pubertal development (Dýrmundsson, 1973) together with a possible effect of photoperiod (Alberio and Colas, 1976; Howles, Webster and Haynes, 1980; Schanbacher and Crouse, 1980). Notwithstanding these variations there are important measurable changes involving maturation of the brain–pituitary–testicular axis during the prepubertal period.

A characteristic of the adult ram is the pattern of circulating LH and testosterone in which pulsatile and rhythmic secretions of LH occur, presumably under the influence of episodic release of GnRH, each release of LH being followed by a marked elevation in circulating testosterone (Purvis, Illius and Haynes, 1974; Katongole, Naftolin and Short, 1974; Sanford *et al.*, 1974; Schanbacher and Ford, 1976; Lincoln, 1976a,b). This is illustrated in *Figure 22.1*. The pituitary–testis axis is functional in the

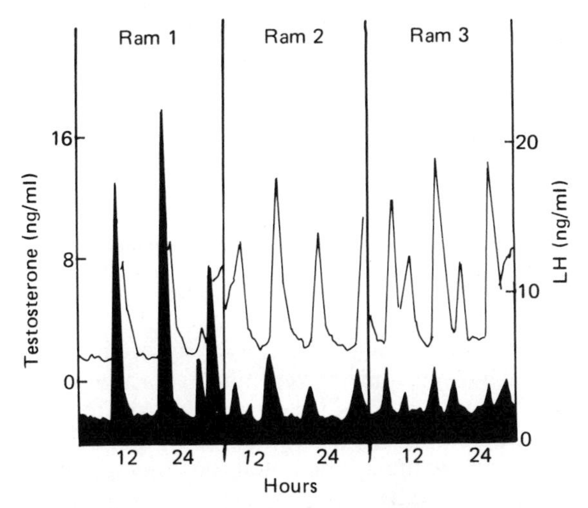

Figure 22.1 Plasma profiles of LH (shaded) and testosterone (unshaded) in three rams. Blood samples were taken from jugular cannulae at 30 minute intervals between 06.30 and 06.00 hours in September. From Schanbacher and Ford (1976)

foetal lamb (with important connotations in regard to subsequent sexual behaviour—a discussion outside the remit of this chapter) but secretion at this stage does not seem to be pulsatile (Foster *et al.*, 1972). The testis will respond to LH injection two or three days after birth with a significant increase in testosterone secretion. Similarly, the pituitary responds to GnRH one day after birth with marked increases in LH production (Lee *et al.*, 1976a) and endogenous pulsatility of LH and testosterone secretion develops from about one week of age, reaching a maximum frequency at around eight weeks but with low hormone concentrations relative to the adult (Foster *et al.*, 1978). Lee *et al.* (1976b) reported that around eight weeks of age there is a relative fall in FSH secretion concurrent with histological changes in the Sertoli cell which may reflect the initial secretion of a non-steroidal FSH inhibiting factor produced by these cells, namely inhibin (cf. Setchell, Davies and Main, 1977; Setchell, 1980; Walton *et al.*, 1980). Subsequent to these events, there is a marked increase in testosterone production at around 30 weeks of age (*Figure 22.2*; Illius *et al.*, 1976). This is concomitant with a rise in LH (Lee *et al.*, 1976b). Hence it seems that higher levels of testosterone are required to suppress LH secretion as pubertal development proceeds, i.e. there is a change in brain–pituitary sensitivity to feedback of testicular secretions. This change is thought to be important in timing the onset of puberty and a similar change is certainly involved in the seasonality of ram breeding (see next

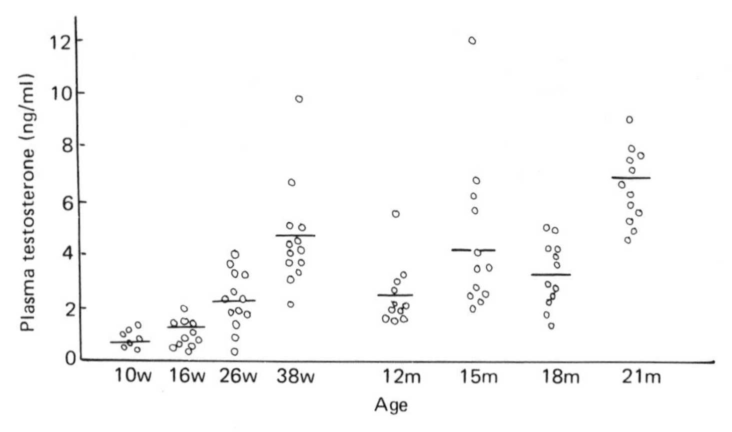

Figure 22.2 The pattern of testosterone secretion in rams up to 21 months of age. Each point represents the mean testosterone concentration for samples taken every 30 minutes for 18 or 24 hours from an individual ram. Group mean for each age shown by ——. From Illius *et al.* (1976)

section on p.438). Artificial manipulation of the sensitivity of this axis represents one possible approach to maximizing reproductive activity and it is important to understand that the process is not straightforward. For instance, there is evidence that testosterone exerts its action at both the hypothalamus and the pituitary; wethers implanted or injected with testosterone have a markedly decreased frequency of LH pulses suggesting that episodic GnRH from the hypothalamus is reduced. Also, wethers treated similarly show a reduced pituitary LH response to exogenous GnRH indicating a direct suppressive action of testosterone on the pituitary (Pelletier, 1974; Galloway and Pelletier, 1975; Garnier, Terqui and Pelletier, 1977; Bremner *et al.*, 1980; D'Occhio, Schanbacher and Kinder, 1982a). Furthermore, compounds other than testosterone and inhibin can exert negative feedback, for instance oestradiol-17β (Riggs and Malven, 1974; Karsch and Foster, 1975; Schanbacher and Ford, 1977) and dihydrotestosterone (Parrott and Davies, 1979) on LH secretion. A physiological role for the latter two compounds has been questioned,

Table 22.1 SERUM LH AND TESTOSTERONE (T) CONCENTRATIONS IN RAMS, WETHERS AND WETHERS IMPLANTED WITH DIFFERENT QUANTITIES OF TESTOSTERONE CAPSULES. WETHERS WERE IMPLANTED AT CASTRATION AND SIX WEEKS LATER ALL ANIMALS WERE BLED AT 10 MINUTE INTERVALS FOR 24 HOURS

Treatment	*No. of animals*	*Mean testosterone* (ng/ml±S.E.M.)	*Mean LH* (ng/ml)	*Number of LH peaks/24 hours*	*Amplitude* (ng/ml)
Rams	3	2.25±0.49	2.1±0.2^{a}	3.3±1.2^{a}	10.2±1.7^{a}
Wethers	3	0.22±0.01^{a*}	13.3±5.2^{b}	29.5±0.5^{b}	9.3±3.6^{a}
0.5T	3	0.54±0.01^{b*}	21.3±4.1^{b}	27.7±1.8^{b}	18.4±2.7^{a}
2T	3	1.11±0.08^{c*}	13.3±1.0^{b}	20.3±0.9^{c}	11.9±1.5^{a}
4T	3	2.01±0.09^{d}	2.2±0.8^{a}	2.7±1.8^{a}	13.6±0.6^{a}

^{a,b,c,d}Means within a column without a common superscript differ significantly ($P<0.05$).
*Significantly different from rams.
From D'Occhio, Schanbacher and Kinder (1982).

however, as well as for inhibin by D'Occhio, Schanbacher and Kinder (1982a,b) with the demonstration that both LH and FSH were suppressed to the levels found in intact animals when wethers were administered implants of testosterone which produced physiological concentrations of testosterone in blood (*Table 22.1*). These workers have also shown that administration of implants which produce low levels of testosterone in wethers (0.54 ng/ml compared with a level of 2.25 ng/ml in normal rams) caused elevations of plasma LH and FSH compared with those found in wethers, implying a positive feedback of testosterone at relatively low blood concentrations (see *Table 22.1*).

Once puberty is reached the level of testosterone is not constant, and despite seasonal fluctuations, continues to rise for at least the first 21 months of life as shown in *Figure 22.2*, a fact which may be related to the greater sperm output, libido and dominance which is often found in older rams (Illius, Haynes and Lamming, 1976).

Seasonality of reproduction in the ram

Many workers have pointed out that the ram is a seasonal breeder. Sexual activity peaks in the autumn to coincide with reproductive cyclicity in the ewe and although most breeds produce semen throughout the year a period of 'summer sterility' or subfertility exists for several months with a seasonal fall in semen production and sexual activity of rams when exposed to oestrous females (Dutt, 1960; Lees, 1965; Pepelko and Clegg, 1965).

HORMONE CHANGES

Since such features are considered to be produced by changes in hormones, the above observations have led to detailed studies into the mechanisms involved in seasonal changes in hormone concentrations in rams. Investigations with a variety of breeds have reported highest concentrations of testosterone during the autumn months with the lowest concentration during the spring. Testosterone is released in a pulsatile fashion, each testosterone peak following a peak of LH, the frequency of which increases with the onset of the breeding season (Johnson, Desjardins and Ewing, 1973; Katongole, Naftolin and Short, 1974; Purvis, Illius and Haynes, 1974; Sanford *et al.*, 1974; Schanbacher and Ford, 1976; Schanbacher and Lunstra, 1976; Davies, Main and Setchell, 1977; Sanford, Palmer and Howland, 1977; Lincoln, 1978a). Typical data from early studies are shown in *Figure 22.3* and *Table 22.2*.

Lincoln and coworkers have described in detail interrelationships between the hypothalamus, gonadotrophins and testicular activity using a highly seasonal primitive breed of sheep, the Soay, as an experimental model. Maintaining these on a 32-week light cycle (16 weeks long days: hours light:dark (LD) 16:8, 16 weeks short days: LD 8:16 with an abrupt change between the two regimes) produced a similar pattern of testis growth and regression and hormone changes as found in the natural environment but compressed into a shorter time period (Lincoln and

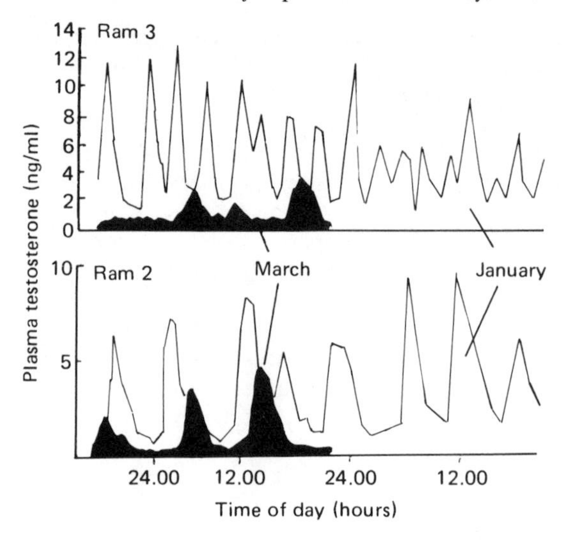

Figure 22.3 Plasma testosterone levels for rams during a 48 hour period in January and a 24 hour period in March. From Purvis, Illius and Haynes (1974)

Table 22.2 SECRETORY PATTERNS OF LH AND TESTOSTERONE[(a)] IN RAMS DURING THE BREEDING AND NON-BREEDING SEASON

	Season	
	Breeding (September)	*Non-breeding* (May)
Mean LH level (ng/ml±S.E.M.)	2.46±0.36	1.81±0.18
LH peaks		
Number/24 hours	5.40±1.21	3.60±0.75
Magnitude (ng/ml)	6.33±2.28	6.50±0.70
Mean testosterone level (ng/ml)	5.22±0.66	1.24±0.24
Testosterone peaks		
Number/24 hours	5.40±0.98	3.20±0.66
Magnitude (ng/ml)	9.81±0.79	3.97±0.80
Interval from LH peak (min)	52.16±3.55	54.50±2.29

[(a)]Plasma values for five rams sampled at 30 minute intervals for 24 hours
From Schanbacher and Ford (1976)

Davidson, 1977). The photoperiod-induced changes are not synchronous. For example, mean plasma FSH concentration was maximum after about six weeks exposure to short days, mean plasma LH, testis diameter and mean plasma testosterone after 13 weeks and sexual and aggressive behaviour after 18 weeks (*Figure 22.4*). This delay between hormonal events and subsequent behavioural changes obviously imposes a limitation on possible methods for treatment of rams to enhance reproductive activity. Benefits could not be obtained in the short term and treatments would have to be relatively chronic. Within the 32-week sexual cycle it was possible to define three phases by measuring testis diameter, namely 'regressed' (testis at minimum size), 'developing' (testes enlarging) and 'active' (testes at maximum size). Associated with these phases, the episodic nature of LH release changed in both frequency and amplitude

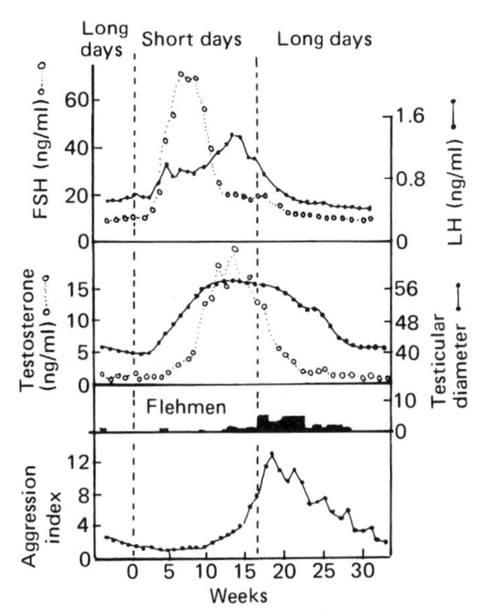

Figure 22.4 Plasma FSH, LH and testosterone levels, testicular diameters and aggression scores (weekly mean) for six Soay rams housed under artificial lighting conditions. The total weekly observations for Flehmen are also shown. From Lincoln and Davidson (1977)

(Lincoln, 1978b). Amplitude was maximum during the developing stage whilst frequency of LH peaks increased with increasing stage of testis development, reaching a maximum in the active phase as shown in *Figure 22.5*. Pituitary responsiveness to exogenous GnRH treatment underwent changes also, the most conspicuous being amplitude of the increase in LH concentration which was greatest during the period of gonadal regression. The situation was, however, complicated in that the duration of the

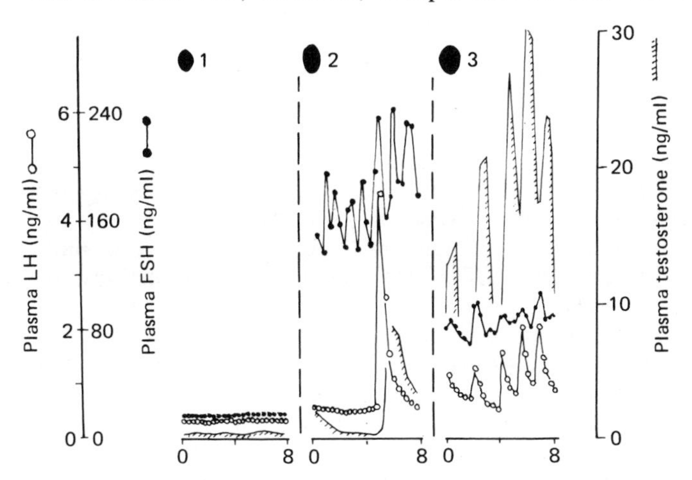

Figure 22.5 Short-term changes in the concentration of LH, FSH and testosterone in the blood plasma of one adult Soay ram sampled at 20 minute intervals for eight hours on three occasions during the seasonal sexual cycle: 1, testes fully regressed; 2, testes redeveloping; 3, testes fully enlarged. From Lincoln (1978b)

response was greatest during the period of maximum testicular activity (Lincoln, 1977), suggesting a change in the pool of releasable LH from the pituitary gland. As discussed previously negative feedback by gonadal steroids forms an integral component of the hypothalamic–pituitary–testicular axis in males. The experiments of Pelletier and Ortavant (1975a;b), Schanbacher (1980a) and others, have demonstrated that one of the consequences of the influence of photoperiod is that there is a change in the sensitivity to steroid negative feedback, sensitivity being greatest in the non-breeding season (long days). The seasonal changes in FSH, LH and testosterone as a result of photoperiodic change are not, however, mediated entirely through changes in steroid feedback since, whilst the levels of LH and FSH in castrates remain much higher than those in intact animals regardless of daylength, the levels are depressed in long days to

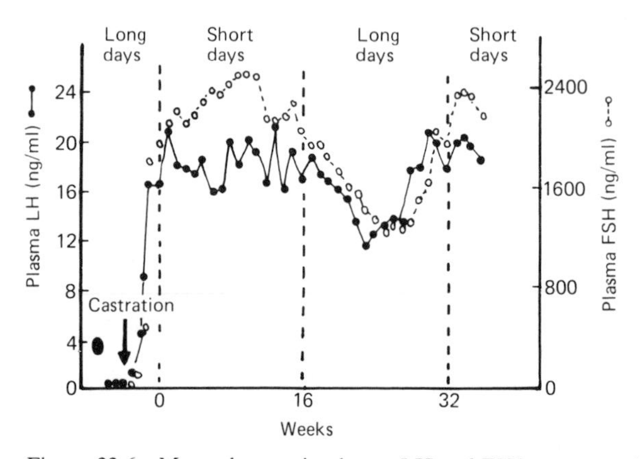

Figure 22.6 Mean changes in plasma LH and FSH concentrations for four adult Soay rams exposed to alternating 16-week periods of long days (16L:8D) and short days (8L:16D). Animals were castrated during a period of long days when the testes were regressed. From Lincoln and Short (1980)

about half the value in short days. This is illustrated in *Figure 22.6* (Pelletier and Ortavant, 1975a,b; Parrott and Davies, 1979; Lincoln and Short, 1980).

As stated earlier, prolactin has been implicated in reproductive activity in other species. Several workers have shown that its blood concentration has a seasonal rhythmicity in rams (Ravault, 1976; Lincoln, McNeilly and Cameron, 1978) but whether or not it has a physiological role in male reproduction in this species is far from clear (Schanbacher, 1980b; Howles, Craigon and Haynes, 1982).

Notwithstanding the fact that the response of the hypothalamo–pituitary–testicular axis to photoperiod involves all facets of the axis and is complicated, the stimulating event at the hypothalamic level leading to increased pituitary and testicular activity would seem to be the increased frequency of release of pulses of GnRH which occurs as daylength shortens. This has been confirmed by Lincoln (1979a) by taking rams in the nadir of the sexual cycle (April–June) and administering small amounts of GnRH at a dose (100 ng) and frequency (2 hours) chosen to mimic the situation existing in rams at the peak of the sexual cycle. The resulting changes in gonadotrophin profiles, testis growth and the secretion of testosterone are comparable to the changes found in rams during the developmental phase of the seasonal sexual cycle (Lincoln, 1979a).

This obviously offers a potential method of stimulating reproductive activity; studies relating to the use of GnRH to do this were in fact carried out earlier (Schanbacher and Lunstra, 1977; Schanbacher, 1978) and are discussed in the last section of this review (p.444).

The higher brain is obviously implicated in seasonal reproduction in the ram, but its role has not been investigated in detail. However, the pineal gland has some part in this control since pinealectomy or superior cervical ganglionectomy, a technique which inactivates the pineal, renders the ram essentially non-periodic and, in fact, in a stimulated state akin to that found in the breeding season as illustrated for testis diameter in *Figure 22.7*

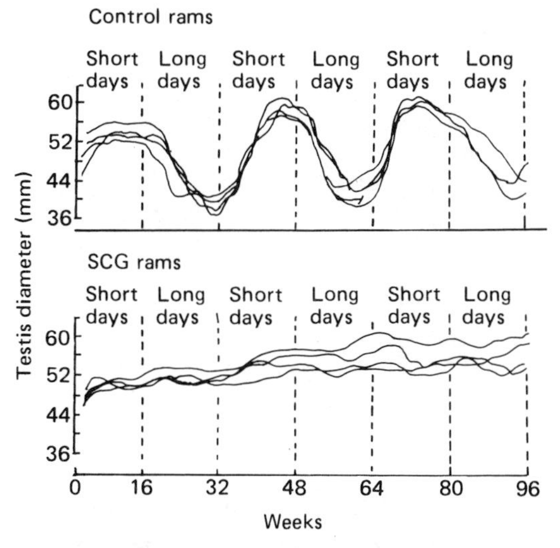

Figure 22.7 Long-term changes in the diameter of the testes for four control Soay rams (upper panel) and four superior cervical ganglionectomized (SCG) rams (lower panel) during a study in which they were exposed to alternating periods of long (16L:8D) and short (8L:16D) days. By courtesy of Lincoln (1979b)

(Barrell and Lapwood, 1978a,b; 1979; Lincoln, 1979b). Melatonin is a major secretory product of the pineal and a diurnal cycle in plasma melatonin concentrations exists with peak values occurring during darkness. There is also a larger nocturnal surge in plasma melatonin in rams during long compared with short days. Melatonin was undetectable in samples from ganglionectomized rams (Lincoln and Short, 1980). The apparent stimulatory effect of pinealectomy and anti-gonadal effects of melatonin suggest that the use of melatonin antagonists could have a stimulatory effect on the reproductive system. To our knowledge, this has not yet been studied.

Seasonality of reproduction in the ram is further complicated by the fact that whilst the ram is considered to be a short day breeder, the involution of the testes and decline in blood hormones tend to occur before the winter solstice and testicular development begins before the summer solstice

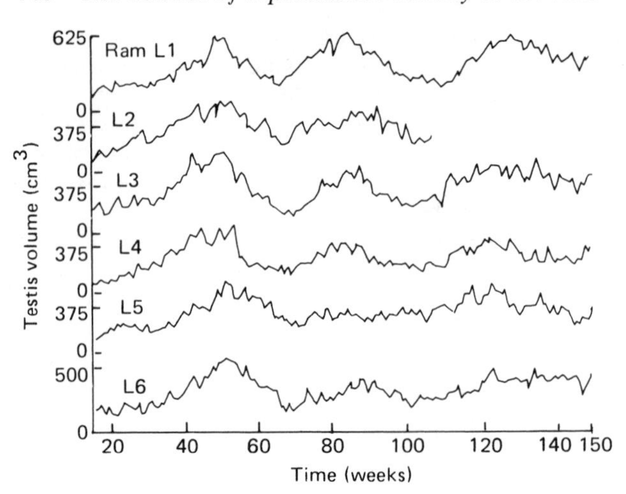

Figure 22.8 Individual testis volumes for six rams kept under constant photoperiod from four months to three years of age. By courtesy of Howles, Craigon and Haynes (1982)

(Purvis, Illius and Haynes, 1974; Lincoln and Davidson, 1977; Lincoln, 1978c; see *Figure 22.4*). This has given rise to the suggestion that photoperiodic change does not drive the seasonal cycle, but merely entrains an endogenous rhythm. There is good evidence for such endogenous rhythms in that cycles of testis growth and hormone concentrations persist for three years with a periodicity of about 36 weeks in rams kept under constant long photoperiod (Howles, Craigon and Haynes, 1982). Data for testis size are shown in *Figure 22.8*.

SEMEN CHANGES

Since LH, FSH and testosterone are involved in spermatogenesis (cf. Courot, 1980; Orgebin-Crist and Hochereau-De Reviers, 1980) and show marked seasonal fluctuations, it is not surprising that seasonal changes in sperm production occur. In the extreme case of the Soay ram, Lincoln and Short (1980) have demonstrated that seasonal regression of the testis involves a reduction in the diameter of the seminiferous tubules with folding of the basement membrane and a marked decrease in germ cell numbers. If these rams are housed, a few cells complete spermatogenesis even during quiescence. However, in the wild, the degree of spermatogenic arrest is greater and most Soay rams fail to produce spermatozoa at all for some months. In domestic breeds the situation is not so severe, but nonetheless, changes can adversely affect fertility (Ortavant, 1956; Colas, Chapter 23). Using a technique for collecting rete testis fluid, sperm production for four breeds was determined in the non-breeding and breeding seasons by Dacheux *et al.* (1981), and the data are illustrated in *Table 22.3*. In all breeds there was a significant increase in sperm production during the breeding season, and the breed differences with the Ile-de-France rams were more susceptible to seasonal effects than the others.

Table 22.3 VARIATIONS IN SPERM PRODUCTION (MEAN±S.E.M.) ACCORDING TO BREED AND BREEDING SEASON (JANUARY–JUNE, NON-BREEDING SEASON, NB; JULY–DECEMBER, BREEDING SEASON, B) IN RAMS

Breed	*Season*	*Sperm production* (10^9/day/testis)
Ile-de-France	NB	1.88±0.15 (53)
	B	4.03±0.19***(47)
Préalpes du Sud	NB	2.37±0.30 (16)
	B	4.62±0.46***(3)
Romanov	NB	2.33±0.58 (6)
	B	3.27±0.35***(6)
Cross-breed Romanov	NB	1.76±0.43 (8)
	B	2.95±0.24***(29)

Values significantly different from those for that breed in the NB season *** $P<0.001$.
Figures in parentheses indicate numbers of animals.
From Dacheux *et al.* (1981)

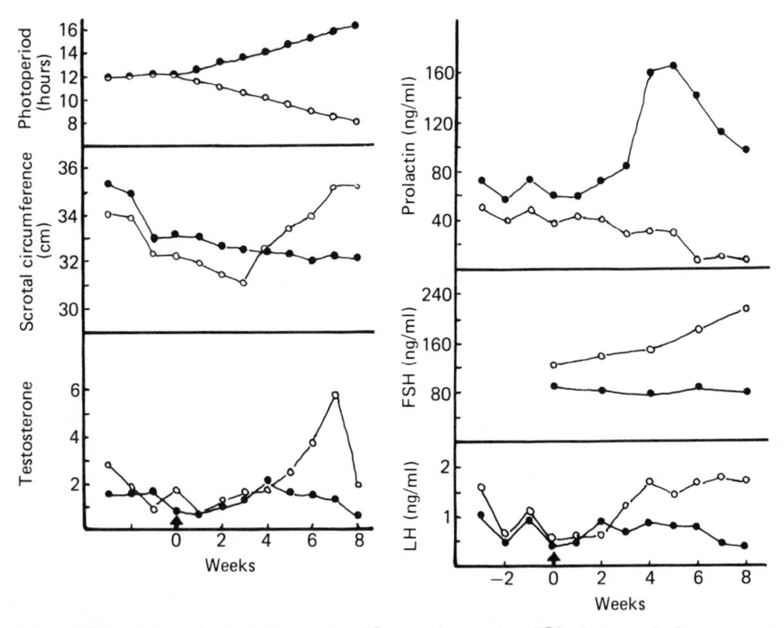

Figure 22.9 The effect of decreasing (○) and increasing (●) photoperiod on mean testicular circumference and mean hormone concentrations in rams (six per group). From Schanbacher and Ford (1979)

Table 22.4 EFFECT OF PHOTOPERIOD ON TESTES WEIGHT, SEMINIFEROUS TUBULE DIAMETERS AND SPERMATOGENESIS IN RAMS (MEANS±S.E.M.)

Treatment	*No. of animals*	*Testis weight* (g)	*Seminiferous tubule diameter* (μm)	*Total daily sperm production* (× 10^9)
Short days	4	477±72*	245±2**	12.9±2.7*
Long days	6	330±111	189±9	6.5±0.8

*$P<0.05$; **$P<0.01$, significantly different from long day rams.
From Schanbacher and Ford (1979)

Schanbacher and Ford (1979) have measured concomitant hormone changes and sperm characteristics in rams artificially exposed to increasing or decreasing daylengths in the non-breeding season for eight weeks. The hormone data are depicted in *Figure 22.9*. In accord with other observations FSH and LH increased in short relative to long day rams and resulted in elevated concentrations of serum testosterone after four weeks of shortening days. The effect of these hormonal changes on semen parameters is shown in *Table 22.4*. Short day treatment of rams in the non-breeding season resulted in testes 45% heavier and sperm production rates two-fold greater than long day rams, the increased sperm production being comparable to the differences found between seasons by Dacheux *et al.* (1981). In addition, seminiferous tubule diameters were larger in short day rams and contained a greater number of pachytene spermatocytes and normal spermatids suggesting that short days increased the probability of germ cells completing the first and second meiotic division.

Semen quantity seems to be affected more by season than does quality, but on the whole, morphological abnormalities are more frequent from January–June compared with July–December in the northern hemisphere (Colas, 1980), probably accounting for the lowered fertility using AI with semen collected in the spring compared with that collected in the autumn. This area is discussed in more detail in Chapter 23.

BEHAVIOURAL CHANGES

It is well established that testicular hormones are involved in the control of sexual behaviour as evinced by the age-old use of castration to abolish sexual behaviour in males. A number of testicular secretory products are involved in the regulation of ram sexual behaviour, namely testosterone, dihydrotestosterone and 17β-oestradiol (Parrott and Davies, 1979; D'Occhio and Brooks, 1980), but the relationship between such hormones and behaviour is far from simple. Prepubertal castration of rams virtually eliminates all adult male sexual behaviour but after postpubertal castration, libido declines slowly and is not attributable to variation in hormone levels (Knight, 1973; Mattner and Braden, 1975). Even within the same breed there can be large variations in mating performance of different rams, with poor correlations between an individual's mating index score and serum testosterone levels (Schanbacher and Lunstra, 1976; Mattner, 1979; Howles, Webster and Haynes, 1980). Furthermore, in the developing ram, photoperiod seems more important *per se* than testosterone, since sexual behaviour was quicker to develop in rams maintained under short compared to long photoperiod, although testosterone levels were equivalent in both groups (Howles, Webster and Haynes, 1980). Although the situation is obviously complex and individual correlations are not good, sexual behaviour in groups of rams has been shown by a number of workers to be affected by season and to be at its highest about six weeks after maximum testosterone secretion. This is demonstrated in *Figure 22.4* for aggressive behaviour and Flehman reactions in Soay rams and in *Figure 22.10* which shows the mating index scores for Finn and Suffolk rams taken through one year (Schanbacher and Lunstra, 1976).

In this study Finn rams were more active overall than Suffolk rams, a situation similar to that found when Finn were compared with Blackface rams (Land, 1970). It was also noted that certain individuals within each breed consistently scored high in the mating index suggesting that regardless of hormone levels, sexual prowess is an inherent trait for which genetic selection may be useful. Lindsay and Signoret (1980) have pointed out that when in constant work, rams produce volumes of semen and numbers of sperm/ejaculate which are well below those considered adequate for artificial insemination. Under such circumstances a ewe must accumulate

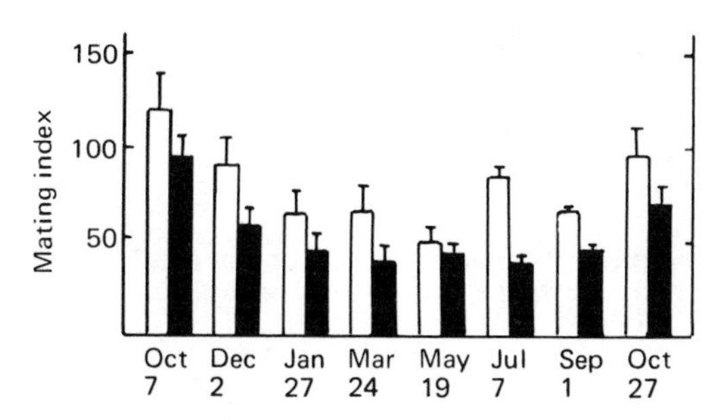

Figure 22.10 Mean mating index scores for five Finn (□) and five Suffolk (■) rams taken through one year. From Schanbacher and Lunstra (1976)

enough sperm from a series of matings either with the same or different rams in order to have a reasonable chance of becoming pregnant (Lindsay, 1979). Rams also show a ewe preference, covering some ewes but not others and it seems probable that rams of low serving capacity may distribute their semen more unequally than very active rams thus accentuating the problem (Synnott, Fulkerson and Lindsay, 1981). On the basis of the above findings, notwithstanding the complexity of factors involved in the control of sexual behaviour, it becomes as important to rectify seasonally depressed sexual behaviour as it is to maximize semen output if rams are needed to be used out of the breeding season.

Manipulation of reproductive activity in the ram

As already described, administration of GnRH in the non-breeding season could be useful in stimulating reproductive characteristics of rams. An earlier study by Schanbacher and Lunstra (1977) demonstrated that a single injection of 50 μg GnRH into rams resulted in significantly higher testosterone levels in blood and an increased mating index in assessments carried out one hour after the injection. Moreover, after chronic treatment with twice daily injections for seven weeks, the testes were larger (cf. *Figure 22.11*) and sperm concentrations, numbers of live sperm and progressive motilities were significantly greater in GnRH treated animals compared with controls. Similar results were obtained in a subsequent

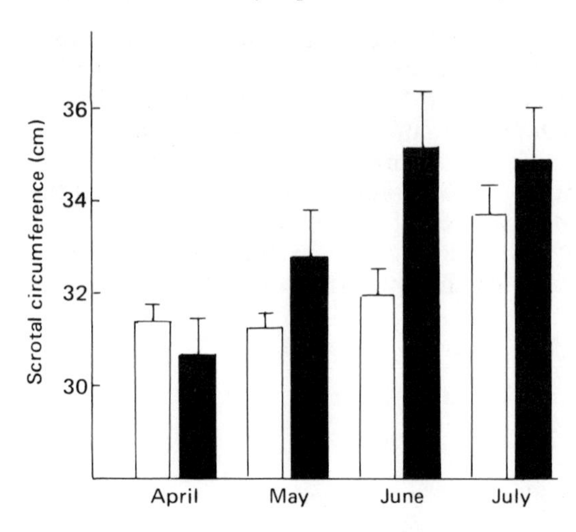

Figure 22.11 Scrotal circumference (mean±S.E.M.) of rams given twice daily intramuscular injections of saline (□; $n = 7$) or 50 µg of GnRH (■; $n = 8$). Treatment began after the April measurement. By courtesy of Schanbacher (1978)

experiment in which fertility was also tested by placing each ram from June 15 to July 15 with 24 ewes in which oestrus and ovulation were induced with progestagen–PMSG treatment (Schanbacher, 1978). Fertility data are shown in *Table 22.5*. The mating activity and percentage of ewes lambing was good compared to previous out-of-season matings in which ewes had

Table 22.5 CHRONIC EFFECTS OF GnRH ON RAM FERTILITY OUTSIDE THE NORMAL BREEDING SEASON

Treatment	*No. of animals*	*Mating activity (% of total ewes ±S.E.M.)*	*Lambing % (% of ewes mated)*	*No. of lambs born/ewe lambing*
Control	7	92.00±0.02	65.6±5.1	2.09±0.08
GnRH	8	93.30±0.02	63.9±3.4	2.02±0.07

Values were determined from lambing records after a 30-day breeding period. Breeding rams received twice daily intramuscular injections of saline or 50 µg of GnRH for two months before and during the sampling period.
From Schanbacher (1978)

been synchronized with progestagens and PMSG, and the rams were untreated. In this particular experiment, however, the control rams also gave good results, perhaps because of the time of the matings (June–July) in relation to the normal breeding season; the testes of the control animals had begun to increase in size by this time (cf. *Figure 22.11*). The results of GnRH therapy are encouraging, however, and should be followed up. In view of the studies of Lincoln (1979a) referred to earlier and Haresign, McLeod and Webster (Chapter 18) it may be advantageous, from a practical standpoint, to investigate the use of slow release implants of GnRH.

A second possible means of stimulating fertility in the ram during the non-breeding season is to increase release of endogenous GnRH by manipulation of the photoperiod. In a study with this aim, rams were exposed to short daylengths (eight hours light) from 28th February to 30th May, and from 9th May to 30th May, each ram was penned with 30 progestagen–PMSG treated ewes (Schanbacher, 1979). Fertility data are shown in *Table 22.6*. Breeding performance by rams on short days was comparable to that during the normal breeding season and all rams sought oestrous females. On the other hand, considerable variation existed in control rams maintained under natural light, with some rams showing very

Table 22.6 LAMBING DATA FOR EWES MATED OUT OF THE NORMAL BREEDING SEASON TO RAMS WHICH WERE EXPOSED TO NATURAL OR ARTIFICIALLY SHORTENED DAYLENGTHS

Treatment	*Oestrus activity (% ±S.E.M.)*	*Mating activity (%)*	*Lambing rate (%)*	*Actual no. of lambs born*
Control rams (5) (natural photoperiod)	95.4±2.3	66.7±18.1	32.0±0.05	81
Short day rams (5) (8L, 16D)	98.0±1.3	89.3±2.9	67.2±0.05**	202

Control rams were maintained out of doors under natural lighting conditions of spring whereas rams on short days were maintained in a closed building under controlled light (8 hours light: 16 hours dark) for 91 days.
** $P<0.01$ significantly different from control rams.
From Schanbacher (1979)

little interest in ewes. Rams on short days not only mated more ewes than controls but produced 2.5 times more lambs than control animals. Furthermore, there was a tendency for more lambs to be born/ewe mated with rams exposed to short days, suggesting a greater number of ova were fertilized. Manipulation of reproductive potential in rams by modification of photoperiod, therefore, is also a potentially useful technique.

Conclusions

More emphasis on lambing throughout the year has led to renewed interest in removing the seasonal depression in ram fertility and making optimally fertile rams available all the year round. In the past decade there have been major advances in knowledge regarding factors controlling reproductive activity in the ram. Although there are breed differences in sperm production and sexual behaviour, and responses to season, common endocrine mechanisms control these functions. Few experiments to date have been carried out using this knowledge to artificially enhance reproductive activity, but those that have, involving treatment with GnRH and modification of photoperiodic cues, have been very encouraging.

References

ALBERIO, R. and COLAS, G. (1976). Influence of photoperiodism on the sexual development of the young Ile-de-France ram. *8th International Congress on Animal Reproduction and Artificial Insemination, Krakow* **3**, 26–29

ALLISON, A.J. (1972). The effect of mating pressure on characteristics of the ejaculate in rams and on reproductive performance in ewes. *Proceedings of the New Zealand Society of Animal Production* **32**, 112–113

ALLISON, A.J. and ROBINSON, T.J. (1971). Fertility of progestagen-treated ewes in relation to the numbers and concentration of spermatozoa in the inseminate. *Australian Journal of Biological Sciences* **24**, 1001–1008

BARRELL, G.K. and LAPWOOD, K.R. (1978a). Seasonality of semen production and plasma LH, testosterone and prolactin levels in Romney, Merino and polled Dorset rams. *Animal Reproduction Science* **1**, 229–244

BARRELL, G.K. and LAPWOOD, K.R. (1978b). Effects of pinealectomy of rams on secretory profiles of LH, testosterone, prolactin and cortisol. *Neuroendocrinology* **27**, 216–227

BARRELL, G.K. and LAPWOOD, K.R.(1979). Effects of pinealectomy on the secretion of LH, testosterone and prolactin in rams exposed to various lighting regimes. *Journal of Endocrinology* **80**, 397–405

BREMNER, W.J., FINDLAY, J.K., LEE, V.W.K., DE KRETSER, D.M. and CUMMING, I.A. (1980). Feedback effects of the testis on pituitary responsiveness to luteinizing hormone releasing hormone infusion in the ram. *Endocrinology* **106**, 329–336

CARMON, J.L. and GREEN, W.W. (1952). Histological study of the development of the testis of the ram. *Journal of Animal Science* **11**, 674–687

COLAS, G. (1980a). Variations saisonnières de la qualité du sperme chez le belier Ile-de-France. 1. Etude de la morphologie cellulaire et de la motilité massale. *Reproduction, Nutrition, Développement* **20**, 1789–1799

COLAS, G. (1980b). Seasonal variations of semen quality in adult Ile-de-France rams. 2. Fertilising ability and its relation to qualitative criteria *in vitro*. *Reproduction, Nutrition, Développement* **21**, 399–407

COUROT, M. (1967). Endocrine control of the supporting and germ cells of the impuberal testis. *Journal of Reproduction and Fertility Supplement* **2**, 89–101

COUROT, M. (1979). Semen quality and quantity in the ram. In *Sheep Breeding*, (G.J. Tomes, D.E. Robertson and R.J. Lightfoot, Eds.), pp. 495–504. London, Butterworths

COUROT, M. (1980). The regulation of testicular function by pituitary gonadotrophins. *9th International Congress on Animal Reproduction and Artificial Insemination, Madrid* **II**, 155–162

CRIM, L.W. and GESCHWIND, I.I. (1972). Patterns of FSH and LH secretion in the developing ram: The influence of castration and replacement therapy with testosterone propionate. *Biology of Reproduction* **7**, 42–46

DACHEUX, J.L., PISSELET, C., BLANC, M.R., HOCHEREAU-DE REVIERS, M.T. and COUROT, M. (1981). Seasonal variations in rete testis fluid secretion and sperm production in different breeds of ram. *Journal of Reproduction and Fertility* **61**, 363–371

DAVIES, R.V., MAIN, S.J. and SETCHELL, B.P. (1977). Seasonal changes in plasma FSH, LH and testosterone in rams. *Journal of Endocrinology* **72**, 12P

D'OCCHIO, M.J. and BROOKS, D.E. (1980). Effects of androgenic and oestrogenic hormones on mating behaviour in rams castrated before or after puberty. *Journal of Endocrinology* **86**, 403–411

D'OCCHIO, M.J., SCHANBACHER, B.D. and KINDER, J.E. (1982a). Relationships between serum testosterone concentration and patterns of luteinizing hormone secretion in male sheep. *Endocrinology* **110**, 1547–1554

D'OCCHIO, M.J., SCHANBACHER, B.D. and KINDER, J.E. (1982b). Testosterone feedback on FSH secretion in male sheep. *Journal of Reproduction and Fertility* **66**, 699–702

DUTT, R.H. (1960). Temperature and light as factors in reproduction among farm animals. *Journal of Dairy Science* **43**, 123–144

DÝRMUNDSSON, O.R. (1973). Puberty and early reproductive performance in sheep. II. Ram lambs. *Animal Breeding Abstracts* **41**, 419–430

DÝRMUNDSSON, O.R. and LEES, J.L. (1972). Pubertal development of Clun Forest ram lambs in relation to time of birth. *Journal of Agricultural Science, Cambridge* **79**, 83–89

FOSTER, D.L., CRUZ, T.A.C., JACKSON, G.L., COOK, B. and NALBANDOV, A.V. (1972). Regulation of luteinizing hormone in the fetal and neonatal lamb. III. Release of LH by the pituitary *in vivo* in response to crude ovine hypothalamic extract or purified bovine gonadotrophin releasing factor. *Endocrinology* **90**, 673–683

FOSTER, D.L., MICKELSON, I.H., RYAN, K.D., COON, G.A., DRONGOWSKI, R.A. and HOLT, J.A. (1978). Ontogeny of pulsatile luteinizing hormone and testosterone secretion in male lambs. *Endocrinology* **102**, 1137–1146

GALLOWAY, D.B. and PELLETIER, J. (1975). Luteinizing hormone release in entire and castrated rams following injection of synthetic luteinizing hormone releasing hormone and the effect of testosterone pretreatment. *Journal of Endocrinology* **64**, 7–16

GARNIER, D.H., TERQUI, M. and PELLETIER, J. (1977). Plasma concentrations of LH and testosterone in castrated rams treated with testosterone or testosterone propionate. *Journal of Reproduction and Fertility* **49**, 359–364

GORDON, I. (1977). Application of synchronisation of oestrus and ovulation in sheep. *Proceedings Symposium on Management of Reproduction in Sheep and Goats*, pp. 15–30. Madison, Wisconsin, University of Wisconsin

HOWLES, C.M., CRAIGON, J. and HAYNES, N.B. (1982). Long term rhythms in testicular volume and plasma prolactin concentrations in rams reared for 3 years in constant photoperiod. *Journal of Reproduction and Fertility* **65**, 439–446

HOWLES, C.M., WEBSTER, G.M. and HAYNES, N.B. (1980). The effect of rearing under a long or short photoperiod on testis growth, plasma testosterone and prolactin concentrations and the development of sexual behaviour in rams. *Journal of Reproduction and Fertility* **60**, 437–447

ILLIUS, A.W., HAYNES, N.B. and LAMMING, G.E. (1976). Effects of ewe proximity on peripheral plasma testosterone levels and behaviour in the ram. *Journal of Reproduction and Fertility* **48**, 25–32

ILLIUS, A.W., HAYNES, N.B., PURVIS, K. and LAMMING,M G.E. (1976). Plasma concentrations of testosterone in the developing ram in different social environments. *Journal of Reproduction and Fertility* **48**, 17–24

JENNINGS, J.T. and CROWLEY, J.P. (1970). The mating of hormone treated sheep. *Animal Production* **12**, 357

JENNINGS, J.T. and CROWLEY, J.P. (1972). The influence of mating management on fertility in ewes following progesterone–PMSG treatment. *Veterinary Record* **90**, 495–498

JOHNSON, B.H., DESJARDINS, C. and EWING, L.L. (1973). Seasonal effects on testis function in rams. *Journal of Animal Science* **37**, 247 (Abstract)

KARSCH, F.J. and FOSTER, D.L. (1975). Sexual differentiation of the mechanism controlling the preovulatory discharge of luteinizing hormone in sheep. *Endocrinology* **97**, 373–379

KATONGOLE, C.B., NAFTOLIN, F. and SHORT, R.V. (1974). Seasonal variations in blood LH and testosterone levels in rams. *Journal of Endocrinology* **60**, 101–106

KNIGHT, T.W. (1973). The effect of the androgen status of rams on sexual activity and fructose concentration in the semen. *Australian Journal of Agricultural Research* **24**, 573–578

LAND, R.B. (1970). The mating behaviour and semen characteristics of Finnish Landrace and Scottish Blackface rams. *Animal Production* **12**, 551–560

LEE, V.W.K., CUMMING, I.A., DE KRETSER, D.M., FINDLAY, J.K., HUDSON, B. and KEOGH, E.J. (1976a). Regulation of gonadotrophin secretion in rams from birth to sexual maturity. I. Plasma LH, FSH and testosterone levels. *Journal of Reproduction and Fertility* **46**, 1–6

LEE, V.W.K., CUMMING, I.A., DE KRETSER, D.M., FINDLAY, J.K., HUDSON, B. and KEOGH, E.J. (1976b). Regulation of gonadotrophin secretion in rams from birth to sexual maturity. II. Response of the pituitary–testicular axis to LH-RH infusion. *Journal of Reproduction and Fertility* **46**, 7–11

LEES, J. (1965). Seasonal variation in the breeding activity of rams. *Nature, London* **207**, 221–222

LINCOLN, G.A. (1976a). Seasonal variation in the episodic secretion of LH and testosterone in the ram. *Journal of Endocrinology* **69**, 213–226

LINCOLN, G.A. (1976b). Secretion of LH in rams exposed to two different photoperiods. *Journal of Reproduction and Fertility* **47**, 351–353

LINCOLN, G.A. (1977). Changes in pituitary responsiveness to luteinizing hormone releasing hormone in rams exposed to artificial photoperiods. *Journal of Endocrinology* **73**, 519–527

LINCOLN, G.A. (1978a). Hypothalamic control of the testes in the ram. *International Journal of Andrology* **1**, 331–341

LINCOLN, G.A. (1978b). The temporal relationship between plasma levels of FSH and LH in the ram. *Journal of Reproduction and Fertility* **53**, 31–37

LINCOLN, G.A. (1978c). The photoperiodic control of seasonal breeding in rams. In *Comparative Endocrinology* (P.J. Galliard and H.H. Boer, Eds.), pp. 149–152. Amsterdam, Elsevier

LINCOLN, G.A. (1979a). Use of a pulsed infusion of luteinizing hormone releasing hormone to mimic seasonally induced endocrine changes in the ram. *Journal of Endocrinology* **83**, 251–260

LINCOLN, G.A. (1979b). Photoperiodic control of seasonal breeding in the ram: participation of the cranial sympathetic nervous system. *Journal of Endocrinology* **82**, 135–147

LINCOLN, G.A. and DAVIDSON, W. (1977). The relationship between sexual and aggressive behaviour and pituitary and testicular activity during the seasonal sexual cycle of rams and the influence of photoperiod. *Journal of Reproduction and Fertility* **49**, 267–276

LINCOLN, G.A. and SHORT, R.V. (1980). Seasonal breeding: nature's contraceptive. *Recent Progress in Hormone Research* **36**, 1–52

LINCOLN, G.A., McNEILLY, A.S. and CAMERON, C.L. (1978). The effects of a sudden decrease or increase in daylength on prolactin secretion in the ram. *Journal of Reproduction and Fertility* **52**, 305–311

LINDSAY, D.R. (1979). Mating behaviour in sheep. In *Sheep Breeding* (G.J. Tomes, D.E. Robertson and R.J. Lightfoot, Eds.), pp. 473–479. London, Butterworths

LINDSAY, D.R. and SIGNORET, J.P. (1980). Influence of behaviour on reproduction. *9th International Congress on Animal Reproduction and Artificial Insemination, Madrid* **1**, 83–92

LOUDA, F., DONEY, J.M., STOLC, L., KRIZEK, J. and SMERHA, J. (1981). The development of sexual activity and semen production in ram lambs of two prolific breeds: Romanov and Finnish Landrace. *Animal Production* **33**, 143–148

MATTNER, P.E. (1979). Gonadal hormone control of male sexual behaviour. University of New England, Armidale, N.S.W., Australia: *Reviews in Rural Science* **IV**, 11–18

MATTNER, P.E. and BRADEN, A.W.H. (1975). Influence of age, hormone treatment, shearing and diet on the libido of Merino rams. *Australian Journal of Experimental Agriculture and Animal Husbandry* **15**, 330–336

ORGEBIN-CRIST, M.C. and HOCHEREAU-DE REVIERS, M.T. (1980). Sperm formation and maturation—role of testicular and epididymal somatic cells. *9th International Congress on Animal Reproduction and Artificial Insemination, Madrid* **1**, 59–82

ORTAVANT, R. (1956). Action de la durée d'eclairement sur les processurs spermatogenètiques chez le belier. *Compte Rendu des Séances de la Société de Biologie* **150**, 471–474

PARROTT, R.F. and DAVIES, R.V. (1979). Serum gonadotrophin levels in prepubertally castrated male sheep treated for long periods with propionated testosterone, dihydrotestosterone, 19-hydroxytestosterone or oestradiol. *Journal of Reproduction and Fertility* **56**, 543–548

PELLETIER, J. (1974). Decrease in the pituitary response to synthetic LH-RF in castrated rams following testosterone propionate treatment. *Journal of Reproduction and Fertility* **41**, 397–402

PELLETIER, J. and ORTAVANT, R. (1975a). Photoperiodic control of LH release in the ram. 1. Influence of increasing and decreasing light photoperiods. *Acta Endocrinologica, Copenhagen* **78**, 435–441

PELLETIER, J. and ORTAVANT, R. (1975b). Photoperiodic control of LH release in the ram. 2. Light–androgen interaction. *Acta Endocrinologica, Copenhagen* **78**, 442–450

PEPELKO, W.E. and CLEGG, M.T. (1965). Influence of season of the year upon patterns of sexual behaviour in male sheep. *Journal of Animal Science* **24**, 633–637

PURVIS, K., ILLIUS, A.W. and HAYNES, N.B. (1974). Plasma testosterone concentrations in the ram. *Journal of Endocrinology* **61**, 241–253

RAVAULT, J.P. (1976). Prolactin in the ram: seasonal variations in the concentration of blood plasma from birth until three years old. *Acta Endocrinologica, Copenhagen* **83**, 720–725

RIGGS, B.L. and MALVEN, P.V. (1974). Spontaneous patterns of LH release in castrate male sheep and the effects of exogenous estradiol. *Journal of Animal Science* **38**, 1239–1244

SANFORD, L.M., PALMER, W.M. and HOWLAND, B.E. (1977). Changes in the profiles of serum LH, FSH and testosterone and in mating performance and ejaculate volume in the ram during the ovine breeding season. *Journal of Animal Science* **45**, 1382–1391

SANFORD, L.M., WINTER, J.S.D., PALMER, W.M. and HOWLAND, B.E. (1974). The profile of LH and testosterone secretion in the ram. *Endocrinology* **95**, 627–631

SCHANBACHER, B.D. (1978). Fertility of rams chronically treated with gonadotrophin releasing hormone during the non-breeding season. *Biology of Reproduction* **19**, 661–665

SCHANBACHER, B.D. (1979). Increased lamb production with rams exposed to short daylengths during the non-breeding season. *Journal of Animal Science* **49**, 927–932

SCHANBACHER, B.D. (1980a). The feedback control of gonadotropin secretion by testicular steroids. *9th International Congress on Animal Reproduction and Artificial Insemination, Madrid* **II**, 177–184

SCHANBACHER, B.D. (1980b). Influence of testicular steroids on thyrotropin releasing hormone-induced prolactin release in mature rams. *Journal of Andrology* **1**, 121–126

SCHANBACHER, B.D. and CROUSE, J.D. (1980). Growth and performance of growing finishing lambs exposed to long or short photoperiods. *Journal of Animal Science* **51**, 943–948

SCHANBACHER, B.D. and FORD, J.J. (1976). Seasonal profiles of plasma luteinizing hormone, testosterone and estradiol in the ram. *Endocrinology* **99**, 661–665

SCHANBACHER, B.D. and FORD, J.J. (1977). Gonadotropin secretion in cryptorchid and castrate rams and the acute effects of exogenous steroid treatment. *Endocrinology* **100**, 387–393

SCHANBACHER, B.D. and FORD, J.J. (1979). Photoperiodic regulation of ovine spermatogenesis: relationship to serum hormones. *Biology of Reproduction* **20**, 719–726

SCHANBACHER, B.D. and LUNSTRA, D.D. (1976). Seasonal changes in sexual activity and serum levels of LH and testosterone in Finnish Landrace and Suffolk rams. *Journal of Animal Science* **43**, 644–650

SCHANBACHER, B.D. and LUNSTRA, D.D. (1977). Acute and chronic effects of gonadotropin releasing hormone on reproductive characteristics of rams during the non-breeding season. *Journal of Animal Science* **44**, 650–655

SCHANBACHER, B.D., GOMES, W.R. and VAN DEMARK, N.L. (1974). Developmental changes in spermatogenesis, testicular carnitine acetyltransferase activity and serum testosterone in the ram. *Journal of Animal Science* **39**, 889–892

SETCHELL, B.P. (1980). The significance of inhibin in the feedback control by the testis of gonadotrophin secretion. *9th International Congress on Animal Reproduction and Artificial Insemination, Madrid* **II**, 163–169

SETCHELL, B.P., DAVIES, R.V. and MAIN, S.J. (1977). In *The Testis*, Volume 4 (A.D. Johnson and W.R. Gomes, Eds.), pp. 189–238. New York, Academic Press

SYNNOTT, A.L., FULKERSON, W.J. and LINDSAY, D.R. (1981). Sperm output by rams and distribution amongst ewes under conditions of continual mating. *Journal of Reproduction and Fertility* **61**, 355–361

WALTON, J.S., EVINS, J.D., HILLARD, M.A. and WAITES, G.M.H. (1980). Follicle stimulating hormone release in hemicastrated prepubertal rams and its relation to testicular development. *Journal of Endocrinology* **84**, 141–152

WATSON, R.H., SAPSFORD, C.S. and McCANCE, I. (1956). The development of the testis, epididymis and penis in the young Merino. *Australian Journal of Agricultural Research* **7**, 574–590

23

FACTORS AFFECTING THE QUALITY OF RAM SEMEN

G. COLAS
INRA, Station de Physiologie de la Reproduction, 37380 Nouzilly, France

Numerous studies have shown that within the same breed, there exists a wide variation in the fertility of rams (Dun, Ahmed and Morrant, 1960; Salamon and Robinson, 1962; Lees 1978; Mickelsen, Paisley and Dahmen, 1981; Colas, 1981). It is also well known that, for the same animal, the fertilizing potential of semen does not remain constant over the year (Colas, 1981).

It is thus important to determine the main factors which may affect the quality of semen and the extent of their effect. Their control would allow rams to be kept under the optimum conditions of sperm production and to eliminate those which produce poor quality semen. This problem is becoming more and more acute with the development of artificial insemination (AI) following oestrus synchronization which makes it possible to use the same sire for a large number of females, at any time of the year and especially in the non-breeding season.

Leaving out the pathological effects on the quality of semen which will not be discussed here, these factors can be divided into three categories related to:

1. Age of the ram,
2. Environment,
3. Semen technology.

However, before tackling this study, it is necessary to give a brief definition of what is meant by the quality of semen as assessed *in vitro*, i.e. to find out the seminal characteristic which is best correlated with fertility *in vivo*.

Relationship between *in vitro* quality and fertility of sperm

The notion of quality applied to male gametes is seen very differently according to individual authors. If the *in vivo* sperm quality is easily measured by pregnancy or lambing rates, its *in vitro* quality remains more difficult to assess since it is estimated by criteria as different as massal motility, percentage of unstained or abnormal spermatozoa, semen density, etc., but not all of them are related in the same way to its fertilizing potential.

The literature available relating to this question shows that in the sheep, correlations between lambing rates and the classical *in vitro* parameters are generally few. In the case of natural mating, however, Hulet, Foote and Blackwell (1965) have reported that the greatest correlation was found with cell morphology ($r = -0.40$ to -0.42). A similar relationship is clearly apparent in the studies of McKenzie and Philips (1934) and Cupps, Laben and Mead (1953). When ewes are artificially inseminated it is even higher ($r = -0.83$) but it is only during the spring months (Colas, 1981) that the morphology and the fertility of semen are significantly different amongst rams. Bearing this in mind, it is interesting to note that authors who have studied the effect of high temperatures on reproduction in the ram, have generally observed an increase in abnormal sperm cells and a reduction in fertility of the treated animals. Both phenomena are thus closely related (Rathore, 1968; 1970). In other words, the morphological control of spermatozoa is, to date, the most accurate test available to evaluate the initial quality of the ejaculate.

By contrast, after cooling and storage, this test has proved to no longer be sufficiently precise in the prediction of the fertility of sperm. A few authors have suggested morphological characteristics of the acrosome to be better, especially for frozen semen (Watson and Martin, 1972; Smorag and Kareta, 1974; Tasseron, Amir and Schindler, 1977). However, the excessive amount of time necessary to carry out this test renders it difficult in practice. For this reason most people prefer to use the viability of spermatozoa following incubation at 37–39 °C (Colas, 1975; Langford *et al.*, 1979; Fiser, Ainsworth and Langford, 1981), at least after freezing sperm in the autumn.

Age of the animal

In sheep, as in most species, the quality of semen is strongly influenced by the age of the animal. Indeed, many studies have shown that the initial ejaculates contain a great number of abnormal cells (Dun, 1955; Louw and Joubert, 1964; Skinner and Rowson, 1968; El Wishy, 1974), but large differences exist between breeds (Colas and Zinszner, 1975). These abnormalities consist, for the most part, of head malformations and proximal cytoplasmic droplets (*Table 23.1*), which indicate incomplete spermatogenetic activity and incomplete epididymal maturation. Quality readily improves with advancing age but the rate of increase seems to depend on daylight environment (Alberio, 1976).

Table 23.1 EVOLUTION OF THE PERCENTAGE OF ABNORMAL SPERMATOZOA WITH AGE IN PUBESCENT RAMS

Abnormalities	*126 days*	*168 days*
Abnormal spermatozoa (%)	70	31
Deformed heads	18	7
Abaxial attachment	4	2
Coiled and looped tails	8	5
Tailless	15	11
Proximal droplets	25	6

From Skinner and Rowson (1968)

The potential to fertilize undergoes a similar evolution. In the Ile-de-France and Lacaune dairy breeds (*Table 23.2*), the percentage of ewes lambing and their litter size is significantly lower following AI with sperm from ram lambs than with sperm from adult rams. This phenomenon has been confirmed by a large number of Lacaune ewes in a dairy AI Centre (60.9% vs. 62.7% lambing rate respectively for semen from ram lambs and adult rams mated to 978 and 13 719 ewes; Briois, 1980, personal communication). However, these differences are smaller than those of Colas and Zinszner (1975) because, in practice, adult rams are 'mated' with only the best dairy ewes, i.e. mainly with the oldest ones which are known to be less fertile than the young females.

Table 23.2 RELATIONSHIP BETWEEN AGE OF RAM AND FERTILITY OF SEMEN FOLLOWING AI

Breed	*Percentage of ewes conceiving*	
	Age of ram (years)	
	<1	*2–4*
Ile-de-France	45.5 (56)	66.2 (65)
Dairy Lacaune	54.0 (354)	61.5 (311)
Total	53.0 (410)*	62.0 (376)

*$P<0.01$
Figures in parentheses indicate fertility in ewes inseminated following treatment with fluorogestone acetate and pregnant mare's serum gonadotrophin.
From Colas and Zinszner (1975)

These data show that a young male is not necessarily a good sire and that its use in AI often results in a lower fertility and prolificacy rate.

Environment

It is well established that in the ram sexual activity is subject to seasonal variations, the intensity of which vary from breed to breed. These variations can affect all the components of the reproductive function, particularly the quality of semen and its fertilizing ability (Salamon and Robinson, 1962; Smyth and Gordon, 1967; Colas, 1980; 1981). They are due to modifications of environmental factors amongst which daylight and temperature are the most important.

PHOTOPERIODISM

Under temperate latitudes, photoperiodism is the main factor controlling the whole reproductive process in the male (Ortavant and Thibault, 1956; Ortavant, 1958; Alberio, 1976; Lincoln, 1976).

In order to know its specific action on the morphology and fertilizing ability of gametes, an experiment has been conducted where two groups of five adult Ile-de-France rams, of different genetic origins, were held for six months (28th October to 30th April) under a controlled ambient temperature (20±3 °C) and submitted to an opposed artificial light regime.

Intensity of light was kept constant (300 lux/m^2). The animals had been preconditioned to these photoperiodic treatments many months before the experiment began and their feeding regime was kept constant over the entire period. All animals were collected regularly (4 ejaculates per week) and semen abnormalities were recorded in the first ejaculate every two weeks according to a method previously described (Colas, 1980). From the third month onwards, fertility of unselected ejaculates of the 10 individual rams was measured in the same flocks. Ewes had been treated with fluorogestone acetate (FGA) and pregnant mare's serum gonadotrophin (PMSG) to induce ovulation. They were inseminated 55±1 hours post-treatment with freshly diluted sperm. Eighteen days later pregnancy was estimated by radioimmunoassay of plasma progesterone (Terqui and Thimonier, 1974).

During the entire experimental period, ejaculated sperm contained significantly more abnormal forms in the rams exposed to an increasing light cycle (long days rams) than in those exposed to a decreasing light duration (short days rams) (*Figure 23.1*). This increase included all types of

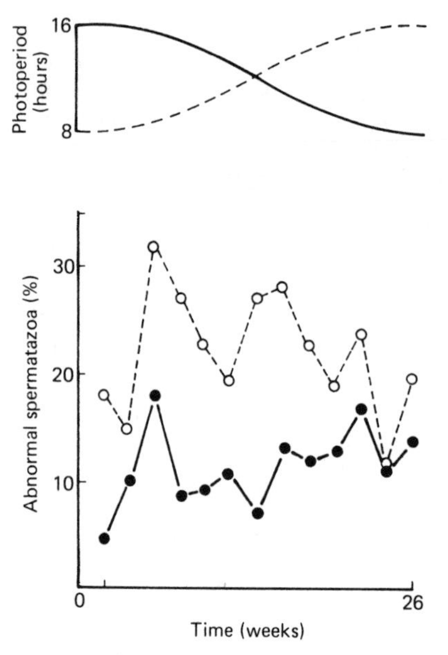

Figure 23.1 Sperm abnormalities in relation to photoperiod in adult Ile-de-France rams (n = 5 per treatment)

abnormalities (tailless spermatozoa, abnormal flagella, abnormal heads, proximal and distal droplets) but differences were significant only for distal (P<0.05) and proximal (P<0.001) droplets. Analysis also revealed that within each group, there were significant variations between rams.

Fertility following AI with the same total number of spermatozoa/ewe (*Table 23.3*) was markedly reduced in the long days rams (51.6% vs. 66.2% lambing rate, P<0.01). It is important to note that significant differences between animals occurred in this group only (40.5% to 64.0%), which confirms previous observations in rams reared under natural conditions

Table 23.3 REPRODUCTIVE PERFORMANCES OF ADULT ILE-DE-FRANCE RAMS[a] ($n = 5$) UNDER CONDITIONS OF ARTIFICIALLY INCREASING OR DECREASING LIGHT (ANNUAL CYCLE)

Ram	*Fertility (%)*	*Fecundity (%)*
Long days rams (increasing light)	51.6 (213)[b]	89.7 (213)
Short days rams (decreasing light)	66.2* (237)	116.9 (237)

*$P<0.01$
[a]Unselected ejaculates.
[b]Number of ewes from the same flocks, inseminated 55 hours following FGA + PMSG treatment.

(Colas, 1981). The total number of lambs born/100 ewes inseminated was also much lower (27 lambs less/100 ewes than in the short days ram group). These results agree well with those reported by Schanbacher (1979) following natural mating.

By way of contrast the percentage of ewes pregnant 18 days post AI was similar in both groups (*Table 23.4*). It differed significantly ($P<0.05$) from the lambing rate only in the long days rams. These results indicate that, as in the case of epididymal sperm (Fournier-Delpech *et al.*, 1979), semen from rams on increasing daylength results in an increase in embryo mortality (13% vs. 26%). They also tend to show that the lower the lambing rate, the higher the embryonic loss, a phenomenon common in the bull (Erb and Flerchinger, 1954; Courot and Tourneur, 1976).

Table 23.4 THE EFFECT OF PHOTOPERIODIC TREATMENT OF THE RAM ON EMBRYO LOSSES

Ram	*Pregnancy rate 18 days post AI (%)*	*Lambing rate (%)*	*Embryo losses (%)*
Long days rams (increasing light)	63.5 (104)[b]*	47.0 (104)	26.0
Short days rams (decreasing light)	66.0 (119)	57.1 (119)	13.0

*$P<0.05$
[a]From unselected ejaculates.
[b]Number of ewes from the same flocks inseminated 55 hours following FGA + PMSG treatment.

In the spring, wide variations exist between animals in both the morphology and fertilizing capacity of semen, whereas in the autumn these differences are less important. From a practical point of view, this means that rams must be selected for semen quality in the spring, not in the autumn. The problem thus lies in knowing how and precisely when to carry out such a selection in the non-breeding season.

It has been shown that, for a single ram, the morphological picture of semen is fairly repeatable from one year to another (Colas, 1980). The maximum percentage of abnormalities occurs every year at about the same time (around the spring equinox in the Ile-de-France breed). A few samples of semen collected at that time (March) thus allows detection of rams which are the most sensitive to photoperiod, i.e. those of low fertility in spring.

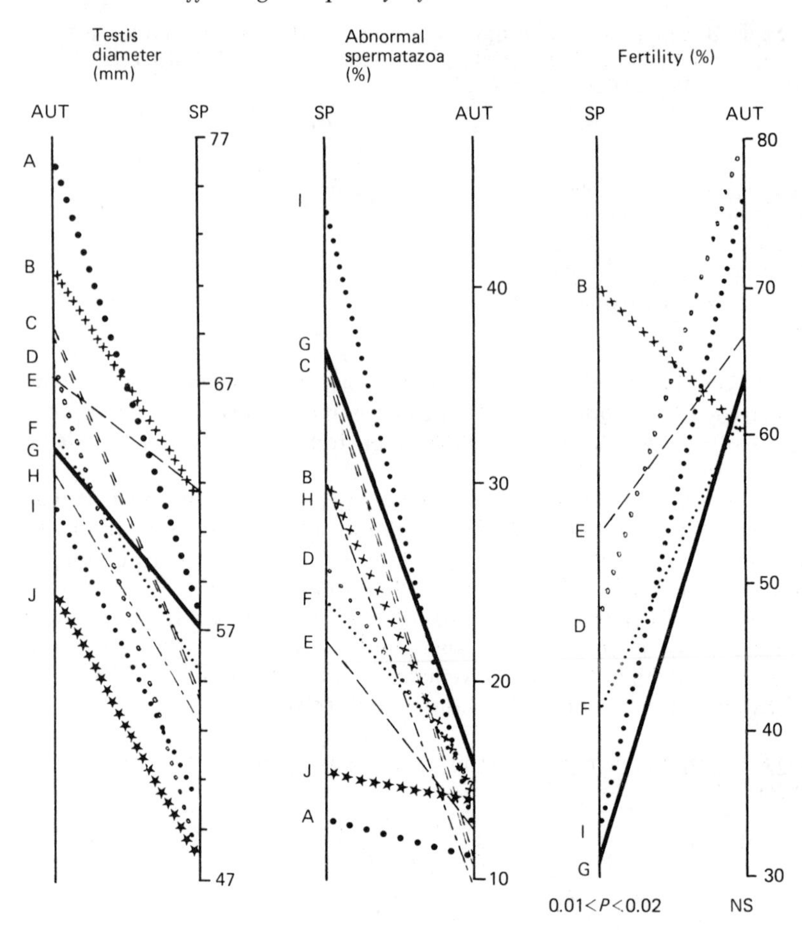

Figure 23.2 Classification of adult Ile-de-France rams according to their seasonal variations in testis diameter, sperm abnormalities and fertility (AUT = Autumn, SP = Spring)

In *Figure 23.2*, 10 unrelated adult Ile-de-France rams have been classified according to variation in testis diameter and semen quality (morphology) in spring and autumn (six of them were also fertility tested *in vivo* following AI). It is difficult at this early stage to see any correlation between morphology and scrotum diameter with season. By contrast, it is clear that seasonal fluctuations in gonadal size are much less marked than those of abnormal cells. If there is also no relationship between testis diameter and conception rates (Mickelsen, Paisley and Dahmen, 1981), then selection of highly fertilizing rams in the spring could be performed with morphological tests only.

TEMPERATURE

The effect of high temperatures on the reproductive performances of the ram is now well established. Exposure of the scrotum alone or of the whole animal to hot temperatures always results in a change in the morphological profile of the semen (Moule and Waites, 1963; Dutt and Hamm, 1957; Rathore, 1968; 1969; 1970; Smith, 1971; Zaba-Branny, 1971), but considerable variation occurs between individuals in the extent of this change (Dun, 1955). These modifications affect every part of the cell, but mainly concern the head. They induce an increase in the proportion of pyriform heads, damaged acrosomes and tailless spermatozoa. Semen degeneration occurs generally in the second week following the beginning of the treatment and complete recovery is not apparent until several weeks after it is completed. Its effect depends upon the severity and the duration of the heating period and also on the breed (Dutt and Hamm, 1957; Lindsay, 1969).

As one might expect, the fertilizing capacity of sperm is also reduced following heat stress (Fowler, 1968; Howarth, 1969; Rathore, 1970). In rams exposed to local or whole body heating, a fall in fertility appears as early as a few days post-treatment, the amplitude of the phenomenon

Table 23.5 REPRODUCTIVE PERFORMANCES OF EWES MATED TO HEAT-TREATED RAMS

Ewe performance	*Duration of heat treatment of rams*		
	Control	*2 days*	*4 days*
23 day non-return rate	14/20	9/20	3/20
40 day pregnant*	12/20	3/20	0/20
Ewes lambed	12/20	3/20	0/20

*All groups significantly different ($P<0.005$)
By courtesy of Rathore (1968)

depending on the duration of treatment. Results reported by Howarth (1969) and Rathore (1970) suggest that high temperatures also affect embryo survival (*Table 23.5*), which indicates that, as for cell morphology, the gametes capable of fertilizing may not be able to maintain a pregnancy.

Semen technology

In the ram many studies have shown that the storage ability of the gametes, generally low in this species (Dauzier, Thibaut and Wintemberger, 1954; Salamon and Robinson, 1962), is further reduced by the fact that the

Table 23.6 EFFECT OF FINAL CONCENTRATION ON THE SURVIVAL RATE AND FERTILITY OF RAM SPERM STORED IN A LIQUID STATE

Final concentration ($\times 10^9$ spermatozoa/ml)	*Motile spermatozoa (%) after incubation*		*Fertility* (% lambing rate)
	3 hours	*5 hours*	
1	38.3±5.2	27.0±4.8	48.9 (88)[a]
2	11.6±2.1**	4.1±1.4**	31.4* (89)

*$P<0.05$; **$P<0.01$
[a] Ewes inseminated (one AI per female)
From Colas *et al.* (1980)

conditions under which AI has to be performed (morphology of the cervix) dictate that sperm must be deposited in small volumes and of a high concentration (Allison and Robinson, 1972). Furthermore it is known that the fertilizing ability and survival of sperm are strongly reduced following storage at high densities (*Table 23.6*). The fall in quality observed during preservation is thus always the result of the interaction between the effect of the dilution rate and that of other parameters of the technique, mainly the composition of the diluents and the temperature at which they are kept.

FRESH DILUTED SEMEN

The decline in fertility is evident as early as the first few hours after dilution (Entwistle and Martin, 1972; Watson and Martin, 1976). However, when fresh sperm is to be used within 10–12 hours after collection, reducing its concentration to 1.6×10^9 spermatozoa/ml and storing it at +15 °C in a skim-milk diluent, gives the best maintenance of fertilizing ability (Smyth, Boland and Gordon, 1978; Colas and Guerin, 1979). When the storage period is longer (24 hours), Maxwell (1978) has shown that the fertility of sperm decreases rapidly (*Table 23.7*). These data also demonstrate an effect of sperm storage on embryonic loss. In practice, fresh diluted semen is generally not used more than 10–12 hours after its preparation.

Table 23.7 FERTILITY AND EMBRYONIC LOSS IN RELATION TO THE AGE OF SEMEN STORED IN A LIQUID STATE

Age of semen (days)	*Pregnancy rate* (%)	*Lambing rate* (%)	*Embryonic loss* (%)
0 (fresh)	70.0 (70)	60.0	14.0
1	45.7 (70)	34.3	25.0
2	46.5 (71)	33.8	27.3
3	32.9 (70)	17.1	47.8

Figures in parentheses indicate the number of ewes inseminated.
From Maxwell (1978)

FROZEN SEMEN

All literature dealing with this subject shows that freezing at very low temperatures greatly reduces the survival and fertilizing capacity of rams' spermatozoa. In fact, this fall in quality occurs chiefly during and after the phase change of the sample, since fertility of liquid semen controlled immediately before freezing is generally high (Colas, unpublished data).

Until now it has been accepted that apart from the importance of sperm concentration, detrimental effects of the freezing process are dependent on three interrelated factors: the nature of the diluents, and the freezing and thawing rates. It was indeed well demonstrated that the percentage of egg yolk (Salamon and Lightfoot, 1969; Watson and Martin, 1975) or glycerol

(Lightfoot and Salamon, 1969; Colas, 1975), and the osmolality of the extender (Salamon and Lightfoot, 1969; Colas, 1975; Fiser, Ainsworth and Langford, 1981) could greatly modify the survival of frozen-sperm. In a similar way it is known that the crystallization phase must be rapid and, consequently, the quality of the semen is always improved when the rate of freezing is rapid as opposed to slow. So far, a few studies have clearly shown that rapid thawing generally reduces the injury caused to the sperm cell (Aamdal and Andersen, 1968; Colas, 1979; Olafsson, 1980). However, recent trials have highlighted the importance of the redilution of thawed semen to improve its fertilizing capacity (Colas and Guerin, 1981). Indeed, the substitution of the freezing medium with an adequate extender, followed by storage of the thawed-rediluted sperm at +15 °C allows a satisfactory lambing rate to be achieved after a single AI, and enables the differences in fertility potential of freshly diluted sperm and frozen-thawed semen to be minimized (*Table 23.8*). It must be added that such semen

Table 23.8 LAMBING AND PROLIFICACY RATES FOLLOWING AI (JULY TO NOVEMBER) WITH FRESH OR FROZEN SEMEN

Sperm	*Fertility* (%)	*Prolificacy* (%)
Fresh	70.3 (798)	164.3 (561)
Frozen	61.1 (242)*	167.6 (148)

*$P<0.01$
Figures in parentheses indicate number of ewes inseminated (one AI/female) on lambing
From Colas and Guerin (1981)

remains fertile after six hours of storage at +15 °C. Resuspension of thawed spermatozoa can thus be performed in the laboratory, which greatly facilitates the use of frozen semen in practice.

These results show that although semen quality is not completely re-established following freezing and thawing, the medium with which the thawed sperm cells are in contact before being deposited into the genital tract of the ewe can be important in minimizing the reduction in fertilizing potential of frozen semen.

Conclusion

Data presented here show that many factors can affect the *in vitro* quality of semen and its fertilizing ability. This also applies to the survival of the embryo and, consequently, to the size of the litter. These results are of real practical importance.

Apart from seasonal variations in fertilizing ability, which one observes in many species, there also exist differences between rams within the same breed, although this occurs to a greater extent during the spring. The choice of ram thus determines to a large extent the fertility level of a flock in spring.

Among the factors which determine the *in vitro* and *in vivo* quality of spermatozoa, photoperiodism is the most important one in Europe where ambient temperatures never reach very high levels. Selecting the best rams amounts to detecting, by means of a morphological test, those which are the less photosensitive. The remaining rams must be either eliminated or exposed to a favourable light treatment. It would be more economically viable to carry out such a selection as early as possible, i.e. when rams are less than one year old, but further work is required to determine whether there is a good correlation between semen characteristics of the ram lamb and of the adult.

Semen quality is also deeply affected by processing treatments, especially by freezing at low temperatures. Changes in the process of the thawing of sperm and its storage under the appropriate conditions is improving conception rates.

References

AAMDAL, J. and ANDERSEN, K. (1968). Freezing of ram semen in straws. *6th International Congress on Animal Reproduction and Artificial Insemination, Paris*, **II**, 977–980

ALBERIO, R. (1976). Rôle de la photopériode dans le développement de la fonction de reproduction chez l'agneau Ile-de-France, de la naissance à 21 mois. Thèse Doct. University of Paris

ALLISON, A.J. and ROBINSON, T.J. (1972). The recovery of spermatozoa from the reproductive tract of the spayed ewe treated with progesterone and oestrogen. *Journal of Reproduction and Fertility* **31**, 215–224

COLAS, G. (1975). Effect of initial freezing temperature, addition of glycerol and dilution on the survival and fertilizing ability of deep frozen ram semen. *Journal of Reproduction and Fertility* **42**, 277–285

COLAS, G. (1979). Fertility in the ewe after artificial insemination with fresh and frozen semen at the induced oestrus and influence of the photoperiod on the semen quality of the ram. *Livestock Production Science* **6**, 153–166

COLAS, G. (1980). Variations saisonnières de la qualité du sperme chez le bélier Ile-de-France. I. Etude de la morphologie cellulaire et de la motilité massale. *Reproduction, Nutrition, Développement* **20**, 1789–1799

COLAS, G. (1981). Variation saisonnière de la qualité du sperme chez le bélier Ile-de-France. II. Fécondance: relation avec les critères qualitatifs observés *in vitro. Reproduction, Nutrition, Développement* **21**, 399–407

COLAS, G. and GUERIN, Y. (1981). A new method for thawing ram semen. *Theriogenology* **6**, 623–631

COLAS, G. and GUERIN, Y. (1979). L'insémination artificielle chez les ovins: acquisitions et perspectives. *5émes Journées de la Recherche Ovine et Caprine, 'L'amélioration génétique des espèces ovine et caprine* (ITOVIC, Ed.), pp. 162–185

COLAS, G. and ZINSZNER, F. (1975). Production spermatique et développement testiculaire chez de l'agneau de race Ile-de-France et Préalpes. *1éres Journées de la Recherche Ovine et Caprine, 'Les Races prolifiques* (ITOVIC, Ed.), pp. 235–243

COLAS, G., TRYER, M., GUERIN, Y. and AGUER, D. (1980). Survival and fertilizing ability of ram sperm stored in a liquid state during 24 hours.

9th International Congress on Animal Reproduction and Artificial Insemination, Madrid, **III**, 315

COUROT, M. and TOURNEUR, J.C. (1976). Qualité du sperme et réussite de l'insémination artificielle. In *Maitrise des cycles sexuels chez les bovins*, pp. 117–125. Paris, INRA-SERSIA-SEARLE

CUPPS, P.T., LABEN, R.C. and MEAD, S.W. (1953). The relation of certain semen quality tests to breeding efficiency and characteristics of semen from low fertility bulls before and after hormone injection. *Journal of Dairy Science* **36**, 422–426

DAUZIER, L., THIBAUT, C. and WINTEMBERGER, S. (1954). Conservation du sperme de bélier après dilution et maintien de son pouvoir fécondant. *Annales d'Endocrinologie* **15**, 341–350

DUN, R.B. (1955). Puberty in Merino rams. *Australian Veterinary Journal* **31**, 104–106

DUN, R.B., AHMED, W. and MORRANT, A.J. (1960). Annual reproductive rhythm in Merino sheep related to the choice of a mating time at Trangie, Central Western New South Wales. *Australian Journal of Agricultural Research* **11**, 805–826

DUTT, R.H. AND HAMM, P.T. (1957). Effect of exposure to high environmental temperature and shearing on semen production of rams in winter. *Journal of Animal Science* **16**, 328–334

EL WISHY, A.B. (1974). Somme aspects of reproduction in fat-tailed sheep in subtropics. IV. Puberty and sexual maturity. *Zeitschrift für Tierzüchtung und Züchtungsbiologie* **91**, 311–316

ENTWISTLE, K.W. and MARTIN, I.C.A. (1972). Effects of the number of spermatozoa and of volume of diluted semen on fertility in the ewe. *Australian Journal of Agricultural Research* **23**, 467–472

ERB, R.E. and FLERCHINGER, F.H. (1954). Influence of fertility level and treatment of semen on non-return decline from 20 to 180 days following artificial service. *Journal of Dairy Science* **37**, 938–948

FISER, P.S., AINSWORTH, L. and LANGFORD, G.A. (1981). Effect of osmolality of skim-milk diluents and thawing rate on cryosurvival of ram spermatozoa. *Cryobiology* **18**, 399–403

FOURNIER-DELPECH, S., COLAS, G., COUROT, M., ORTAVANT, R. and BRICE, G. (1979). Epididymal sperm maturation in the ram: motility, fertilizing ability and embryonic survival after uterine insemination in the ewe. *Annales de Biologie Animale, Biochimie et Biophysique* **19**, 597–605

FOWLER, D.G. (1968). Skin folds and Merino breeding. 7. The relations of heat applied to the testis and scrotal thermoregulation to fertility in the Merino ram. *Australian Journal of Experimental Agriculture and Animal Husbandry* **8**, 142–148

HOWARTH, B. (1969). Fertility in the ram following exposure to elevated ambient temperature and humidity. *Journal of Reproduction and Fertility* **19**, 179–183

HULET, C.V., FOOTE, W.C. and BLACKWELL, R.L. (1965). Relationship of semen quality and fertility in the ram to fecundity in the ewe. *Journal of Reproduction and Fertility* **9**, 311–315

LANGFORD, G.A., MARCUS, G.J., HACKETT, A.J., AINSWORTH, L., WOLYNETZ, M.S. and PETERS, H.F. (1979). A comparison of fresh and frozen semen in the insemination of confined sheep. *Canadian Journal of Animal Science* **59**, 685–691

LEES, J.L. (1978). Functional infertility in sheep. *Veterinary Record* **18**, 232–236

LIGHTFOOT, R.J. and SALAMON, S. (1969). Freezing ram spermatozoa by the pellet method. II. The effects of method of dilution, dilution rate, glycerol concentration, and duration of storage at 5 °C prior to freezing on survival of spermatozoa. *Australian Journal of Biological Science* **22**, 1547–1560

LINCOLN, G.A. (1976). Secretion of LH in rams exposed to different photoperiods. *Journal of Reproduction and Fertility* **47**, 351–353

LINDSAY, D.R. (1969). Sexual activity and semen production of rams at high temperature. *Journal of Reproduction and Fertility* **18**, 1–8

LOUW, D.F. and JOUBERT, D.M. (1964). Puberty in the male Dorper sheep and Boer goat. *South African Journal of Agricultural Science* **7**, 509–520

McKENZIE, F.F. and PHILLIPS, R.W. (1934). The effects of temperature and diet on the onset of the breeding season (estrus) in sheep. *Journal of American Veterinary Medical Association* **84**, 189–195

MAXWELL, W.M.C. (1978). Studies on the survival and fertility of chilled-stored ram spermatozoa and frozen stored boar spermatozoa. PhD Thesis. University of Sydney

MICKELSEN, W.D., PAISLEY, L.G. and DAHMEN, J.J. (1981). The effect of scrotal circumference, sperm motility and morphology in the ram on conception rates and lambing percentage in the ewe. *Theriogenology* **16**, 53–59

MOULE, G.R. and WAITES, G.M. (1963). Seminal degeneration in the ram and its relation to the temperature of the scrotum. *Journal of Reproduction and Fertility* **5**, 433–446

OLAFSSON, T. (1980). Insemination of sheep with frozen semen: Results obtained in a field trial in Norway. *Zuchthygiene* **15**, 50–59

ORTAVANT, R. (1958). Le cycle spermatogénétique chez le bélier. Thèse Doct. Sci. University of Paris

ORTAVANT, R. and THIBAUT, C. (1956). Influence de la durée d'éclairement sur les productions spermatiques du bélier. *Compte Rendu des Seances de la Société de Biologie* **150**, 358–361

RATHORE, A.K. (1968). Effects of high temperature on sperm morphology and subsequent fertility in Merino sheep. *Proceedings of the Australian Society of Animal Production* **7**, 270–274

RATHORE, A.K. (1969). Mid-piece sperm abnormality due to high temperature exposure of rams. *British Veterinary Journal* **125**, 534–538

RATHORE, A.K. (1970). Fertility of rams heated for 1, 2, 3 and 4 days, mated to superovulated ewes. *Australian Journal of Agricultural Research* **21**, 355–358

SALAMON, S. and LIGHTFOOT, R.J. (1969). Freezing ram spermatozoa by the pellet method. I. The effect of diluent composition on survival of spermatozoa. *Australian Journal of Biological Science* **22**, 1527–1546

SALAMON, S. and ROBINSON, T.J. (1962). Studies on artificial insemination of Merino sheep: I. The effect of frequency and season of insemination, age of the ewe, rams and milk diluents on lambing performance. *Australian Journal of Agricultural Research* **13**, 52–68

SCHANBACHER, B.D. (1979). Increased lamb production with rams exposed to short daylengths during the non-breeding season. *Journal of Animal Science* **49**, 927–932

SKINNER, J.D. and ROWSON, L.E.A. (1968). Puberty in Suffolk and cross-bred rams. *Journal of Reproduction and Fertility* **16**, 479–488

SMITH, J.F. (1971). The effect of temperature on characteristics of semen of rams. *Australian Journal of Agricultural Research* **22**, 481–490

SMORAG, Z. and KARETA, W. (1974). Fertility evaluation of the frozen semen of ram on the basis of acrosomal cap state. *Medycyna Weterynaryjna* **30**, 689–691

SMYTH, P.A. and GORDON, I. (1967). Seasonal and breed variations in the semen characteristics of rams in Ireland. *Irish Veterinary Journal* **21**, 222–223

SMYTH, P.A., BOLAND, M.E. and GORDON, I. (1978). Conception rate in ewes: effect of method of breeding and number of insemination. *Journal of Agricultural Science, Cambridge* **91**, 511–512

TASSERON, F., AMIR, D. and SCHINDLER, H. (1977). Acrosome damage of ram spermatozoa during dilution cooling and freezing. *Journal of Reproduction and Fertility* **51**, 461–462

TERQUI, M. and THIMONIER, J. (1974). Nouvelle méthode radioimmunologique rapide pour l'estimation du niveau de progestérone plasmatique: application chez la brebis et chez la chèvre. *Comptes Rendus de l'Academie des Sciences, Paris* **279**, 1109–1112

WATSON, P.F. and MARTIN, I.C.A. (1972). A comparison of changes in the acrosomes of deep-frozen ram and bull spermatozoa. *Journal of Reproduction and Fertility* **28**, 99–101

WATSON, P.F. and MARTIN, I.C.A. (1975). The influence of some fractions of egg yolk on the survival of ram spermatozoa at 5 °C. *Australian Journal of Biological Science* **28**, 145–152

WATSON, P.F. and MARTIN, I.C.A. (1976). Artificial insemination of sheep: the fertility of semen extended in diluents containing egg yolk and inseminated soon after dilution or stored at 5 °C for 24 or 48 hours. *Theriogenology* **6**, 559–564

ZABA-BRANNY, A. (1971). Effect of exposure to high environmental temperature on production and characteristics of semen in the ram. *Acta Agraria et Silvestria* **11**, 69–92

24

MANAGEMENT OF THE FREQUENT LAMBING FLOCK

W.M. TEMPEST
Harper Adams Agricultural College, Newport, Shropshire, UK

Introduction

The technology of sheep production has advanced considerably over the last ten years. Methods for the controlled breeding of ewes which were initiated in the mid 1960s by Robinson in Australia (Robinson, 1965) have been refined, particularly by Gordon in Ireland (Gordon, 1975) into a standard technique. The application of such a technique in breeding ewes has received considerable attention from Robinson and his co-workers at the Rowett Research Institute (Robinson, Fraser and Gill, 1972; Robinson, 1974; 1978; Robinson, Fraser and McHattie, 1977), and by French research workers at Nouzilly (Thimonier *et al.*, 1975), where a continuous system of lamb production has been developed. At the same time, it was necessary to develop nutritional regimes to support the frequently bred ewe (Robinson, 1974; 1978; Robinson, Fraser and McHattie, 1977). All of these advances are reviewed elsewhere in this volume (Haresign, Chapter 18; Mauleon, Chapter 19; Robinson, Chapter 6; Treacher, Chapter 7; Orskov, Chapter 8), and need not be detailed again here. The purpose of this chapter is to show how these techniques can be synthesized into a frequent lambing system of production, to detail the management factors to which particular attention must be paid and to assess the performance of such a system.

Frequent lambing intervals

Frequent lambing is lambing at intervals more frequent than annually. With a gestation period of five months, the minimum theoretical lambing interval would therefore appear to be six months. Although individual ewes may lamb successfully at a six-month interval without artificial aids (Robinson and Orskov, 1975), such an interval has not been continuously achieved on a flock basis even with the use of exogenous hormones on a breed with a long breeding season (Speedy, Black and Fitzsimons, 1976). This is because the average time necessary for uterine involution is too long (up to two months) (Robinson, 1978).

The minimum lambing interval on a flock basis is therefore likely to be about seven months. Continuous lambing at this interval has been successfully achieved at the Rowett Research Institute (Robinson, Fraser and McHattie, 1975; Robinson and Orskov, 1975). In France (Thimonier and

Table 24.1 THE EIGHT-MONTH LAMBING SEQUENCE

1 month	Preparation for mating
5 months	Gestation
2 months	Lactation

Cognie, 1971; Thimonier *et al.* 1975), and at Harper Adams Agricultural College the lambing interval is eight months (*Table 24.1*).

The sequence of events is one month for mating preparation, which includes any hormone administration, five months of pregnancy and two months of lactation (including drying-up). The period of lactation can be flexible, going from as little as four weeks to as high as ten weeks, thus fluctuating the sequence slightly to avoid a high labour input at the peak Easter holiday period (which moves from year to year). The frequent lambing systems so far developed involve early curtailment of lactation—at four weeks for a seven-month interval (Robinson, 1974) and at seven to ten weeks for an eight-month interval as at Harper Adams College. However, there is some evidence to suggest that lack of postpartum oestrous activity may be more related to the presence of the lamb than to lactation *per se* (Cognie *et al.*, 1981), and it is interesting to observe that the sheep dairying practitioners claim that oestrus occurs soon after lamb removal but whilst machine milking is being continued (Mills, 1981).

Within the system operating at Harper Adams College, mating takes place in December, August and April. This particular timing was chosen for two reasons: (a) because it provides two out of the three mating times within the normal breeding season, with a high chance of success; and (b) because it provides events which fit in with the College's farming pattern. Other possible mating times are November, July, March; or October, June, February.

At Harper Adams, the eight-month sequence is operated with two sub-flocks (A and B) (*Table 24.2*) so that if a ewe fails to conceive at one

Table 24.2 THE INTEGRATION OF TWO FLOCKS IN THE HARPER ADAMS COLLEGE SYSTEM

Flock	*Mating*	*Lambing*	
A	DEC	MAY	First year
B	Apr	Sep	
A	AUG	JAN	
B	Dec	May	Second year
A	APR	SEP	
B	Aug	Jan	

mating, she is transferred to the alternate sub-flock, and has the opportunity to mate four months later. The eight-month lambing interval, operated in sub-flocks, provides a regular sequence of events on an annual basis, which is an added advantage in developing the system for commercial practice. In France the sequence is even more streamlined, with a series of seven flocks each separated in physiological state by seven weeks. Ewes which do not conceive at one mating are then moved to the next flock for re-mating.

Management factors

BREED OF EWE

From a comprehensive review by the Scottish Agricultural Colleges (1977), it would appear that the more suitable breeds for frequent lambing are those with a long duration breeding season, coupled with high prolificacy, in order to achieve high annual production. The most common British and Irish prime lamb mother, the halfbred ewe, has a restricted breeding season, and gives poor breeding results in early autumn and spring (Gordon, 1975; Speedy and Fitzsimons, 1977).

In the UK the proven ewe breed for frequent lambing is the Finn-Dorset (Finnish Landrace × Dorset Horn) (Robinson and Orskov, 1975), and in France the Romanov and Prealpe-du-Sud are successful. The Friesland × Dorset Horn and the Cambridge have also successfully lambed at a continuous eight-month interval at Harper Adams College. All of these breeds, however, are in numerically short supply. There is thus a strong case for a wider use of the Finnish Landrace in Dorset Horn flocks to produce in the one mating, Finn-Dorset ewes for very intensive frequent lambing systems, and Finn-Dorset rams for use as crossing rams on hill ewes or ewes of low prolificacy, the latter use having been recently successfully demonstrated by ADAS in the south-west (P. Stone, personal communication). The Finn-Dorset has considerable potential for both commercial breed development and for genetic study by a breeding organization.

Because of its large contribution (70%) to the Cambridge breed (Scottish Agricultural Colleges, 1977) and its responsiveness to artificial daylength patterns (Ducker and Bowman, 1972), and because it is the most numerous purebred lowland breed (National Sheep Association, 1979), the Clun Forest would seem to be worthy of further investigation for frequent lambing. Another breed with the potential for frequent lambing because of its less restricted breeding season, is the Suffolk. The Suffolk is the most widely used terminal sire for prime lamb production, and because of its repeated use for two or more generations, makes a high contribution to the breeding ewe population (Meat and Livestock Commission, 1972). Many Suffolk crossbred ewes already produce a December/January lamb crop, thus offering the prospect of a February/March/April re-breeding after early weaning and hormone treatment (Northern Ireland Agricultural Trust, 1974). Such an April remating with Suffolk × Mule and Suffolk × Welsh Halfbred ewes produced a successful September lambing at the Ministry of Agriculture's Rosemaund Experimental Husbandry Farm in 1981 (S. Meadowcroft and D. Brown, personal communication). Because of the ready availability of Suffolk crossbred females, this breed may offer the easiest opportunity for the commercial development of frequent lambing.

BREED OF RAM

The first requirement of the ram in a frequent lambing system is to maintain libido and semen production throughout the year. Colas has reviewed the factors affecting quality of ram semen in Chapter 23, but reported earlier that although semen of acceptable quality could be collected for AI throughout the year (Colas *et al.*, 1972), the problem with some breeds may be the total absence of libido in the spring (Colas, Brice and Guerin, 1974). Suffolk and other Down breeds have not presented any problems of mating out of season at Harper Adams College, but rams of the Texel breed have shown low libido in spring.

The second requirement of the ram is to influence the type of carcass produced, and this is discussed later.

OESTRUS SYNCHRONIZATION

A standard controlled breeding technique (after Gordon, 1975) is followed. This involves intravaginal administration by sponges of 30 mg fluorogestone acetate (Cronolone, Intervet Ltd) or 60 mg medroxyprogesterone acetate (MAP, Upjohn Ltd) for 12 days, followed by a single intramuscular injection of 400 iu pregnant mare's serum gonadotrophin (PMSG; Serum Gonadotrophin, Upjohn Ltd; Folligon, Intervet Ltd) at sponge withdrawal (*Table 24.3*).

Table 24.3 CONTROL OF BREEDING FOR FREQUENT LAMBING FLOCKS

Day	*Operation*
–14	Sponges inserted
–2	Sponges withdrawn PMSG injected
0	Rams joined

Despite this being a 'standard' technique, several points may be raised. Firstly, Gordon (1975) reported no differences in effect between Cronolone and MAP at the levels indicated, but reported higher conception rates with 45 mg than with 30 mg Cronolone in out-of-season breeding; and Colas *et al.* (1973) reported similarly in favour of 40 mg Cronolone in AI trials in the breeding season. The recommendations for the length of sponge insertion and the timing of PMSG injection in relation to sponge withdrawal vary from company to company and may be important, but there appears to be a lack of critical data in this respect.

The need for, and level of PMSG, is debatable, as cyclic sheep show oestrus after sponge withdrawal in the absence of PMSG. Colas *et al.* (1973) found more precise and reliable synchronization when low levels of PMSG were given; Gordon (1975) reported no difference between 375 and 750 iu PMSG at an August mating and Robinson, J.J. (personal communication) feels that PMSG may only be necessary for non-cyclic ewes mating during normal anoestrum. There is, however, a distinct lack of dose response data for Finn-Dorset ewes mated regularly every eight months at varying times of the year. Although in its early stages, results at Harper Adams College show a response in conception rate up to 1000 iu PMSG at spring matings, and up to 500 iu PMSG at early autumn matings.

MATING MANAGEMENT

Adequate preparation for mating is essential in a frequent lambing system. The ewe is in a 'production state' for all but one month of the eight month cycle—the month immediately prior to mating, and it has been shown by Robinson and Orskov (1975) that provided the ewes are well fed in this month (1.5–1.75 × maintenance) to adjust body condition to a score of 3 at mating, then there is little problem in maintaining ewe body weight for the rest of the cycle, and from cycle to cycle. The procedure adopted at Harper Adams College is to body condition score the ewes at weaning, to adjust that body condition accordingly on hay or grass for the next two weeks until sponge insertion when condition is reassessed and concentrates are introduced if necessary.

In developing frequent lambing for commercial practice, a simple but effective mating system, which is realistic under most farming conditions, is needed. Forty-eight hours after sponge withdrawal, fertile rams are introduced. The value of this delayed introduction was shown by Joyce (1972) and Bryant and Tomkins (1976). Although Colas, Brice and Guerin (1974) recommend one ram for every five ewes, Bryant and Tomkins (1975) have shown acceptable conception rates (73%) at up to 12 synchronized ewes/ram. The ratio adopted for practical conditions at Harper Adams College is one ram to ten ewes, in separate mating groups in December and August, but in a multi-ram group for matings in April. Joyce (1972) showed that this paddock mating method gave comparable conception results with those achieved by hand mating. Neither recourse to this technique nor to set-time AI (Colas *et al.* 1968) has been necessary at Harper Adams College, although there is a strong argument for bringing the ram's role under greater control for frequent lambing. The ratio of one ram to ten ewes does not mean that a formidable array of rams is needed to ensure adequate mating power (Gordon, 1975) when the flock is split into staggered mating sub-flocks; indeed Harper Adams College carry four rams for 120 ewes. With such regular and frequent active mating periods, it has not been found necessary to expose the rams to teaser ewes at regular intervals, and rams are kept under natural daylength and grazing conditions.

The rams are removed from ewes after the synchronized oestrus, and reintroduced 14 days later to mate at the first repeat heats, after which they are finally removed.

PREGNANCY MANAGEMENT

The success of frequent lambing is influenced mainly by the level and pattern of nutrition throughout the cycle, factors which also apply under any circumstances. Nutrition has been adequately reviewed in this volume by Gunn (Chapter 5), Robinson (Chapter 6) and Treacher (Chapter 7), and requirements are further detailed by the Agricultural Research Council (1980) and the Meat and Livestock Commission (1981). There is some discrepancy between these sources and this, together with differences in the way the requirements are expressed, adds confusion to the farming practitioner.

Robinson and Orskov (1975) have adequately demonstrated that the feeding of ewes lambing every seven months can be easily controlled to give no cause for concern. Their recommendations are shown in *Figure 24.1*. The pre-mating level of nutrition is maintained for the first month of

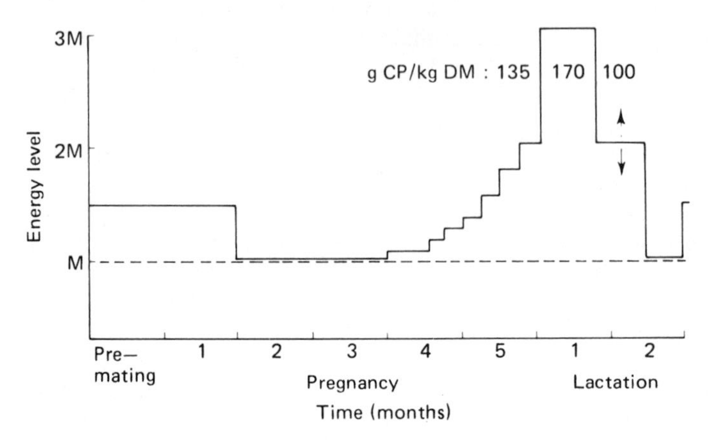

Figure 24.1 Pattern of feeding throughout an eight-month cycle (M = maintenance energy requirement)

gestation, but were it to be maintained until late pregnancy, inappetance and hypoglycaemia could result. In mid pregnancy, it is critical to restrict food intake so that the ewe apparently maintains her weight, but ensuring a carefully controlled mobilization of body fat for foetal growth. In late pregnancy, increased feeding is necessary to achieve high lamb birth-weights, udder tissue development and to prevent excessive mobilization of body fat in the ewe. A 20% weight increase over the last two months is generally the target for lowland ewes carrying twins. In order to help achieve these targets, the ewes are shorn in mid pregnancy (Rutter, Laird and Broadbent, 1971; 1972; Morgan and Broadbent, 1980) and fed concentrate diets in late pregnancy based on whole barley and pelleted fishmeal (Orskov, 1977) formulated to give 135 g crude protein/kg dry matter of the total diet when fed in conjunction with hay (concentrate composition in kg/tonne: whole barley 875, pelleted fishmeal 100, minerals/vitamins 25).

At approximately three months of gestation, the ewes are pregnancy tested using ultrasound (Medata Ltd) and barren ewes are 'slipped' into the alternate sub-flock and sponge-treated as normal.

CONTROL OF LAMBING

Because lambing occurs every four months, there is a need to avoid extended lambing periods. Even ewes which are synchronized in oestrus may have an extended lambing over six days (Gordon, 1975). It is therefore useful to control parturition in these ewes, which Bosc (1973)

Table 24.4 CONTROL OF LAMBING

Day	*Operation*
+140	Ewes penned
+141	Ewes induced with betamethasone
+143	Ewes lamb

Table 24.5 CONTROL OF BREEDING AND LAMBING

Operation	*Day*	*Weekday*
Sponges in	–14	Friday
Sponges out	–2	Wednesday
Rams in	0	Friday
Pen ewes	+140	Friday
Induce	+141	Saturday 18.00 hours
Lamb	+143	Monday 06.00 hours to Tuesday 06.00 hours

showed could be done by administration of a corticosteroid. There appear to be several factors involved in the response of the ewe—especially the type of corticosteroid, the dose level, the time of treatment in gestation, and the breed of sheep. One-day lambing at Harper Adams is achieved with a single intramuscular injection of 16 mg betamethasone (Betsolan, Glaxo Ltd), administered at 18.00 hours on day 141 of gestation, to achieve lambing between 06.00 hours on day 143 and 06.00 hours on day 144 (*Tables 24.4* and *24.5*).

LACTATION MANAGEMENT

The management imposed in lactation depends on the growth rate required in the lambs (see *Lamb management and marketing*). For maximum growth rate maximum milk output from the ewe is needed, and this is closely related to the level and type of dietary crude protein (Robinson *et al.*, 1974). The optimum level of dietary crude protein would appear to be about 170 g CP/kg DM with a protein source of low degradability and the diet fed at 3 × maintenance in terms of energy. This would still leave the ewe in negative energy balance, and she would draw upon her own body fat reserves to supplement her energy requirements (concentrate diet composition in kg/tonne for feeding in conjunction with hay: whole barley 775, pelleted fishmeal 200, minerals/vitamins 25). In later lactation, it is necessary in the frequently bred ewe to avoid excessive depletion of body reserves before re-breeding and to begin drying-up the ewe thus encouraging a greater solid food intake in the lamb before its abrupt weaning. Calderon Cortes *et al.* (1977) showed that the withdrawal of the low degradability protein supplement reduced milk yield, and decreased the mobilization of body fat. This technique is followed at Harper Adams College by removal of the protein supplement at three weeks of lactation, with regulation of energy intake from 3–6 weeks of lactation to restore body condition while in the final stages of drying-up, thus easing the restoration of body condition in the pre-mating month, which can be readily achieved at 1.5 × maintenance level of feeding (*Figure 24.1*).

LAMB MANAGEMENT AND MARKETING

Despite the regularity of lambing which can be achieved, the seasonal nature of the UK price structure has precluded regular marketing. The three different lamb crops are reared according to the availability of farm resources and the market (*Table 24.6*). May-born lambs are stored at grass and finished traditionally on sugar beet tops at 8–10 months of age. September-born lambs are reared indoors on diets of low nutrient density to finish towards Easter at six months of age, and January-born lambs are finished on concentrates for the Easter market at four months of age.

Table 24.6 LAMB REARING METHODS

Born	*Method*	*Sell*
May	Grazed—stores	January–March
September	Housed—medium growth	Easter
January	Housed—concentrates	Easter

On concentrate finishing, success has not been achieved with the whole barley/pelleted fishmeal diets of Orskov (1977), but highest performance has been achieved on complete pelleted diets which include a proportion of roughage incorporated into the pellet (Lamlac Start-to-Finish, Volac Ltd), similar to those developed for calves by Thomas, Hinks and Gilchrist-Shirlaw (1980) and Thomas and Hinks (1981) (*Table 24.7*).

As there is variation in the type, weight, age and rearing method of the different lamb crops, so the optimum choice of breed of ram may vary from lamb crop to lamb crop (*Table 24.8*).

Table 24.7 LAMB FATTENING 'BEST PERFORMANCES'

	B/FM	*Creep*	*Complete*
Daily liveweight gain (g/d)	182	336	375
Food conversion rate (kg food/kg LWG)	5.60	3.64	3.50
Daily feed intake (kg/d)	1.00	1.19	1.27
Highest daily feed intake (kg/d)	1.00	1.36	1.45

B/FM = Barley/fishmeal mix
Creep = proprietary lamb creep pellet
Complete = complete lamb rearing pellet

Table 24.8 RAM BREEDS IN RELATION TO LAMB REARING

Lambs born	*Rearing method*	*Ram breed*
May	Stores	Oxford, Texel, Texel × Oxford
September	Medium growth	Suffolk
January	Concentrates	Dorset Down

EWE REPLACEMENT POLICY

No accurate assessment of the annual replacement rate can be made, as the ewes have not been replaced on a regular basis. Ewes have been culled only as a result of death, barrenness at two consecutive matings, mastitis, or a persistent decline in lamb output. Robinson and Orskov (1975) came to the conclusion that there was little point in extending ewe life beyond five pregnancies, although many of their ewes went on to complete eight pregnancies. The longer the life of the ewe, provided lamb output is maintained, the greater will be the biological efficiency, and ewes at Harper Adams have been retained as long as they are highly productive (average of two lambs reared/lamb crop). Consequently, an absolute replacement rate cannot be given, but up to the present it has been between 20% and 25% per annum.

Replacement ewes are purchased as mature two-crop ewes. The Rowett policy is to introduce ewe lambs to the frequent lambing programme at eight months of age (Robinson, Fraser and McHattie, 1977).

Performance and potential

PHYSICAL PERFORMANCE

The levels of physical performance achieved in the Harper Adams flock in its first four years of operation are shown in *Tables 24.9* to *24.13*.

Conception rates (*Table 24.9*) show a seasonal trend for matings at different times of the year. The level of conception and the seasonal variation is largely in agreement with that obtained by Speedy and

Table 24.9 CONCEPTION RATES (%) IN HARPER ADAMS FLOCK

Mating	*Flock*			*Mean*
	A	*B*	*A*	
December	98	98	94	97
August	94	93	87	91
April	76	79		78
Mean	89	90		89

Fitzsimons (1977) of 88, 82 and 73% respectively (average 81%) for November, August and February matings of Finn-Dorset ewes kept under conditions of summer grazing and winter housing. Robinson, Fraser and McHattie (1975) reported conception rates varying between 81 and 98% (average 88%) in Finn-Dorset ewes kept under controlled conditions of daylength and hormones, but with no evidence of seasonal variation. The Rowett flock kept under natural daylength conditions, in which PMSG replaced daylength manipulation, achieved conception rates of 96, 95 and 93% in mature ewes mated in October, January and May (Fraser *et al.*, 1976).

The number of lambs born/ewe lambing is shown in *Table 24.10*. Again a distinct seasonal trend is evident. Speedy and Fitzsimons (1977) reported

Table 24.10 LAMBS BORN/EWE LAMBING IN HARPER ADAMS FLOCK

Mating	*Flock*			*Mean*
	A	*B*	*A*	
December	2.82	2.71	2.52	2.68
August	2.30	2.33	2.12	2.25
April	1.75	1.64		1.70
Mean	2.29	2.23		2.21

2.51, 2.25 and 2.22 for November, August and February matings (average 2.33); Robinson, Fraser and McHattie (1975) 2.1 lambs born/ewe at seven-month intervals with no seasonal effect under controlled conditions; and Fraser *et al.* (1976) 2.2, 2.3 and 2.6 for October, January and May matings under natural daylength (average 2.37).

Lamb mortality is shown in *Table 24.11*. There is no obvious seasonal effect, and the figure of 16% records total lamb deaths from conception to sale. Speedy and Fitzsimons (1977) recorded a mortality of 27% of all

Table 24.11 LAMB MORTALITY (%) IN HARPER ADAMS FLOCK

Mating	*Flock*			*Mean*
	A	*B*	*A*	
December	10	16	22	16
August	30	10	21	20
April	14	8		11
Mean	18	11		16

lambs; but it is difficult to determine a mortality figure for the Rowett flock, because the lambs were required for experimental work rather than being reared under field conditions. For normal lowland flocks, a lamb mortality of 14% between birth (including stillbirths) and sale is given by the Meat and Livestock Commission (1980).

The number of lambs sold/ewe lambing is shown in *Table 24.12*. Annual production would appear to depend on two major variables when frequent

Table 24.12 LAMBS SOLD/EWE LAMBING IN HARPER ADAMS FLOCK

Mating	*Flock*			*Mean*
	A	*B*	*A*	
December	2.55	2.28	1.97	2.27
August	1.60	2.10	1.67	1.79
April	1.54	1.50		1.52
Mean	1.90	1.96		1.86

lambing is extended into the field. These are the replacement policy, i.e. whether ewe lambs are brought into the flock and the facility for slipping ewes into other subsequent flocks at two or four-month intervals. With mature ewes kept under commercial conditions, and the slipping facility, the Harper Adams flock has sold 270 lambs/100 ewes in the flock per annum (*Table 24.13*).

Table 24.13 ANNUAL FLOCK PERFORMANCE AT HARPER ADAMS

Number of ewes to ram/tupping	50
Total number of ewes in flock	100
Ewe tuppings expected @ 1.5 times/year	150
89% conceive/tupping, second chance ewes	11
Therefore, actual number of ewe tuppings	161
At 89% conception, number of lambings	145
Lambs born/ewe lambing	2.21
Lambs born/100 ewes in flock	320
Mortality	16%
Lambs sold/ewe lambing	1.86
Lambs sold/100 ewes in flock	270

Speedy and Fitzsimons (1977) recorded an annual production of 2.13 lambs weaned/ewe from Finn-Dorsets, but this figure was a reflection of the low annual conception rate (81%) because of the lack of slipping (3.51 lambs born/ewe lambing/year; 2.84 lambs born/ewe to ram/year) and the high lamb mortality. The Nouzilly flock averaged three lambs/ewe/year (Thimonier *et al.*, 1975) at an eight-month interval with slipping; and the Rowett flock with a seven-month interval achieved a mean annual production of 3.45 lambs/ewe after adjusting for lamb mortality (Robinson and Orskov, 1975).

ECONOMIC PERFORMANCE

There are very few reports of the economic performance of frequent lambing flocks. The University of Reading (1973) study, based on a number of technical assumptions, concluded that 'frequent lambing systems do not appear sufficiently attractive (gross margins too low) to warrant practical testing'; and the Scottish Agricultural Colleges (1977) study showed that increasing the production frequency was not reflected in higher net margins. Although total financial output was much increased the variable costs were approximately doubled. The most noticeable increase in costs was in purchased feedstuffs, reflecting in particular the high proportion of lambs fattened on a concentrate diet.

Economic success would therefore appear to depend on integrating the frequent lambing flock with the available resources of a particular farm,

Table 24.14 INTEGRATION OF FREQUENT LAMBING FLOCK WITH FARM RESOURCES

Stage of cycle	*December mating*		*August mating*		*April mating*	
	Month	*Location*	*Month*	*Location*	*Month*	*Location*
Flush	Nov	Cow leys	Jul	Sheep ley	Mch	Housed
Tup	Dec	Cow leys	Aug	Sheep ley	Apr	Sheep ley
Pregnancy month 1	Jan	Cow leys	Sep	Sheep ley	May	Sheep ley
Pregnancy month 2	Feb	Housed	Oct	Cow ley	Jun	Sheep ley
Pregnancy month 3	Mch	Housed	Nov	Cow ley	Jul	Sheep ley
Pregnancy month 4	Apr	Housed	Dec	Housed	Aug	Sheep ley
Lamb	May	Housed	Jan	Housed	Sep	Housed
Lactate	Jun	Sheep ley	Feb	Housed	Oct	U/S ley

U/S ley = undersown ley

especially for lamb finishing. The way in which this is achieved for the ewes in the Harper Adams flock is shown in *Table 24.14* and for lamb finishing in *Table 24.6*. In achieving such integration, the flock has produced extremely high margins (*Table 24.15*) relative to annual lambing lowland flocks recorded by the Meat and Livestock Commission (1980). The

Table 24.15 FINANCIAL PERFORMANCE (£/EWE IN FLOCK)

	Annual[(a)]	*Frequent*
Lamb sales	34	87
+ Wool	3	3
— Flock replacements	6	6
OUTPUT	31	84
Fertilizer + forage	5	1
Purchased forage	<1	7
Ewe concentrates	5	6
Lamb concentrates	<1	12
Veterinary and medical	2	4
Hormones	0	3
Miscellaneous	1	1
VARIABLE COSTS	14	34
GROSS MARGIN	17	50
Grass stocking rate (ewes/hectare)	11.7	22.5
Gross margin/hectare of grass	205	1125

[(a)]Figures from MLC recorded flocks (1980) for comparison

purchased forage costs include purchased hay and an allowance for grazing cow leys and sugar beet tops. Thus high stocking rates can be practised and high gross margins/hectare are achieved, but it is to be emphasized that purchasing forage is purchasing hectares.

Conclusion

Robinson in 1977 had reached the stage where it was possible to predict with reasonable certainty the levels of production that could be achieved in a frequent lambing system. In conjunction with the Scottish Agricultural Colleges (1977), a specification for the integration of frequent lambing systems on British farms was produced, based on estimates of the level of production, the availability of grass and the nutrient requirements of the ewe and lamb. He pointed out that the estimates needed to be validated by the provision of information from development projects. The Harper Adams system is such a development, and can be taken as an example of successful 'Science into Practice', but caution is still needed because information is still required over a longer period of time to determine if performance can be maintained and to determine optimum ewe replacement policy.

Acknowledgement

I would wish to record that the Harper Adams system is largely the development of the fundamental research of Dr J.J. Robinson, and his advice and encouragement has been very much appreciated.

References

AGRICULTURAL RESEARCH COUNCIL (1980). *The Nutrient Requirements of Ruminant Livestock.* Technical Review by an Agricultural Research Council Working Party. Slough, Commonwealth Agricultural Bureaux

BOSC, M.J. (1973). A review of methods of inducing parturition in the ewe and cow. *Revue Médicale et Véterinaire* **149**, 1463–1480

BRYANT, M.J. and TOMKINS, T. (1975). The flock-mating of progestagen-synchronised ewes. 1. The influence of ram-to-ewe ratio upon mating behaviour and lambing performance. *Animal Production* **20**, 381–390

BRYANT, M.J. and TOMKINS, T. (1976). The flock-mating of progestagen-synchronised ewes. 2. The influence of time of ram introduction upon mating behaviour and lambing performance. *Animal Production* **22**, 379–384

CALDERON CORTES, J.F., ROBINSON, J.J., McHATTIE, I. and FRASER, C. (1977). The sensitivity of ewe milk yield to changes in dietary crude protein concentration. *Animal Production* **24**, 135 (Abstract)

COGNIE, Y., GAYERIE, F., OLDHAM, C.M., POULIN, N. and MAULEON, P. (1981). Frequent lambing: underlying philosophy. *32nd Annual Meeting of the EAAP, Zagreb*

COLAS, G., BRICE, G. and GUERIN, Y. (1974). Recent progress in sheep artificial insemination. *Bulletin Technique d'Information de Ministère d'Agriculture* **294**, 795–800

COLAS, G., LASZCZKA, A., BRICE, G. and ORTAVANT, R. (1972). Seasonal variations in semen production in the ram. *Acta Agraria et Silvestria: Seria Zootechniczna* **12**, 3–15

COLAS, G., THIMONIER, J., COUROT, M. and ORTAVANT, R. (1973). Fertility, prolificacy and fecundity during the breeding season of ewes artificially inseminated after treatment with fluorogestone acetate. *Annales de Zootechnie* **22**, 441–451

COLAS, G., DAUZIER, L., COUROT, M., ORTAVANT, R. and SIGNORET, J.P. (1968). Results obtained while investigating some important factors in AI in the sheep. *Annales de Zootechnie* **17**, 47–57

DUCKER, M.J. and BOWMAN, J.C. (1972). Photoperiodism in the ewe. 5. An attempt to induce sheep of three breeds to lamb every eight months by artificial daylength changes in a non-light proofed building. *Animal Production* **14**, 323–334

FRASER, C., ROBINSON, J.J., McHATTIE, I. and GILL, J.C. (1976). Field studies on the reproductive performance of Finnish Landrace × Dorset Horn ewes. *Animal Production* **22**, 162 (Abstract)

GORDON, I. (1975). Hormonal control of reproduction in sheep. *Proceedings of the British Society of Animal Production* **4**, 79–93

JOYCE, M.J.B. (1972). A comparison of three different mating systems.

Proceedings of the 7th International Congress of Animal Reproduction and Artificial Insemination, Munich **2**, 935–938

MEAT AND LIVESTOCK COMMISSION (1972). *Sheep Improvement*. Scientific Study Group Report. Bletchley, Meat and Livestock Commission

MEAT AND LIVESTOCK COMMISSION (1980). *Commercial Sheep Production Yearbook 1979–80*. Bletchley, Meat and Livestock Commission

MEAT AND LIVESTOCK COMMISSION (1981). *Feeding the Ewe*. Sheep Improvement Services. Bletchley, Meat and Livestock Commission

MILLS, O. (1981). In *Practical Sheep Dairying*, pp. 58, 61. Wellingborough, Thorsons

MORGAN, H. and BROADBENT, J.S. (1980). A study of the effects of shearing pregnant ewes at housing. *Animal Producion* **30**, 476 (Abstract)

NATIONAL SHEEP ASSOCIATION (1979). *British Sheep*. Tring, National Sheep Association

NORTHERN IRELAND AGRICULTURAL TRUST (1974). *Lamb Production. The Development of a New System.*

ØRSKOV, E.R. (1977). Nutrition of lambs from birth to slaughter. In *Sheep Nutrition and Management*, pp. 35–46. London, US Feed Grains Council

ROBINSON, J.J. (1974). Intensifying ewe productivity. *Proceedings of the British Society for Animal Production* **3**, 31–40

ROBINSON, J.J. (1978). Techniques and systems for very intensive sheep production. In *Sheep on Lowland Grass*, pp. 51–60. Paper presented to the Summer Meeting of the British Society of Animal Production, 1978.

ROBINSON, J.J. and ORSKOV, E.R. (1975). An integrated approach to improving the biological efficiency of sheep meat production. *World Review of Animal Production* **11**, 63–76

ROBINSON, J.J., FRASER, C. and GILL, J.C. (1972). Preliminary observations on the performance of Finnish Landrace × Dorset Horn ewes in an intensive system. *Proceedings of the British Society of Animal Production* (1972). p.134 (Abstract)

ROBINSON, J.J., FRASER, C. and McHATTIE, I. (1975). *Annales de Biologie Animale, Biochimie et Biophysique* **15**, 345

ROBINSON, J.J., FRASER, C. and McHATTIE, I. (1977). Development of systems for lambing sheep more frequently than once per year. In *Sheep Nutrition and Management*, pp. 5–33. London, US Feed Grains Council

ROBINSON, J.J., FRASER, C., GILL, J.C. and McHATTIE, I. (1974). The effect of dietary crude protein concentration and time of weaning on milk production and body weight change in the ewe. *Animal Production* **19**, 331–339

ROBINSON, T.J. (1965). Use of progestagen-impregnated sponges inserted intravaginally or subcutaneously for the control of the oestrus cycle in sheep. *Nature, London* **206**, 39–41

RUTTER, W., LAIRD, T.R. and BROADBENT, P.J. (1971). The effects of clipping pregnant ewes at housing and of feeding different basal roughages. *Animal Production* **13**, 329–336

RUTTER, W., LAIRD, T.R. and BROADBENT, P.J. (1972). A note on the effects of clipping pregnant ewes at housing. *Animal Production* **14**, 127–130

SCOTTISH AGRICULTURAL COLLEGES (1977). *A Study of High Lamb Output Production Systems*. Technical Note No. 16

SPEEDY, A.W. and FITZSIMONS, J. (1977). The reproductive performance of Finnish Landrace × Dorset Horn and Border Leicester × Scottish Blackface ewes mated three times in two years. *Animal Production* **24**, 189–196

SPEEDY, A.W., BLACK, W.J.M. and FITZSIMONS, J. (1976). The performance of Finnish Landrace × Dorset Horn ewes mated every six months. *Animal Production* **22**, 138 (Abstract)

THIMONIER, J. and COGNIE, Y. (1971). Increasing lambing frequency and flock management. *Bulletin Technique d'Information des Ingénieurs des Services Agricoles* **257**, 187–196

THIMONIER, J., COGNIE, Y., CORNU, C., SCHNEBERGER, J. and VERNUSSE, G. (1975). *Annales de Biologie Animale Biochimie et Biophysique* **15**, 365

THOMAS, D.B. and HINKS, C.E. (1981). The use of complete diets for early weaned calves. *Animal Production* **32**, 354 (Abstract)

THOMAS, D.B., HINKS, C.E. and GILCHRIST-SHIRLAW, D.W. (1980). The importance of roughage in relation to performance, feed intake and rumen fermentation patterns in early weaned calves. *Animal Production* **30**, 468 (Abstract)

UNIVERSITY OF READING (1973). *An Assessment of Continuous Lamb Production in the United Kingdom*. Report of Working Party from the Department of Agriculture, University of Reading and The Grassland Research Institute, Hurley

VI

Genetic Improvement

25

DEFINING SELECTION OBJECTIVES

D.E. ROBERTSON
Muresk Agricultural College, Western Australia

The definition of selection objectives is critically important for every sheep breeder since it determines the success of the sheep breeding enterprise. Therefore the topic merits careful consideration and extensive research before commencement, and frequent review during a breeding programme. Yet it is a topic which is very difficult to discuss usefully in general terms. Every sheep breeder is faced with a unique set of circumstances which will influence his choice of selection objectives. Different environments, different markets, access to different sheep genotypes, different personal goals all lead to sheep breeders having a near infinite range of selection objectives.

Instead of focusing on any particular sheep breeding programmes this chapter will highlight some of the variety of selection objectives being pursued in different sheep industries, and will review the factors which should be considered in defining selection objectives. Following chapters will provide information relevant to selection for carcass quality, reproduction and wool production.

Principal selection objectives

All sheep breeding plans are intended to enhance the efficient production of one or more of the following products: wool, meat, milk, pelts, exhibition sheep. The last is perhaps contentious because breeders may not agree that the production of exhibition sheep has become an end in itself. Many still believe that the showing of sheep and the presentation of especially handsome sheep for sale is an aid to selection for productivity. This is not so. Performance records obtained under appropriate conditions should be the only criteria for selection. The appearance of a sheep at a sale or show is all but worthless as a guide to its genetic potential for production. Yet strong demand in the market place is a very persuasive incentive for breeders to make the breeding of exhibition sheep one of their principal selection objectives. Most sheep breeders are distracted to some degree from selection for productivity. Very few breeders of pedigree sheep in any of the sheep producing countries can yet afford to ignore the appearance of their sheep and select solely on performance. Their markets compel them to make appearance one of their principal selection objectives.

Diverse production systems

Although the list of principal selection objectives is short and apparently straightforward, a multitude of production pathways is used to attain these end products. Sheep are maintained under a wider range of conditions than any other domesticated species. At one extreme in the pastoral areas of Australia, sheep are given little more husbandry than wild animals, being mustered sometimes only once a year from free range, where they graze natural, uncultivated and unfertilized native plants. At the other extreme some sheep in Western Europe are kept almost in battery conditions, fed formulated rations, their reproduction is controlled by injected hormones, and their lambs reared artificially. Between these extremes all levels of intensive and extensive husbandry systems are practised. To meet these diverse requirements, many sheep genotypes have been evolved and are still being selected.

The sheep is a versatile animal and the following examples illustrate the great diversity of sheep and sheep production systems.

Wool production in tropical Australia

At places like Julia Creek in north-western Queensland the harsh environment severely limits production, and imposes stress on sheep. Very poor nutritional conditions prevail for more than six months of the year because of the prolonged 'dry' season. Temperatures are very high, maximum daily temperatures in the summer are mostly above 40 °C and often exceed 45 °C. The naturally treeless plains of the area offer no protection. Under the combined burdens of heat and poor nutrition, sheep productivity is low; Merino ewes average 3.73 kg of wool per head per annum and only 40% of ewes mated rear a lamb (G. O'B. Roberts, personal communication).

Under these conditions the breeder can apply no selection in the ewes and very little in the rams; any that is operated must be for adaptation and wool production in this very difficult environment.

Super-fine wool production from housed sheep in Australia

Also in Australia, a new industry is developing, producing wool in a system that is at the opposite end of the husbandry spectrum from the Queensland pastoral industry. In Victoria fine-wool Merinos are being housed and hand-fed throughout the year to produce a clean, fine fibre of uniform diameter and strength throughout its length. These qualities attract a premium price from specialty wool spinners and the industry, though small, is expanding (R. Beggs, personal communication).

More typical Australian sheep production

Between these two extremes, the vast majority of Australia's sheep graze improved pastures to produce wool and meat with more income derived from wool than meat. Breeders serving this industry have traditionally selected for high wool cuts and traits associated with wool production, and the evidence would seem to indicate that the industry is healthy. Recently an Australian Merino ram was sold for a record price of Australian $79 000 (about £48 000 sterling). In addition, the Australian Merino Society operates a very large cooperative breeding scheme for traits associated with wool production in which more than a million ewes are mated annually (Shepherd, 1976).

More emphasis is now being placed on meat as a selection objective, largely because of the good markets for sheep meats in the oil-rich Arab states of the Middle East. Large-framed sheep are preferred by this market and so breeders are selecting large, fast-growing animals.

More slowly Australian sheep breeders are adopting greater reproductive rate as a selection goal. Although average lambing percentages have been around 70% for many years, breeders did not perceive this as a problem until meat started to contribute more to sheep farmers' returns. Quarantine restrictions prevent fecund breeds like the Finn, Romanov and East Friesian entering Australia, so improvement will have to come from within Australia's existing genetic resources, at least for the time being. For this purpose, the Booroola Merino will be invaluable, because its fecundity is so readily transferred in crosses (Robertson, 1974; 1976; Piper, 1980; Davis *et al.*, 1981).

New Zealand wool and meat production

New Zealand sheep farmers have long derived a very significant part of their income from the export of prime lambs as well as wool, from their Romney and Romney-based flocks, grazing highly improved pastures. Their selection objectives therefore emphasize both wool and reproduction. In recent years New Zealand breeders and scientists have pursued increased reproductive rate with great vigour. They are already widely exploiting the Booroola's capacity to lift lambing percentages in crosses with other breeds, and have also successfully developed breeds like the Coopworth by rigid adherence to clearly defined selection objectives.

A new market has recently been opened, selling very young lightweight New Zealand lambs to Greek communities. If this market persists, some breeders will have to modify their selection objectives accordingly.

The Drysdales, Tukidales and similar 'hairy' Romney strains were bred in New Zealand, specifically for the carpet wool production. The development of these breeds was achieved by adherence to predetermined and well defined selection objectives.

Western European sheep meat production

In Western Europe the value of wool produced is small beside the value of sheep meat. The fleece is sometimes worth less than its cost of removal from the sheep. Here, meat production is the principal selection objective. Given high and relatively stable lamb prices, it may be profitable to apply high technology, inducing ewes to lamb out of season, perhaps more than once a year, to have litters of lambs, and to rear lambs artificially. New genotypes are being developed which are more responsive to such systems of intensive husbandry. However, these sheep production systems are not widespread; traditional systems are still the norm.

Chinese wool and meat production

China affords one example where the trend from wool to meat production has been reversed. In the Sinkiang Province starting with the Kazakh Fat-rumped sheep, a primitive, coarse-woolled breed shedding its fleece in Spring, Chinese breeders developed the Sinkiang Finewool. This was achieved by crossing the Kazakh with Russian and Australian Merinos, then selecting over some forty years for body size and fleece characteristics. The result is a dual purpose sheep which grows a 5 kg fleece of fine white wool (estimated to be of 64–70 spinning count), while mature ewes weigh 50 kg and mature rams 100 kg. Furthermore the Sinkiang Finewool copes with nomadic systems of production, walking up to 1000 km annually (Morris, Armstrong and Howe, 1979).

Libyan sheep meat production

Although the Libyan Barbary sheep produces a fleece of coarse wool, its carcass is worth much more than its fleece. Selection in these sheep should therefore be aimed primarily at increasing meat production.

West African sheep meat production

The Fulani sheep of West Africa are used to produce meat from sparse grazing, often in nomadic flocks. Adaptation and ability to thrive in an environment which would be regarded as being unkind to sheep, are the important attributes of this breed.

South-east Asian sheep meat production

The Kelantan in Malaysia and the Priangan in Indonesia produce meat in a wet, tropical environment where other sheep breeds fail to survive because of the heavy endoparasite burden.

Karakul fur production

Karakul fur production offers an especially interesting example of a sheep system made to fit a particular environmental niche. The Karakul is run in marginal country where grazing is so poor that lamb rearing would impose a heavy stress on ewes. Karakul lambs are therefore harvested for their pelts soon after birth. Selection is for colour and quality of the pelt on the newborn lamb.

Thus a 'sheep production system' can mean many things, and the range of sheep genotypes is even more diverse than the range of husbandry systems. Breeders setting selection goals can consider the many working sheep production models operating throughout the world. Methods or ideas gleaned from an examination of other systems may lead to modified goals.

Defining selection objectives

In defining their selection objectives, breeders should be prepared to assess their entire operation from its beginning in a systematically planned way. Each assumption and decision should be queried, and breeders not doing this risk missing opportunities. Too often breeders impose unnecessary constraints on themselves by applying only traditional husbandry methods, by producing for established markets, or by confining their choice of breeding stock to local strains. A step-by-step analysis of the whole operation should be undertaken.

Review of resources available

Resources to be considered include the climate, soils, feed supplies, labour, markets and sheep breeds available. All are important in determining what sort of enterprise can be run and hence what selection objectives will be.

Decide upon product(s)

Given the mix of resources available, it must be decided whether mutton, lamb, wool, milk, pelts or combinations of these are to be marketed. These are principal selection objectives.

Consider adaptive traits

It is obvious in the above mentioned examples of sheep production systems that adaptation to environment can be crucial in sheep husbandry. Adaptive traits which might be important in the particular environment must be considered for inclusion as secondary selection objectives. Examples of these are heat tolerance, parasite resistance, susceptibility to mineral deficiency or overload, resistance to fleece rot, resistance to dust penetration of the fleece, tolerance to footrot, and having a long breeding season.

Some of these may be more or less easily dealt with by amending management. Whether or not it is worthwhile selecting for these adaptive traits will depend in part on the cost of alternative management practices.

Determine relative economic weightings

Breeders always do have some feelings for the relative value of the sheep's products and adaptive traits. It is implied in a breed standard which allocates points and in an index or any other system used to rank animals. Yet when breeders are asked to nominate relative economic weightings they often react by saying it is too difficult. Nevertheless, no matter how unstable markets may be, it is necessary if a rational breeding programme is to be developed that realistic relative economic weightings be found for products and adaptive traits.

Review genotypes available

Breeders must ascertain the breeds or crosses that most nearly approximate the specified requirements. They should be prepared to exploit the tremendous diversity of sheep genotypes already existing. Of course not all of these will be readily available or may ever be available because of quarantine restrictions. Nevertheless, it is highly probable that genotypes or genes outside the traditional local gene pool could be used with advantage. It is unlikely that any existing gene pool already includes all of the genes which could be useful in that environment. Unfortunately too many breeders unnecessarily confine their selection within a breed or even within strain boundaries.

Develop a breeding plan

The breeder may decide to use a single breed, or to use a crossbred which must be reproduced by repeated crossing, or to pool two or more breeds as a foundation for the development of a new breed.

If the choice is a single breed, then the likely rate of improvement under selection can be predicted. To do this accurately it is necessary to know the heritabilities, phenotypic variances, and the phenotypic and genetic correlations of the traits to be selected as well as achievable selection pressures for the particular population. In practice, most breeders base their predictions on parameters estimated in similar populations and achieve acceptable precision. Efficient within-breed selection programmes should lead to cumulative responses of between 1% and 2% per annum.

The breeder should also calculate how much his proposed improvement programme is likely to accelerate if he utilizes techniques such as artificial insemination, superovulation, embryo transfers, correcting records for twinning and young dams, early weaning, recording feed intake, yield of clean wool, fibre diameter and so on. However, against the expected increased genetic gains must be balanced the extra costs of these techniques. When the expected gains and costs have been evaluated it would be sensible to review again other breeds available. A breed or combination of breeds may be discovered which will require less selection effort to attain the chosen selection objectives.

Where a combination of breeds is selected, precise predictions of performance levels cannot be made, unless the particular cross has been previously evaluated. Levels of heterosis resulting from a specific cross cannot be predicted, nor can the level of response to selection be anticipated in newly formed synthetic breeds, although it is not unreasonable to hope for fairly rapid responses to selection in a population created by recent crossbreeding.

The future

In all breeding programmes there is an inevitable time lag between the plan formulation and the attainment of goals, and with sheep it may be decades. Therefore the breeder is forced to anticipate market conditions and technology changes well into the future. This is probably the most difficult part in defining breeding objectives, yet it is crucial.

Lerner and Donald (1966) forecast the ruminant would retreat to regions where it utilized only stubbles or rough grazing, that human food needs would prevent grazing animals competing with crops on arable land. This has not happened yet. Probably more sheep are fed concentrated rations now than when Lerner and Donald were writing. While many in the world cannot afford to eat much meat, others can pay a high price for meat, and at the present time some of those able to pay have a liking for sheep meat. This is likely to continue.

The future would seem to hold as much diversity as the present. In some countries, the trend is towards more intensive sheep production. Yet at the same time in Australia most sheep growers are adopting less intensive husbandry practices and so reducing labour costs. Some farmers are now managing 15 000 sheep per man. For the same reasons in New Zealand the 'easy care' sheep is a popular concept and a selection objective accorded high priority.

Breeders must therefore make their own assessments of what sheep will be needed in the future, and base their selection objectives on these.

References

DAVIS, G.H., MONTGOMERY, G.W., ALLISON, A.J., KELLY, R.W. and BRAY, A.R. (1981). Fecundity in Booroola Merino sheep—further evidence of a major gene. *Proceedings of 13th Annual Conference of the Australian Society of Reproductive Biology*, p.5

LERNER, I. and DONALD, H.P. (1966). *Modern Developments in Animal Breeding*. London and New York, Academic Press

MORRIS, R.S., ARMSTRONG, J.M. and HOWE, R.R. (1979). *Report on Animal Husbandry and Animal Health Aspects of the Australian Agricultural Mission to the Peoples' Republic of China, 21st June to 12th July, 1978.* Australian Bureau of Animal Health, Department of Primary Industry, Canberra

PIPER, L.R. (1980). *Proceedings of a Workshop on the Booroola Merino.* CSIRO Publication, Melbourne (in print)

ROBERTSON, D.E. (1974). *Sheep Fertility. Recent Research and its Application in Western Australia*, pp. 25–30. Perth, Univesity of Western Australia

ROBERTSON, D.E. (1976). Ovulation and lambing results with Booroola and Booroola-cross Merino ewes in Western Australia. In *Sheep Breeding Proceedings of the 1976 International Congress, Muresk Agricultural College, Western Australia*, pp. 372–376

SHEPHERD, J.H. (1976). The Australian Merino Society nucleus breeding scheme. In *Sheep Breeding Proceedings of the 1976 International Congress, Muresk Agricultural College, Western Australia*, pp. 188–199

26

SELECTION FOR CARCASS QUALITY

B.T. WOLF
Welsh Agricultural College, Aberystwyth, UK
and
C. SMITH
Animal Breeding Research Organisation, Edinburgh, UK

The main long-term objective in the genetic improvement of sheep in the UK is seen as the efficient production of lean meat from herbage and plant residues. Improvement concerns several aspects of production which may affect the quantity and the quality of the product, including fertility, prolificacy, survival, liveweight growth, mature size, carcass shape, composition and palatability. Thus, selection for carcass quality cannot be considered in isolation from liveweight growth and, in some situations, reproductive traits.

The emphasis upon improvement in carcass quality is seen against the background of a post-war decline in lamb meat consumption in the UK. This decline in consumption is related to a problem of overfatness in the lamb carcass (Kempster, 1979), although other factors such as price relative to alternative meats, unattractive fat characteristics, limited potential for manufacturing processes and social changes in meat eating habits are also involved. Defined in terms of consumer requirements the selection objectives for improved carcass quality would be for maximum lean, minimum bone and optimum levels of fat. Other selection objectives may include carcass weight or killing-out percentage, favourable tissue weight distribution, carcass conformation, eye-muscle area, lean:bone ratio and the ratio of subcutaneous:intermuscular fat. This last trait is directly related to the problem of overfatness since a minimum subcutaneous fat may be required by the butchering trade to minimize post-slaughter moisture loss from the carcass. At a given carcass weight and level of subcutaneous fat cover a high ratio of subcutaneous:intermuscular fat is indicative of a low percentage carcass fat. Furthermore, fat in excess of consumer requirement may be removed from the subcutaneous fat depot without risk of damage to the joint.

In the UK the sheep industry is based upon the complementary use of breeds specialized for either performance in extreme environments, reproductive and maternal ability or growth and carcass quality (King, 1976). Since heterosis and specific combining ability for liveweight growth and carcass composition are generally considered to be small (Nitter, 1978), this review will be limited to a consideration of the potential for selection between and within the specialized meat breeds.

Sheep breed comparisons

Evaluation of breed differences in carcass characteristics may be made at a constant age, weight, level of subcutaneous fat in the carcass (constant 'finish'), degree of mature size or over defined ranges of any of these variables. The interpretation of the results of experiments will depend upon the basis chosen for comparison. Where breeds of similar mature size are compared at commercially acceptable liveweights, valid inferences may be made about the biological and commercial significance of differences in breed means. However, when breeds differ widely in mature size the differences observed at constant weights may be due to differences in stage of development rather than to variation in growth patterns. McClelland, Bonaiti and Taylor (1976) have explored the implications of making breed comparisons at constant degree of expected mature size. Their results suggest that breed comparisons made at constant 'finish' may approximate to 'constant degree of mature size' as well as being readily interpreted in commercial terms.

Indirect evidence for breed differences in carcass composition can be found in the survey data of Kempster and Cuthbertson (1977) and Wood *et al.* (1980). At constant levels of subcutaneous fat the breed types surveyed by Kempster and Cuthbertson (1977) gave a wide range of carcass weights which ranked in accordance with the expected mature size of the lambs (*Table 26.1*). Differences in percentage lean in the carcass reflected breed type differences in the ratios of subcutaneous:internal fat and of lean:bone. In practice, however, Southdown crosses and British Longwool lambs were

Table 26.1 CARCASS CHARACTERISTICS OF SOME TYPES OF BRITISH LAMB: MARKET SURVEY DATA REPORTED BY KEMPSTER AND CUTHBERTSON (1977) AND ADJUSTED TO A CONSTANT LEVEL OF SUBCUTANEOUS FAT IN THE SIDE

Breed type	*Side weight* (kg)	*Conformation score*[4]	*Percentage lean*	*Percentage intermuscular fat*	*Percentage KKCF*	*Lean:bone ratio*
Welsh Mountain	6.39^{a}	3.28^{a}	54.7^{a}	11.9^{cd}	4.5^{d}	3.71^{c}
Blackfaced Mountain	8.26^{cd}	3.60^{ab}	55.6^{b}	10.9^{a}	3.7^{c}	3.52^{b}
British Longwool crosses[1]	9.43^{d}	3.80^{bc}	56.7^{d}	10.5^{a}	3.4^{bc}	3.64^{bc}
Suffolk crosses	9.43^{d}	4.08^{cd}	54.5^{a}	12.1^{d}	3.0^{a}	3.37^{a}
Intermediate[2]	8.20^{c}	4.30^{d}	56.4^{cd}	11.4^{bc}	3.2^{ab}	3.80^{c}
Southdown crosses	7.15^{b}	3.73^{bc}	55.4^{b}	11.9^{cd}	3.3^{bc}	3.68^{c}
British Longwool[3]	10.26^{e}	3.76^{bc}	55.9^{bc}	11.2^{bc}	3.7^{c}	3.76^{c}

Note: Numbers with different superscript letters within a column are significantly different from each other.
[1]Border Leicester × Scottish Blackface, Blueface Leicester × Swaledale, Teeswater × Dalesbred.
[2]Dorset Down or Hampshire Down rams with ewes of intermediate mature size.
[3]Purebred Romney Marsh, South Devon, Devon Longwool.
[4]Subjectively assessed on a 7-point scale, 1 = very poor to 7 = very good.

slaughtered to yield heavier, fatter carcasses than shown in *Table 26.1*. Conversely, the Blackfaced Mountain types and Suffolk crosses were slaughtered to give slightly lighter but leaner carcasses than shown. Although these data give an interesting insight into the utilization and performance of the various breed types under commercial conditions it is unwise to make definitive statements about genetic differences in carcass composition since environmental causes of variation are involved. Unbiased breed comparisons require that random samples of sires from each breed produce progeny which are reared in the same environment.

Selected results from four experimental sire breed comparisons are presented in *Table 26.2*. Breeds differed in carcass weight at constant liveweight (killing-out percentage) although the ranking of the breeds was not always consistent between experiments (More O'Ferrall and Timon, 1977a; Wolf, Smith and Sales, 1980; Wolf and Smith, unpublished). Percentage lean in the carcass at constant liveweights varied widely between breeds with a low value being recorded for the relatively mature Dorset Down and an outstandingly high value being recorded for the Texel, a result which is in good agreement with the reports of Osikowski and Borys (1976), Flamant and Perret (1976), Latif and Owen (1979;1980). Adjustment to constant levels of subcutaneous fat in the carcass gave a wider range in carcass weight and a reduction in breed differences for percentage lean in the carcass (Croston, Jones and Kempster, 1979; Wolf, Smith and Sales, 1980). However, even at constant finish the Texel maintains an advantage of 1.5–2.5 percentage units lean (Wolf, Smith and Sales, 1980; Kempster, Croston and Jones, 1981). The data of Croston, Jones and Kempster (1979) suggest that there is potential for breed substitution to increase carcass weight at a given level of 'finish' but, with the exception of the Texel, there is limited potential to increase percentage lean at a given carcass weight. However, sire breeds may differ in the ratios of subcutaneous:intermuscular fat and lean:bone. These factors combine in the Texel to give a high percentage lean (Wolf, Smith and Sales, 1980; Kempster, Croston and Jones, 1981).

Wood *et al.* (1980) have suggested that the ratio of internal:carcass fat may be higher in breeds used as ewe breeds than in meat sire breeds, the internal fats having an important metabolic role. Supportive evidence for breed differences in fat partitioning between depots may be drawn from a number of sources (Donald, Read and Russell, 1970; McClelland and Russel, 1972; Tempest and Boaz, 1977; Geenty, Clarke and Jury, 1979) whilst a high level of kidney knob and channel fat (KKCF) has been identified as a problem in some strains of the Welsh Mountain breed (Kempster and Cuthbertson, 1977; Kempster, 1979).

THE EFFECT OF BREED ON THE GROWTH OF CARCASS TISSUES

Having made breed comparisons at constant endpoints the researcher is often required to predict how the breeds would compare at different weights or ages to those considered. This leads to a problem of extrapolation from limited information. One approach is to slaughter at two or more slaughter points and test for breed of sire × slaughter group interaction.

Table 26.2 A SUMMARY OF BREED COMPARISONS FOR SLAUGHTER AGE AND CARCASS COMPOSITION

Breed	*Slaughter age* (days)			*Carcass weight* (kg)					*% Lean*				
References	*1*	*3a*	*3b*	*1*	*2*	*3a*	*3b*	*4*	*1*	*2*	*3a*	*3b*	*4*
Border Leicester								19.7^{b}					55.9^{ab}
Cotswold					17.4^{c}					4.9^{a}			
Dorset Down	231^{d}	151^{b}	148^{ab}	18.8^{a}	18.2^{ab}	16.2	14.5^{a}	17.5^{d}	54.7^{c}	4.9^{a}	54.7^{a}	55.4^{a}	55.8^{c}
Dorset Horn	193^{ab}			16.8^{cde}					57.4^{b}				
Hampshire Down	207^{c}			16.9^{cd}				17.6^{d}	55.7^{bc}				55.6^{c}
Ile-de-France	227^{d}	153^{b}	153^{bc}	17.5^{b}	17.7^{ac}	16.1	14.8^{ab}	18.5^{c}	55.1^{bc}	5.0^{a}	55.8^{ab}	55.4^{a}	56.7^{a}
Lincoln	205^{abc}			15.3^{e}					56.1^{bc}				
North Country Cheviot								18.4^{c}					56.3^{ab}
Oldenburg		151^{b}	163^{cd}		17.8^{ac}	15.4^{a}	14.8^{ab}			5.3^{b}	57.5^{c}	55.5^{a}	
Oxford Down	195^{ab}	129^{a}	139^{a}	16.6^{e}	17.7^{ac}	16.1	15.1^{b}	20.3^{a}	56.1^{bc}	4.9^{a}	56.3^{bc}	55.2^{a}	55.8^{c}
Southdown					19.2^{d}			16.2^{e}		5.0^{a}			56.3^{ab}
Suffolk	195^{a}	137^{a}	147^{ab}	17.4^{c}	18.0^{ab}	16.0	15.2^{b}	19.7^{b}	55.9^{bc}	5.0^{a}	56.3^{bc}	55.0^{a}	56.1^{ab}
Texel	203^{bc}	150^{b}	166^{d}	16.7^{e}	18.8^{bd}	16.2	16.0^{c}		59.9^{a}	5.5^{b}	60.5^{d}	57.5^{b}	(57.1)
Wensleydale								20.4^{a}					56.6^{a}

Table 26.2 (CONTINUED)

Breed	*% Fat*				*Lean:bone ratio*				*Subcutaneous:intermuscular fat*		
References	*1*	*2*	*3a*	*3b*	*1*	*3a*	*3b*	*4*	*3a*	*3b*	*4*
Border Leicester								3.40[c]			1.03[b]
Cotswold		2.2[b]									
Dorset Down	29.8[a]	2.6[a]	27.9[a]	26.8[a]	3.58[a]	3.47[a]	3.45[a]	3.46[c]	1.13[a]	1.11[a]	1.06[b]
Dorset Horn	25.8[b]				3.50[abcd]						
Hampshire Down	27.6[b]				3.44[bc]			3.42[c]			1.04[b]
Ile-de-France	26.8[b]	2.2[b]	26.3[b]	26.8[a]	3.34[bcde]	3.44[a]	3.44[a]	3.57[b]	1.08[abc]	1.11[a]	1.04[b]
Lincoln	26.2[b]				3.22[e]						
North Country Cheviot								3.40[c]			1.05[b]
Oldenburg		1.9[d]	23.2[c]	26.1[b]		3.29[b]	3.34[a]		1.07[bc]	1.18[a]	
Oxford Down	26.6[b]	2.2[b]	24.6[cd]	26.2[b]	3.32[de]	3.24[b]	3.27[a]	3.40[c]	1.10[ab]	1.17[a]	1.04[b]
Southdown		3.2[e]						3.66[a]			1.03[b]
Suffolk	27.4[b]	2.5[ac]	25.0[bd]	26.8[a]	3.39[bcd]	3.30[b]	3.34[a]	3.46[c]	1.03[abc]	1.11[a]	1.05[b]
Texel	22.7[c]	2.3[bc]	21.5[e]	25.8[b]	3.58[a]	3.72[c]	3.80[b]	(3.73)	1.08[abc]	1.23[b]	
Wensleydale								3.44[c]			1.11[a]

Note: Numbers with different superscript letters within a column are significantly different from each other.

References: 1. More O'Ferrall and Timon (1977a,b): lambs slaughtered at either 35.6 kg or 41.1 kg liveweight.
2. B.T. Wolf and C. Smith (unpublished): serial slaughter between 13 weeks and 15 months of age. Data presented as tissue weights in half carcass at 41.6 kg liveweight.
3a. Wolf, Smith and Sales (1980): lambs slaughtered at either 35.7 kg or 41.6 kg liveweight.
3b. Wolf, Smith and Sales (1980): data adjusted to 11.3% subcutaneous fat in the side.
4. Croston, Jones and Kempster (1979): preliminary results of MLC Ram Breed Comparisons. Data adjusted to 11.7% subcutaneous fat in the carcass. Data for the Texel are taken from Kempster, Croston and Jones (1981).

This method was used by More O'Ferrall and Timon (1977b) and Wolf, Smith and Sales (1980). Neither experiment gave any evidence of change in breed rank for carcass composition, over a limited range of liveweights between 35 kg and 41 kg.

Serial slaughter of groups of lambs over a range of age, liveweight or finish yields information about breed differences in the differential growth patterns of carcass tissues, is statistically efficient and is also capable of giving breed comparisons at defined levels of the independent variate. In studies of the growth of carcass tissues of domestic animals the allometric equation has commonly been used. This equation takes the form:

$$y = ax^b$$

where y represents the weight of a part, x the weight of the whole animal or some other part, a is a constant which defines the elevation of the curve and b is a constant which represents the ratio of the specific growth rates of x and y. The proportion of y relative to x remains constant for $b = 1$, increases for $b>1$ and decreases for $b<1$. For descriptions of alternative methods of analysis and discussion of the problems encountered see Seebeck (1968a) and Finney (1978).

McClelland, Bonaiti and Taylor (1976) compared the Soay, Southdown, Finnish Landrace and Oxford Down (estimated mature weights 24.7, 60.2, 62.8 and 110.5 kg respectively) at four stages of maturity between 39.5% and 72.6% of mature weight. Breed did not significantly affect muscle weight expressed as a percentage of fleece-free empty body weight. The percentage total chemical fat in the fleece-free empty body was particularly low in the Soay (15.3%) compared with values of 22.3, 25.7 and 25.6% for the Southdown, Finnish Landrace and Oxford Down respectively. However, when carcass tissue weights were expressed as a percentage of carcass weight the variation between breed means was increased, although this was largely a function of the low killing-out percentage and low percentage of dissectible carcass fat in the Soay. Significant breed × maturity interactions occurred only for percentage lean in the carcass. Thus over the range of maturity considered, and with the exception of the Soay breed, only small differences between breeds were recorded for the percentage composition of the fleece-free empty body and carcass and there was little evidence of change in breed rank between stages of maturity.

Similar results were presented by Prud'hon (1976) for the growth of muscle and bone relative to empty body weight for lambs of the Romanov, Merinos d'Arles and Berrichon du Cher breeds slaughtered between 25% and 60% of their mature size. However, allometric coefficients for the growth of total fat relative to empty body weight ranged from 1.37 to 1.83 although neither the significance of breed effects nor the standard errors of the estimates were reported.

Further evidence to support the near constancy of percentage lean in the liveweight can be drawn from the work of Knapman (1976) who found that for Southdown, Suffolk and Cotswold cross lambs of 34 kg liveweight, dissectible lean tissue constituted 25% of the liveweight and 22% of each subsequent unit of liveweight gain up to 70% of mature size. Where the growth of dissectible lean tissue has been studied relative to liveweight or empty body weight over weight ranges encompassing commercial slaughter weights, allometric coefficients of 1.0 or slightly below have been recorded (Tulloh, 1964; Fourie, Kirton and Jury, 1970; Wolf and Smith, unpublished). Breeds did not significantly affect allometric coefficients for the growth of lean relative to liveweight (Fourie, Kirton and Jury, 1970; Wolf and Smith, unpublished). However at a constant liveweight of 41.6 kg the Texel and Oldenburg yielded significantly greater weights of lean than the other breeds considered (Wolf and Smith, unpublished; see *Table 26.2*). It is therefore possible that the Texel and Oldenburg are unusual in that they may contain a greater proportion of lean tissue in the empty body than other breeds or, alternatively, they may contain a lower proportion of some non-carcass component of the liveweight. However, no major differences were observed between the Texel and Suffolk crossbred lambs for weights of non-carcass components or daily food intake (Latif and Owen, 1980).

A number of authors have studied breed effects upon the differential growth of carcass tissues for lambs slaughtered over similar ranges of degree of maturity. Fourie, Kirton and Jury (1970) reported that for lambs slaughtered between birth and maturity the rate of muscle and bone growth relative to carcass weight was significantly higher in Romney than in Southdown and Southdown × Romney lambs. The rate of fat deposition relative to carcass weight was not significantly affected by breed. However, the rates of body fat (carcass fat + internal fat) deposition relative to starved liveweight, and of carcass fat deposition relative to muscle plus bone weight were significantly greater ($P<0.05$) in the Southdown and Southdown × Romney than in the Romney. Similarly, Wilson (1975) and Knapman (1976) have presented results which indicate that breed may have a significant effect upon the relative growth of carcass tissues. In contrast, other authors have been unable to demonstrate significant breed differences in the allometric coefficients of carcass tissue growth relative to a range of independent variates and for a wide range of breeds (Seebeck, 1966; Thompson, Atkins and Gilmour, 1979a,b; Geenty, Clarke and Jury, 1979; Wolf and Smith, unpublished).

There is little evidence to suggest that breeds may differ in the rates of subcutaneous and intermuscular fat deposition relative to carcass weight although breeds may differ in patterns of KKCF deposition (Wood *et al.*, 1980; Wolf and Smith, unpublished). In conclusion it is suggested that breed differences in tissue growth prior to the lamb's achievement of 30% of mature size may make an important contribution to differences in breed means for carcass tissue weights at constant live or carcass weight. The importance of breed effects on differential carcass tissue growth over the commercial weight range is less clear.

LEAN TISSUE DISTRIBUTION

Detailed anatomical studies of the growth and development of the musculature of sheep have revealed a relationship between muscle distribution and total muscle weight in the carcass (Lohse, Moss and Butterfield, 1971; Jury, Fourie and Kirton, 1977). There is, however, considerable evidence for breed differences in muscle or dissectible lean tissue distribution, although the commercial importance of these differences is often questioned. Jury, Fourie and Kirton (1977) found significant differences between the Southdown, Romney and their cross for the allometric coefficients for the growth of a number of individual muscles and standard muscle groups relative to total muscle weight. However, at defined weights of total muscle within the commercial range the breed differences in muscle distribution between standard muscle groups were small and considered to be of little commercial importance. Seebeck (1968b) found significant differences between the Merino and Dorset Horn cross (Border Leicester × Merino) for the percentage of total lean in the neck and thorax joints. Subsequently, Thompson, Atkins and Gilmour (1979a,b) reported that the progeny of Dorset Horn and Border Leicester rams out of Merino, Corriedale and Border Leicester × Merino ewes did not differ significantly for lean tissue distribution between standard joints.

The results of a number of surveys and experimental investigations involving a wide range of breeds suggest that although many breeds may differ little for percentage total lean in the high-priced cuts, individual breeds may differ significantly from the overall mean (*Table 26.3*). The

Table 26.3 A SUMMARY OF PUBLISHED BREED MEANS FOR THE PERCENTAGE OF TOTAL CARCASS LEAN FOUND IN THE HIGH-PRICED JOINTS

Breed-type group	A			
Welsh Mountain	57.3^{c}			
Blackfaced Mountain	56.0^{b}			
British Longwool crosses	55.7^{b}			
Suffolk crosses	56.2^{b}			
Intermediate	56.3^{b}			
Southdown crosses	57.2^{c}			
British Longwool	54.7^{a}			
Breed (of sire)	*B*	*C*	*D*	*E*
Border Leicester	53.5^{c}			
Clun			56.4^{b}	
Colbred			54.9^{a}	
Dorset Down	54.7^{abc}	54.9		
Finnish Landrace				40.5
Hampshire Down	54.9^{ab}		55.8^{b}	
Ile-de-France	55.4^{a}	55.1		
North Country Cheviot	54.1^{bc}			
Oldenburg		54.4		
Oxford Down	54.5^{abc}	55.1		39.8
Soay				40.8
Southdown	55.2^{ab}			43.4
Suffolk	55.2^{ab}	54.9	56.1^{b}	
Texel		54.3		
Wensleydale	54.6^{abc}			

Note: Numbers with different superscript letters within a column are significantly different from each other.
A Kempster and Cuthbertson (1977)
B Croston, Jones and Kempster (1979)
C Wolf (1982)
D Wood *et al.* (1980)
E Taylor, Mason and McClelland (1980)

maximum differences recorded between breed means for percentage lean in the high-priced cuts were 2.6 percentage units (Kempster and Cuthbertson, 1977) and 1.9 percentage units of total lean weight (Croston, Jones and Kempster, 1979). Although breed did not significantly affect percentage lean in the high-priced cuts (Wolf, 1982), a number of breed differences were recorded for lean tissue distribution between eight standard joints. The results of Prud'hon (1976) and Taylor, Mason and McClelland (1980) indicate that breed differences in lean tissue distribution may occur even when comparisons are made at the same stage of maturity.

DISTRIBUTION OF SUBCUTANEOUS AND INTERMUSCULAR FAT

Early evidence for breed differences in the distribution of both intermuscular and subcutaneous fat at constant weight of depot fat was presented by Seebeck (1968a,b). Gaili (1978) reported significant differences between the Dorset Horn, Clun and Hampshire in intermuscular fat distribution while Thompson, Atkins and Gilmour (1979a,b) were unable to detect a significant effect of breed on the distribution of either subcutaneous or intermuscular fat. Our own results (*Table 26.4*) are in good agreement with those of Kempster (1981) in showing small but significant effects of breed on the distribution of both subcutaneous and intermuscular fat at constant weights of total depot fat. In practice, lambs of each breed will differ in

Table 26.4 LEAST SQUARES MEANS[(a)] (DECAGRAMS) FOR THE WEIGHT OF SUBCUTANEOUS AND INTERMUSCULAR FAT IN STANDARD JOINTS OF SIRE BREEDS IN THE ABRO TERMINAL SIRE BREED COMPARISON

	Sire breed						
	Dorset Down	*Ile-de-France*	*Oldenburg*	*Oxford Down*	*Suffolk*	*Texel*	*Significance of breed effect*
Subcutaneous fat							
Leg	24.4	23.1	24.6	24.6	25.0	23.7	**
Chump	10.9	11.2	10.6	11.0	10.8	10.7	NS
Loin	15.5	16.0	15.7	16.0	16.4	15.4	*
Breast	15.5	15.0	15.5	16.1	15.4	16.1	NS
Best-end neck	13.1	12.9	12.5	12.5	12.3	12.4	NS
Shoulder	22.3	23.3	22.8	21.8	22.0	23.3	**
Scrag	2.9	3.2	2.8	2.6	2.7	2.9	***
Intermuscular fat							
Leg	10.2	10.1	10.1	10.3	10.6	10.0	NS
Chump	5.1	5.2	5.1	5.3	5.3	5.4	NS
Loin	7.6	7.9	7.9	8.2	7.8	7.6	NS
Breast	17.2	15.6	17.3	17.0	16.7	17.2	*
Best-end neck	9.5	9.6	9.1	9.7	9.3	9.0	**
Middle-neck	15.7	15.8	15.3	14.9	15.1	15.4	NS
Shoulder	24.1	25.2	24.5	24.3	25.0	25.2	*
Scrag	2.6	2.6	2.4	2.3	2.2	2.6	**

[(a)]Comparisons made at constant weight of subcutaneous (104.5 dg) and intermuscular (92.0 dg) fat in the half carcass.
From B.T. Wolf (unpublished)

weight of depot fat when slaughtered at constant finish but it is unlikely that the differences between extreme breed means would be more than double those shown in *Table 26.4*.

BONE DISTRIBUTION

Breed differences in the distribution of bone between standard joints have been reported by Seebeck (1968b) and Thompson, Atkins and Gilmour

Table 26.5 LEAST SQUARES MEANS[(a)] (DECAGRAMS) FOR BONE WEIGHTS IN STANDARD JOINTS OF SIRE BREEDS IN THE ABRO TERMINAL SIRE BREED COMPARISON

	Sire breed						
	Dorset Down	*Ile-de-France*	*Oldenburg*	*Oxford Down*	*Suffolk*	*Texel*	*Significance of breed effect*
Leg	32.8	34.2	33.8	34.0	33.5	34.0	***
Chump	9.8	9.8	9.7	9.7	9.6	9.5	NS
Loin	10.7	10.8	10.7	10.8	11.0	10.5	NS
Breast	12.6	12.3	12.4	12.1	12.5	12.3	NS
Best-end neck	11.9	11.4	11.4	11.4	11.4	11.5	NS
Middle-neck	21.9	21.0	21.5	21.0	21.2	21.3	*
Shoulder	26.6	27.2	27.0	27.7	27.5	27.3	***
Scrag	5.7	5.3	5.2	5.3	5.3	5.4	*

[(a)]Comparisons made at constant weight of bone (132 dg) in the half carcass.
From B.T. Wolf (unpublished)

(1979b). The results presented in *Table 26.5* show small but economically unimportant effects of breed on bone distribution.

LIVEWEIGHT GROWTH RATE

Rate of gain and age at slaughter are important factors influencing the choice of sire breed although relatively more emphasis may be placed on these traits for production systems which are limited by time, seasonality of price and grass supply. The results of More O'Ferrall and Timon (1977a) and Wolf, Smith and Sales (1980) were consistent in showing rapid growth rates to constant liveweights for the Oxford Down and Suffolk. There was, however, some disagreement in the values of the Texel, Ile-de-France and Dorset Down for slaughter age (*Table 26.2*). The crossbred progeny of Texel and Ile-de-France sires were of similar age at 45 kg liveweight (Osikowski and Borys, 1976) or of similar weight at 35 weeks of age (B.T. Wolf and C. Smith, unpublished).

Statistical adjustment of the data to a constant percentage subcutaneous fat in the carcass altered breed rank for slaughter age (*Table 26.2*) with the relatively lean Texel cross being most markedly affected. The increased age at slaughter of the Texel relative to the Suffolk, combined with a lack of an adequate premium payment for the leaner carcass produced, is a major factor inhibiting the use of the Texel for the production of finished lambs off summer grass in the UK. The comparison of the Texel with the Suffolk is particularly interesting since they are of similar mature size but differ widely in liveweight growth rate and carcass composition. In comparisons made at constant slaughter weights the differences between these breeds in carcass weight for age have generally been in proportion to differences in liveweight growth rate. However, the two breeds have similar lean tissue growth rates (lean weight/day of age), the higher carcass growth rate of the Suffolk being almost entirely explained by a higher rate of fat deposition (More O'Ferrall and Timon, 1977b; Latif and Owen, 1979; 1980; Wolf, Smith and Sales, 1980; Wolf and Smith, unpublished). It is unclear whether this factor would explain the observed differences in liveweight growth rate. McClelland (1975) has suggested that the reduced rate of fat deposition in the Texel may be the result of a reduction in voluntary food intake. However, published evidence is limited and does not support this suggestion (Osikowski and Borys, 1976; Latif and Owen, 1980).

Selection within breeds

In planning genetic improvement schemes decisions must be made about the choice of selection objectives and the method of evaluating breeding stock. Selection objectives of carcass quality were previously described in terms of consumer satisfaction although other aspects of performance, notably food conversion efficiency and growth rate will also be considered by the producer. Since the goal of the producer is to maximize profit all traits which influence the economic return to the production enterprise must be considered for inclusion in the selection objectives. The selection index (Hazel, 1943) is regarded as the best method of selecting for multiple objectives. Information which is required for the construction of the selection index includes estimates of the heritability, phenotypic standard deviation and relative economic value of each trait and of the phenotypic and genetic correlations between traits. The relative economic value indicates the relative importance of the contribution of each trait to profit and represents the amount by which net financial return from the animal is increased as a result of one unit of improvement in each trait.

Economic aspects of the problems involved in determining selection objectives and methods for the calculation of relative economic values have been reviewed by Miller and Pearson (1979). Few European countries use index selection in sheep improvement and estimates of economic weights for carcass quality traits are rare and vary widely between countries (Croston *et al.*, 1980). In the UK there is no short-term financial incentive to include carcass quality traits in the selection objective since the market value of the lamb is closely related to weight, with only small price differentials between lambs of similar weight but varying fatness. There is, therefore, a clash of interest between short-term financial considerations and the long-term maintenance of market demand. A similar problem was documented by Bradford (1967) who showed that selection for yield of lean cuts in the carcass on the basis of progeny test results was unlikely to be cost effective in the USA. The limited number of progeny produced by selected rams in natural service and the small price differential paid for the carcasses did not cover the cost of the progeny test. The problem of setting objectives for the improvement of carcass quality is therefore one of forecasting future trends in demand and of predicting the relative emphasis to be placed on live animal and carcass traits.

Genetic parameters of carcass composition

Estimates of the heritability of carcass traits are shown in *Tables 26.6* and *26.7*. Variability in heritability estimates between studies is shown for a number of traits reflecting differences in population and environments. In general, percentage lean and fat in the carcass are moderately heritable with lower values being reported for the heritability of percentage bone, lean:bone ratio and subcutaneous:intermuscular fat ratio. Estimates of the phenotypic standard deviations of each trait have been presented in a number of the papers summarized in *Tables 26.6* and *26.7*. The combination of moderate heritability estimates and phenotypic standard deviations of *ca.* 3.0 percentage units indicates a potentially useful response to direct selection for percentage lean in the carcass.

Table 26.6 SUMMARY OF PUBLISHED PATERNAL HALF-SIB ESTIMATES OF THE HERITABILITY OF LAMB CARCASS TRAITS EVALUATED AT CONSTANT LIVEWEIGHT

Trait	*Heritability*	*Authors*
Carcass weight	0.02	Bowman, Marshall and Broadbent (1968)
Carcass weight	0.11±0.18	Bowman and Hendy (1972)
Carcass weight	0.33±0.11	Botkin *et al.* (1969)
Carcass weight/day of age	0.35±0.11	Botkin *et al.* (1969)
Killing-out %	0.16±0.07	Wolf *et al.* (1981)
Killing-out %	0.16±0.17	Cotterill and Roberts (1976)
Killing-out %	0.41±0.12	Botkin *et al.* (1969)
Carcass composition		
% Lean	0.40±0.12	Botkin *et al.* (1969)
% Fat	0.54±0.13	Botkin *et al.* (1969)
% Bone	0.23±0.09	Botkin *et al.* (1969)
% Lean	0.41±0.13	Wolf *et al.* (1981)
% Fat	0.37±0.13	Wolf *et al.* (1981)
% Bone	0.16±0.10	Wolf *et al.* (1981)
Lean:bone ratio	0.13±0.09	Wolf *et al.* (1981)
Lean:bone ratio	0.19±0.09	Botkin *et al.* (1969)
Subcutaneous:intermuscular fat ratio	0.12±0.09	Wolf *et al.* (1981)
Eye-muscle area	0.14±0.10	Wolf *et al.* (1981)
Eye-muscle area	0.14±0.20	Bowman and Hendy (1972)
Eye-muscle area	0.34±0.11	Botkin *et al.* (1969)
Eye-muscle area	0.53	Bowman, Marshall and Broadbent (1968)
Back-fat depth	0.21±0.11	Wolf *et al.* (1981)
Back-fat depth	0.37±0.21	Cotterill and Roberts (1976)
Back-fat depth	0.40±0.26	Bowman and Hendy (1972)
Back-fat depth	0.51±0.13	Botkin *et al.* (1969)
Conformation score	0.18±0.12	Wolf *et al.* (1981)

Table 26.7 SUMMARY OF PUBLISHED PATERNAL HALF-SIB ESTIMATES OF THE HERITABILITY OF LAMB CARCASS TRAITS EVALUATED AT CONSTANT AGE

Trait	*Heritability*	*Authors*
Carcass weight	0.35±0.14	Olson *et al.* (1976)
Carcass weight	0.53±0.25	Timon (1968)
Killing-out %	0.06±0.11	Olson *et al.* (1976)
Carcass composition		
% Protein	0.51±0.26	Timon (1968)
% Ether extract	0.50±0.26	Timon (1968)
% Bone	0.32±0.22	Timon (1968)
% Bone	0.04±0.11	Olson *et al.* (1976)
Weight of bone	0.43±0.14	Olson *et al.* (1976)
Lean:bone ratio	0.36±0.23	Timon (1968)
Eye-muscle area	0.12	Smith *et al.* (1968)
Eye-muscle area	0.56±0.26	Timon (1968)
Back-fat depth	0.27	Smith *et al.* (1968)
Back-fat depth	0.28±0.13	Olson *et al.* (1976)

Estimates of the genetic correlations between carcass traits have been presented by Botkin *et al.* (1971), Bradford and Spurlock (1972), Olson *et al.* (1976) and Wolf *et al.* (1981). In general no unfavourable correlations have been reported although a negative correlation is shown for the relationship between percentage lean and subcutaneous:intermuscular fat ratio (*Table 26.9*).

Within-breed genetic variation in carcass weight distribution has been reported by a number of authors. Watson and Broadbent (1968) recorded differences between the progeny of six Suffolk rams for tissue weight and individual muscle weights in the leg joint. Bowman, Marshall and Broadbent (1968) and Bowman and Hendy (1972) reported zero to moderate heritabilities for the weight of standard joints for lambs slaughtered on the basis of liveweight plus handling appraisal. Timon (1968) recognized a need to make adjustments for carcass weight and fatness. Although joint weights were highly heritable (lambs slaughtered at constant age), joint weights expressed as a percentage of carcass weight had low heritability. Adjustment for fatness gave low or zero heritability suggesting that there was little scope for genetic selection to change carcass weight distribution at constant levels of fatness. However, in studies of muscle distribution a number of authors have remarked upon variation which was not explained by any of the factors included in their statistical models and have suggested that this may be of genetic origin (Jackson, 1969; Taylor, Mason and McClelland, 1980). In cattle, Andersen (1977) (quoted by Berg, Andersen and Libroriussen, 1978) reported a heritability of 0.29 for the percentage of total lean found in the pistol joint and Bergström (1978) cited evidence from experiments with twins in support of within-breed genetic variation in muscle distribution.

Studies of lean tissue (Wolf, 1982) and subcutaneous fat distribution (B.T. Wolf, unpublished data) at constant weights of total dissectible tissue showed low to moderate heritabilities for these traits (*Table 26.8*). In general coefficients of variation for lean tissue distribution traits were about 3–10%, the lower values being found for joints (leg, shoulder and

Table 26.8 LEAST SQUARES MEANS ($\bar{x}$), PHENOTYPIC STANDARD DEVIATIONS (σ_P) AND HERITABILITY (h^2) ESTIMATES FOR THE PERCENTAGE DISTRIBUTION OF DISSECTIBLE LEAN AND SUBCUTANEOUS FAT

Joint	*Lean*			*Subcutaneous fat*		
	$\bar{x}$	σ_P[a]	h^2±S.E.M.	$\bar{x}$	σ_P[a]	h^2±S.E.M.
Leg[b]	29.8	1.16	0.21±0.10	24.0	3.31	0.31±0.12
Chump[b]	7.9	0.70	0.28±0.11	10.3	1.64	0.21±0.10
Loin[b]	10.6	1.03	0.46±0.14	14.6	2.35	0.10±0.08
Best-end neck[b]	6.5	0.54	0.07±0.08	11.8	1.91	0.21±0.10
Breast	9.4	1.22	0.35±0.12	14.7	3.16	0.10±0.08
Middle-neck	12.6	0.74	0.15±0.09			
Shoulder	20.8	1.01	0.33±0.12	21.7	3.59	0.08±0.08
Scrag	2.4	0.38	0.28±0.11	2.8	0.99	0
High-priced joints	54.8	1.60	0.65±0.15			

[a] Calculated after adjustment for fixed effects of breed of sire and dam, year, ewe age, rearing type, sex of lamb and total weight of tissue in the half carcass.
[b] Joint included in the high-priced joints.

high-priced joints combined), which represented the largest proportions of the carcass. Coefficients of variation for traits measuring the distribution of subcutaneous fat were generally higher than for lean tissue distribution and heritability estimates were lower.

Selection for liveweight growth traits

Since liveweight growth rate can readily be measured on the individual at an early age, selection on the basis of individual performance is possible. However, estimates of the heritability of early growth traits in Down breeds have generally been close to zero (Bichard and Yalçin, 1964; Bowman and Broadbent, 1966; Bowman and Hendy, 1972; Wolf *et al.*, 1981). Bowman (1968) considered that the heritability of liveweight growth was lower in the Down breeds than in the Merino and American range breeds and suggested that this may be due either to a reduction in additive genetic variation due to selection or to differences in environment. Analysis of large numbers of pedigree flock records gave a reliable pooled heritability estimate of lamb adjusted eight-week weight of 0.16±0.02 (P.R. Bampton, personal communication). It is considered that much of the variability in lamb performance at early ages may be attributable to variation in dam's milk supply rather than to differences in the genetic potential of the lamb (Owen, 1971). Indirect evidence to support this contention is drawn from reports in which heritability estimates have been higher when calculated within populations of single lambs than for twin lambs (Vogt, Carter and McClure, 1967; Gjedrem, 1967; Hallgrimsson, 1971 (quoted by Jónmundsson, 1977)) although other studies have shown no significant differences (Jónmundsson, 1977; Martin *et al.*, 1980).

In the United Kingdom the MLC Pedigree Flock Recording Scheme makes use of the finding that the heritability of weight for age increases with time and thus distance from the early maternal influence (Bowman, 1968). Litter weight at eight weeks is used as a measure of maternal performance and breeders are advised to make selection decisions between lambs on the basis of weights at 21 weeks or on average daily gain (ADG) from eight to 21 weeks. Schemes such as this allow selection on the basis of individual performance with a minimum generation interval of one year in males. Further progress may be made through the use of sire summaries and progeny test assessments. Methods of optimizing genetic progress within Down flocks using performance and/or progeny testing for liveweight growth rates have been studied by Eikje (1978), the utilization of a crossbred ewe flock for progeny testing has been considered by Bichard and Yalçin (1964) and the benefits and costs of selection have been evaluated (Morris, 1980a,b).

Early weaning, followed by artificial rearing and selection for weight at 90 days of age has been examined as a method of improving the response to selection for early growth rate in Suffolk rams (Owen *et al.*, 1978). Ram lambs selected for either high or low weight for age were subsequently progeny tested with progeny either artificially reared or reared under field conditions. Estimates of the doubled regression of progeny on sire for 90-day weight were 0.20±0.10 and 0.30±0.06 respectively. These results have encouraged further investigation in an attempt to demonstrate the efficiency of the method both in terms of the genetic parameters and cost effectiveness.

Even if selection after artificial rearing is shown to be cost effective it is expected that breeders will show some resistance to its adoption. Alternative methods of selection may involve evaluation of individual performance to older ages than 21 weeks. Although such methods will increase the generation interval for selected rams to two years, a greatly increased heritability estimate of the selected trait and a high genetic correlation with weight for age in early life could lead to a greater genetic gain/year. Croston *et al.* (1983) have investigated the method using Suffolk sires selected for high and low 18-month weight. Estimates of the effective heritability ($r_{g12}h_1h_2$) of 12-week weight were 0.18±0.07 and −0.02±0.06 in two different flocks. On the basis of this evidence it seems unlikely that selection for 18-month weight will be a useful method of improving early lamb growth rate.

Correlated responses to selection for live animal traits

There are few published estimates of genetic correlations between carcass traits and growth traits in sheep and generally published estimates have high standard errors attached. However, correlations between traits which can be measured in the live animal and carcass traits are of particular interest because they may indicate methods of selection for carcass quality which are not dependent upon the slaughter of sibs or progeny. These latter methods will incur higher costs than a simple performance test and due to a limitation of testing facilities the intensity of selection which can be achieved will be lower and the generation interval increased.

The genetic correlations between liveweight growth rate from birth to slaughter and carcass composition reported by Wolf *et al.* (1981) are given in *Table 26.9*. These estimates suggest that selection for liveweight growth

Table 26.9 CORRELATIONS[a] (× 100) ±S.E.M. BETWEEN SOME CARCASS TRAITS AND DAILY GAIN (ADG) TO SLAUGHTER

		(1)	*(2)*	*(3)*	*(4)*	*(5)*	*(6)*	*(7)*	*(8)*	*(9)*	*(10)*	*(11)*	*(12)*	*(13)*	*(14)*	*(15)*
ADG birth to slaughter	(1)		8	3	–7	3	–7	–27	–0	4	8	8	3	94	–0	20
% Lean	(2)	15±29		41	–93	–83	–77	–60	22	88	–26	–61	9	21	–47	–38
% Bone	(3)	80±37	67±22		–71	–64	–47	–54	–79	63	–34	–50	–26	–1	–43	–52
% Total fat	(4)	–32±29	–98±2	–82±12		91	79	68	14	–93	34	68	4	–15	53	49
% Subcutaneous fat	(5)	–47±41	–98±7	–31±42	92±6		50	50	12	–83	67	69	8	–5	53	52
%Intermuscular fat	(6)	–27±24	–75±9	–77±21	83±8	57±19		38	–1	–74	–29	43	–0	–16	33	33
% KKCF	(7)	–27±27	–67±14	–84±14	77±10	74±19	35±18		19	–62	23	42	–1	–28	32	26
Lean/bone ratio	(8)	–89±39	36±31	–45±26	–15±31	–64±46	0±26	15±25		–9	17	12	33	13	14	28
Lean/fat ratio	(9)	28±26	94±3	81±12	–97±2	–88±9	–84±7	–70±9	12±24		–35	–57	–5	13	–50	–48
Subcutaneous: intermuscular fat ratio	(10)	–57±39	–58±26	41±44	38±25	31±33	–60±21	58±21	–119±47	–34±20		37	10	7	33	32
Subcutaneous fat depth	(11)	–30±39	–80±15	–14±45	74±16	80±14	50±17	55±23	–61±42	–64±18	18±34		5	4	45	40
Eye-muscle area	(12)	38±31	53±29	–12±44	–42±33	–56±41	–7±26	–56±30	64±30	38±29	–54±38	–47±39		44	13	29
Lean weight/day of age	(13)	95±3	41±24	93±30	–56±22	–75±32	–42±20	–33±20	–75±32	55±3	–80±7	–47±32	49±24		–1	21
Subcutaneous fat score	(14)	–23±29	–85±12	–59±19	85±10	78±18	39±18	73±13	–30±28	–81±9	82±33	73±20	–19±32	–34±21		64
Conformation score	(15)	7±32	–79±18	–60±20	81±14	66±30	46±26	49±19	–21±33	–77±11	26±32	44±36	40±38	–16±25	100±7	

[a] Phenotypic correlations above diagonal, genetic correlations below

By courtesy of Wolf *et al.* (1981)

rate to fixed weights might be expected to give a correlated response in lean weight/day of age but little reduction in percentage total fat. Evidence to support this view can be drawn from the results of Botkin *et al.* (1971). Although none of the estimates of the genetic correlations were significantly different from zero, Olson *et al.* (1976) found that post-weaning growth rates and weight for age tended to be negatively correlated to back-fat depth but positively correlated to percentage kidney fat at a constant age. In two-way selection for weaning weight in Merino sheep Pattie and Williams (1966) found that at a constant slaughter age the high weaning weight lines were slightly fatter, although the divergence recorded was proportional to differences in liveweight.

The genetic relationships between measures of liveweight growth rate and carcass weight or killing-out percentage are unclear. The results of Botkin *et al.* (1971) suggest zero or slightly negative correlations whilst Wolf *et al.* (1981) reported a positive correlation of 0.53±0.21 between average daily gain from birth to slaughter and killing-out percentage. Post-weaning growth rates and weight for age were positively and highly correlated to carcass weight at constant age but had zero correlations with killing-out percentage (Olson *et al.*, 1976).

Genetic correlations between ultrasonic back-fat depth (Botkin *et al.*, 1971) or subcutaneous fat depth and eye-muscle area (Wolf *et al.*, 1981) and carcass traits were favourable to overall selection objectives. Thus selection for increased eye-muscle area or for reduced back-fat depth would be expected to give correlated increases in percentage lean, lean:bone ratio and perhaps lean weight/day of age. However, caution is required in the interpretation of the data of Wolf *et al.* (1981) since correlated responses from ultrasonically predicted measurements may not be as large as those expected from a consideration of genetic correlations with measurements made on the carcass. However, the evidence of Botkin *et al.* (1971) would suggest that carcass fat depth and ultrasonic fat depth are essentially the same trait, with similar genetic correlations with carcass composition.

Carcass conformation at constant liveweight had positive genetic and phenotypic correlations with measures of carcass fatness (*Table 26.9*), a result which is in broad agreement with those of Kempster, Croston and Jones (1981). These authors showed that conformation is a poor indicator of carcass lean content and proportion of lean in the higher priced joints among carcasses of equal weight and subcutaneous fat cover. We agree with their conclusion that the use of conformation as an objective in selection schemes could have a deleterious effect on progress in more important traits.

Conclusions

Genetic selection to change lamb carcass composition may involve both selection between and within breeds. Comparisons of potential terminal sire breeds have been reviewed and these show that breed substitution may give rapid changes in carcass weight at a given level of subcutaneous fat cover. However, with the exception of the Texel, breed differences in percentage lean in the carcass are not large and long-term selection within breeds will be necessary to improve this trait.

The objectives of within-breed selection for carcass quality have not been well defined or evaluated. Estimates of the heritability and phenotypic standard deviation of carcass traits would suggest that useful responses may be expected from within-breed selection. Although there is not enough available information to suggest that the genetic parameter estimates are entirely reliable, they should provide a suitable base for use in the construction of selection indices and the evaluation of alternative breeding plans. There is little evidence of effective response to direct selection for liveweight growth rate in terminal sire breeds in the United Kingdom, the nature of the correlated responses in carcass composition are unclear and there are no adequate systems for measuring genetic change.

References

BERG, R.T., ANDERSEN, B.B. and LIBRORIUSSEN, T. (1978). Growth of bovine tissues. 2. Genetic influences on muscle growth and distribution in young bulls. *Animal Production* **27**, 51–62

BERGSTROM, P.L. (1978). Sources of variation in muscle weight distribution. In *Patterns of Growth and Development in Cattle* (H. de Boer and J. Martin, Eds.), pp. 91–131. The Hague, Martinus Nijhoff

BICHARD, M. and YALCIN, B.C. (1964). Crossbred sheep production. III. Selection for growth rate and carcass attributes in the second-cross lamb. *Animal Production* **6**, 179–187

BOTKIN, M.P., FIELD, R.A., RILEY, M.L., NOLAN, J.C. and ROEHRKASSE, G.P. (1969). Heritability of carcass traits in lambs. *Journal of Animal Science* **29**, 251–255

BOTKIN, M.P., RILEY, M.L., FIELD, R.A., JOHNSON, C.L. and ROEHRKASSE, G.P. (1971). Relationship between productive traits and carcass traits in lambs. *Journal of Animal Science* **32**, 1057–1061

BOWMAN, J.C. (1968). Genetic variation in body weight in sheep. In *Growth and Development of Mammals* (G.A. Lodge and G.E. Lamming, Eds.), pp. 291–308. London, Butterworths

BOWMAN, J.C. and BROADBENT, J.S. (1966). Genetic parameters of growth between birth and sixteen weeks in Down Cross sheep. *Animal Production* **8**, 129–135

BOWMAN, J.C. and HENDY, C.R.C. (1972). Study of retail requirements and genetic parameters of carcass quality in polled Dorset Horn sheep. *Animal Production* **14**, 189–198

BOWMAN, J.C., MARSHALL, J.E. and BROADBENT, J.S. (1968). Genetic parameters of carcass quality in Down Cross sheep. *Animal Production* **10**, 183–191

BRADFORD, G.E. (1967). Genetic and economic aspects of selecting for lamb carcass quality. *Journal of Animal Science* **26**, 10–15

BRADFORD, G.E. and SPURLOCK, G.M. (1972). Selection for meat production in sheep—results of a progeny test. *Journal of Animal Science* **34**, 737–745

COTTERILL, P.P. and ROBERTS, E.M. (1976). Preliminary heritability estimates of some lamb carcass traits. *Proceedings of the Australian Society of Animal Production* **11**, 53–56

CROSTON, D., JONES, D.W. and KEMPSTER, A.J. (1979). A comparison of the performance and carcass characteristics of lambs by nine sire breeds. *Animal Production* **28**, 456–457 (Abstract)

CROSTON, D., READ, J.L., JONES, D.W., STEANE, D.E. and SMITH, C. (1983). Selection on ram 18-month weight to improve lamb growth rate. *Animal Production* (in press).

CROSTON, D., DANELL, O., ELSEN, J.M., FLAMANT, J.C., HANRAHAN, J.P., JAKUBEC, V., NITTER, G. and TRODAHL, S. (1980). A review of sheep recording and evaluation of breeding animals in European Countries: A group report. *Livestock Production Science* **7**, 373–392

DONALD, H.P., READ, J.L. and RUSSELL, W.S. (1970). Influence of litter size and breed of sire on carcass weight and quality of lambs. *Animal Production* **12**, 281–290

EIKJE, E.D. (1978). Genetic progress from performance and progeny test selection in Down sheep. *Proceedings of the New Zealand Society for Animal Production* **38**, 161–173

FINNEY, D.J. (1978). Growth curves: their nature, uses and estimation. In *Patterns of Growth and Development in Cattle* (H. de Boer and J. Martin, Eds.), pp. 658–672. The Hague, Martinus Nijhoff

FLAMANT, J.C. and PERRET, G. (1976). Le croisement et la production de viande d'agneaux. Comparison et séléction des races de mâles. In *2èmes Journées de la Recherche Ovine et Caprine*, pp. 110–134. Paris, INRA and ITOVIC

FOURIE, P.D., KIRTON, A.H. and JURY, K.E. (1970). Growth and development of sheep. II. Effect of breed and sex on the growth and carcass composition of the Southdown, Romney and their cross. *New Zealand Journal of Agricultural Research* **13**, 753–770

GAILI E.S.E. (1978). A note on the effect of breed-type and sex on the distribution of intermuscular fat in carcasses of sheep. *Animal Production* **26**, 217–219

GEENTY, K.G., CLARKE, J.N. and JURY, K.E. (1979). Carcass growth and development of Romney, Corriedale, Dorset and crossbred sheep. *New Zealand Journal of Agricultural Research* **22**, 23–32

GJEDREM, T. (1967). Phenotypic and genetic parameters for weight of lambs at five ages. *Acta Agriculturae Scandinavica* **17**, 199–216

HAZEL, L.N. (1943). The genetic basis for constructing selection indexes. *Genetics* **28**, 476–490

JACKSON, T.H. (1969). Relative weight changes in the tissues of the gigot joint as Scottish Blackface castrated male lambs develop from weaning to maturity and an analysis of the observed individual variation. *Animal Production* **11**, 409–417

JÓNMUNDSSON, J.V. (1977). A study of data from the sheep recording associations in Iceland. I. Sources of variation in weight of lambs. *Journal of Agricultural Research, Iceland* **1**, 16–30

JURY, K.E., FOURIE, P.D. and KIRTON, A.H. (1977). Growth and development of sheep. IV. Growth of the musculature. *New Zealand Journal of Agricultural Research* **20**, 115–121

KEMPSTER, A.J. (1979). Variation in the carcass characteristics of commercial British sheep with particular reference to overfatness. *Meat Science* **3**, 199–208

KEMPSTER, A.J. (1981). Fat partition and distribution in the carcasses of cattle, sheep and pigs. A review. *Meat Science* **5**, 83–98

KEMPSTER, A.J. and CUTHBERTSON, A. (1977). A survey of the carcass characteristics of the main types of British lamb. *Animal Production* **25**, 165–179

KEMPSTER, A.J., CROSTON, D. and JONES, D.W. (1981). Value of conformation as an indicator of sheep carcass composition within and between breeds. *Animal Production* **33**, 39–49

KING, J.W.B. (1976). National Sheep Breeding Programmes—Great Britain. In *Sheep Breeding*, (G.J. Tomes, D.E. Robertson and R.J. Lightfoot, Eds.), pp. 67–76. London, Butterworths

KNAPMAN, P.W. (1976). A growth study of young lambs. PhD Thesis. University of Reading

LATIF, M.G.A. and OWEN, E. (1979). Comparison of Texel- and Suffolk-sired lambs out of Finnish Landrace × Dorset Horn ewes under grazing conditions. *Journal of Agricultural Science, Cambridge* **93**, 235–239

LATIF, M.G.A. and OWEN, E. (1980). A note on the growth performance and carcass composition of Texel- and Suffolk-sired lambs in an intensive feeding system. *Animal Production* **30**, 311–314

LOHSE, C.L., MOSS, F.P. and BUTTERFIELD, R.M. (1971). Growth patterns of muscles of Merino sheep from birth to 517 days. *Animal Production* **13**, 117–126

McCLELLAND, T.H. (1975). *The Texel Breed of Sheep*, Technical Note 8, East of Scotland Agricultural College, Edinburgh

McCLELLAND, T.H. and RUSSEL, A.J.F. (1972). The distribution of body fat in Scottish Blackface and Finnish Landrace lambs. *Animal Production* **15**, 301–306

McCLELLAND, T.H., BONAITI, B. and TAYLOR, St. C.S. (1976). Breed differences in body composition of equally mature sheep. *Animal Production* **23**, 281–294

MARTIN, T.G., SALES, D.I., SMITH, C. and NICHOLSON, D. (1980). Phenotypic and genetic parameters for lamb weights in a synthetic line of sheep. *Animal Production* **30**, 261–269

MILLER, R.H. and PEARSON, R.E. (1979). Economic aspects of selection. *Animal Breeding Abstracts* **47**, 281–290

MORE O'FERRALL, G.J. and TIMON, V.M. (1977a). A comparison of eight sire breeds for lamb production. 1. Lamb growth and carcass measurements. *Irish Journal of Agricultural Research* **16**, 267–275

MORE O'FERRALL, G.J. and TIMON, V.M. (1977b). A comparison of eight sire breeds for lamb production. 2. Lamb carcass composition. *Irish Journal of Agricultural Research* **16**, 277–284

MORRIS, C.A. (1980a). Some benefits and costs of genetic improvement in New Zealand's sheep and beef cattle industry. 1. The annual selection response within closed flocks or herds. *New Zealand Journal of Experimental Agriculture* **8**, 331–340

MORRIS, C.A. (1980b). Some benefits and costs of genetic improvement in New Zealand's sheep and beef cattle industry. 2. Discounted costs and returns on a farm basis following selection. *New Zealand Journal of Experimental Agriculture* **8**, 341–345

NITTER, G. (1978). Breed utilisation for meat production in sheep. *Animal Breeding Abstracts* **46**, 131–143

OLSON, L.W., DICKERSON, G.E., CROUSE, J.D. and GLIMP, H.A. (1976). Selection criteria for intensive market lamb production: carcass and growth traits. *Journal of Animal Science* **43**, 90–101

OSIKOWSKI, M. and BORYS, B. (1976). Effect on production and carcass quality characteristics of wether lambs of crossing Blackheaded Mutton, Ile de France and Texel rams with Polish Merino ewes. *Livestock Production Science* **3**, 343–349

OWEN, J.B. (1971). *Performance Recording in Sheep*. Farnham Royal, Buckinghamshire, Commonwealth Agricultural Bureaux

OWEN, J.B., BROOK, L.E., READ, J.L., STEANE, D.E. and HILL, W.G. (1978). An evaluation of performance-testing of rams using artificial rearing. *Animal Production* **27**, 247–259

PATTIE, W.A. and WILLIAMS, A.J. (1966). Growth and efficiency of post-weaning gain in lambs from Merino flocks selected for high and low weaning weight. *Proceedings of the Australian Society of Animal Production* **6**, 305–309

PRUD'HON, M. (1976). La croissance globale de l'agneau: ses characteristiques et ses lois. In *2èmes Journées de la Recherche Ovine et Caprine*, pp. 6–26. Paris, INRA and ITOVIC

SEEBECK, R.M. (1966). Composition of dressed carcasses of lambs. *Proceedings of the Australian Society of Animal Production* **6**, 291–297

SEEBECK, R.M. (1968a). Developmental studies of body composition. *Animal Breeding Abstracts* **36**, 167–181

SEEBECK, R.M. (1968b). A dissection study of the distribution of tissues in lamb carcasses. *Proceedings of the Australian Society of Animal Production* **7**, 297–302

SMITH, R.H., KEMP, J.D., MOODY, W.G. and CUNDIFF, L.V. (1968). Heritability estimates of some lamb carcass traits. *Progress Report of the Kenyan Agricultural Experimental Station*, No. 176, 24–25

TAYLOR, St. C.S., MASON, M.A. and McCLELLAND, T.H. (1980). Breed and sex differences in muscle distribution in equally mature sheep. *Animal Production* **30**, 125–133

TEMPEST, W.M. and BOAZ, T.G. (1977). The influence of the Tasmanian Fine Woolled Merino on carcass characteristics of lambs. *Livestock Production Science* **4**, 191–202

THOMPSON, J.M., ATKINS, K.D. and GILMOUR, A.R. (1979a). Carcass characteristics of heavy weight crossbred lambs. 2. Carcass composition and partitioning of fat. *Australian Journal of Agricultural Research* **30**, 1207–1214

THOMPSON, J.M., ATKINS, K.D. and GILMOUR, A.R. (1979b). Carcass characteristics of heavyweight crossbred lambs. III. Distribution of subcutaneous fat, intermuscular fat, muscle and bone in the carcass. *Australian Journal of Agricultural Research* **30**, 1215–1221

TIMON, V.M. (1968). Genetic studies of growth and carcass composition in sheep. In *Growth and Development of Mammals* (G.E. Lodge and G.E. Lamming, Eds.), pp. 400–415. London, Butterworths

TULLOH, N.M. (1964). The carcass compositions of sheep, cattle and pigs as functions of body weight. In *CSIRO Symposium: Carcass Composition and Appraisal of Meat Animals*, Paper 5 (D.E. Tribe, Ed.), Melbourne, CSIRO

VOGT, D.W., CARTER, R.C. and McCLURE, W.H. (1967). Genetic and phenotypic parameter estimates involving economically important traits in sheep. *Journal of Animal Science* **26**, 1232–1238

WATSON, J.H. and BROADBENT, J.S. (1968). Inherited variation in carcass composition of Suffolk × Welsh lambs. *Animal Production* **10**, 257–264

WILSON, A. (1975). Carcass studies in crossbred lambs. PhD Thesis. University of Newcastle upon Tyne

WOLF, B.T. (1982). An analysis of the variation in the lean tissue distribution of sheep. *Animal Production* **34**, 1–8

WOLF, B.T., SMITH, C. and SALES, D.I. (1980). Growth and carcass composition in the crossbred progeny of six terminal sire breeds of sheep. *Animal Production* **31**, 307–313

WOLF, B.T., SMITH, C., KING, J.W.B. and NICHOLSON, D. (1981). Genetic parameters of growth and carcass composition in crossbred lambs. *Animal Production* **32**, 1–7

WOOD, J.D., MACFIE, H.J.H., POMEROY, R.W. and TWINN, D.J. (1980). Carcass composition in four sheep breeds: the importance of type of breed and stage of maturity. *Animal Production* **30**, 135–152

27

GENETIC IMPROVEMENT OF REPRODUCTIVE PERFORMANCE

R.B. LAND, K.D. ATKINS† and R.C. ROBERTS
ARC Animal Breeding Research Organisation, Edinburgh, UK

Genetic improvement offers the opportunity to increase output from existing resources. The basis of such improvement is the exploitation of genetic variation, by increasing the frequency of favourable alleles. For reproduction in sheep, the variation among breeds has long been recognized and utilized in programmes of crossbreeding. Variation within populations has only recently been incorporated into breeding programmes. The first part of this chapter will therefore review the evidence for the presence of genetic variation for the reproductive performance of sheep, the extent of that variation and the opportunities for genetic change.

The implementation of genetic methods of improvement is governed by the contribution of reproduction to overall merit and the speed of the improvement which can be offered. While the presence of genetic variation is a prerequisite of genetic change and while the extent of that variation influences the change which might be achieved, the rate of improvement is influenced by the recognition of relevant variation. The more accurate the recognition of genetic merit, the faster the rate of improvement and hence the greater the relative role of genetics in any programme of improvement. The second part of this chapter will therefore consider the possibilities and opportunities to recognize genetic merit for reproductive performance more accurately.

Knowledge itself, however, is only the first step to improvement; equally, present knowledge is the only basis of future research. In the final section of the chapter consideration will be given to the implementation of improvement programmes and the prospects for future research, including the benefits of international collaboration. In addition, most attention will be paid to the improvement of female traits.

Genetic variation in aspects of reproduction

There are two main methods available for exploiting genetic variation. The first is by crossing different genotypes, normally breeds, utilizing additive

†Usual address: N.S.W. Department of Agriculture, Agricultural Research Station, Trangie, N.S.W., Australia.

effects and any heterosis that may be available. The second is based on genetic variation within breeds, leading to selection responses. Both are related to systems of sheep husbandry. For lamb production, crossing features prominently in most commercial systems. It is necessary therefore to first review the effects of crossing on fertility, and then to examine the scope for improvement within breeds.

DIFFERENCES BETWEEN BREEDS

Differences in fertility between breeds are widely accepted, but cannot be quantified unless the breeds have been kept under the same conditions, and for a length of time sufficient to preclude carry-over effects from previous treatments. This still leaves a plethora of documented evidence revealing large breed differences. Timon (1974) claimed that breed differences in reproductive performance exceeded those among other production traits. It is well known that breeds like the Finnish Landrace and Romanov produce a larger number of offspring than, for example, the Merino or the British hill breeds. The reader is referred to reviews by Rae (1956), Bowman (1966), Bradford (1972) and Timon (1974) for access to the extensive literature on breed differences in fertility.

Breed differences are evident for several aspects of reproductive performance. The Dorset Horn breed, for example, is valued for its ability to secure out-of-season lambing, as it does not share the dependency of other breeds on reducing daylength to ovulate. This has long been known, but an understanding of genetic effects on physiological mechanisms governing reproductive perfomance is of more recent origin. Much of this work stems from the interest in the reproductive capacity of the Finnish Landrace and the Romanov. Briefly, genetic variance has been found in various aspects of ovarian activity, and its hormonal control, and these aspects will be discussed in more detail in a later section.

CROSSES BETWEEN BREEDS

As mentioned earlier, breeds are frequently crossed for reasons of improving general performance, and particularly to increase production of hill and upland sheep when moved to lowland pastures. Crossbreeding exploits both the additive and non-additive genetic components, the additive through the complementarity of traits effecting economic merit, and the non-additive through expressions of heterosis. Examples of complementarity from the use of the highly fertile Finn and Romanov breeds are numerous (Donald, Read and Russell, 1968; Meyer *et al.*, 1977). The more traditional use of the Longwool breeds exemplifies the heterotic aspects of crossbreeding.

Crossbreeding can affect fertility through different routes: crossbreeding in the dam, in the offspring and in the sire. There is only limited evidence that crossbred rams lead to higher fertility than purebred rams when mated to the same type of ewe, though individual ram effects can be large. Nitter (1978) reviewed the evidence, showing mostly small ram effects on number reared, and never more than 10%.

The main findings of crossbreeding in the dam and the offspring have been discussed comprehensively by Nitter (1978), and are summarized in *Table 27.1*. It should be emphasized that these data come from separate studies, and should therefore be interpreted with caution. But taking them at face value, it appears that dam effects are the main cause of heterosis in number of ewes lambing, while offspring effects are largely responsible for heterosis in lamb survival to weaning. There is no direct evidence on the separate effects of dam and offspring on any heterosis in embryo survival. Nitter (1978) concluded that in general, heterotic effects are greater, though more variable, for reproductive traits than they are for growth. The greater amount perhaps accords with general expectation, but the variability renders prediction of heterosis in any particular cross uncertain.

Table 27.1 HETEROSIS IN THE COMPONENTS OF REPRODUCTIVE PERFORMANCE—UNWEIGHTED MEANS FROM STUDIES REVIEWED BY NITTER (1978)

Trait	*Effect of heterosis in:*	
	Lamb[a]	*Ewe*[b]
Number of ewes lambing	+2.6%	+8.7%
Number of lambs born	+2.8%	+3.2%
Lamb survival to weaning	+9.8%	+2.7%
Lambs weaned/ewe mated	+15.2%	+14.7%

[a] Heterosis in the lamb estimated as the difference between reciprocal F_1 matings and purebred matings of both parental breeds.
[b] Heterosis in the ewe estimated as the difference between F_1 crossbred ewes and purebred ewes mated to rams of a third breed.

Age effects on heterosis in fertility have also been reported, being greatest in younger ewes (Nitter, 1978). Heterosis for sexual maturity has been observed (Land, Russell and Donald, 1974), as also has age at puberty (Dickerson and Laster, 1975).

Heterotic effects on the number born are rather small (*Table 27.1*), with the exception of two studies involving Border Leicester crosses in Australia and New Zealand (McGuirk, 1967; Hight and Jury, 1970). One important determinant of number born—ovulation rate—shows no heterosis at all, the unweighted mean estimate from a number of studies not differing from zero (Nitter, 1978).

GENETIC VARIATION WITHIN BREEDS

Perhaps the most striking example of variation in fertility within a breed is what has become known as the 'Booroola gene'. Its effect was first reported by Turner (1968), among a strain of Merinos from New South Wales. It became a matter of conjecture whether it was a single gene or not, and the issue was not resolved until Piper and Bindon (1982) applied their own diagnostic methods, and found the acceptable segregation demanded of the hypothesis. They classified the ewes according to whether they had produced at least one litter of triplets (or more) over their lifetime production. The gene behaves as a single dominant gene with respect to litter size, which explains its perpetuation in the flock of origin despite the continual purchase of rams from other flocks. The effect of the 'Booroola gene' was to increase litter size from 1.37 to 2.42 in the data of Piper and Bindon (1982). There are some indications (Piper and Bindon, 1982) that the effects of the gene are additive for ovulation rate, but that the increased number of ova in the homozygote is not matched by an increase in the number of lambs born, when compared to the heterozygote.

Table 27.2 HERITABILITY AND REPEATABILITY ESTIMATES FOR COMPONENTS OF REPRODUCTIVE PERFORMANCE

Trait	*Herita-bility*	*Repeata-bility*	*Breed*	*Reference*[a]
Ewes lambing	0	0.09	UK hill breeds	1
	0.07	0.05	Clun Forest	2
	0.07	0	Finn × Dorset	3
	0.12	0.13	Australian Merino	4
	0	0.05	Australian Merino	5
	0.17	0.04	Rambouillet	6
Lambs born/ewe lambing	0.15	0.26	UK hill breeds	1
	0.12	0.14	Clun Forest	2
	0.12		Australian Merino	4
	0.08	0.15	Australian Merino	7
	0.10	0.19	Australian Merino	8
	0.12	0.14	Rambouillet	6
	0.04	0.18	Various British breeds	9
	0.12	0.10	Swedish and British breeds	10
	0.08	0.06	Rahmari	11

[a]References: 1. Purser (1965); 2. Forrest and Bichard (1974); 3. Martin *et al.* (1981); 4. McGuirk (1973); 5. Turner (1966); 6. Shelton and Menzies (1970); 7. Mann, Taplin and Brady (1978); 8. Hanrahan, J.P., personal communication; 9. Mechling and Carter (1969); 10. Rendel (1956); 11. Karam (1957)

There is also overwhelming evidence of general genetic variation within breeds, and the literature is summarized in *Table 27.2*. The summary covers ewe fertility (percentage lambing) and number of lambs born/ewe, and all the values are based on single records.

Estimates of heritability and repeatability of the percentage of ewes lambing are somewhat variable, but average 0.07 and 0.06 respectively. Thus, the genetic component is not a strong one, and progress under selection would be slow. In addition, there is an upper limit of 100% for this trait, and because of its binomial nature, the approach to that maximum would become even slower.

The corresponding estimates for number of lambs born (litter size) are somewhat higher though still low, being 0.10 and 0.15 respectively. These values are the unweighted means from all the studies cited, which in turn are remarkably consistent, despite being conducted on many breeds in very diverse environments and this testifies to their likely accuracy. The comparison of heritability and repeatability is of some interest, since the latter is the upper limit of the former. In Falconer's (1981) terminology, the repeatability (r) is:

$$r = \frac{V_G + V_{Eg}}{V_P}$$

which may be rewritten as:

$$r = \frac{V_A}{V_P} + \frac{V_{NA} + V_{Eg}}{V_P}$$

where V represents variance, and the subscripts, P, A, NA and Eg are phenotypic, additive genetic, non-additive genetic and general environmental, respectively. The ratio V_A/V_P is of course the heritability, and by substituting the values quoted above, the V_{NA} and V_{Eg} components together are only half as great as the additive component. This does not allow for the fact that some of the V_{Eg} component may be a genetic maternal effect, which reduces further the relative importance of V_{NA}. This explains why, in the previous section, the heterotic effect on numbers born was rather small. The inescapable conclusion from this is that selection offers a more powerful method of improving numbers born than by the exploitation of heterosis from crossbreeding. The fact remains that the heritability of 10% is low, and ways of increasing this value will be discussed later in this chapter (p.523).

A more immediate way of increasing the heritability is to base it on more than one record, which are then combined. For this procedure to be valid, it carries the implicit assumption that the genetic correlation between successive measurements is always unity, which may well not be true. On the further assumption that the absolute value of the additive genetic component remains unchanged over successive parities, which is also arguable, we can calculate the heritability of multiple records (h_n^2):

$$h_n^2 = h^2 \frac{n}{1 + (n - 1)r}$$

where h^2 is the heritability, n is the number of records and r the repeatability. Using this formula, the calculated heritability for one to five records is 0.10, 0.17, 0.23, 0.28 and 0.31, respectively. Even if the assumptions noted above do not fully hold, it still allows for worthwhile increases in the heritability by using successive records. The effect of this on genetic gain becomes somewhat complicated, since it affects the generation interval in the opposite direction. Under many systems (e.g. hill sheep) the generation interval matches fairly accurately the length of time required to obtain an adequate number of female replacements, with considerable overlapping of generations. The implication is still that as a predictor of genetic merit, multiple records would offer faster gains than selection based on single records, but bearing in mind that generations should be turned over as quickly as is practically feasible, there will be an optimum beyond which the extra accuracy would be outweighed by the loss of time incurred in collecting the additional information.

To pursue these considerations further, there is a need for more experimental information on the following points: (i) What are the empirical values of the heritabilities of multiple records? (ii) How does the additive genetic component change, in absolute terms, over successive parities? (iii) What is the value of the genetic correlation between successive records? Only when these questions have been adequately answered can the optimal breeding system for increasing the number of lambs be properly designed. With these qualifications, the recommendation to use multiple records in a selection programme seems to be valid. Eikje (1975) addressed the same issue and came to a similar conclusion.

No discussion of increasing lambing rates would be complete without the mention of lamb survival after birth. The estimates of heritability, considered as a trait of the lamb, vary from zero to 0.16 (Piper and Bindon, 1977; Shelton and Menzies, 1970). There are other reports quoting values somewhere in between. Such variation among estimates should be accepted since, biometrical considerations apart, they have been collected under a range of environments giving wide differences in mortality. Nevertheless, the possibilities of genetic differences in survival under given conditions cannot be ignored. J.P. Hanrahan (personal communication) states that where total number of lambs weaned is the objective, it is more efficient to select for this trait than for the correlated trait of numbers born.

REALIZED RESPONSES TO SELECTION

There are five selection experiments reported in the literature. All have been for increased number of lambs born, and the results are summarized in *Table 27.3*. As can be seen from the table, the selection criteria have varied somewhat between experiments. Four of the studies have, to varying degrees, used multiple records. Mann, Taplin and Brady (1978) selected rams only, and on a system involving only minimal recording. The Turner (1978) study is also different because it began after screening a large population for prolific ewes, which procedure itself improved performance by 0.1 of a lamb. In this context the study of Owen (1976) should also be noted, though the selection aspects have not yet been fully documented. He also adopted a screening procedure, among several breeds, to establish the prolific Cambridge sheep. Similarly, Hanrahan (1976) screened a population of Galway ewes for high lambing rates, and the top 3.4% of the screened ewes yielded an increase of 0.1 in number of lambs born by their progeny.

Except for the report of Mann, Taplin and Brady (1978) all studies show a remarkably uniform rate of response, at about 1.5% per annum, which furthermore accord well with the rates predicted from the genetic parameters of the base populations. The responses reported are all linear, and none showed any sign of falling off. Thus, although the total responses in *Table 27.3* seem small in absolute terms, over 20 years or so, as in the study of Clarke (1972), they can be substantial. The responses in percentage terms are also similar to those in other production traits in sheep, like fleece weights and body weights. This means that the low heritability is counterbalanced by the relatively large variance of number born. Forrest and Bichard (1974) reported a coefficient of variation of 30% for number born alive, as against the more usual value of 12% for growth rate. In summary, selection responses are clearly an efficacious way of improving the number of lambs, and it compares very favourably with any benefits in this respect stemming from crossbreeding, despite the popular assumption that fertility traits profit from heterosis.

Table 27.3 REALIZED GENETIC RESPONSES IN NUMBER OF LAMBS BORN/EWE LAMBING

Breed	*Selection criteria*	*Length of experiment* (years)	*Total response*	*Annual rate of response* (%)	*Predicted response*	*Reference*
Romney	Rams, Ewes: pooled dam records	22	0.40	0.018(1.5%)	0.021	Clarke (1972)
Merino	Rams, Ewes: pooled dam records	14	0.38	0.02[a]	—	Turner (1978)
Galway	Rams, Ewes: pooled dam records	10	0.22	0.022(1.5%)	—	Hanrahan and Timon (1978)
Merino	Rams: pooled dam records Ewes: single dam record	10	0.18	0.018(1.3%)	0.015	Atkins (1980)
Merino	Rams: single dam record Ewes: no selection	10	0.02	0.002	0.004	Mann, Taplin and Brady (1978)

[a] Excluding initial effect of screening

As in all traits in all species, selection for fertility in sheep can lead to correlated responses in other traits. Estimates of genetic correlations by their nature are erratic, and the published estimates reflect this, as shown in *Table 27.4*. The mean unweighted estimate for adult body weight is +0.25, showing expected increases in body size following selection for more lambs. Clarke (1972) reported a small correlated response in this direction. In the study of Hanrahan (1979) the correlated response was even smaller, and not significant, though still positive. Weaning weights show no detectable genetic correlation with number born, though there

Table 27.4 GENETIC CORRELATIONS BETWEEN REPRODUCTION RATE AND SOME PRODUCTION CHARACTERS

Genetic correlations between reproduction and:			*Reference*
Body weight	*Weaning weight*	*Fleece weight*	
0.29		0.27	Young, Turner and Dolling (1963)
0.61		–0.02	Purser (1965)
	–0.12	–0.25	Gjedrem (1966)
0.20		–0.52	Kennedy (1967)
	–0.03	–0.14	Shelton and Menzies (1968)
0.30	0.24		Ch'ang and Rae (1972)
0.21			Forrest and Bichard (1974)
	0.06	0.05	Eikje (1975)
–0.07		–0.15	More O'Ferrall (1976)

will be of course the obvious negative environmental correlation. The genetic correlation with fleece weight yields mostly negative estimates, but on average are very small, their unweighted mean being –0.1. The correlated response in hogget fleece weight by Clarke (1972) fully substantiates this, both in direction and magnitude. To conclude, selection for number of lambs, in terms of both expectation and practice, has only minor genetic consequences for other production traits. However, there are environmental and management consequences which cannot be ignored, as anyone familiar with the practicalities of rearing twins, especially in adverse environments, will readily appreciate.

GENETIC VARIATION IN OVULATION RATE

It has been argued by Hanrahan (1980) that since there are no demonstrable differences between ewes in their uterine capacity to support a given number of fertilized eggs, it follows that the genetic correlation between ovulation rate and litter size is unity. From these premises, he argues further that litter size is only an index of ovulation rate but an inaccurate one because of random loss. The accuracy declines as the number of ova shed increases, because of the consequent increased loss. Hanrahan concluded that by selecting on ovulation rate rather than on litter size, the annual genetic gain could be as much as three times greater.

The basis of this conclusion is that the heritability of ovulation rate is much higher than that of litter size, both in the Finn (0.45 vs. 0.10) and Galway (0.57 vs. 0.06) breeds (Hanrahan, 1980). However, the only other study quoting the heritability of ovulation rate, that of Piper *et al.* (1980), reported a very much lower value of 0.05 at 18 months of age, this time in Merinos. The authors considered then at that age there would be little advantage is using ovulation rate instead of litter size as the selection criterion. Different breeds from different environments might give different answers. Without more data, therefore, the issue cannot be pursued further.

The recognition of genetic merit

As already indicated genetic variation in the reproductive performance of sheep occurs both among and within populations, and selection may lead to improvement of reproductive performance. The value of genetic methods to the improvement of sheep reproduction is, however, dependent upon the rate of that improvement. This in turn is profoundly affected by our ability to measure the trait. Basic genetic theory shows that with given genetic variation, the rate of response to selection is influenced by the heritability, the selection differential and the generation interval. Of these, both the heritability and the selection differential are partly a reflection of the accuracy with which reproductive performance may be measured, while improved reproductive rates may shorten the generation interval.

The suggestion that reproductive performance may be difficult to measure may seem incompatible with the apparent precision of recording reproductive performance. Despite practical difficulties of recording under some conditions, it is very evident whether a ewe lambs or not, and how many lambs she produces. Fertility and prolificacy may be recorded precisely, but this does not mean that reproductive merit can be measured precisely. Although a ewe may be fertile, or produce twins on one occasion, this record of performance is a poor indicator of subsequent performance. Because the repeatability of reproductive traits is low, one record of performance is a poor estimate of phenotypic worth, and an even poorer estimate of genetic worth. With a low repeatability, the heritability of reproductive traits is inevitably lower, as shown in the preceding section. This, however, does not mean that genetic variation in reproduction is not important, but that it is masked by environmental variation, which by appropriate methodologies it may be possible to circumvent.

For females it is very dificult to recognize genetically superior individuals from their phenotype. For males, there is no expression of litter size or conception rate, and hence no opportunity for direct selection, with the result that the selection differential must be exercised on one sex only. The difficulty of recognizing genetic merit has a serious adverse effect on the rate of response to genetic selection. Might reproduction be measured more accurately?

The possibility that reproductive merit may be measured more accurately is based partly on the presumption that performance is a function of the value of some underlying continuous variable, and of thresholds for the discrete characteristics of reproductive traits, rather like the model of Falconer (1965) of underlying liability to disease and a threshold beyond which the disease is expressed. On this model, one may ask whether an underlying variable may exist in some tangible form, and whether reproductive merit may be recognized more effectively from knowledge of the physiology of reproduction. The general principles that both sexes carry virtually the same genes, that the same hormones control reproduction in both sexes has been argued before (e.g. Land, 1973). The present question is the likelihood that a useful underlying variable might exist. To be useful, an indirect measure of reproductive merit has to be easy to measure, and associated sufficiently with performance itself.

THE VALUE OF INDIRECT CRITERIA

The heritability of a trait indicates the extent to which the phenotype of an individual indicates genetic merit for that trait. With the use of indirect traits, the heritability as a single concept is replaced by a combination of the square roots of the heritabilities of both the desired and indirect traits, and of the genetic correlation between the two. Instead of h^2, one must use $(h_D r_g h_I)$, where r_g is the genetic correlation between the desired and indicator traits, and h_D and h_I the square roots of their respective heritabilities. The higher r_g and h_I, the more valuable the indicator trait. To be better than the desired trait itself, $(r_g\ h_I)$ must be higher than h_D, unless the indicator trait also allows the selection intensity to be increased. It is possible in principle that even if $(r_g\ h_I)$ is less than h_D, indicator traits may still make a useful contribution to progress through their effect on the selection intensity.

To quantify the advantages of indicator traits which can be measured in young animals of both sexes, Walkley and Smith (1980) calculated the expected rates of genetic change for a range of heritabilities and genetic correlations and some of these are given in *Table 27.5*. Using a male trait alone with a heritability of 0.4 and a genetic correlation with the desired trait of 0.4, the rate of improvement in female performance would be similar to that when selecting for female performance itself. The combination of such a trait with the female trait would increase the response by 30%. The higher r_g and h_I, the greater the benefits, but vice versa with respect to h_D.

Is it likely that a readily measurable underlying variable might be sufficiently associated with any of the important components of reproduction to secure worthwhile gains?

Table 27.5 THE RELATIVE (%) RATE OF CHANGE IN PROLIFICACY FOLLOWING SELECTION ONLY ON A TRAIT MEASURED IN MALES BEFORE MATING OR ON SUCH MALES AND PROLIFICACY ITSELF WHEN THE HERITABILITY OF THE MALE TRAIT IS 0.1 OR 0.4, THE GENETIC CORRELATION BETWEEN IT AND PROLIFICACY 0.3 OR 0.7 AND THE HERITABILITY OF PROLIFICACY 0.1

Heritability of male trait	0.1		0.4	
Genetic correlation with prolificacy	0.3	0.7	0.3	0.7
Male trait alone	57	110	80	170
Male trait and prolificacy	115	145	125	195

From Walkley and Smith (1980)

PHYSIOLOGICAL CRITERIA OF REPRODUCTIVE MERIT

The possibility that a physiological criterion might be related to reproductive performance has been argued recently (Land, 1981; Land *et al.*, 1982a); it is based on knowledge of the physiology of reproduction and the results of a very limited number of experiments specifically designed to investigate this possibility.

In brief, the number of eggs shed is taken as the limit to reproduction (Hanrahan and Quirke, 1982). The question then becomes the endocrine control of the ovary, or more detail on which hormones control the ovary. What evidence is there for genetic variation in the activity of those hormones, and what are the actual correlations between the hormonal traits and reproduction, i.e. do the hormones considered operate quantitatively?

Follicle stimulating hormone

All the evidence for variation in the activity of FSH is based on the comparison of different breeds or strains within breeds.

Female, but not male, lambs of the highly prolific Booroola strain of Merino sheep have higher concentrations of FSH than control lambs (Findlay and Bindon, 1976). There is therefore genetic variation associated with the Booroola gene. Bindon and Piper (1976) concluded that despite its high repeatability of 0.63, at least in adult ewes (Land *et al.*, 1980), the limitation of the expression of that variation of FSH to females restricts its utility for selection. More extensive studies are required in the young lamb and its relationship to adult levels.

In the adult female, Cahill *et al.* (1981) indicated two very interesting differences between high ovulation rate Romanov and low ovulation rate Ile-de-France ewes. The second FSH peak in the Romanov was twice as great as that of the Ile-de-France, and the drop in FSH from the luteal to the preovulatory phase of the oestrous cycle was much less in Romanov than Ile-de-France ewes. The authors drew attention to a potential role of the second FSH peak in the recruitment of follicles, but since ovulation rate of the remaining ovary doubles after hemicastration late in the oestrous cycle, it seems unlikely that this peak is closely related to ovulation rate. The maintenance of high FSH levels virtually until ovulation is possibly of greater significance. Feedback hormones from the preovulatory follicle are presumed to have a much smaller effect on the release of FSH in prolific than in the non-prolific ewes. This could be the basis of the maintenance of a large number of follicles through ovulation.

Variation in the concentration of FSH in the plasma of sheep is therefore partly of genetic origin, and is in turn associated with variation in ovarian activity. The general principle that variation in FSH is under genetic control is supported by the study of correlated changes following selection of mice for characteristics of growth and reproduction (Bradford, Berkley and Spearow, 1980), and extended by that study to the demonstration of variation within, as well as among, populations. The extent of such variation within populations of sheep has not been evaluated nor has the covariation with ovulation rate.

Luteinizing hormone

The concentration of this hormone in peripheral plasma has been studied much more extensively than that of FSH and some of the earlier work has been reviewed by Land and Carr (1979). It was concluded that there is no evidence to suggest that the concentration of LH during the oestrous cycle is higher in prolific than in non-prolific ewes, but that the interval from oestrus to the LH peak is greater. This is supported by the later studies of Bindon *et al.* (1979) and Cahill *et al.* (1981).

Because of the low repeatability of LH, Bindon and Piper (1976) concluded that even if the concentrations of LH in the plasma of male lambs were related to the ovulation rate of their female relatives, it would be difficult to alter the trait by selection. One alternative is to use the LH response to LH releasing hormone. Land and Carr (1979) have shown this characteristic to vary extensively both between and within populations. Their results from comparisons of lambs of the prolific Finn breed, its crosses and lambs of low prolificacy breeds, indicated a positive correlation with the LH response to LH releasing hormone, but the study of Hanrahan, Quirke and Gosling (1981) does not support this. For practical utility, one needs to know the extent of covariation between LH releasing hormone and ovulation within a population and, as with FSH, this has not yet been evaluated, but preliminary results of the study of Land and Carr (1979) show the trait to be heritable ($h^2 = 0.53$).

Prolactin

Many roles have been suggested for prolactin. In the control of the ovary, it is postulated to be part of the luteotrophic stimulus but much of the recent interest in prolactin arises from the quite remarkable degree to which it varies with daylength in both males (Ravault, 1976) and females (Walton *et al.*, 1977). With Ravault (1976) showing that this variation is expressed in young as well as in mature males, the obvious implication was to use the concentration of prolactin as a criterion of selection for seasonality.

This prediction has not withstood the study of covariation with seasonality. Perturbation of seasonality by artificial changes of daylength dissociated ovarian cyclicity from prolactin levels in females (Thimonier, Ravault and Ortavant, 1978; Walton *et al.*, 1980). Conversely, pharmacological reduction of the concentration of prolactin did not lead to a resumption of cyclicity (Land *et al.*, 1980). Of particular relevance, the Dorset Horn breed with its early end to anoestrus did not show a correspondingly earlier decline in the concentration of prolactin (G.M. Webster and W. Haresign, 1981, personal communication).

Equally, in males, artificial light patterns dissociated cyclical variation in prolactin and testicular activity (Howles, Webster and Haynes, 1980), and again genetic variation in reproductive activity among breeds was independent of variation in prolactin concentration (Carr and Land, 1982). Of the three breeds studied, only the Scottish Blackface showed synchrony between the change in prolactin concentration and the expected time of onset and end of female reproductive activity. Prolactin rose at the same time in Finnish Landrace males, despite the later onset of oestrous cycles in that breed; prolactin fell at the same time in Tasmanian Merino as in the Blackface males despite the earlier end of anoestrus in that breed. The evidence to date suggests little genetic variation among breeds in the prolactic response to photoperiodicity, despite the abundant genetic variation in the reproductive response. Genetic covariation has not been studied within breeds, which might yet prove useful, though the expectation that it might is now low.

The marked seasonal variation in the concentration of prolactin, and its possible covariation with seasonality of reproduction may have distracted attention from other associations of possible importance. Carr and Land (1982) found the concentration to be much higher during the summer in Finnish Landrace than Tasmanian Merino males. If prolactin blocks the induction of aromatase by FSH in the sheep as it does in the rat (Dorrington and Gore-Langton, 1981), high prolactin could lower oestrogen production, and hence an allele for high prolactin could be an allele for high ovulation rate, through a reduction in oestrogen feedback on the release of FSH and LH.

Prolactin has yet to be shown to vary with ovulation rate, but new knowledge may show it to be important. Depression of prolactin with bromocriptine during the oestrous cycle increased the ovulation rate of Blackface ewes, over two consecutive oestrous cycles, from 1.0 to 1.5 whereas it decreased the ovulation rate of Finn ewes from 3.8 to 3.2 (R.B. Land and W.R. Carr, unpublished). The high variability of ovulation rate coupled with small sample sizes indicate caution, but even so, the results suggest the possibility that the high ovulation rate of the Finn may be dependent upon high prolactin and should be pursued. This possibility is supported by the approximately two-fold greater concentration of prolactin during the two days before the preovulatory discharge of LH in Romanov compared with Ile-de-France ewes (Cahill *et al.*, 1981). The story of prolactin is by no means finished.

Ovarian activity

The interrelations between prolactin and the ovary introduce the possibility that genetic variation in ovarian activity may reside in the intrinsic characteristics of the ovary itself. Certainly there are vast differences between breeds and strains in the number of follicles of all sizes, and there is a tendency for the numbers in various development classes to be associated with the numbers which finally develop and ovulate. Sheep of the prolific Finn and Romanov breeds have less primordial follicles but more growing follicles (Cahill, Mariana and Mauléon, 1979). By contrast the prolific Booroola strain of Merino sheep has less growing follicles (Bindon and Piper, 1982). Another problem is knowing whether variation follows from different hormone patterns or whether it itself is the genetic basis of some of the differences in the fecundity of sheep. This has not yet been resolved. There is extensive variation of oocyte number within breeds, and also overlap between breeds with very different ovulation rates. Romanov ewes in the study of Cahill, Mariana and Mauléon (1979) were found to have between 12 and 58 × 10^3 oocytes, while Ile-de-France ewes with less than half the ovulation rate had from 20 to 86 × 10^3. Together with the possibility of changing the number of eggs shed by endocrine intervention shortly before ovulation, this shows that the number of oocytes does not control the 'fine tuning' of the ovulation rate.

Knowledge of genetic variation in oocyte populations does, however, suggest that it may be possible to develop a physiological criterion for selection. The failure of ovarian anatomy to form a basis for the genetic variation in ovulation rate indicates the control is extra-ovarian, possibly through systemic endocrine levels. Knowledge that differences may be detected in the newborn lamb (Land, 1970; Trounson *et al.*, 1974) indicates that variation is common to a wide range of ages.

The intrinsic characteristics of the ovary might nevertheless have a role to play in the determination of the number of eggs shed. The concentration of progesterone in peripheral plasma during the luteal phase is at the most only 50% greater in the prolific Finn and Romanov ewes, despite an ovulation rate two to three times as high (Wheeler and Land, 1977; Bindon *et al.*, 1979; Quirke, Hanrahan and Gosling, 1979; Cahill *et al.*, 1981). The linear relationship between the number of corpora lutea and the concentration of progesterone in peripheral plasma is even weaker in a population where the Booroola gene is segregating; indeed Bindon *et al.* (1981) found the two to be independent. As with the relationship between ovarian anatomy and ovulation rate, this independence, and also an apparently similar total feedback on the release of gonadotrophins, could be a consequence rather than a cause of the differences in ovulation rate. However, the lack of generality in the system is illustrated by the finding that there is a direct relationship between the number of corpora lutea and the concentration of progesterone in D'man sheep (M. Marie and A. Lahlou-Kassi, personal communication).

The possibility that reduced production of feedback hormones might contribute to prolificacy is supported by the recent finding of Cummins *et al.* (1982) that Booroola Merino follicles produce less inhibin. These authors also suggest that Booroola Merinos may be more sensitive to the inhibin but their results indicate that this might not be so in the natural range of FSH values of the intact ewe.

Differences in prolificacy might then arise from differences in the biochemical basis of ovarian activity, but this has yet to be demonstrated. Only if it were, could its use as a selection criterion be assessed.

Negative feedback

Reduced production of feedback hormones is one way in which a high ovulation rate may be achieved. Equally, increased degradation or perception of these hormones may lead to an increase in the number of eggs shed. Finnish Landrace ewes, for example, are less sensitive to the negative feedback effects of oestrogen than are Scottish Blackface (Land, 1976), but such a characteristic is likely to be based on variation in neural biochemistry, inaccessible and hence inappropriate for selection.

Testis size

A particularly simple potential criterion in the male of ovarian activity in related females is the size of the testis. In mice, a positive genetic correlation is clear (Eisen and Johnson, 1981); in sheep most breed and cross comparisons indicate the presence of a similar correlation, but not all (Land *et al.*, 1982a). The covariation within a population has been studied by comparison of male and female Romanov sibs by Ricordeau *et al.* (1979) who found the genetic correlation to be 0.43. Selection for testis size (Land *et al.*, 1982a) was complicated by the decision to correct for body weight; the ovulation rate and number of lambs born/unit of body weight changed in the direction of selection, both principally through changes in body weight. The relationships between body and gonadal growth, and genetic merit for reproduction, offer a rewarding area for study.

The heritability of testis size is of the order of 0.5 in both mice and sheep. The genetic correlation with ovulation rate in mice was 0.5. If it were even 0.3 in sheep, the incorporation of such a male character into a programme of genetic selection for prolificacy would, as shown earlier, increase the rate of genetic change to 170% of that possible with selection in the female alone, or alternatively allow the same rate to be maintained without any direct selection for prolificacy in the female.

Prospects

The study of genetic variation in reproductive performance has shown the opportunity for genetic improvement in reproduction. This opportunity is utilized extensively by choice among populations, but with a few conspicuous exceptions the opportunity to select within populations remains to be grasped in practice. One such exception is the Waihora group breeding scheme in New Zealand where prolificacy has increased by 28% in less than 10 years (Hight, 1977).

Research has shown that the rate of change within populations may be increased in the future by the development and use of criteria of reproductive performance which are neither sex- nor age-limited. There is the physiological framework for the development of such criteria but values of genetic parameters are lacking. Most studies have been based on the comparisons of breeds. These might give valuable clues but it is the genetic covariation within populations which determines the usefulness of possible criteria. Even then the prolific breeds principally studied—the Booroola, D'man, Finnish Landrace and Romanov—have yet to be compared in a single environment or against a common reference breed. There is an urgent need to do so.

Covariation within populations has tended to be measured in pedigree, unselected populations or by selection for a potentially appropriate trait. Both have the drawback that they are relevant only for the traits measured at the time. Any new or even modified physiological characteristic has to be assessed afresh. An alternative would be selection for a key component of reproduction, say ovulation rate, so that lines would be available for the assessment of covariation at any time. This possibility is developed further by Land *et al.* (1982a). It is important, however, to make sure that populations are of adequate size or that they are adequately replicated, to obtain estimates of the required accuracy.

Genetic studies of the physiological control of reproduction have also led to greater understanding of the importance of various physiological components in the determination of reproductive performance. Such knowledge is the basis of the development of procedures to manipulate reproduction artificially. Sensitivity of gonadotrophin release to the negative feedback effects of gonadal hormones was postulated to be an important component of the genetic superiority of Finnish Landrace sheep. The manipulation of feedback equilibria by immunization promises to be an appropriate practical route to greater fecundity. Trials in Australia show that active immunization might be appropriate; equally, a single injection of antiserum at the start of mating has been found to raise the prolificacy of Welsh Mountain sheep from 1.0 to 1.5, without regard to the stage of the oestrous cycle (Land *et al.*, 1982b).

Genetics has a considerable contribution to make to physiological understanding. Physiology may contribute to the identification of particular genes, their biochemical manipulation and the introduction of molecular biology to animal breeding in the future.

References

ATKINS, K.D. (1980). Selection for skin folds and fertility. *Proceedings of the Australian Society of Animal Production* **13**, 174–176

BINDON, B.M. and PIPER, L.R. (1976). Assessment of new and traditional techniques of selection for reproduction rate. In *Sheep Breeding* (G.J. Tomes, D.E. Robertson and R.J. Lightfoot, Eds.), pp. 357–371. London, Butterworths

BINDON, B.M. and PIPER, L.R. (1982). Physiological characteristics of high fecundity sheep and cattle. *Proceedings of the World Congress on Sheep and Beef Cattle Breeding, New Zealand* (R.A. Barton and W.C. Smith, Eds.) (in press)

BINDON, B.M., CUMMINS, L.J., PIPER, L.R. and O'SHEA, T. (1981). Relation between ovulation rate and plasma progesterone in Merinos with natural and induced high fecundity. *Proceedings of Australian Society of Reproductive Biology* **13**, 79

BINDON, B.M., BLANC, M.R., PELLETIER, J., TERQUI, M. and THIMONIER, J. (1979). Periovulatory gonadotrophin and ovarian steroid patterns in sheep of breeds with differing fecundity. *Journal of Reproduction and Fertility* **55**, 15–25

BOWMAN, J.C. (1966). Meat from sheep. *Animal Breeding Abstracts* **34**, 293–319

BRADFORD, G.E. (1972). Genetic control of litter size in sheep. *Journal of Reproduction and Fertility, Supplement* **15**, 23–41

BRADFORD, G.E., BERKLEY, M.S. and SPEAROW, J.L. (1980). Physiological selection for aspects of efficiency of reproduction. In *Selection Experiments in Laboratory and Farm Animals* (A. Robertson, Ed.), pp. 161–175. Slough, Commonwealth Agricultural Bureaux

CAHILL, L.P., MARIANA, J.C. and MAULÉON, P. (1979). Total follicular populations in ewes of high and low ovulation rates. *Journal of Reproduction and Fertility* **55**, 27–36

CAHILL, L.P., SAUMANDE, J., RAVAULT, J.P., BLANC, M.R., THIMONIER, J., MARIANA, J.C. and MAULÉON, P. (1981). Hormonal and follicular relationships in ewes of high and low ovulation rates. *Journal of Reproduction and Fertility* **62**, 141–150

CARR, W.R. and LAND, R.B. (1982). Seasonal variation in the plasma concentration of prolactin in castrate rams of breeds of sheep with different seasonality of reproduction. *Journal of Reproduction and Fertility* **66**, 231–235

CH'ANG, T.S. and RAE, A.L. (1972). The genetic basis of growth, reproduction and maternal environment in Romney ewes. II. Genetic covariation between hogget characters, fertility and maternal environment of the ewe. *Australian Journal of Agricultural Research* **23**, 149–165

CLARKE, J.N. (1972). Current levels of performance in the Ruakura Fertility flock of Romney sheep. *Proceedings of the New Zealand Society of Animal Production* **32**, 99–111

CUMMINS, L.J., O'SHEA, T., BINDON, B.M., LEE, V.M.K. and FINDLAY, J.K. (1982). Ovarian inhibin content and sensitivity to inhibin in Booroola and control strain Merino ewes. *Journal of Reproduction and Fertility* (in press)

DICKERSON, G.E. and LASTER, D.B. (1975). Breed, heterosis and environmental influences on growth and puberty in ewe lambs. *Journal of Animal Science* **41**, 1–9

DONALD, H.P., READ, J.L. and RUSSELL, W.S. (1968). A comparative trial of crossbred ewes by Finnish Landrace and other sires. *Animal Production* **10**, 413–421

DORRINGTON, J. and GORE-LANGTON, R.E. (1981). Prolactin inhibits oestrogen synthesis in the ovary. *Nature, London* **290**, 600–602

EIKJE, E.D. (1975). Studies on sheep production records VII. Genetic phenotypic and environmental parameters for productivity traits of ewes. *Acta Agriculturae Scandinavica* **25**, 242–252

EISEN, E.J. and JOHNSON, B.M. (1981). Correlated responses in male reproductive traits in mice selected for litter size and body weight. *Genetics* **99**, 513–524

FALCONER, D.S. (1965). The inheritance of liability to certain diseases estimated from the incidence among relatives. *Annals of Human Genetics* **29**, 51–76

FALCONER, D.S. (1981). *Introduction to Quantitative Genetics,* Second edition. London and New York, Longman

FINDLAY, J.K. and BINDON, B.M. (1976). Plasma FSH in Merino lambs selected for fecundity. *Journal of Reproduction and Fertility* **46**, 515

FORREST, P.A. and BICHARD, M. (1974). Analysis of production records from a lowland sheep flock. 3. Phenotypic and genetic parameters for reproductive performance. *Animal Production* **19**, 33–45

GJEDREM, T. (1966). Selection index for ewes. *Acta Agriculturae Scandinavica* **16**, 21–29

HANRAHAN, J.P. (1976). Response to selection for litter size in Galway sheep. *Irish Journal of Agricultural Research* **15**, 291–300

HANRAHAN, J.P. (1979). Genetic and phenotypic aspects of ovulation rate and fecundity in sheep. Paper presented to the 21st British Poultry Breeders Roundtable, Glasgow

HANRAHAN, J.P. (1980). Ovulation rate as the selection criterion for litter size in sheep. *Proceedings of the Australian Society of Animal Production* **13**, 405–408

HANRAHAN, J.P. and QUIRKE, J.F. (1982). Results of superovulation egg transfer and selection for ovulation rate in sheep. *Proceedings of World Congress on Sheep and Beef Cattle Breeding, New Zealand.* (In press)

HANRAHAN, J.P. and TIMON, V.M. (1978). Response to selection for increased litter size in Galway sheep. *Animal Production* **26**, 372

HANRAHAN, J.P., QUIRKE, J.F. and GOSLING, J.P. (1981). Effect of lamb age, breed and sex on plasma LH after administration of GnRH. *Journal of Reproduction and Fertility* **61**, 281–288

HIGHT, G.K. (1977). *Lands and Survey Romney breeding scheme.* Annual Report, Research Division of New Zealand Ministry of Agriculture and Fisheries 1975–6, pp. 88–89

HIGHT, G.K. and JURY, K.E. (1970). Hill country sheep production. I. The influence of age, flock and year on some components of reproduction rate in Romney and Border Leicester × Romney ewes. *New Zealand Journal of Agricultural Research* **13**, 641–659

HOWLES, C.M., WEBSTER, G.M. and HAYNES, N.B. (1980). The effect of rearing under a long or short photoperiod on testis growth, plasma testosterone and prolactin concentrations, and the development of sexual behaviour in rams. *Journal of Reproduction and Fertility* **60**, 437–447

KARAM, H.A. (1957). Multiple birth and sex ratio in Rahmani sheep. *Journal of Animal Science* **16**, 990–997

KENNEDY, J.P. (1967). Genetic and phenotypic relationships between fertility and wool production in 2-year-old Merino sheep. *Australian Journal of Agricultural Research* **18**, 515–522

LAND, R.B. (1970). Number of oocytes present at birth in the ovaries of pure and Finnish Landrace cross Blackface and Welsh sheep. *Journal of Reproduction and Fertility* **21**, 517–521

LAND, R.B. (1973). The expression of female sex-limited characters in the male. *Nature, London* **241**, 208–209

LAND, R.B. (1974). Physiological studies and genetic selection for sheep fertility. *Animal Breeding Abstracts* **42**, 155–158

LAND, R.B. (1976). The sensitivity of the ovulation rate of Finnish Landrace and Blackface ewes to exogenous oestrogen. *Journal of Reproduction and Fertility* **48**, 217–218

LAND, R.B. (1981). Physiological criteria and genetic selection. *Livestock Production Science* **8**, 203–213

LAND, R.B. and CARR, W.R. (1979). Reproduction in domestic mammals. In *Genetic Variation in Hormone Systems*, Volume 1 (J.G.M. Shire, Ed.), pp. 89–112. Boca Raton, Fla., C.R.C. Press

LAND, R.B., RUSSELL, W.S. and DONALD, H.P. (1974). The litter size and fertility of Finnish Landrace and Tasmanian Merino sheep and their reciprocal crosses. *Animal Production* **18**, 265–271

LAND, R.B., CARR, W.R., McNEILLY, A.S. and PREECE, R.D. (1980). Plasma FSH, LH, the positive feedback of oestrogen, ovulation and luteal function in the ewe given bromocriptine to suppress prolactin during seasonal anoestrus. *Journal of Reproduction and Fertility* **59**, 73–78

LAND, R.B., GAULD, I.K., LEE, G.J. and WEBB, R. (1982a). Further possibilities for manipulating the reproductive process. In *Future Developments in the Genetic Improvement of Animals* (J.S.F. Barker, K. Hammond and A.E. McClintock, Eds.). Academic Press (in press)

LAND, R.B., MORRIS, B.A., BAXTER, G., FORDYCE, F. and FORSTER, J. (1982b). The improvement of sheep fecundity by treatment with antisera to gonadal steroids. *Journal of Reproduction and Fertility* **66**, 625–634

McGUIRK, B.J. (1967). Breeding for lamb production. *Wool Technology and Sheep Breeding* **14**, 73–75

McGUIRK, B.J. (1973). The inheritance of production characters in Merino sheep. PhD Thesis. University of Edinburgh

MANN, T.L.J., TAPLIN, D.E. and BRADY, R.E. (1978). Response to partial selection for fecundity in Merino sheep. *Australian Journal of Experimental Agriculture and Animal Husbandry* **18**, 635–642

MARTIN, T.G., NICHOLSON, D., SMITH, C. and SALES, D.I. (1981). Phenotypic and genetic parameters for reproductive performance in a synthetic line of sheep. *Journal of Agricultural Science, Cambridge* **96**, 107–113

MECHLING, E.A. and CARTER, R.C. (1969). Genetics of multiple births in sheep. *Journal of Heredity* **60**, 261–266

MEYER, H.H., CLARKE, J.N., BIGHAM, M.L. and CARTER, A.H. (1977). Reproductive performance, growth and wool performance of exotic breeds and their crosses with the Romney. *Proceedings of the New Zealand Society of Animal Production* **37**, 220–229

MORE O'FERRALL, G.J. (1976). Phenotypic and genetic parameters of productivity in Galway ewes. *Animal Production* **23**, 295–304

NITTER, G. (1978). Breed utilisation for meat production in sheep. *Animal Breeding Abstracts* **46**, 131–143

OWEN, J.B. (1976). The development of a prolific breed of sheep. *Proceedings of the 27th Annual Meeting, EAAP, Zurich*, Abstract 543

PIPER, L.R. and BINDON, B.M. (1977). The genetics of early viability in Merino sheep. *Proceedings of the Third International Congress of the Society for the Advancement of Breeding Researches in Asia and Oceania (SABRAO), Canberra.* Animal Breeding Papers, pp. 10.17–10.22

PIPER, L.R. and BINDON, B.M. (1982). Genetic segregation for fecundity in Booroola Merino sheep. *Proceedings of the World Congress on Sheep and Beef Cattle Breeding, New Zealand* (R.A. Barton and W.C. Smith, Eds.). (in press)

PIPER, L.R., BINDON, B.M., ATKINS, K.D. and McGUIRK, B.J. (1980). Genetic variation in ovulation rate in Merino ewes aged 18 months. *Proceedings of the Australian Society of Animal Production* **13**, 409–412

PURSER, A.F. (1965). Repeatability and heritability of fertility in hill sheep. *Animal Production* **7**, 75–82

QUIRKE, J.F., HANRAHAN, J.P. and GOSLING, J.P. (1979). Plasma progesterone levels throughout the oestrous cycle and release of LH at oestrus in sheep with different ovulation rates. *Journal of Reproduction and Fertility* **55**, 37–44

RAE, A.L. (1956). The genetics of the sheep. *Advances in Genetics* **8**, 189–265

RAVAULT, J.P. (1976). Prolactin in the ram: seasonal variations in the concentration of blood plasma from birth until three years old. *Acta Endocrinologica, Denmark* **83**, 720–725

RENDEL, J. (1956). Heritability of multiple birth in sheep. *Journal of Animal Science* **15**, 193–201

RICORDEAU, G., PELLETIER, J., COUROT, M. and THIMONIER, J. (1979). Phenotypic and genetic relationships between endocrine criteria and testicular measurements of young Romanov rams and the ovulation rates at 8 months of their half sisters. *Annales de Genetique et de Selection animale* **II**, 145–159

SHELTON, M. and MENZIES, J.W. (1968). Genetic parameters of some performance characteristics of range fine-wool ewes. *Journal of Animal Science* **27**, 1219–1223

SHELTON, M. and MENZIES, J.W. (1970). Repeatability and heritability of components of reproductive efficiency in fine-wool sheep. *Journal of Animal Science* **30**, 1–5

THIMONIER, J., RAVAULT, J.P. and ORTAVANT, R. (1978). Plasma prolactin variations and cyclic ovarian activity in ewes submitted to different light regimes. *Annales de Biologie Animale, Biochimie, Biophysique* **18**, 1229–1235

TIMON, V.M. (1974). The evaluation of sheep breeds and breeding strategies. *Proceedings of the Working Symposium on Breed Evaluation and Crossing Experiments with Farm Animals, Zeist*, pp. 367–387

TROUNSON, A.O., CHAMLEY, W.A., KENNEDY, J.P. and TASSELL, R. (1974). Primordial follicle numbers in ovaries and levels of LH, FSH in pituitaries and plasma of lambs selected for and against multiple births. *Australian Journal of Biological Science* **30**, 229–241

TURNER, H.N. (1966). Selection for increased reproduction rate. *Wool Technology and Sheep Breeding* **13**, 69–79

TURNER, H.N. (1968). The effect of selection on lambing rates. *Proceedings of a Symposium on Physiology of Reproduction in Sheep, Stillwater, Oklahoma*, pp. 67–93

TURNER, H.N. (1978). Selection for reproduction rate in Australian Merino sheep: Direct responses. *Australian Journal of Agricultural Research* **29**, 327–350

WALKLEY, J.R.W. and SMITH, C. (1980). The use of physiological traits in genetic selection for litter size in sheep. *Journal of Reproduction and Fertility* **59**, 83–88

WALTON, J.S., EVINS, J.D., FITZGERALD, B.P. and CUNNINGHAM, F.J. (1980). Abrupt decrease in daylength and short term changes in the plasma concentrations of FSH, LH and prolactin in anoestrous ewes. *Journal of Reproduction and Fertility* **59**, 163–171

WALTON, J.S. McNEILLY, J.R., McNEILLY, A.S. and CUNNINGHAM, F.J. (1977). Changes in concentrations of follicle stimulating hormone, luteinising hormone, prolactin and progesterone in the plasma of ewes during the transition from anoestrus to breeding activity. *Journal of Endocrinology* **75**, 127–136

WHEELER, A.G. and LAND, R.B. (1977). Seasonal variation in oestrus and ovarian activity of Finnish Landrace, Tasmanian Merino and Scottish Blackface Ewes. *Animal Production* **24**, 363–376

YOUNG, S.S.Y., TURNER, H.N. and DOLLING, C.H.S. (1963). Selection for fertility in Australian Merino sheep. *Australian Journal of Agricultural Research* **14**, 440–482

28

FACTORS AFFECTING THE PRODUCTION AND QUALITY OF WOOL

J.M. DONEY
Hill Farming Research Organisation, Bush Estate, Penicuik, Midlothian, UK

Although the remit does not include either the economic background to wool production or the role within the wide range of sheep farming systems in the UK and throughout the world, it is obvious that an academic review of the biological factors involved would have little relevance out of the context of whole farm systems. Some of the practical questions which have been addressed to the research worker range from the hopeful 'how can we manage our sheep to increase the financial returns from the wool crop' to the somewhat cynical 'how do we breed sheep without fleeces'.

The volume of scientific work on the biology of wool growth has declined very considerably in the past decade, apart perhaps from the major wool-producing states such as Australia and the USSR. But, despite this fall in interest, there is a considerable body of basic research which can be applied to most potential questions of sheep management application. For this reason a comprehensive list of references to biological aspects of wool growth will not be given; all such can be found within existing reviews such as those of Ryder and Stephenson (1968) and Black and Reis (1979).

It can be accepted that, as an economic crop, wool is best suited for low-cost extensive systems with low labour and capital inputs, although, in some situations, for example in eastern Europe, it has been regarded as of sufficient importance to the economy to develop quite intensive systems with wool as the major product. At the present time in the UK wool is not a primary product in any of the varied environments in which sheep husbandry is practised. These differences in approach have quite significant consequences on the basic product/value relationships as determined either by market forces or by deliberate intervention. Thus the relative value per kilogram of fleece wool and lamb meat may vary from around 1:1 or less to as high as 5:1 or greater. Obviously, then, the application of basic biological knowledge to management decisions will give different answers depending on the specific circumstance to which it is applied.

In this chapter the biological basis of wool growth regulation will be discussed and, finally, the possibilities for application of this knowledge will be considered in terms of different sheep husbandry systems in which wool may be regarded as a valuable component of the system.

Biological factors affecting wool production

DEVELOPMENT OF FLEECE STRUCTURE

Wool fibres are produced by epidermal structures, the follicles. These develop during foetal life in characteristic groups made up of two types, known as primary and secondary. The first-formed primaries develop in rows, usually of three follicles, and the later, more numerous secondaries develop in association with these trios to form the follicle group. The type of fleece produced in adult life, in terms of fibre density, staple length and relative fineness and uniformity of fibres within the fleece, depends on two attributes of the follicle group—the relative size and the relative number of secondary and primary follicles within the group. In fine-wool sheep such as the Merino, there is no difference in size between the two follicle types and the ratio of secondary to primary fibre number is high (*c.* 20:1) whereas in the carpet-wool breeds, characterized by distinct inner and outer coats, the primaries may have a diameter up to five times greater than the secondaries and the secondary:primary ratio is low (*c.* 3:1). These differences are largely determined by genotype, although there is some evidence to suggest that poor nutrition during late pregnancy and in early postnatal life can delay or prevent the maturation of a proportion of the secondary follicles, thus reducing the fibre type ratio and, possibly, changing the adult fleece structure.

The wool fibres of the adult fleece can be classified into three types—fine wool, hair and kemp. The true wool fibres are the finest, ranging from 15–50 μm in diameter, depending on breed and other factors; they are usually associated with secondary follicles. Kemp fibres are very coarse (100–200 μm) and tend to be short because of their limited growth period; they are shed seasonally, forming the typical 'brush-ends'. Where present, kemp fibres are usually associated with the central primaries. Hair fibres, or heterotypes, are intermediate in diameter and like kemps may be heavily medulated for part of their length. They have the continuous growth habit of the wool fibres but are more susceptible to marked changes in diameter associated with environmental factors. Hair fibres are generally associated with the primary follicles.

The distribution of fibre types within the fleece determines fleece quality and, like the follicle development on which it depends, quality is very largely determined by genetic factors associated with breed. Some modifications of fibre growth rate and relative dimensions may be caused by environmental factors but, whilst the effects of these on quantitative wool production may be considerable, the effect on fleece structure and, hence, quality, are marginal at the most. Definitive work on the development of fleece structure can be found in Ryder and Stephenson (1968).

ENVIRONMENTAL EFFECTS ON ADULT WOOL GROWTH

There is conflicting evidence on the long-term effect of undernutrition during the follicle development stage (late foetal and early postnatal period) on the structure and potential growth rate of the adult fleece

(Ryder and Stephenson, 1968; Corbett, 1979). A limitation may be imposed directly by the possible reduction in secondary follicle development already discussed or indirectly by a permanent reduction in body size which could increase the competition between adjacent follicles for substrate by decreasing skin surface area or which may have a continuing influence on potential feed intake associated with body size. In practice it can be concluded that the potential reduction in follicle number may only be of significance in the high secondary:primary ratio breeds such as the Merino strains and, even then, may not be expressed except in cases of unusually severe nutritional limitation. The positive correlations between adult size and fleece weight certainly suggest that any residual effects of early growth retardation could result in a reduction of potential wool growth throughout adult life.

By far the most important factor affecting adult wool growth, however, is the current level of nutrition (Ryder and Stephenson, 1968; Allden, 1979). Increases in energy intake, except at very low levels of protein content in the diet, can have an immediate effect on wool growth, detectable within a few days. The full effects of such increases in energy intake are, however, cumulative and may not be reached until up to three months after the change in diet level. Whilst it is generally accepted that wool growth rate is dependent on energy rather than protein intake, this may be due to the normal degradation of protein in the rumen since experiments in which the rumen is bypassed or the diet contains protected proteins have shown very significant responses in wool growth to protein levels. Wool protein contains a high proportion of the high sulphur amino acid, cystine, and it has been shown that variation in the availability of the sulphur amino acids to the follicle can affect both fibre growth rate and fibre composition. However, supplementation of the diet has proved ineffective because of rumen degradation. In general, nutritional limitations to wool growth mainly affect quantitative production. Both length and diameter are reduced in individual fibres to the extent that, in the whole fleece, a period of severe undernourishment would be represented by a marked thinning down of all fibres at the corresponding part of the staple. In many cases this thinning could seriously influence the manufacturing value of the fleece and, in severe cases, could result in a physical break in the staple with consequent loss of fleece before shearing. Undernourishment does not appear to increase the number of fibres which completely cease to grow.

Thus, the potential growth rate of wool fibres is set by the follicle structure (genotype and, perhaps, early nutrition) and by current nutritional status. The achieved growth rate, however, can be further limited by other factors in the external and internal environment.

Rate of wool growth varies considerably throughout the year. The distinct seasonal rhythm of production was, at first, attributed to seasonal fluctuation in temperature or intake. Controlled experimental studies have shown that wool growth follows a sinusoidal curve irrespective of nutrition or temperature and it is now considered that the seasonal pattern is regulated by an endogenous rhythm influenced by photoperiod and other factors (Ryder and Stephenson, 1968; Nagorcka, 1979; Panaretto, 1979). The amplitude of the growth curve varies with breed. In the fine-wool

Merino, winter growth rate has been found to be around 85% of that in summer whereas in British hill breeds the winter rate may be as low as 30% of that in summer. In the more primitive breeds growth may stop completely, with many fibres forming brush-ends, especially those of the outer coat.

Interaction of endogenous rhythm and nutrition varies with breed. In the Merino strains there may be a positive response to nutrition at any time of the year, whereas in breeds such as the British hill breeds there is a positive response in summer and no response in winter (Doney, 1966). Hence, since the pattern of grazing intake also follows a seasonal pattern the amplitude of the wool growth curve may be increased beyond that established by the endogenous rhythm alone. Other UK breeds show an intermediate response but, in general, are closer to the example of the hill breeds than they are to the Merino.

Ambient temperature has been shown to influence fibre growth rate, possibly by its effect on skin temperature and blood flow through the skin (Ryder and Stephenson, 1968; Bottomley, 1979). Whilst this may be part of the mechanism involved in the reduction of wool growth during winter, variation in temperature is unlikely to have any direct effect of practical significance. Other aspects of weather such as wind and rain exposure may influence wool growth in relation to their cooling effect but, indirectly, all aspects can interact with the nutritional effect either by their influence on food availability or, through behavioural and physiological responses, on variation in food intake.

Physiological state of the animal has been found to have a significant effect on wool growth. In general entire males produce more wool than castrated males which produce more than females. There is no real evidence, however, to suggest that these differences are determined by anything other than differences in size and nutritional state. Effects of age are also found; rate of wool production in most breeds increases to a maximum at between four and five years and then declines, often rapidly. The reproductive cycle in females can reduce wool growth significantly (Corbett, 1979). During late pregnancy, depression of wool growth rate in the range of 20–40% has been found and, even in breeds which exhibit a marked seasonal depression, pregnancy can depress the rate still further. Where there are no restrictions to intake, voluntary feed intake during pregnancy is not reduced, except, perhaps, for a short period before parturition, so it is accepted that the depression of wool growth is likely to be associated with hormonal effects on the partition of ingested nutrients. In lactating ewes, wool growth is also reduced by up to 30% or more, this despite the accepted increase in voluntary intake associated with lactation. In all breeds investigated, including those with a high seasonal amplitude, improved nutrition in the form of concentrate supplements or better pasture quality, can reduce the difference between dry and lactating ewes considerably, providing the lactation period occurs in spring and early summer. It can be accepted that, as in pregnancy, the reduction in wool growth during lactation is influenced by hormonal factors affecting nutrient partition. It can be seen that there are possibilities for interaction effects of season, available nutrition and physiological state.

During both pregnancy and lactation several workers have found that the depression in wool output is associated not only with a reduction in the length and diameter of individual fibres but also with a reduction in the number of active follicles. Thus, during this period, qualitative aspects of the fleece may be affected in terms of a change in fibre type ratio, partial break or shedding of fibres leading to fleece cotting or, in some cases, the complete casting of fleeces, delayed until after the period of restricted growth but often occurring before the normal time for shearing.

It is evident that the growth and quality of the adult fleece is regulated by an interaction of genetic factors (follicle structure and endogenous rhythm) and, mainly, nutritional factors (level of intake, diet composition and priority in partition of nutrients according to physiological demand). A knowledge of the genetic limitations of a given breed and of the nutritional limitations of a given management system can, therefore, lead to an ability both to predict potential levels of production and to modify management to achieve optimum, cost-effective improvement in wool production.

HORMONAL REGULATION OF WOOL GROWTH

The hormonal basis of variation in wool growth has been studied both from the point of view of understanding the mechanisms by which the environmental responses are mediated and with the objective of developing methods of obtaining regulation and improvement in practice by the use of exogenous control (Ryder and Stephenson, 1968; Wallace, 1979). The types of hormone response which can be considered are regulatory responses in the natural physiological state and pharmacological responses to non-physiological levels of administered hormones.

It is generally accepted that hormones secreted by the pituitary have a permissive effect on wool growth since excision of this gland results in a virtual cessation of growth. The role of all identified pituitary hormones has not been established but thyroid stimulating hormone (TSH) appears to be essential, in that daily administration of thyroid hormones can restore normal wool growth in hypophysectomized sheep. Administration of thyroid hormone to normal sheep has been found to increase wool growth in association with an increase in feed intake and/or a decrease in storage of tissue protein and liveweight. Growth hormone, which has a direct action, does not appear to have a permissive effect on wool growth in that it does not permit growth in hypophysectomized sheep nor does it supplement the response to thyroid hormone in these sheep. It has, however, been found to have a depressive effect on wool growth when administered to normal sheep although this may be followed by enhanced growth after treatment ceases. Other pituitary hormones such as prolactin, gonadotrophic hormones or oxytocin do not appear to have any significant permissive or regulatory effect on wool. Apart from hormones secreted by the thyroid, only the adrenal cortex has been found to influence wool growth in that administration of either adrenocorticotrophic hormone (ACTH) or cortisol and its analogues can depress wool growth, leading to complete cessation and formation of brush-ends in extreme cases.

It would appear that, apart from the permissive role of certain hormones, the regulatory effect is likely to be determined by hormone balance leading to control of tissue storage or depletion and the partition of nutrients. The endogenous balance achieved at different levels in the different breed types at different times of year may account for the seasonal rhythms largely entrained to light stimuli. It is unlikely that the disruptive effect of corticosteroids could be responsible for the observed seasonal depression of growth although it could play an important role in relation to fibre shedding and fleece loss as a response to specific acute stress from climatic or nutritional factors occurring during the winter period.

At this time, there are no indications of a suitable systematic hormone intervention which would consistently improve wool growth rates in commercial flocks without incurring unacceptable penalties in other aspects of production.

The application of wool knowledge to system development

The foregoing discussion has shown that much of the basic biological knowledge of factors affecting wool production has been available for many years. Very few new factors have been reported since the intensive period of investigation in the 1960s and early 1970s. It can be said, with confidence, that the agriculturalist has sufficient information available to him which can be applied in any system of sheep production to determine the economic role of wool production in that system and to assess the scope for management change designed to increase or improve that production. It is evident, however, that no single recommendation can be made. The answers depend on economic factors, the nature of the nutritional resources available to the sheep enterprise, the genotype of the sheep currently involved and the alternative product of the system. Whilst the general magnitude of the effects of the major factors is known, accurate prediction of absolute levels of wool production cannot be made in any single situation (White, Nagorcka and Birrell, 1979).

The crucial question which must be faced is how the existing knowledge can be applied in a realistic way. The question has been discussed, with specific reference to Australia where wool production is a primary objective, by McLaughlin (1979). Only the general principles of application can be discussed here. Sheep are normally considered as dual purpose animals with meat and wool being the most common combination. In some situations, for example in the extensive conditions of Australian agriculture, wool may be the sole product of significance, whilst, in other situations dairy production may replace meat or may even be considered as a third primary product. Very different answers, in terms of management change, would result from the application of the basic principles in these situations.

In a two-product system, the most important factors to be taken into account are the relative cash values and existing levels of production of the two products, the genetic and nutritional relationships between factors affecting their production and the relative return of investment aimed at improvement of either product. In the UK, the relative price/kilogram of lamb liveweight or greasy wool has been changing consistently over the past decade in favour of meat and there is no reason to suppose this trend will not continue. Individual meat production efficiency depends on reproductive rate and lamb growth; increases in both of these components would tend to reduce individual fleece weights and fleece quality in a pasture-based system unless the increased primary production was supported by increased nutrient provision during the lactation period. Flock output depends on stocking rate and increases in this could increase the output of both meat and wool simultaneously. Application of these principles to the development of a system designed to increase the economic efficiency of lamb production from hill land in the UK, has been discussed by Russel, Doney and Maxwell (1976). In two examples of the application of principles designed to reduce the nutritional limitations to lamb production, increases in flock output of 51% and 94%, respectively, were obtained. These increases were accompanied by increases in wool output of 31% and 84%, respectively. In both cases, the increased production could be attributed to a combination of increases in both stocking rate and individual fleece weights. Conversely, it has been shown (HFRO, 1979) that a management system designed solely to increase wool production in these situations may confer some benefit in terms of lamb growth but would have little effect on reproductive performance—a major component of total lamb production. In other intensive or semi-intensive systems it is likely that management changes initiated solely in regard to the maximization of wool production would result in an overall reduction of economic return, whereas changes initiated to maximize lamb production could have a small but positive influence on wool, even though the relative return to the enterprise from wool may decline further.

It is obvious that the value of the wool crop, both in terms of quantity and quality, as defined by its suitability for specific manufacturing processes and its freedom from faults, can be increased by change of flock size, by within-breed selection or change of breed, by adjustment of the nutritional inputs to maximize response when the potential for wool growth permits and by adjustment of other aspects of management including change of lambing time, change in management of the growing lamb, elimination of climatic stress factors, etc. It is essential to recognize that unless wool is the primary and, indeed, perhaps the only, economic product, additional input of resources, specifically based on a knowledge of factors affecting wool growth, can only be justified when they could be expected to show a clear return from the wool clip, without detracting from other more important products. Alternatively, the additional resources should be directed towards the primary products (reproductive performance, lamb growth, milk production, etc.) in such a way that realization of improvement in these products carries with it a bonus in either the quantity or quality of the wool produced. In other words, if it is clearly recognized that, in most sheep systems, wool is a valuable by-product, the producer will have the best chance of improving the output and value of that by-product by applying existing knowledge of the biological basis of wool growth.

References

ALLDEN, W.G. (1979). Feed intake, diet composition and wool growth. In *Physiological and Environmental Limitations to Wool Growth* (J.L.

Black and P.J. Reis, Eds.), pp. 61–78. Armidale, University of New England Publishing Unit

BLACK, J.L. and REIS, P.J. (Editors) (1979). *Physiological and Environmental Limitations to Wool Growth.* Armidale, University of New England Publishing Unit

BOTTOMLEY, G.A. (1979). Weather conditions and wool growth. In *Physiological and Environmental Limitations to Wool Growth* (J.L. Black and P.J. Reid, Eds.), pp. 115–126. Armidale, University of New England Publishing Unit

CORBETT, J.L. (1979). Variation in wool growth with physiological state. In *Physiological and Environmental Limitations to Wool Growth* (J.L. Black and P.J. Reis, Eds.), pp. 79–98. Armidale, University of New England Publishing Unit

DONEY, J.M. (1966). Breed differences in response of wool growth to annual nutritional and climatic cycles. *Journal of Agricultural Science, Cambridge* **67**, 25–30

HILL FARMING RESEARCH ORGANISATION (1979). Science and hill farming. Twenty five years of work at the Hill Farming Research Organisation. Penicuik, Scotland, HFRO

McLAUGHLIN, J.W. (1979). Application of current knowledge for improving the wool production of grazing sheep. In *Physiological and Environmental Limitations to Wool Growth* (J.L. Black and P.J. Reis, Eds.), pp. 355–361. Armidale, University of New England Publishing Unit

NAGORCKA, B.N. (1979). The effect of photoperiod on wool growth. In *Physiological and Environmental Limitations to Wool Growth* (J.L. Black and P.J. Reid, Eds.), pp. 127–138. Armidale, University of New England Publishing Unit

PANARETTO, B.A. (1979). Effects of light on cyclic activity of wool follicles and possible relationships to changes in the pelage of other mammals. In *Physiological and Environmental Limitations to Wool Growth* (J.L. Black and P.J. Reis, Eds.), pp. 327–336. Armidale, University of New England Publishing Unit

RYDER, M.L. and STEPHENSON, S.K. (1968). *Wool Growth.* London, Academic Press

RUSSEL, A.J.F., DONEY, J.M. and MAXWELL, T.J. (1976). The effect on wool production of changes in management designed to increase output of lamb from hill land in the United Kingdom. *Livestock Production Science* **3**, 178–182

WALLACE, A.L.C. (1979). The effect of hormones on wool growth. In *Physiological and Environmental Limitations to Wool Growth* (J.L. Black and P.J. Reis, Eds.), pp. 257–268. Armidale, University of New England Publishing Unit

WHITE, D.H., NAGORCKA, B.N. and BIRRELL, H.A. (1979). Predicting wool growth of sheep under field conditions. In *Physiological and Environmental Limitations to Wool Growth* (J.L. Black and P.J. Reis, Eds.), pp. 139–161. Armidale, University of New England Publishing Unit

29

GENETIC SELECTION FOR WOOL PRODUCTION

B.J. McGUIRK
CSIRO Division of Animal Production, Blacktown, New South Wales, Australia

This chapter will review the available information on the inheritance of fleece weight and other fleece characters, and on the consequences of selecting for increased fleece weight. These topics are relevant to sheep improvement programmes in many countries and for many breeds of sheep, since wool is produced by the great majority of the world's estimated population of approximately 990 million sheep (Australian Wool Corporation, 1981). That is not to say that the improvement of wool production is equally important in all such programmes. For example, fleece weight and other wool traits receive little attention in sheep performance recording schemes in Europe (Croston *et al.*, 1980) where wool makes only a small contribution to total flock returns and the major concern is to increase lamb production. By contrast the collection of information on fleece weights and, more recently, on average fibre diameter has been the overriding aim in performance recording schemes for specialized wool-producing breeds such as the Merino (McGuirk, 1982).

This review will draw heavily on the results of genetic investigations of fleece characters in Merino sheep, and especially on responses observed in selection flocks in which the sole or major aim was to increase fleece weight. Some of these selection flocks were established before 1950, making them among the earliest selection experiments with a domestic species of livestock. They have provided invaluable information on both direct and correlated responses to selection, and especially on the physiological and biochemical mechanisms underlying genetic differences in wool production (Williams, 1979; McGuirk, 1980a). These findings will be summarized here and their relevance to future industry improvement programmes will also be discussed.

The inheritance of fleece weight

Heritability estimates for both greasy and clean fleece weight are summarized in *Table 29.1*. Both measures of wool production have been included for, while the amount of clean wool is the character of ultimate concern in wool processing, the genetic and phenotypic correlations between the two weights are high (+0.7 to +0.9; Turner and Dunlop, 1974) and so the cost

Table 29.1 HERITABILITY (h^2) ESTIMATES FOR YEARLING AND HOGGET GREASY AND CLEAN FLEECE WEIGHTS

Sheep type	*Breed*	*Greasy fleece weight*		*Clean fleece weight*		*References*
		h^2	*Adjustment*[a]	h^2	*Adjustment*[a]	
Fine wools (Merinos)	US Rambouillet	0.38	—	0.28	—	Terrill and Hazel (1943)
	Australian Merino (fine)	0.30[b]	—	0.29[b]	—	Mullaney *et al.* (1970)
	Australian Merino (medium)	0.40	—	0.47	—	Morley (1955)
	Australian Merino (medium)	0.42	DB	0.40	DB	Brown and Turner (1968)
	Australian Merino (strong)	0.46	—	0.28	—	Gregory (1982)
	South African Merinos	—	—	0.31	DWS	Heydenrych *et al.* (1977)
Breeds derived from Merino types	Australian Polwarth	0.14[b]	—	0.30[b]	—	Mullaney *et al.* (1970)
	Australian Corriedale	0.22[b]	—	0.28[b]	—	Mullaney *et al.* (1970)
	Romnolet	—	—	0.48	DBAS	Veseley and Slen (1961)
General purpose wool breeds	NZ Romney	0.28	DB	—	—	Blair (1981)
	NZ Romney	0.41	DBAS	—	—	Baker *et al.* (1979)
	Norwegian breeds (including Cheviot)	0.36	D	—	—	Eikje (1975a)
	NZ Perendale	0.36	DB	—	—	Elliott *et al.* (1979)
Breeds used for carpet wool production	Welsh Mountain	0.61	B	—	—	Doney (1958)
	Welsh Mountain	0.58	—	—	—	Dalton (1962)
	Scottish Blackface	0.41	NK	—	—	Purser (unpublished data)

[a]The adjustments considered were as follows: D = dam age; B = birth type or birth status; S = sex; A = animal's own age at shearing; NK = not known; — = none.

[b]In these studies, relationship estimated between offspring performance as hoggets and parent performance, adjusted for reproductive performance, at various ages.

of estimating clean wool production is probably only justified when comparing potential sire replacements in the specialized wool-producing breeds.

The estimates included in *Table 29.1* indicate that fleece weight is at least moderately heritable in a wide range of breeds. This same conclusion was reached in earlier reviews (see for example Turner and Dunlop (1974) who presented a more exhaustive review of published estimates). The broad agreement between estimates for such a diverse group of breeds and wool types is striking. With estimates in the general area of 0.3–0.4, an animal's own performance is a reasonably good guide to its breeding value and mass selection should be effective, provided that animals being compared have been run together and managed similarly. Offspring–dam and offspring–sire regression estimates would appear to be similar (Baker *et al.*, 1979).

Mass selection programmes are most effective if selection is practised before animals are mated for the first time, and if the measurements have been adjusted to remove the effect that identifiable environmental factors may have on performance. The estimates in *Table 29.1* are for yearling or hogget records (12–18 months of age) and are thus based on information collected before most specialized wool-producing breeds are first mated. This is not always so for breeds used primarily for meat production which are often mated earlier (Eikje, 1971). Factors such as an animal's birth type, the age of its dam and even its own age at shearing may affect fleece weights. In principle, failure to adjust for such factors will lead to lower heritability estimates but the effect of these adjustments on heritability estimates for yearling or hogget fleece weights appears to be small (Baker *et al.,* 1979; Gregory, 1982), at least where the animals have previously been shorn as lambs. This finding is of considerable importance for the specialized wool-producing breeds, such as the Merino, where large flocks are often run under relatively unsupervised conditions and where, as a consequence, the identity of a lamb's dam is generally unknown.

Little information is available on the inheritance of fleece weight for ewes after they enter the breeding flock. While selection for increased hogget fleece weight has been found to increase production at later shearings (Brown *et al.*, 1966), estimates for the heritability of these records and of their genetic correlations are necessary in order to establish optimum selection procedures for improving lifetime production. The available evidence suggests that the heritability of adult ewe fleece weights, if adjusted for such environmental effects as her lambing performance, is similar to that for yearling or hogget records, though the genetic correlation between yearling and later fleece weights is only of the order of +0.5 to +0.6 (Eikje, 1975a).

Selection for increased fleece weight

Flocks selected for increased fleece weight have been established in a number of breeds and those to be discussed here are described in *Table 29.2*. The flocks have been classified as either single character selection flocks, in which selection was solely for increased greasy or clean fleece weight (Experiments 1–3), or flocks in which selection was primarily for

Table 29.2 SUMMARY OF SELECTION EXPERIMENTS FOR INCREASED FLEECE WEIGHT

Experiment No.	*Breed*	*Flocks*	*Size of flocks*		*Duration of experiment*	*References*
			Rams	*Ewes*		
(a) Flocks selected solely on fleece weight:						
1	Australian Merino	High clean fleece weight	5	100	1951–	Pattie and Barlow (1974)
		Control	10–25	100	1951–	McGuirk (1980b)
2	Australian Merino	High clean fleece weight	1	32–50	1954–1976	Turner, Brooker and Dolling (1970)
		Control	5–10	200–250	1948–1976	
3	New Zealand Romney	High greasy fleece weight	4	80	1958–	Blair (1981)
		Control	4	80	1958–	
(b) Multi-trait selection flocks:						
4	Australian Merino	High clean fleece weight with checks on crimp frequency, skin wrinkle and face cover	6	200	1947–1972	Dun and Eastoe (1970)
		Control–as for Experiment 1				
5	Australian Merino	High clean fleece weight with checks on skin wrinkle and fibre diameter	5–15	200–600	1950–1978	Turner, Dolling and Kennedy (1968) Turner and Jackson (1978)
		Control–as for Experiment 2				
6	Australian Merino	High clean fleece weight with checks on skin wrinkle and either fibre diameter (1950–52) or crimp frequency (1961–76)	5–8	100–250	1950–1976	Turner, Dolling and Kennedy (1968) Turner and Jackson (1978)
		Control–as for Experiment 2				
7	South African Merino	Increased clean fleece weight with checks on fibre diameter and crimp frequency	5	160	1969–	Heydenrych *et al.* (1977)
		Control	5	160	1969–	

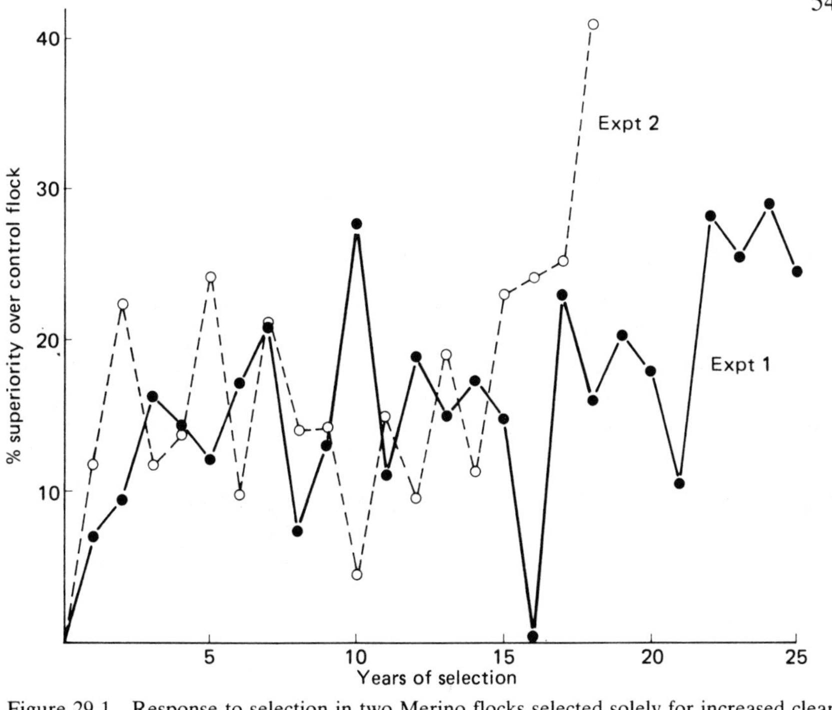

Figure 29.1 Response to selection in two Merino flocks selected solely for increased clean fleece weight. Response measured as a percentage deviation from an unselected control flock

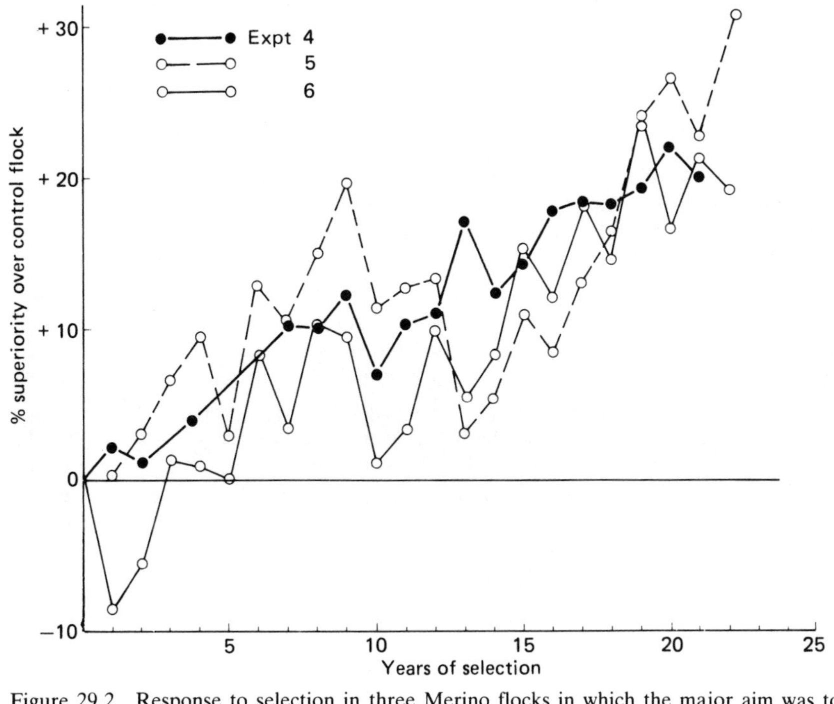

Figure 29.2 Response to selection in three Merino flocks in which the major aim was to increase clean fleece weight. Response measured as a percentage deviation from an unselected control flock

increased fleece weight but where other characters were taken into account (Experiments 4–7). This was done in each case by independent culling levels to remove animals with extreme values for the wool or body faults listed. In each experiment response to selection was measured as the deviation from an unselected control flock, drawn from the same base population as the selection flock. In Experiments 1 and 2, flocks were also selected for low clean fleece weight, and these will be referred to later when considering correlated responses to selection.

All of the selection programmes were effective in that, over the course of the experiments, average fleece weights in the selection flocks were higher than those in the control populations. Patterns of response in the five selected flocks of Australian Merinos are illustrated in *Figures 29.1* and *29.2*. Annual rates of improvement for these five flocks and for Experiment 7 are given in *Table 29.3*; annual performance figures were not

Table 29.3 ANNUAL PERCENTAGE RATES OF IMPROVEMENT IN MERINO FLOCKS SELECTED FOR INCREASED FLEECE WEIGHT

Experiment[(a)]	*Period of response*	*Estimated annual rate of improvement* (Mean % ±S.E.M.)
1	1952–1977	1.09±0.09
2	1954–1976	1.58±0.20
4	1947–1967	1.03±0.03
5	1950–1976	1.05±0.09
6	1950–1976	0.85±0.25
7	1969–1975	1.22±0.21

[(a)]See *Table 29.2* for details of flocks and selection procedures.

available for Experiment 3. For each flock the percentage superiority of the selection flock hoggets over corresponding control flock animals was calculated each year and the regression then calculated for these percentage gains against years of selection. Regression lines were forced through the origin as both selection and control flocks had been drawn from the same base population. Annual rates of gain varied between 0.85 and 1.66% gain/year. As might be expected, the results in *Table 29.3* suggest that gains were more rapid in the single character Merino flocks (Experiments 1 and 2) than in the flocks drawn from the same base populations but in which characters other than fleece weight were considered when selecting replacement breeding stock (Experiments 4–6).

A more satisfactory procedure for examining responses to selection is to estimate the realized heritability for fleece weight, defined as the response obtained expressed as a percentage of the cumulated selection differential. This enables a more direct comparison to be made between the response achieved and that expected from the heritability of fleece weight in the base population or control flock. It also enables the pattern of response to be examined having first removed variation attributable to fluctuations in selection differentials achieved in different years. Unfortunately it is only possible to estimate the realized heritability of fleece weight in single character selection flocks or when selection is based on an index, but not for the selection procedures followed in Experiments 4–7.

Realized heritability estimates have been calculated for Experiments 1 and 3. In Experiment 1, where the heritability of clean fleece weight in the base population was 0.47 (Morley, 1955), estimates of the realized heritability of clean fleece weight after one, four and six generations of selection were respectively 0.49, 0.50 and 0.30 (McGuirk, 1973). Blair (1981) estimated the realized heritability in Experiment 3 over 21 years of selection as 0.15, which was close to the average half-sib estimate for rams and ewes.

There has been considerable discussion in Australia over the pattern of response in the Merino selection flocks. This was sparked initially by the apparent declining rate of response in Experiment 1. Pattie and Barlow (1974) suggested that response had 'plateaued' in this flock after the first two generations of selection. Ferguson (1976) then further claimed that there was evidence of a marked slackening of response to selection for increased fleece weight in Experiments 2, 5 and 6. In subsequent papers, Turner and Jackson (1978) and McGuirk (1980b) have argued that further responses have been obtained in Experiments 5 and 6 and 1 respectively, using essentially the same information as that presented for these flocks in *Figures 29.1* and *29.2*. Blair (1981) has also claimed that there was no evidence of a decline in the rate of progress in Experiment 3, over the 21-year period of response which he examined.

One of the noticeable features of the fleece weight selection experiment with Australian Merino sheep is the wide variation in performance among successive groups of hoggets. For example, Robards (1979) estimated that the between-year coefficient of variation for fleece weight in control flock animals in some of these experiments was in excess of 20%. This variability complicates the measurement of progress because the absolute level of superiority of animals from high fleece weight selection flocks would appear to be directly related to the level of performance of the control group. This has been observed both under paddock and pen conditions (McGuirk, 1980a). The effect of this genotype × environment interaction would be reduced by expressing differences between the flocks in percentage rather than in absolute terms, which is why responses have been presented as percentage changes in *Figures 29.1* and *29.2*. An adequate analysis of responses in Experiments 1–3 would involve the calculation of realized heritability estimates for various time periods during the experiment and allow for possible genotype × environment interactions (Robards and Pattie, 1967; Turner and Jackson, 1978).

Predicted and observed correlated responses to selection for increased fleece weight

Characters which are correlated with fleece weight are of interest for a number of reasons. Because all breeds are to some extent dual purpose, it is important to know if there are genetic associations between fleece weight and characters such as reproductive performance and growth. Secondly, a genetic correlation with a fleece or skin character can throw some light on the mechanisms by which selection can bring about an increase in wool production, and might also suggest useful indirect selection criteria in improvement programmes. These two topics have been jointly examined through a study of the components of fleece weight (Turner, 1958), in which gains in fleece weight can be initially attributed to increases in wool production/unit area of skin or by an increase in the wool growing surface area. Wool production/unit area of skin can in turn be described as a function of follicle or fibre density, average fibre diameter and fibre length, which is often measured as staple length. The wool growing surface area is influenced both by the size of the animal, a function of body weight, and by the degree of skin wrinkle or fold.

The availability of animals from fleece weight selection flocks has enabled the physiological mechanisms underlying genetic differences in wool production to be examined much more directly than was possible from the study of fleece components, and wool from these flocks can be processed to assess its manufacturing potential. All of these topics will be discussed in the following sections, using estimates of genetic correlations as well as responses observed in selection flocks, notably those in which selection was for increased fleece weight (*Table 29.2*).

GENETIC CORRELATIONS WITH OTHER PRODUCTION CHARACTERS AND WITH SKIN, FLEECE AND BODY TRAITS

Genetic associations between fleece weight and measures of reproductive performance were reviewed by Turner (1972) and will not be considered in detail here. Heritability estimates for reproductive traits are generally low (0 to 0.2) and correlations with fleece weight are low and variable in sign. Consequently, changes observed in characters such as lambs born/ewe joined in flocks selected for increased fleece weight have been small (Turner *et al.*, 1972; Barlow, 1974).

Estimates of the heritability of fleece, skin and body characters listed in *Table 29.4* are essentially an updated version of those published by Turner and Dunlop (1974) for apparel wool breeds. Not all fleece, skin and body characters have been considered; emphasis has been placed on those of interest in wool biology or to the processor. Resistance to compression, the degree of medullation, post-scouring colour and staple crimp frequency are all thought to have some effect on the processing performance of wools, or their suitability for different end uses, as will be discussed later.

Most of the characters in *Table 29.4* have a moderate to high heritability. Estimates tend to be consistent across sheep types (see *Table 29.1*), although it is difficult to be sure for the follicle characteristics and for characters such as resistance to compression and post-scouring colour, where there are so few estimates.

The genetic correlations listed in *Table 29.4* can conveniently be grouped to represent the average degree of correlation found with fleece weight. The bounds suggested by Brown and Turner (1968) have been used, namely:

–0.6 and higher	high negative
–0.4 to –0.6	moderate negative
–0.2 to –0.4	low negative
and 0 to –0.2	negligible

Table 29.4 ESTIMATES OF THE HERITABILITY OF FLEECE, SKIN AND BODY CHARACTERS AND OF THEIR GENETIC CORRELATIONS WITH FLEECE WEIGHT

Character	*Heritability*	*Genetic correlations with:*		*Reference*
		Greasy fleece weight	*Clean fleece weight*	
Greasy fleece weight (GFW)	See *Table 29.1*	—	+0.6 to +1.0	1
Clean fleece weight (CFW)	See *Table 29.1*	+0.6 to +1.0	—	1
Yield	0.3 to 0.8	−0.2 to +0.8	+0.5 to +0.8	1
Yearling/hogget liveweight	0.3 to 0.7	−0.3 to +0.6	−0.2 to +0.6	1
Skin folds or wrinkle	0.2 to 0.8	0.0 to +0.7	−0.4 to +0.1	1
Average fibre diameter	0.2 to 0.6	+0.1 to +0.5	0 to +0.4	1
Staple length	0.4 to 0.7	0.0 to +0.7	+0.3 to +0.9	2
Follicle density	0.3 to 0.6	0.2, −0.1	0.3, 0.1, 0.0	2, 3
S/P ratio	0.5, 0.4	0.5	0.3, 0.3	3, 4
Follicle depth	0.4	0.3	0.4	3
Follicle bending	0.4, 0.5	−0.1, +0.1	−0.5, −0.2	3, 5
Resistance to compression	0.8	0.0	−0.2	5
Medullation (hairyness)	0.3 to 0.7	0.3, 0.3, 0.1	0.1	1, 6
Post-scouring colour	0.3		+0.4	7
Crimp frequency	0.3 to 0.6	−0.2 to −0.1	−0.2 to −0.6	2

References: 1 Turner and Dunlop, 1974
2 Brown and Turner, 1968
3 Jackson, Nay and Turner, 1975
4 Heydenrych, Vosloo and Meissenheimer, 1977
5 Watson, Jackson and Whiteley, 1977
6 Blair, 1981
7 Whiteley and Jackson, 1982

This same system of grouping was then repeated for the positive correlations. Using this grouping procedure, genetic correlations with fleece weight would be classified as follows, and with the correlation holding for both greasy and clean fleece weight unless otherwise specified:

Moderate negative:	crimp frequency with clean fleece weight.
Low negative:	crimp frequency with greasy fleece weight; follicle curvature;
Low positive:	fibre or follicle number, staple length and skin wrinkle with greasy fleece weight, ratio of secondary/primary follicles, follicle depth;
Moderate positive:	yield and staple length with clean fleece weight, post-scouring colour, percent medullated fibres;
High positive:	greasy with clean fleece weight.

All other correlations could be classed as negligible, although correlations between fleece weight and fibre diameter are almost always positive. It must be stressed that this grouping of correlations represents the average degree of relationship across published estimates, and that the picture in individual flocks and especially in different breeds may not conform to this average situation.

Correlated responses in fleece, skin and body characters have been published for Experiments 1 and 2. Barlow (1974) compared estimates of the realized genetic correlations with two generations of divergent selection for fleece weight and found good overall agreement with estimates for the same base or control flock. He employed two methods for estimating rates of correlated change, to allow for the declining rate of response in the high fleece weight selection flock. Both of his sets of published regression coefficients have been used here to estimate the realized genetic correlations to six generations of selection for increased fleece weight and are given in *Table 29.5*. A figure of 0.39 has been used as the realized

Table 29.5 ESTIMATES OF REALIZED GENETIC CORRELATIONS (r_g) TO SIX GENERATIONS OF SELECTION FOR INCREASED FLEECE WEIGHT

Character	*Estimation*	*Procedure*[(a)]	*Previous estimates*		
	1	*2*	r_g	h^2	*Source*
Greasy fleece weight	0.55	0.67	0.65	0.40	1
Yield	0.77	0.83	0.56	0.39	1
Hogget body weight	−0.04	0.06	−0.12	0.36	1
Skin wrinkle	0.18	0.20	0.12	0.50	1
Staple length	0.32	0.36	0.39	0.56	1
Average fibre diameter	0.26	0.36	0.16	0.47	2
Follicle density	0.20	0.14	0.30	0.40	2
S/P ratio	0.19	0.23	0.32	0.45	3
Crimp frequency	−0.60	−0.64	−0.53	0.47	1

[(a)]See Barlow (1974). Estimates are averages for ewes and rams except for average fibre diamater, follicle density and S/P ratio, which are for ewes only.

References

1 Morley (1955)

2 Brown and Turner (1968)

3 Jackson, Nay and Turner (1975)

Table of figure derived by Barlow (1974) from the above three sources

heritability of fleece weight over this period of the experiment (McGuirk, 1973).

The estimates of realized genetic correlations in *Table 29.5* are in overall agreement with predictions for this and similar Merino populations. Moderate to high positive correlations were found for greasy fleece weight and yield, and there was a moderate negative correlation with crimp frequency. Correlations with the components of fleece weight were mostly positive, but low or negligible in magnitude. The only exception was with body weight where the estimated (Morley, 1955) and the realized correlations were slightly negative. Barlow (1974) concluded that the gains in wool production/head were in components of wool production/unit area, principally in staple length, fibre diameter and in fibre density, and not in the components of the wool growing area of the sheep. This general conclusion is fully supported by direct observations on wool production per unit area made under pen conditions (Williams, 1979).

A less formal analysis of correlated changes in Experiment 2 was published by Turner, Brooker and Dolling (1970). They drew essentially the same conclusions. Selection for increased fleece weight led to increases in staple length and follicle density, with smaller changes in body weight, skin folds and fibre diameter, and to an appreciable reduction in crimp frequency. Taken together, the results of Experiments 1 and 2 indicate that correlated changes in fleece, skin and body characters are in the direction expected, and of the same order of magnitude as predicted by estimates of genetic correlations for those populations. There is an obvious need to extend the observations on these flocks to the other characters listed in *Table 29.4*.

CSIRO also established a number of flocks in which selection was for increased and decreased expression of the components of fleece weight. Turner, Brooker and Dolling (1970) described the direct responses in the flocks selected for body weight, skin wrinkle, follicle density, fibre diameter and staple length, and for measured wool production/unit area and yield, as well as the changes in fleece weight in these flocks. Selection for increased wool production/unit area produced the greatest gains in fleece weight/head, although selection for increased body weight, staple length and yield also led to statistically significant gains in fleece weight. Selection for increased fibre diameter, fibre density and skin wrinkle were less effective or even ineffective in increasing fleece weight. Rendel and Nay (1978) described responses to selection for increased and reduced follicle density and secondary/primary follicle ratios, and claimed that selection for increased density or S/P ratio did not increase fleece weight. Hence it is not clear if the components of fleece weight, and especially the follicle characteristics, are causally related to fleece weight and could be used as indirect selection criteria in attempting to improve fleece weight. The possible value of follicle depth and curvature as indirect selection criteria is currently being examined (Jackson, Nay and Turner, 1975).

Brief mention should also be made of Merino flocks selected for increased or reduced crimp frequency (Robards and Pattie, 1967). Crimp frequency and fleece weight have been found to have a negative correlation of moderate size in a number of flocks, and the observed changes in wool production in flocks selected for high and low crimp frequency were as predicted (Robards, Williams and Hunt, 1974). Wool from these flocks will be referred to later when discussing the physiological and biochemical consequences of selecting for increased fleece weight.

PHYSIOLOGICAL AND BIOCHEMICAL CONSEQUENCES OF SELECTION

There have been a number of studies of the efficiency of conversion of feed to wool, using animals from flocks selected for increased fleece weight. This is a topic of considerable importance, for unless gains in wool production/head are due at least in part to improved efficiency, no economic advantage may follow. Previously it had been observed within Merino flocks that animals which produced the most wool also tended to be the most efficient (Schinckel, 1960; Williams, 1979), an observation that had also been recorded for Romneys and Corriedales (Clark *et al.*, 1965; Wodzicka-Tomaszewska, 1966). However, comparisons of animals from high and low fleece weight selection flocks, or from high fleece weight flocks and control flock animals, provided direct evidence that the differences in wool production were due almost entirely to improved efficiency, and not to increased intake (Williams, 1979; McGuirk, 1980a). Pen studies over a number of generations of selection indicate that for animals in Experiment 1 fed at maintenance, there has been a trend of continuing gains with time both in wool growth/head and in feed efficiency (Robards, Williams and Hunt, 1974).

A further point to emerge from these pen studies was that the absolute advantage in wool production to animals from the high fleece weight selection flock over either control flock or low fleece weight animals, increased as the average level of production increased (Williams, 1979; McGuirk, 1980a). This finding has now been observed for animals from Experiments 1, 4 and 5. The varying levels of production were generally produced by altering intake, although Piper and Dolling (1969) also varied the nutritive value of the diets fed.

Studies to further examine differences in production and feed efficiency have focused on the availability and utilization of cystine and on related changes in the sulphur content of wool. Wool is composed predominantly of keratin which has a high content of sulphur. Individual wool samples have been found to have between 2.7% and 4.2% sulphur. Most of the sulphur is present as cystine, with smaller amounts of cysteine and methionine (Reis, 1979). Keratin is not homogeneous and the wool proteins are usually grouped as high sulphur, low sulphur or high tyrosine proteins. There is considerable heterogeneity within these major groupings, especially in the high sulphur and high tyrosine proteins where there are many sub-fractions (Gillespie, 1982). For convenience these sub-fractions are generally grouped into families, based on similarities in chemical composition. The three major protein groups are thought to be associated with different structural components of the cortical cells of the wool fibre, with the low sulphur proteins concentrated in the microfibrils and the high sulphur and probably also the high tyrosine proteins concentrated in the surrounding non-fibrous matrix.

Both the rate of wool growth and its sulphur content are influenced by the availability of sulphur-containing amino acids. When supplements of cystine, methionine or casein are infused into the abomasum of sheep, both wool production and the sulphur content of the wool are also increased (Reis, 1979). The increase in sulphur content is due to an increased yield of high sulphur proteins, and the fact that their sulphur content is also increased due to a change in the proportions of the sub-fractions.

Selection for increased fleece weight has been shown to reduce the sulphur content of wool in Experiments 1 (Reis *et al.*, 1967) and 5 (Piper and Dolling, 1966). The effect is not simply a dilution of the same amount of sulphur in a greater amount of wool, as selection for increased fleece weight increased total sulphur output. More extensive chemical analyses have been undertaken on wools from high and low crimp frequency flocks. In summary, selection for reduced crimp frequency increased clean wool production and the efficiency of conversion of feed to wool and the wool had a lower sulphur content (McGuirk, 1980a). Campbell, Whiteley and Gillespie (1972; 1975) isolated the high sulphur wool proteins from these wools and found both the overall yield and sulphur content was reduced in wool from the low crimp frequency flock. The amino acid composition of the two types of wool also differed, with selection for low crimp frequency reducing the percentage content of cystine, and more generally reducing the content of those amino acids which are associated with the non-fibrous protein matrix within the cortical cells (threonine, serine, proline and cystine; Kulkarni, 1977). Not all of these changes produced by selection for low crimp frequency have as yet been looked for in high fleece weight selection flocks, but given the strong agreement between the crimp and fleece weight selection flocks in other characters (McGuirk, 1980a), they appear more than likely. If this proves to be true, then the effect of selection for increased fleece weight on the chemical composition of the wool would be exactly counter to that produced by infusing sulphur-containing amino acids.

Selection for increased fleece weight does not appear to have altered feed digestibility or the proportion of dietary nitrogen reaching the abomasum (Piper and Dolling, 1969; Williams, 1979). Differences in production and efficiency would appear to be due to differences in the utilization of nutrients after absorption from the alimentary tract. Cystine metabolism is presumably altered although precisely what these effects are is not clear (Williams, 1979). Blood plasma cystine levels are reduced and this change would appear to be responsible for the lower concentration of sulphur in wool of sheep from high fleece weight selection flocks.

Perhaps the most exciting finding in this field was the observation by Williams, Robards and Saville (1972) that animals from the high fleece weight selection flock in Experiment 1 were much more responsive to infusions of cystine or methionine than animals from the corresponding low fleece weight selection flock (*Table 29.6*). Sheep from the high fleece weight selection flock showed a greater response both in wool production and in sulphur output. The responses in wool production show the same pattern as in the previously described studies in which intake was increased, and indicate that the availability of sulphur-containing amino is limiting the ability of animals with a high genetic potential for wool

Table 29.6 RESPONSE OF HIGH AND LOW FLEECE WEIGHT GROUPS (EXPERIMENT 1) TO AMINO ACID SUPPLEMENTATION

Flock	*Treatment*	*Wool production* (μg/cm²/day)	*Sulphur content* (%)	*Sulphur output* (μg/cm²/day)
High fleece weight	No supplementation	743	2.99	22.2
	+ cystine	1114	3.45	38.4
	+ methionine	1190	3.53	42.0
Low fleece weight	No supplementation	639	3.48	22.2
	+ cystine	714	3.77	26.9
	+ methionine	757	3.74	28.3

Adapted from Williams, Robards and Saville (1972)

production to express that potential. This finding points to the need to develop special nutritional strategies if we are to reap the full benefits of our improvement programmes.

PROCESSING PERFORMANCE

Apparel wool

Whiteley and Jackson (1982) recently reviewed the raw wool characteristics of importance in the production of fine, uniform yarns and concluded that average fibre diameter had a much greater effect on processing performance than any of the other characters considered. Average fibre diameter affects the spinning potential of a given quantity of wool which is why, within a breed such as the Merino, fine wools usually demand a price premium. In the past crimp frequency has been used to predict average fibre diameter, but now diameter is measured directly when testing samples from individual animals and lots prior to sale. Average fibre diameter is moderately heritable and positively correlated with fleece weight (see *Table 29.4*). Most breeders wish to maintain average fibre diameter in their flocks at its traditional level and would attempt to do this by excluding rams with average diameter measurements well above the flock average when choosing sires. Some genetic progress in fleece weight will be foregone by imposing this restriction on average fibre diameter.

Many of the other fibre characteristics which influence processing performance are strongly influenced by environmental factors and are thus unlikely to be of much interest in a breeding programme. This would include such characters as the degree and type of vegetable fault contamination and staple strength. Whiteley and Jackson (1982) suggested that resistance to compression, or 'bulk' as it is sometimes called, did require possible inclusion in breeding programmes, especially in view of the higher spinning speeds now used in industry. Resistance to compression reflects the degree of fibre crimping (Chaudri and Whiteley, 1968), is positively correlated with staple crimp frequency (Watson *et al.*, 1975) and it is probably positively correlated with sulphur content of the wool. A high degree of fibre crimp is associated with less fibre entanglement during scouring, but leads to greater wastage during carding and poorer spinning performance. To avoid some of these difficulties, Australian superfine wool is processed at lower speeds then coarser wool and this contributes to the much higher costs of processing superfine wools (Australian Wool Corporation, 1977). As selection for increased fleece weight reduces crimp and thereby resistance to compression (*Table 29.4*), it could thus lead to lower processing costs in some circumstances.

Selection for increased fleece weight could change processing properties and fibre and fabric characteristics either by altering the physical dimensions of the wool fibres or their chemical composition. Lipson and Walls (1962) examined the worsted manufacturing performance of wools from the high fleece weight selection flock in Experiment 5 and from its control flock. The gains in fleece weight in this experiment were achieved without increasing average fibre diameter but there was a reduction in staple crimp frequency. Wool from the high fleece weight flock was superior in carding and top-making performance and in yarn and fabric properties. Card waste was lower, the ratio of top to noil increased, average fibre length in the top was greater, there were fewer end-breaks during spinning and both the yarn and cloth were of greater strength. Possible explanations for these differences include the slightly lower average fibre diameter and higher average fibre length in both the raw wool and top from the high fleece weight wools. With hindsight, the differences may have also been due to lower resistance to compression in these wools, associated with lower crimp frequency.

Selection for increased fleece weight may produce wools with some poorer processing characteristics, due to the expected reduction in sulphur content. Perhaps the most important of these is the ability of the wool to take and retain set, a property made use of in permanent pleating of worsted materials (skirts, trousers, etc.). To date only wools from the high and low crimp frequency flocks (Robards and Pattie, 1967) have been compared (Campbell, Whiteley and Gillespie, 1972) but, as setting characteristics are associated with sulphur content (Whiteley *et al.*, 1976), they are likely to be found between wools from high and low fleece weight groups.

Other end uses

The effects of selection for increased fleece weight on the processing performance of wools for woollen and carpet manufacture are less easy to specify. Both systems utilize a wide variety of wool types, which can be blended to produce the desired result. For example, carpet blends usually consist of three major types of wool, traditional carpet wools such as the Scottish Blackface, Swaledale and New Zealand Drysdale, general purpose wools which are generally thought to have the virtues of good colour and strength and spin well, and 'filler' wools which are used because they are cheap (Ross, Wickham and Elliot, 1982). The composition of a blend will be determined by price and availability as well as by its anticipated performance.

Changes in three wool characters, namely the degree of medullation, resistance to compression and average fibre diameter would be of some importance for both woollen and carpet manufacture. The presence of medullated fibres is apparently critical to the appearance of a carpet, but if the proportion of medullated fibres is too high, the wearing performance of a carpet, and the uptake of dye and final appearance of the dyed product can all be adversely affected. A high resistance to compression will improve the appearance of carpets, especially after use, and would also improve the shape retention of woollen garments. Average fibre diameter will affect the softness of woollen goods.

Information on the relationships between fleece weight and either fibre diameter or resistance to compression is limited in breeds producing wool for the woollen or carpet industries. Carnaby and Elliott (1980) reported generally negative estimates for the phenotypic correlations between greasy fleece weight and resistance to compression, which were similar in magnitude to the genetic correlation of 0 to –0.2 reported for the Merino (Watson *et al.*, 1977). The incidence of medullated fibres would be expected to increase following selection for increased fleece weight, as the characters are positively correlated.

In summary selection for increased fleece weight could affect the suitability of individual wools for particular end uses, through possible correlated changes in the percentage of medullated fibres, resistance to compression or average fibre diameter. However, given the wide variety of wools used in the woollen and carpet industries, and the substantial variation in their characteristics, selection for increased fleece weight in such breeds is unlikely to be of any great significance to either processing performance or the quality of the final product.

Industry application of selection programmes using fleece measurement

The results summarized in the previous sections have shown that fleece weight is moderately heritable and will respond to mass selection, that the gains achieved are likely to reflect increased efficiency of conversion of feed to wool, and that they are unlikely to be associated with major undesirable effects on other production characters or on the processing characteristics of the wool. The relevance of these findings to sheep performance recording programmes will now be discussed briefly.

In breeds where wool is relatively unimportant as a source of return, selection for increased fleece weight should remain an unimportant breeding objective. The emphasis given to fleece weight in a selection programme should reflect its relative economic importance, ideally in an index of overall genetic merit (see for example Gjedrem, 1966), and the expected gains in fleece weight would be small. In Norway, where wool contributes only 20% of total returns from sheep (Gjedrem, 1969), it is estimated that there has been a genetic decline in fleece weight in both lambs (Eikje, 1975b) and ewes (Eikje and Steine, 1976), resulting from their new sheep breeding programme introduced in the 1960s.

There are no estimates of genetic progress in fleece weight in commercial flocks in specialized wool-producing breeds, and only limited information on sire selection differentials achieved for production characters

(McGuirk, 1982). Because pedigree information is generally not recorded in these flocks, and systematic progeny testing little used, estimates of genetic change cannot readily be obtained from an analysis of production data. The use of frozen semen would be one way of estimating genetic progress on a widespread scale (Smith, 1977). Such a scheme would enable breeders to make a direct comparison of the progeny of successive groups of sires.

Improvement programmes for the specialized wool-producing breeds will continue to aim to increase clean wool production, while maintaining average fibre diameter. The recommended procedure has been to select ram replacements on clean fleece weight, having removed those with an average fibre diameter above acceptable limits, while ewes are selected on greasy fleece weight. In the past most Merino breeders have made little use of the fleece measurement laboratories established to determine yield and average fibre diameter (McGuirk, 1982). In Australia and South Africa, for example, only a minority of breeders have used the laboratories and generally only samples from selected groups of rams were then tested. This situation is changing slowly as ram breeders and ram buyers become more aware of the value of measurement as a selection aid, and as individual breeders successfully incorporate measurement into their breeding, management and ram selling operations.

The fleece measurement laboratories would appear to be the obvious focus for more comprehensive performance recording schemes for the specialized wool-producing breeds, given that yield and fibre diameter measurements will still be required for individual ram samples. At present little attention is given to the recording of reproductive data or growth rates in many Merino flocks, even though the number of surplus stock for slaughter can be an important source of income for commercial producers (Ponzoni, 1979). There is growing interest among breeders in characters which protect the fleece while it is being grown, from dust and water penetration, and the measurement of appropriate fleece constituents could easily be accommodated within the existing fleece testing services. As the number of characters increases, there will be a greater need to use selection indexes in these improvement programmes, so as to make most efficient use of the information recorded.

Acknowledgements

Dr Neville Jackson kindly provided the data from which genetic gains in the CSIRO selection flocks were estimated. I am also grateful to Dr Jackson and to Drs Helen Newton Turner and K.J. Whiteley for their helpful comments on this paper. The results presented in *Table 29.6* are reproduced with permission of the CSIRO, publishers of the *Australian Journal of Biological Sciences*.

References

AUSTRALIAN WOOL CORPORATION (1977). *Superfine Wool and Stud Industries*. Australian Wool Corporation, Melbourne

AUSTRALIAN WOOL CORPORATION (1981). *Statistical Data on Wool* No. 27, December 1981, p. 3

BAKER, R.L., CLARKE, J.N., CARTER, A.H. and DIPROSE, G.D. (1979). Genetic and phenotypic parameters in New Zealand Romney sheep. I. Body weights, fleece weights and oestrous activity. *New Zealand Journal of Agricultural Research* **22**, 9–21

BARLOW, R. (1974). Selection for clean fleece weight in Merino sheep. II. Correlated responses to selection. *Australian Journal of Agricultural Research* **25**, 973–994

BLAIR, H.T. (1981). Response to selection for open face and greasy weight in Romney sheep. PhD Thesis. Massey University

BROWN, G.H. and TURNER, H.N. (1968). Response to selection in Australian Merino sheep. I. Estimates of phenotypic and genetic parameters for some production traits in Merino ewes and an analysis of the possible effects of selection on them. *Australian Journal of Agricultural Research* **19**, 303–322

BROWN, G.H., TURNER, HN., YOUNG, S.S.Y. and DOLLING, C.H.S. (1966). Vital statistics for an experimental flock of Merino sheep. III. Factors affecting wool and body characteristics including the effect of age of ewe and its possible interaction with method of selection. *Australian Journal of Agricultural Research* **17**, 557–581

CAMPBELL, M.E., WHITELEY, K.J. and GILLESPIE, J.M. (1972). Compositional studies of high- and low-crimp wools. *Australian Journal of Biological Sciences* **25**, 977–987

CAMPBELL, M.E., WHITELEY, K.J. and GILLESPIE, J.M. (1975). Influence of nutrition on the crimping rate of wool and the type and proportion of constituent proteins. *Australian Journal of Biological Sciences* **28**, 389–397

CARNABY, G.A. and ELLIOTT, K.H. (1980). Bulk: a wool trait of importance to the carpet industry. *Proceedings of the New Zealand Society of Animal Production* **40**, 196–204

CHAUDRI, M.A. and WHITELEY, K.J. (1968). The influence of natural variation in fibre properties on the bulk compression of wool. *Textile Research Journal* **38**, 897–906

CLARK, V.R., KESHARY, K.R., COOP, I.E. and HENDERSON, A.E. (1965). The relationship between fleece weight and efficiency in Romney and Corriedale sheep. *New Zealand Journal of Agricultural Research* **8**, 511–522

CROSTON, D., DANELL, O., ELSEN, J.M., FLAMANT, J.C., HANRAHAN, J.P., JAKUBEC, V., NITTER, G. and TRODAHL, S. (1980). A review of sheep recording and evaluation of breeding animals in European countries: a group report. *Livestock Production Science* **7**, 373–392

DALTON, D.C. (1962). Characters of economic importance in Welsh Mountain sheep. *Animal Production* **4**, 269–278

DONEY, J.M. (1958). The role of selection in the improvement of Welsh Mountain hill sheep. *Australian Journal of Agricultural Research* **9**, 819–829

DUN, R.B. and EASTOE, R.D. (1970). *Science and the Merino Breeder.* Sydney, NSW Government Printer

EIKJE, E.D. (1971). Studies on sheep production records. I. Effects of environmental factors on weight of lambs. *Acta Agriculturae Scandanavica* **21**, 26–32

EIKJE, E.D. (1975a). Studies on sheep production records. VII. Genetic, phenotypic and environmental parameters for productivity traits of ewes. *Acta Agriculturae Scandanavica* **25**, 242–252

EIKJE, E.D. (1975b). Studies on sheep production records. VIII. Estimation of genetic change. *Acta Agriculturae Scandanavica* **25**, 253–260

EIKJE, E.D. and STEINE, T.A. (1976). Realized genetic change in ewe productivity traits. *Meldinger fra Norges Landbrukshøgskole* **55**, 12p

ELLIOTT, K.H., RAE, A.L. and WICKHAM, G.A. (1979). Analysis of records of a Perendale flock. 2. Genetic and phenotypic parameters for immature body weights and yearling fleece characteristics. *New Zealand Journal of Agricultural Research* **22**, 267–272

FERGUSON, K.A. (1976). Australian sheep breeding programmes—aims, achievements and the future. In *Sheep Breeding* (G.J. Tomes, D.E. Robertson and R.J. Lightfoot, Eds.), pp. 13–25. London, Butterworths

GILLESPIE, J.M. (1982). Structural proteins of hair: isolation, characterization and regulation of biosynthesis. In *Biochemistry and Physiology of Skin* (L.A. Goldsmith, Ed.) Oxford, Oxford University Press (In press)

GJEDREM, T. (1966). Selection index for ewes. *Acta Agriculturae Scandanavica* **16**, 21–29

GJEDREM, T. (1969). Phenotypic and genetic parameters for fleece weight and some wool quality traits. *Acta Agriculturae Scandanavica* **19**, 103–115

GREGORY, I.P. (1982). Genetic studies of South Australian Merino sheep. III. Heritabilities of various wool and body traits. *Australian Journal of Agricultural Research* **33**, 355–362

HEYDENRYCH, H.J., VOSLOO, L.P. and MEISSENHEIMER, D.J.B. (1977). Selection response in South African Merino sheep selected either for high clean fleece mass or for a wider S/P ratio. *Agroanimalia* **9**, 67–73

JACKSON, N., NAY, T. and TURNER, H.N. (1975). Response to selection in Australian Merino sheep. VII. Phenotypic and genetic parameters for some wool follicle characteristics and their correlation with wool and body traits. *Australian Journal of Agricultural Research* **26**, 937–957

KULKARNI, V.G. (1977). The isolation and characterization of cortical cells and their constituent protein fractions from high and low crimp wool fibres. PhD Thesis. University of New South Wales

LIPSON, M. and WALLS, G.W. (1962). Processing of wool from a flock selected for high fleece weight. *Journal of the Textile Institute* **53**, 416–422

McGUIRK, B.J. (1973). The importance of production characters in Merino sheep. PhD Thesis. University of Edinburgh

McGUIRK, B.J. (1980a). Selection for wool production in Merino sheep. In *Selection Experiments in Laboratory and Domestic Animals*, pp. 176–197. Slough, Commonwealth Agricultural Bureau

McGUIRK, B.J. (1980b). The effects of selection for increased fleece weight. *Proceedings of the Australian Society of Animal Production* **13**, 171–174

McGUIRK, B.J. (1982). Recording and the use of records in sheep selection programmes overseas. *Proceedings of World Congress on Sheep and Beef Cattle Breeding* (W.C. Smith and R.A. Barton, Eds.), Volume 2, pp.49–55

MORLEY, F.H.W. (1955). Selection for economic characters in Australian

Merino sheep. 5. Further estimates of phenotypic and genetic parameters. *Australian Journal of Agricultural Research* **6**, 873–881

MULLANEY, P.D., BROWN, G.H., YOUNG, S.S.Y. and HYLAND, P.G. (1970). Genetic and phenotypic parameters for wool characteristics in fine-wool Merino, Corriedale and Polwarth sheep. II. Phenotypic and genetic correlations. *Australian Journal of Agricultural Research* **21**, 527–540

PATTIE, W.A. and BARLOW, R. (1974). Selection for clean fleece weight in Merino sheep. I. Direct response to selection. *Australian Journal of Agricultural Research* **25**, 643–655

PIPER, L.R. and DOLLING, C.H.S. (1966). Variation in the sulphur content of wool of Merino sheep associated with genetic differences in wool-producing capacity. *Australian Journal of Biological Sciences* **19**, 1179 –1182

PIPER, L.R. and DOLLING, C.H.S. (1969). Efficiency of conversion of food to wool. IV. Comparison of sheep selected for high clean wool weight with sheep from a random control group at three levels of dietary protein. *Australian Journal of Agricultural Research* **20**, 561–578

PONZONI, R.W. (1979). Objectives and selection criteria for Australian Merino sheep. *Proceedings of the Inaugural Conference of the Australian Association of Animal Breeding and Genetics*, pp. 320–336

REIS, P.J. (1979). Effects of amino acids on the growth and properties of wool. In *Physiological and Environmental Limitations to Wool Growth* (J.L. Black and P.J. Reis, Eds.), pp. 223–242. Armidale, University of New England Publishing Unit

REIS, P.J., TUNKS, D.A., WILLIAMS, O.B. and WILLIAMS, A.J. (1967). A relationship between sulphur content of wool and wool production by Merino sheep. *Australian Journal of Biological Sciences* **20**, 153–163

RENDEL, J.M. and NAY, T. (1978). Selection for high and low ratio and high and low primary density in Merino sheep. *Australian Journal of Agricultural Research* **29**, 1077–1086

ROBARDS, G.E. (1979). Regional and seasonal variation in wool growth throughout Australia. In *Physiological and Environmental Limitations to Wool Growth* (J.L. Black and P.J. Reis, Eds.), pp. 1–42. Armidale, University of New England Publishing Unit

ROBARDS, G.E. and PATTIE, W.A. (1967). Selection for crimp frequency in wool of Merino sheep. I. Direct response to selection. *Australian Journal of Experimental Agriculture and Animal Husbandry* **7**, 552–558

ROBARDS, G.E., WILLIAMS, A.J. and HUNT, M.H. (1974). Selection for crimp frequency in the wool of Merino sheep. 2. Efficiency of conversion of feed to wool. *Australian Journal of Experimental Agriculture and Animal Husbandry* **14**, 441–448

ROSS, D.A., WICKHAM, G.A. and ELLIOT, K.H. (1982). Breeding objectives to improve wool used in carpets. *Proceedings of the World Congress on Sheep and Beef Cattle Breeding* (W.C. Smith and R.A. Barton, Eds.), Volume 1, pp.37–45

SAVILLE, D.G. and ROBARDS, G.E. (1972). Efficiency of conversion of food to wool in selected and unselected Merino types. *Australian Journal of Agricultural Research* **23**, 117–130

SCHINCKEL, P.G. (1960). Variation in feed intake as a cause of variation in wool production of grazing sheep. *Australian Journal of Agricultural Research* **11**, 585–594

SMITH, C. (1977). Use of frozen semen and embryos to measure genetic trends in farm livestock. *Zietschrift für Tierzüchtung und Züchtungsbiologie* **94**, 119–127

TERRILL, C.E. and HAZEL, L.N. (1943). Heritability of yearling fleece and body traits of range Rambouillet ewes. *Journal of Animal Science* **2**, 358–359

TURNER, H.N. (1958). Relationships among clean wool weight and its components. I. Changes in clean wool weight related to changes in the components. *Australian Journal of Agricultural Research* **9**, 521–552

TURNER, H.N. (1972). Genetic interactions between wool, meat and milk production in sheep. *Animal Breeding Abstracts* **40**, 621–634

TURNER, H.N. and DUNLOP, A.A. (1974). Selection for wool production. *Proceedings of the First World Congress on Genetics Applied to Livestock Production.* Vol. 1, 739–756

TURNER, H.N. and JACKSON, N. (1978). Response to selection in Australian Merino sheep. VIII. Further results on selection for high clean wool weight with attention to quality. *Australian Journal of Agricultural Research* **29**, 615–629

TURNER, H.N., BROOKER, M.G. and DOLLING, C.H.S. (1970). Response to selection in Australian Merino sheep. III. Single character selection for high and low values of wool weight and its components. *Australian Journal of Agricultural Research* **21**, 955–984

TURNER, H.N., DOLLING, C.H.S. and KENNEDY, J.F. (1968). Response to selection in Australian Merino sheep. I. Selection for high clean wool weight, with a ceiling on fibre diameter and degree of skin wrinkle. Response in wool and body characteristics. *Australian Journal of Agricultural Research* **19**, 79–112

TURNER, H.N., McKAY, E. and GUINANE, F. (1972). Response to selection in Australian Merino sheep. IV. Reproduction rate in groups selected for high clean wool weight with a ceiling on degree of skin wrinkle and either fibre diameter or crimp frequency. *Australian Journal of Agricultural Research* **23**, 131–148

VESELEY, J.A. and SLEN, S.B. (1961). Heritabilities of weaning weight, yearling weight and clean fleece weight in range Romnolet sheep. *Canadian Journal of Animal Science* **41**, 109–114

WATSON, N., JACKSON, N. and WHITELEY, K.J. (1977). Inheritance of the resistance to compression property of Australian Merino wool and its genetic correlation with follicle curvature and various wool and body characters. *Australian Journal of Agricultural Research* **28**, 1083–1094

WHITELEY, K.J. and JACKSON, N. (1982). Breeding for apparel wool. *Proceedings of the World Congress on Sheep and Beef Cattle Breeding*, November 1980 (W.C. Smith and R.A. Barton, Eds.), Volume 1, pp.47–55

WHITELEY, K.J., CAMPBELL, M.E., SIU, C. and KULKARNI, V.G. (1976). Mechanical and chemical properties of wool fibres exhibiting extreme variations in chemical composition and crimp form. *Proceedings of the 5th International Wool Textile Research Conference, Aachen, 1975*, Vol. 4, 57–66

WILLIAMS, A.J. (1979). Speculation on the biological mechanisms responsible for genetic variation in the rate of wool growth. In *Physiological and Environmental Limitations to Wool Growth* (J.L. Black and P.J. Reis,

Eds.), pp. 337–354. Armidale, University of New England Publishing Unit

WILLIAMS, A.J., ROBARDS, G.E. and SAVILLE, D.G. (1972). Metabolism of cystine by Merino sheep genetically different in wool production. II. The responses in wool growth to abomasal infusion of L-cystine or DL-methionine. *Australian Journal of Biological Sciences* **25**, 1269–1276

WODZICKA-TOMASZEWSKA, M. (1966). Efficiency of wool growth. 1. Comparison of differences between high- and low-producing sheep under restricted and *ad libitum* feeding. *New Zealand Journal of Agricultural Research* **9**, 909–915

LIST OF PARTICIPANTS

Aguer, Mr D.	Intervet S.A. 43, Avenue Joxe, 49000 Angers, France
Allen, Dr J.D.	Colborn-Dawes Nutrition Ltd., Heanor Gate Industrial Estate, Heanor, Derby, UK
Alliston, Dr J.C.	Animal Breeding Research Organisation, West Mains Road, Edinburgh, UK
Ammann, Mr P.B.	Schweiz Zentralstelle Fur, Kleinviehzucht, Belpste, 16, CH-3007, Bern, Switzerland
Appleton, Mr M.	Liscombe EHF, Dulverton, Somerset, UK
Appleyard, Mr W.T.	Animal Diseases Research Association, Moredun Institute, 408 Gilmerton Road, Edinburgh, UK
Atkins, Mr K.D.	Animal Breeding Research Organisation, Kings Buildings, Edinburgh, UK
Awa, Prof. O.A.	Animal Science Division, Arab Centre for the Studies of Arid Zones and Dry Lands, Damascus, Syria
Awgichew, Mr K.	Institute of Agricultural Research -H.R.S., P.O. Box 2003, Addis Ababa, Ethiopia
Baker, Mr H.K.	Meat and Livestock Commission, PO Box 44, Queensway House, Bletchley, Milton Keynes MK2 2EF, UK
Barker, Mr J.D.	Animal Breeding Research Organisation, West Mains Road, Edinburgh, UK
Bath, Dr I.H.	Faculty of Veterinary Medicine, University College, Dublin, Ireland
Bebbington, Mr F.	Lancashire College of Agriculture. Myerscough Hall, Bilsborrow, Preston, Lancs., UK
Beck, Dr N.F.G.	Department of Agriculture, University College of Wales, Penglais, Aberystwyth SY23 3DD, UK
Black, Dr J.L.	CSIRO, Division of Animal Production, Prospect, PO Box 239, Blacktown, 2198, Australia
Booth, Miss A.M.	RHM Agriculture Ltd, Deans Grove, Colehill, Wimborne, Dorset, UK
Bouffault, Mons J.C.	Roussel Uclaf, Division Agrovet, 163 Ave. Gambetta, 75020 Paris, France
Boundy, Mr T.	Kilaganoon, Montgomery, Powys SY15 6HW, UK
Bouvard, Hans Ing.C.G.	Ciba Geigy SA, Research Station Les Barges, CH-1896, Voury, Switzerland

Box, Mr P.G.	Glaxo Animal Health, Breakspeare Road South, Harefield, Uxbridge, Middx, UK
Broadbent, Dr J.S.	ADAS/MAFF, Block 3, Government Buildings, Burghill Road, Westbury on Trymm, Bristol, UK
Butler, Ms G.	ADAS, Government Buildings, Kenton Bar, Newcastle upon Tyne, UK
Butler-Hogg, Dr B.W.	Animal Physiology Division, ARC Meat Research Institute, Langford, Bristol BS18 7DY, UK
Campbell, Mr C.	US Feed Grains Council, 11 College Green, Gloucester, UK
Cave-Penney, Mr A.L.	Hampshire College of Agriculture, Sparsholt, Winchester, Hants, UK
Chalmers, Mr D.A.	Pauls Agriculture Ltd, Research and Advisory Department, New Cut West, Ipswich, Suffolk, UK
Chapman, Mr D.G.	Northumberland College of Agriculture, Kirkley Hall, Ponteland, Newcastle upon Tyne, UK
Clark, Miss C.F.S.	Animal Husbandry Division, North of Scotland College of Agriculture, 581 King Street, Aberdeen AB0 1UD, UK
Cole, Dr D.J.A.	University of Nottingham School of Agriculture, Sutton Bonington, Loughborough, Leics, UK
Cooper, Mr A.	Seale Hayne College, Newton Abbot, Devon TG12 6NQ, UK
Coulson, Mr A.	Ag/Vet Technical Services Manager, Upjohn Ltd, Fleming Way, Crawley, W. Sussex RH10 2NJ, UK
Crosby, Dr F.	University College of Dublin, Dept of Agriculture, Lyons, Newcastle, Co. Dublin, Eire
Cunningham, Prof. J.M.M.	The West of Scotland Agriculture College, Auchincruive, Ayr KA6 5HW, UK
Cuthbert, Mr N.H.	BOCM Silcock, Basing View, Basingstoke, Hants, UK
Davis, Dr S.R.	Ruakura Animal Research Station, Private Bag, Hamilton, New Zealand
Deaville, Mr S.P.	Rumenco Limited, Stretton House, Derby Road, Burton on Trent, Staffs, UK
Deeley, Ms S.M.	Butterworth & Co (Publishers) Ltd, Borough Green, Sevenoaks, Kent TN15 8PH, UK
Dickson, Mr I.A.	The West of Scotland Agricultural College, Pathfoot Building, University of Stirling, Stirling FK9 4LA, UK
Dingwall, Mr W.S.	Dept of Applied Nutrition, Rowett Research Institute, Bucksburn, Aberdeen AB2 9SB, UK
Doney, Dr J.M.	Hill Farming Research Organisation, Bush Estate, Penicuik, Midlothian, UK
Dove, Dr H.	CSIRO, Division of Plant Industry, PO Box 1600, Canberra City, A.C.T. 2601, Australia
Dýrmundsson, Dr O.R.	The Agricultural Society of Iceland, Baendahollin, PO Box 7080, Reykjavik, Iceland
Eales, Mr F.A.	Moredun Research Institute, 408 Gilmerton Rd, Edinburgh EH17 7JH, UK
Eddie, Mr J.	Intervet Laboratories Ltd., Science Park, Melton Road, Cambridge CB4 4BH, UK

Ellis, Mr C.A.	Agriculture Department, University College of North Wales, Bangor, Gwynedd, UK
Emmerson, Mr S.A.	ADAS/MAFF, Shardlow Hall, Shardlow, Derby, UK
Fitzgerald, Mr S.	Agricultural Institute, Belclare, Tuam, Co. Galway, Eire
Foglini, Dr A.	Istituto Zooprofilattico, Sperimentale, Via Campo Boario-64100 Teramo, Italy
Francis, Mr R.B.	M & J Ranch, General Delivery, Priddis, Alberta, Canada T0L 1W0
Garnsworthy, Dr P.C.	University of Nottingham School of Agriculture, Sutton Bonington, Loughborough, Leics., UK
Geary, Mr M.	Intervet Laboratories Ltd., Science Park, Milton Road, Cambridge BB4 4BH, UK
Giudice, Dr G.	Corso Cavour 92, 06100 Perugia, Italy
Gunn, Dr R.G.	Hill Farming Research Organisation, Bush Estate, Penicuik, Midlothian, UK
Hale, Prof. N.	Animal Industries Department, University of Connecticut, Storrs, Connecticut 06268, USA
Hankey, Dr M.S.	Department of Agriculture, University of Newcastle upon Tyne, UK
Hannagan, Mr M.J.	Dalgety Spillers, Feed Division, Dalgety House, The Promenade, Clifton, Bristol BS8 3NJ, UK
Hanrahan, Dr J.P.	The Agricultural Institute, Belclare, Tuam, Co. Galway, Eire
Harris, Mr C.I.	MAFF/ADAS, Government Buildings, Coley Park, Reading RG1 6DT, UK
Haynes, Dr N.B.	Dept. Physiology and Environmental Sciences, University of Nottingham School of Agriculture, Sutton Bonington, Loughborough, Leics, UK
Hinxman, Mr R.	Shuttleworth Agricultural College, Old Warden Park, Biggleswade, Beds SG18 9DX, UK
Howles, Dr C.M.	Dept of Animal Physiology, University of Nottingham School of Agriculture, Sutton Bonington, Loughborough, Leics, UK
Hughes, Mr G.J.	ADAS, SE Regional Sheep Specialist, Ministry of Agriculture, Coley Park, Reading, Berks, UK
Irwin, Mr J.H.D.	Department of Agriculture, Livestock Husbandry Farm, Manor House, Loughgall, Armagh BT61 8JB, UK
Kelly, Mr P.	ADAS, Nutrition Chemistry Dept, Olantigh Road, Wye, Ashford, Kent, UK
Kempster, Dr A.J.	Meat and Livestock Commission, PO Box 44, Queensway House, Bletchley, Milton Keynes, MK2 2EF, UK
Kilpatrick, Dr M.J.	Glaxo Animal Health Ltd, Breakspear Road, South, Harefield, Middx, UK
Kimberlin, Dr R.H.	ARC & MRC Neuropathogenesis Unit, West Mains Road, Edinburgh EH9 3JQ, UK
Kirk, Dr J.A.	Seale Hayne College, Newton Abbot, Devon TQ12 6NQ, UK
Lamming, Prof. G.E.	University of Nottingham School of Agriculture, Sutton Bonington, Loughborough, Leics, UK

Land, Dr R.B.	Animal Breeding Research Organisation, Edinburgh EH9 3JQ, UK
Lazenby, Prof. A.	The Grassland Research Institute, Hurley, Maidenhead, Berks SL6 5LR, UK
Lewis, Mr T.	Beecham Animal Health, Broadmead Lane, Keynsham, Bristol, UK
Lindemann, Mr M.A.	BOCM Silcock Ltd, Basing View, Basingstoke, Hants, UK
Lloyd, Miss M.D.	Animal Production Advisory Dept, East of Scotland College of Agriculture, Bush Estate, Penicuik, Midlothian, UK
Lush, Mr M.H.	Lincolnshire College of Agriculture and Horticulture, Caythorpe Court, Grantham, Lincs, UK
Manfredini, Prof. M.	Istituto di Alomentazione Animale, via S. Giacomo 11, 40126 Bologna, Italy
Manson, Mr J.D.	Bryson & Manson, 32 Belfast Road, Antrim, N. Ireland
Marine, A.	Ariestolas (Huesca), Spain
Mauleon, Dr P.G.	Institut National de la Recherche Agronomique, 149 Rue de Grenelle, 75341 Paris, Cedex 07, France
Maxwell, Dr T.J.	Hill Farming Research Organisation, Bush Estate, Penicuik, Midlothian, EH26 0PY, UK
McKeown, Mr D.	136 Knock Road, Dervock, Ballymoney BT53 8AA, Northern Ireland
McKergow, Mr P.R.W.	Ministry of Agriculture, Fisheries and Food, Southgate, Bury St Edmunds, Suffolk IP33 2BD, UK
McLeod, Mr B.J.	University of Nottingham School of Agriculture, Sutton Bonington, Loughborough, Leics, UK
Melrose, Dr D.R.	Meat and Livestock Commission, PO Box 44, Queensway House, Milton Keynes, UK
Mitchell, Mr C.C.	Warwickshire College of Agriculture, Moreton Hall, Moreton Morrell, Warwicks, UK
Morgan, Mr H.	ADAS Regional Office, Woodthorne, Wergs Road, Wolverhampton, UK
Noble, Miss J.E.	J. Bibby Agriculture Ltd, Feeds and Seeds Division, Adderbury, Nr. Banbury, Oxon, UK
Nuttall, Mr B.R.	ADAS Trawsgoed, Aberystwyth, Dyfed, UK
Oliver, Mr S.A.C.	ADAS, Government Buildings, Kenton Bar, Newcastle-upon-Tyne, UK
Outhwaite, Mr J.R.	Ministry of Agriculture, Fisheries and Food, The Lodge, Stamford Road, Oakham, Leics, UK
Papasolomontos, Dr S.	Dalgety Spillers Agriculture Ltd, Dalgety House, The Promenade, Clifton, Bristol BS8 2NJ, UK
Parrett, Mr J.H.	Lancashire College of Agriculture and Horticulture, Myerscough Hall, Bilsborrow, Preston, Lancashire, UK
Phillips, Mr A.	Pen-y-gelli Farm, Ffordd, Bethel, Caernarvon, Gwynedd, UK
Pickard, Mr N.	Ministry of Agriculture, Fisheries and Food, Copthall House, Potter Street, Worksop, Notts, UK
Pollott, Mr G.E.	Meat and Livestock Commission, PO Box 44, Queensway House, Bletchley, Milton Keynes, UK

Prache, Ms Sophie	CRZV-INRA, Laboratoire Production Orine, Theix, 63110 Beaumont, France
Quirke, Dr J.F.	The Agricultural Institute, Belclare, Tuam, Co. Galway, Eire
Roberts, Mr J.	Harper Adams Agricultural College, Edgmond, Shropshire, UK
Read, Mr J.L.	Head of Sheep Improvement Services, Meat and Livestock Commission, PO Box 44, Queensway House, Bletchley, Milton Keynes MK2 2EF, UK
Roberts, Dr R.C.	ARC Animal Breeding Research Organisation, West Mains Road, Edinburgh EH9 3JQ, UK
Robertson, Dr D.	Muresk Agricultural College, Muresk, Western Australia 6401
Robertson, Mrs J.	Gidgegannup, Western Australia 6555
Robinson, Dr J.J.	Rowett Research Institute, Bucksburn, Aberdeen AB2 9SB, UK
Ruiter, Mr T.	Research and Advisory Institute for Cattle Husbandry, Runderweg 6, 8219 PK Lelystad, The Netherlands
Russel, Dr A.J.F.	Hill Farming Research Organisation, Bush Estate, Penicuik, Midlothian, UK
Rutter, Mr W.	East of Scotland College of Agriculture, School of Agriculture, Edinburgh, UK
Savery, Mr C.R.	MAFF/ADAS, Staplate Mount, Starcross, Exeter, Devon, UK
Schanbacher, Dr B.	Roman L. Hruska US Meat Animal Research Center, US Dept Agriculture, Clay Centre, NE 68933, USA
Scheer, Mr H.D.	Alberta Agriculture, 9718-107 Street, Edmonton, Alberta, Canada T5K 2C8
Schackell, Mr G.H.	Invermay Agricultural Research Centre, Private Bag, Mosgiel, New Zealand
Sharp, Mr M.	Ministry of Agriculture, Fisheries and Food, Veterinary Investigation Centre, Block C, Government Buildings, Whittington Road, Worcester, UK
Shrestha, Dr J.N.B.	Animal Research Centre, Agriculture Canada, Ottawa, Ontario K1A 0C6, Canada
Smith, Mr A.D.M.	Hill Farming Research Organisation, Bush Estate, Penicuik, Midlothian, UK
Speedy, Dr A.W.	University of Oxford, Department of Agricultural & Forest Sciences, Parks Road, Oxford OX1 3PF, UK
Stafford, Mr J.R.	Marks and Spencer Ltd, Michael House, 57 Baker Street, London W1, UK
Stephen, Mr T.G.	North of Scotland College of Agriculture, Hunter Hall, Bucksburn, Aberdeen, UK
Suttle, Dr N.F.	Moredun Research Institute, 408 Gilmerton Road, Edinburgh EH17 7JH, UK
Sykes, Prof. A.R.	Department of Animal Science, Lincoln College, University of Canterbury, New Zealand
Tait, Dr A.J.	Glaxo Animal Health Ltd, Breakspear Road, South, Harefield, Uxbridge, UK

Tait, Dr R.M.	Department of Animal Science, University of British Columbia, Vancouver, BC Canada V6T 2A2
Targett-Adams, Dr R.M.	Welsh Agricultural College, Llanbadarn Fawr, Aberystwyth, Dyfed, UK
Tasker, Mr C.D.	Ministry of Agriculture, Fisheries and Food, The Avenue, Bakewell, Derbyshire, UK
Tempest, Dr W.M.	Harper Adams Agricultural College, Newport, Shropshire, UK
Thompson, Dr F.	Rumenco Ltd, Stretton House, Derby Road, Burton-on-Trent, Staffs, UK
Thompson, Mr W.	ICI PLC Agricultural Division, Jealott's Hill Research Station, Bracknell, Berkshire, UK
Thomson, Dr E.F.	Farming Systems Program, International Centre for Agricultural Research in the Dry Areas (Karda), PO Box 5466, Aleppo, Syria
Treacher, Dr T.T.	Grassland Research Institute, Hurley, Maidenhead, Berkshire, UK
Tweed, Mr C.W.	Ballycoose Farm, Ballycally, Larne, Co. Antrim, N. Ireland
Valderrabano, Mr J.	I.N.I.A., Apartado 202, Zaragoza, Spain
Vipond, Dr J.	School of Agriculture, 581 King Street, Aberdeen AB2 1UD, UK
Wallace, Mr D.N.	117 Lylehill Road, Templepatrick, Co. Antrim, N. Ireland
Wallace, Mr D.	117 Lylehill Road, Templepatrick, Co. Antrim, N. Ireland
Walters, Mr B.R.	Redesdale EHF, Rochester, Otterburn, Newcastle-upon-Tyne, UK
Ware, Mr M.A.	Butterworth & Co (Publishers) Ltd, Borough Green, Sevenoaks, Kent TN15 8PH, UK
Watson, Mr D.E.	Fough Farm, Longnor, Nr. Buxton, SK17 0RP, UK
Webster, Mr G.M.	Intervet International BV, PO Box 31, 5830 AA Boxmeer, The Netherlands
Wildig, Mr J.	Pwllpeiran EHF, Cwmystwyth, Aberystwyth, Dyfed, UK
Williams, Dr H.L.	The Royal Veterinary College, Boltons Park, Potters Bar, Herts, UK
Wilson, Prof. P.N.	BOCM Silcock, Basing View, Basingstoke, Hants, UK
Wiseman, Dr J.	University of Nottingham School of Agriculture, Sutton Bonington, Loughborough, Leics, UK
Wolf, Mr B.T.	Welsh Agricultural College, Llanbadarn Fawr, Aberystwyth SY23 3AL, UK

INDEX